Mathematics in Mind

The monographs and occasional textbooks published in this series tap directly into the kinds of themes, research findings, and general professional activities of the **Fields Cognitive Science Network**, which brings together mathematicians, philosophers, and cognitive scientists to explore the question of the nature of mathematics and how it is learned from various interdisciplinary angles. Themes and concepts to be explored include connections between mathematical modeling and artificial intelligence research, the historical context of any topic involving the emergence of mathematical thinking, interrelationships between mathematical discovery and cultural processes, and the connection between math cognition and symbolism, annotation, and other semiotic processes. All works are peer-reviewed to meet the highest standards of scientific literature.

Melanija Mitrović •
Mahouton Norbert Hounkonnou
Editors

Algebra Without Borders

Classical and Constructive Semigroups and Applications

Springer

Editors

Melanija Mitrović
Center for Applied Mathematics of the
Faculty of Mechanical Engineering Niš,
CAM-FMEN
University of Niš
Niš, Serbia

Mahouton Norbert Hounkonnou
International Chair in Mathematical Physics
and Applications (ICMPA-UNESCO Chair)
University of Abomey-Calavi
Cotonou, Benin

ISSN 2522-5405 ISSN 2522-5413 (electronic)
Mathematics in Mind
ISBN 978-3-031-86476-6 ISBN 978-3-031-86477-3 (eBook)
https://doi.org/10.1007/978-3-031-86477-3

Mathematics Subject Classification: 20Mxx, 03F65, 03Gxx, 92Bxx, 60Bxx

This Springer imprint is published by the registered company Springer Nature Switzerland AG
The registered company address is: Gewerbestrasse 11, 6330 Cham, Switzerland

Foreword

The great analyst Einar Hille wrote in the preface to his 1948 classic book "Functional Analysis and Semi-Groups," "I hail a semigroup when I see one and I seem to see them everywhere." The omnipresence of semigroups, obvious in 1948, when semigroup theory was still in its infancy, becomes even more noticeable in 2024, for which the present volume serves as another manifestation. The reader will find here diverse samples of impressive applications of semigroups in natural, social, and life sciences as well as their usage in various parts of mathematics such as algebra, geometry, logic, dynamical systems, and Markov processes. Along with the contributions elaborating on classical semigroup theory, the volume also includes a survey of recent achievements in a novel direction on the edge of semigroups and constructive mathematics. In addition, the volume contains a thorough yet easy-to-read overview of the origins of algebraic semigroup theory and its early development.

I envisage that many scholars may find some specific chapters in the volume immediately useful for their work. At the same time, the volume goes beyond this direct utility as it provides a valuable opportunity to see at once many facets of this magic crystal that the realm of semigroups constitutes. It is really the case when the whole is greater than the sum of its parts!

Many thanks to the editors, Melanija Mitrović and Mahouton Norbert Hounkonnou, for coming up with the idea of the volume and making this idea a reality, to all contributors, and last but not least—to the publishers, Springer Nature, who keeps producing literature on semigroups, including monographs, conference volumes, and "Semigroup Forum," the journal covering all aspects of semigroup theory!

Mikhail Volkov
Managing Editor of "Semigroup Forum"

Preface

"Algebra is beautiful. It is so beautiful that many people forget that algebra can be very useful as well." [R. Lidl, G. Pilz, Applied abstract algebra, Springer, 1998]

Following S. Shapiro, 2002, "Everything we do is based on some mathematical structure, and although mathematics is often considered abstract, it is fundamental to how we understand nature, the larger universe, with its time and space dimensions and a myriad of uncertainties". Algebraic structures, as often considered, represent templates for solving problems. Within a scope of an algebraic structure can be found details enabling us with possibility to think clearly about the bigger picture. This can give us a precise way to communicate with other (AI) scientists providing clarity of thought and precise communication. Going through the literature written by renowned scientists, one can find that the whole of modern or abstract algebra has been designed to be elegantly simple. It can be described in a couple of sentences: the story starts with sets (collections of objects) accommodated with certain relations on them, and proceeds to the concept of a semigroup, a set with an associative binary operation defined on it. Each new concept is based on the previous ones, and, ultimately, the whole multi-story edifice rests on the sparse foundation of sets and relations with semigroups being the building blocks for the structures comprising the subject of modern algebra. In the history of classical mathematics the algebraic theory of semigroups is a relative newcomer, with the theory proper developing only in the second half of the twentieth century. Nowadays, classical semigroup theory is an enormously broad topic with applications which have advanced on a very broad front. A prominent semigroup theoretician, B. M. Schein, states that "One meets ... semigroups much more often than groups (and much more often than one thinks he does), and only the existing polarization of mind on the topic of groups and other classical algebraic structures prevents one from seeing semigroups in various processes and phenomena of mathematics and the universe". On the other hand, it is only recently that constructive mathematics has paid attention to semigroup theory. The use of intuitionistic logic in constructive mathematics facilitates the extraction of computational content from constructive proofs, something almost impossible within classical mathematics. In this way, constructive mathematics is closely related to theoretical computer science. Following J. M. Howie, links are clearly very important in mathematics. Links to other areas of mathematics persuade us that the ideas we are exploring are important. Applications outside mathematics are a bonus. As the first in a series of books on Algebra without Borders, "Classical & Constructive Semigroups and Applications" aims to blend all these aspects together in order to give an opportunity to mathematicians and all those from other sciences who deal with algebraic structures to experience both (classical and constructive) semigroup theory and its numerous applications. The path toward this book started from our meeting at a conference in 2019 in Sweden where we discussed and developed a plan for a series of books based on conferences on Algebra without Borders. The first edition of this series of conferences titled *Classical & Constructive Semigroups and Applications*, ACaCS 2021, was scheduled for August 2020, but was unfortunately postponed to September 2021 due to the awful COVID-19 pandemic. In meantime, we organized a webinar on Mathematics for Human Flourishing in the Time of COVID-19 and Post COVID-19, 21 October 2020, an international scientific conference on Mathematics for Social Sciences and Arts-Algebraic Modeling, 24–26 May 2021, in collaboration with Professor Philippa Pattison from Australia and Professor Dragana Martinovic from Canada: https://mathsocart.masfak.ni.ac.rs/, and an international conference on Classical and Constructive Nonassociative Algebraic Structures: Foundations and Applications, CaCNAS:FA 2021, 30 June–2 July 2021 in collaboration with Professor Mujahid Abbas and Professor Madad Khan from Pakistan: https://cacnas.masfak.ni.ac.rs/. The *ACaCS 2021* conference (https://acacs2020.masfak.ni.ac.rs/) was organized by higher education institutions from Europe (Center for Applied Mathematics of the Faculty of Mechanical Engineering, *CAM-FMEN*, University of Niš, Serbia: https://camfmen.masfak.ni.ac.rs/), and Africa (International Chair in Mathematical Physics and Applications, *ICMPA-UNESCO Chair*, University of Abomey-Calavi, Benin: http://www.icmpa.net). It gathered the world's leading scientists in semigroups and mathematics (more precisely in algebra), and showcased the participation of prominent invited experts interested in applications of these fields. The participants came from around 35 different countries from all continents. The conference ACaCS 2021 took place in Niš, Serbia, at the Faculty of Mechanical Engineering, 6–8 September.

In addition to the General Session dedicated to semigroup theory (with classical or intuitionistic logic as the background, which went on in the first part of every conference day), the conference comprised a set of colloquiums chaired by leading scientists as described on the website: The general session and the colloquiums focused on recent developments in semigroups and other basic algebraic structures as well as on some of their potential applications in other fields.

The title of the book *Classical & Constructive Semigroups and Applications* suggests the main novelty comparing with already existing literature. To our best knowledge, this is the very first book that presents the framework of constructive semigroups with apartness, a novel algebraic structure—that is theory of semigroups developed within intuitionistic logic as background. As already pointed above, our book has the advantage to relate new trends in classical and constructive semigroups with growing application developments in other fields. Regarding novelties in such applications, connections to other areas of mathematics (logics, biomathematics, analysis, geometry, etc.), natural sciences, engineering as well as interconnections between semigroups, cognitive sciences, linguistics, and humanities will be presented. Let us finish paraphrasing E. Frenkel in his citation, which we took as the motto of our Algebra without Borders series: Algebra (like mathematics) is as much part of our cultural heritage as art, literature, and music. Algebra is the source of timeless profound knowledge, which goes to the heart of all matter and unites us across cultures, continents, and centuries.

The Editors wish to express their gratitude to the Professor Mikhail Volkov for the time he devoted to read the book and write the Foreword. They also thank all the contributors who accepted to be part of this first volume, and to the anonymous referees.

The Editors would like to express their thanks to the Faculty of Mechanical Engineering, University of Niš, Serbia, and *ICMPA-UNESCO Chair*, University of Abomey-Calavi, Benin Republic, for their various supports. Special thanks go to the *CAM-FMEN* Technical supporting team for successful organization of *ACaCS 2021* conference, and to the *ICMPA-UNESCO Chair* team, especially to Robinson Degbegni, for their efforts in preparations of this book.

In conclusion, we would like to express our sincere thanks to the Springer team for the remarkable help they have given us during this exciting work.

We also dedicate this book to our children Cornelia, Mladen, Jelena, Doriane, Vladimir, Mehdy, and Nikola.

Niš, Serbia
Cotonou, Benin
July 2024

Melanija Mitrović
Mahouton Norbert Hounkonnou

References

1. Frenkel, E.: Love and Math: The Hearth of Hidden Reality. Basic Books, New York (2013)
2. Howie, J.M.: Semigroups, Past, Present and Future. In: Proceedings of the International Conference on Algebra and Its Applications (2002)
3. Lidl, R., Pilz, G.: Applied Abstract Algebra. Springer, Berlin (1998)
4. Schein, B.M.: Relation algebras and function semigroups. Semigroup Forum **1**(1), 1–62 (1970)
5. Shapiro, S.: Philosophy of Mathematics: Structure and Ontology, 296 p. Oxford University Press, Oxford (2000)

Contents

Author Biographies

Melanija Mitrović is a Full Professor at the Department of Mathematics and Informatics of the Faculty of Mechanical Engineering, University of Niš.

She received her BSc and MSc degrees at the Faculty of Philosophy, University of Niš, and her PhD degree at the Faculty of Mathematics and Sciences, University of Niš, in 2000. She lectures at all three levels of higher education in Serbia, as well as abroad. She is a member of the Faculty Council (2015–2022), Quality Board, Committee for the Student Evaluation of Educational Quality (2009–2015), Deputy Vice head of the Department of Mathematics and Informatics (2005–2010). She is the head of the Center for Applied Mathematics of the Faculty of Mechanical Engineering Niš, CAM-FMEN (https://camfmen.masfak.ni.ac.rs), around which, among other activities, she develops an interdisciplinary research group investigating applications of algebraic structures to problems in engineering science. She holds the status of Permanent Full Professor at the International Chair in Mathematical Physics and Applications (ICMPA-UNESCO Chair), University of Abomey-Calavi, Benin Republic (https://www.cipma.net/spip.php?page=rubrique&id_rubrique=27). Her membership includes the Editorial Board of the Book series Mathematics in Mind, Springer (https://www.springer.com/series/15543?detailsPage=aboutTheEditor), the Fields Cognitive Science Network (http://www.fields.utoronto.ca/generalinfo/Fields-Cognitive-Science-Network), Peer Community In (PCI) Neuroscience. Major directions of her research, professional work, and expertise focus on foundations of constructive mathematics, classical and constructive algebraic structures, especially within areas of semigroup and semiring theory, Witt and Virasoro algebras, integrable systems, groups, and representations. Her innovating work within the foundations of constructive mathematics (theory of sets with (non-tight) apartness, theory of ordered sets with (non-tight) apartness) and its applications to constructive semigroups with apartness positions her among the pioneers of the constructive mathematics in Serbia. Outside of Serbia, she is recognized as the mother of the novel theory of constructive semigroups with apartness, considered as a new algebraic theory. She publishes her works in renowned outstanding international journals. She is reviewer for renowned journals in mathematics. She is invited as plenary speakers at highly ranking international conferences in a regular basis to communicate about the results of her research. She often acts as the chair of organizing committees and as program committee member of prestigious international conferences. She was the Chair of the Organizing Committee and member of the Program Committee of the Workshop “Theoretical Computer Science - from Foundation to Applications,” TCS-FP 2009, as well as “Constructive Mathematics: Foundations and Practice,” CMFP 2013—both organized at the Faculty of Mechanical Engineering, University of Niš. In connection with the CMFP 2013 organization, she got IMU/CDC Support (Conference Grant Support) as the first Serbian mathematical conference ever co-sponsored by IMU (International Mathematical Union, Berlin). In addition, CM:FP 2013 was the first conference on constructive mathematics (mathematics with intuitionistic logic) organized at a developing country. In the period from October 2020 till November 2021 she was the Chair of the Organizing Committee and member of the Program Committee of the following online and/or hybrid international scientific conferences: “Mathematics for human flourishing in the time of COVID-19 and post Covid-19” (https://camfmen.masfak.ni.ac.rs/Webinar_Covid.html), MS^2A^2M 2021 (https://mathsocart.masfak.ni.ac.rs/), CaCNAS:FA 2021 (http://cacnas.masfak.ni.ac.rs/), ACaCS 2021 (https://acacs2020.masfak.ni.ac.rs/). She is a guest speaker in leading universities worldwide. She serves as an international jury member for doctoral theses. She is author of a well-known scientific monograph, at national level, on semigroup theory. Her works in the field of algebra so far has made a significant contribution to its development, which is reflected in the expansion of existing knowledge (classical semigroup theory and semiring theory) and in the establishment of completely new theory of constructive semigroups with apartness. She received the Award Povelja of the Faculty of Mechanical Engineering Niš in 2019 for her permanent work connected to the promotion of the faculty and University of Niš algebraic group (https://www.melanijamitrovic.net/).

Dr. Sc in 1992 from catholic University of Louvain in Belgium, **Mahouton Norbert Hounkonnou** is a full Professor of Mathematics and Physics at the University of Abomey-Calavi, Benin Republic.

His works deal with noncommutative and nonlinear mathematics, and complex systems. He authors and reviews books. Further, he serves as member and associate editor of editorial boards for renowned journals and books in mathematics and mathematical physics, including the Editorial Boards of Mathematics in Mind, Springer (https://www.springer.com/series/15543?detailsPage=aboutTheEditor), Fields Cognitive Science Network (http://www.fields.utoronto.ca/generalinfo/Fields-Cognitive-Science-Network), the Peer Community In (PCI) Neuroscience, etc. He has published over 200 main research papers in outstanding ISI-ranked peer-reviewed journals and international conference proceedings in the fields of mathematics, mathematical physics, and complexity. He is a visiting professor at African, Asian, European, and North American universities. Together with his peers at the international level, he founded the International Chair in Mathematical Physics and Applications (ICMPA-UNESCO Chair) of the University of Abomey-Calavi where he created multi-university master degree and PhD programs in mathematics with connections, motivations, or applications to physics, and in physics with important relations to mathematics. The best African students from over 13 French- and English-speaking countries are selected to follow these graduate programs, which attract prominent and leading mathematicians and mathematical physicists worldwide who come to give lectures and supervise student's research, what has substantially increased international collaboration with African, Asian, American, European, and Indian mathematicians. The ICMPA-UNESCO Chair presently hosts the international conference and school series on Contemporary Problems in Mathematical Physics, which are held every two years and each year since 1999 and 2005, respectively. These activities have led to a significant network of researchers connected with the ICMPA-UNESCO Chair, which benefits from the resources available in mathematics and mathematical physics in Africa. Professor Hounkonnou supervised over 35 PhD and 40 MSc students from various countries and continents, including Belgium, Benin, Burkina-Faso, Burundi, Cameroun, Democratic Republic of Congo, Niger, Nigeria, Senegal, Togo, Zambia, etc.

Professor Hounkonnou was awarded a series of recognitions for his work such as the 2023 Yang-Hui Award *for seminal contributions to the deformed quantum algebras*, the American Institute of Physics 2023 Tate Medal *for leadership in building and maintaining an enduring transnational African mathematical physics research and education community, in particular the COPROMAPH conferences and schools and Academy level international networks*, the 2016 World Academy of Sciences C. N. R. Rao Prize for Scientific Research *for his incisive work on noncommutative and nonlinear mathematics and his contributions to world-class mathematics education*, the Tokyo University of Science President Award in 2015, and the Prize of the Third World Academy of Sciences (TWAS) in 1996. Professor Hounkonnou is the Chair of The World Academy of Sciences (TWAS) Independent Expert Committee (IEC) for the review of "TWAS Awards" candidates in the field of Mathematical Sciences (present–). He was a member of UNESCO Scientific Board for International Basic Sciences Programme (IBSP), NANUM 2014 Award Committee Member of the International Congress of Mathematicians (ICM 2014) as reviewer for region Africa, and member of the InterAcademy Partnership working group on *Harnessing Science, Engineering and Medicine to Address Africa's Challenges*, etc. He was TWAS research Professor in Zambia, the chair of the African Academy of Sciences Commission on Pan-African Sciences Olympiad (2014), and chair of the African Academy of Sciences Membership Advisory Committee (MAC) in Mathematical Sciences (2013–present). Professor Hounkonnou is the co-chair of the network of African, European and Mediterranean Academies for Science Education (AEMASE III), the President of the Network of African Science Academies (NASAC), former President of the Benin National Academy of Sciences, Arts and Letters (ANSALB, 2015–2020). His membership extends to InterAcademy Partnership Advisory Committee and Science Education Programme (IAP SEP), the International Association of Mathematical Physics, American Mathematical Society, London Mathematical Society, Society for Industrial and Applied Mathematics (SIAM), Academy of Science of South Africa (ASSAf), Hassan II Academy of Sciences and Technology, Morocco, African Academy of Sciences (AAS), The World Academy of Sciences (TWAS), Scientific Council of the Centre International de Mathématiques Pures et Appliquées (CIMPA), Scientific Committee of the International Centre for Advanced Training and Research in Physics (CIFRA, Magurele-Bucharest, Romania), as well as other scientific and learned societies. He is the President of the ANSALB Permanent Commission *Education and Ethics*, and a member of the Presidential Body *Delegation for Control and Ethics in Higher Education* in Benin. He is Doctor Honoris Causa of the University Toulouse III Paul Sabatier, France, and representative for Africa of the International Mathematical Union (IMU) Commission for Developing Countries (CDC). In May 2023, he was elected Chair of the Associate Members of the Organisation of the Southern Cooperation (OSC) (2023–2024). He is a Knight of the Benin National Order.

Part I
General Theory

Three Themes in the Development of Classical Semigroup Theory

Christopher D. Hollings

Abstract The algebraic theory of semigroups emerged in the mid-twentieth century in a form that is sometimes referred to as the 'classical' theory, first brought together coherently in the textbooks of Lyapin (*Semigroups*, 1960) and Clifford and Preston (*The algebraic theory of semigroups*, 1961–67). In this chapter, we survey aspects of three of the major themes that can be found in that early theory: the study of arbitrary transformations of a set, efforts to mimic familiar results about rings in a semigroup setting, and the questions arising from differential geometry that gave rise to the notion of an inverse semigroup.

Keywords A. K. Sushkevich · Transformation · Generalised group · Kernel · David Rees · Completely 0-simple semigroup · Rees matrix semigroup · Rees theorem · V. V. Wagner · G. B. Preston · Inverse semigroup

1 Introduction

One of the most notable themes in the mathematics of the nineteenth century was the gradual shift towards greater abstraction. This can be seen right across the mathematical spectrum, from the emergence of new geometries [80] to the formalisation of the notion of a set [61]. This latter development in particular had been underway since the early years of the century, with mathematicians realising that it was often useful to study *collections* of objects alongside the more direct study of the objects themselves. One of the most famous examples of this in the first half of the century was Évariste Galois' gathering of certain permutations into collections that he labelled 'groups' [176]. Here of course it was not just the elements of the collection that mattered but the way in which they interacted with

C. D. Hollings (✉)
Mathematical Institute, University of Oxford, Oxford, UK

The Queen's College, Oxford, UK
e-mail: christopher.hollings@maths.ox.ac.uk

M. Mitrović, M. N. Hounkonnou (eds.), *Algebra Without Borders*, Mathematics in Mind, https://doi.org/10.1007/978-3-031-86477-3_1

each other under composition. Beginning in the writings of Arthur Cayley in the 1850s, the essential properties of Galois' groups of permutations were identified and used to repackage the notion as the abstract definition of a group that we know today [250]. Moreover, the burgeoning theory of groups, with its emphasis on the rules of interaction between arbitrary elements, provided a model for how other mathematical objects might be handled. Thus, by the beginning of the twentieth century, structural algebra, dealing in other such abstractly defined notions as fields and integral domains, was already emerging as a major branch of mathematics [45, 81]. It would go on to be promoted in particular in the works of Emmy Noether, whose approach to algebra was subsequently spread more widely by B. L. van der Waerden's 1930 textbook *Moderne Algebra* [240].

Alongside the growing structural approach to algebra, axiomatic methods were also in vogue in mathematics at the turn of the twentieth century, having been promoted by David Hilbert in his *Grundlagen der Geometrie* of 1899 [90]. Hilbert's influence is seen in particular in the works produced by the group of so-called postulate analysts who formed around E. H. Moore in Chicago and who took collections of axioms as their objects of study, investigating questions of consistency, minimality, and independence [45, §3.5]. It is out of their work that the word 'semigroup' emerged in a mathematical context in 1904, although a proper *theory* of such objects was still some way off. A finite group having usually been defined up to this point by the conditions of closure, associativity, and cancellation, the question naturally arose of whether this definition also serves to define an infinite group. It was easily seen that it does not, and the term 'semigroup' was coined for the infinite object that is defined by the three axioms just listed: in modern terminology, a cancellative semigroup [107, §1.2]. This axiomatic tinkering, however, can hardly be considered as the beginning of the theory of semigroups, nor did it ultimately have much impact on the theory of groups, aside perhaps from some pedagogical value [110, 111, 175].

What eventually provided the spark for the development of a *theory* of semigroups was a series of problems emerging from other branches of mathematics. In some cases, questions arose from the realisation that problems posed elsewhere contained redundant conditions and that the problems might therefore be tackled instead within a simpler framework. For instance, it was noted that the solutions of certain problems relating to rings did not call upon any of the properties of addition—it was arguably more natural therefore to consider these problems in a purely multiplicative context. Thus, early aspects of semigroup theory were developed as generalisations or extensions of ideas from elsewhere in abstract algebra. These were nevertheless well-motivated investigations and were rarely generalisations for generalisation's sake. This was certainly true of the study of arbitrary transformations of a set that underpinned much of the early investigation of semigroups: it seemed entirely natural to step away from the special case of bijections, at the heart of the study of groups, in order to determine what happens in a wider setting.

During the 1920s and 1930s, these semigroup-theoretic investigations gathered pace, although they remained largely disjoint and were not necessarily recognised as aspects of a broader theory. This began to change, however, with the appearance in 1940 of a paper by David Rees, which not only provided a proof of one of the classical results of semigroup theory, but also fixed the term 'semigroup' in its modern sense [193]. This latter point is one of the more trivial aspects of Rees' paper, and yet this anchoring of terminology provided a core around which a coherent theory of semigroups could begin to coalesce. The developments continued throughout the 1940s and 1950s and were then brought together in textbook form first by E. S. Lyapin in his *Semigroups* of 1960 [153] and then by A. H. Clifford and G. B. Preston in the first volume of their *Algebraic theory of semigroups* in 1961 [42]. Both texts remained the standard references on the subject for many years and secured semigroup theory as a valid branch of mathematical study.[1] They also reflect a type of semigroup theory that has sometimes been referred to as the 'classical' theory, which might be (very) broadly characterised as having an overall flavour of ring theory;[2] it is the history of this phase of the theory that we deal with here.

In the present chapter, we provide a survey of selected early features of the development of (algebraic) semigroup theory. First of all, we consider the study of arbitrary transformations of a set, which was a particular occupation of the semigroup pioneer A. K. Sushkevich during the 1920s and 1930s (Sect. 2). After this, in Sect. 3, we turn to an example of an early semigroup-theoretic study that arose through efforts to mimic results about rings, namely, the Rees theorem. In Sect. 4, we look to differential geometry and the questions asked there that eventually motivated the definition of perhaps the most-studied class of semigroups, inverse semigroups. We conclude the main part of the chapter in Sect. 5 with a few brief comments on other aspects of the theory. In the interests of saving space, we focus on mathematics and keep biographical details to a minimum. One of the side effects of this policy is that we will say only a little about the way in which many aspects of semigroup theory were developed independently in the East and West during the years of the Cold War; for more on these aspects, the reader is referred to [107] and also to [109] for a wider context. At the end of the present chapter, we include an appendix that gives references to other writings on the history of semigroup theory and to biographies of prominent contributors to the theory. Throughout the chapter, we shall assume (we hope not unreasonably) that the reader will already be well versed in the theory of semigroups, and so the majority of the semigroup-theoretic terminology that we employ will remain undefined; any reader seeking clarification is referred to [116].

[1] A huge number of further textbooks have of course appeared over the following decades, and we do not attempt to enumerate them here. A natural successor to [42] was [114, 116]. A text written by one of the editors of the present volume [171] appears to have been the first book on (algebraic) semigroups to have been written by a woman (the corresponding such book for topological semigroups is [182]).

[2] For one characterisation of the 'classical' theory of semigroups, see [197].

2 Transformations

In the interests of brevity, our discussion of semigroups of transformations will be confined solely to the works of A. K. Sushkevich on this topic in the 1920s and 1930s. Born in Borisoglebsk in southern Russia, Sushkevich spent most of his working life in Kharkiv in Ukraine, and, aside from an interval in Voronezh during the 1920s, it was here that he completed most of his work on semigroups.[3] One major theme of his investigations was the study of minimal ideals ('kernels') of semigroups, which led him to what we now recognise as a finite version of the Rees theorem (see Sect. 3). The development of different versions of his 'kernel' construction in a range of settings was what occupied Sushkevich throughout his career in the more abstract reaches of his work [107, §3.3]. But in a more concrete direction, the other significant theme of his investigations was semigroups of arbitrary transformations.

Sushkevich's work in this area started with his doctoral dissertation, *The theory of operations as the general theory of groups*, begun in Kharkiv around 1918 and completed in Voronezh in 1922 [226]. Most of the content of the dissertation eventually found a wider (though not necessarily wide) readership through its recycling in his published papers of the following decade and then in his monograph, *Theory of generalised groups*, of 1937 [228]. Throughout these works, Sushkevich's goal, simply put, was to develop a theory of 'generalised groups' as an abstract version of that of 'generalised substitutions' (i.e. arbitrary transformations of a set) by direct analogy with the abstraction of the theory of groups from the study of permutations.[4] In seeking to develop such a theory, Sushkevich saw himself as addressing a natural problem and filling a conspicuous gap in the existing literature.

Foreshadowing the work on the history of mathematics that Sushkevich would publish later in his career,[5] his dissertation begins with a short historical account of group theory, with a focus on the emergence of the axiomatic definition. Of particular interest to Sushkevich was a set of postulates that had been laid down for a finite group by F. G. Frobenius in a paper of 1895 entitled simply 'Über endliche Gruppen' [66].[6] We record that definition here, in something close to Frobenius' original terms, for future reference:

[3] See the appendix for a list of biographies of Sushkevich. For English translations of the six papers that he published in Ukrainian, see [180].

[4] Since the early investigations of the permutations of a set by Galois and Cauchy in the first half of the nineteenth century, the term 'substitution' had been employed for what we would now term a permutation; see, for example, [124].

[5] He wrote, for example, on the history of algebra in Russia [230] and on mathematics in Kharkiv [231].

[6] Sushkevich had first learnt group theory as a student of Frobenius in Berlin during the first decade of the twentieth century [112].

Definition 1 (Frobenius' Definition of a Finite Group) By a *group* we mean a finite system of elements subject to an operation which assigns to each pair of elements A, B a third element AB subject to the following conditions:

(i) The result of the operation is unique [i.e. it is well-defined].
(ii) The operation is 'uniquely invertible' [i.e. cancellative on both sides].
(iii) The operation is associative but not necessarily commutative.
(iv) The operation is 'limited in its effect' in the sense that only a finite number of elements is generated via its repeated application.

Sushkevich's approach to 'generalised groups' was to weaken Frobenius' definition in particular ways. In line with his goal of studying transformations, he repeatedly emphasised the role played by the associative law in these investigations.[7] Also of great interest was Frobenius' second condition, which Sushkevich termed the 'law of unique invertibility'. A one-sided version of this condition would later be key to Sushkevich's most influential semigroup-theoretic paper, as we shall see in Sect. 3.

Having given this background to his investigations, Sushkevich spent much of the following chapter of his dissertation engaged in theory-building. Confusingly to modern eyes and perhaps also to the eyes of mathematicians at the time, he took as his fundamental notion that of a 'group', by which he meant simply a set upon which there is defined a binary operation; for clarity, he often termed this a 'generalised group', as opposed to an 'ordinary group'. Staying largely in the finite case, his groups could be defined by multiplication tables, and Sushkevich placed so-called group properties any property of the group that can be deduced from the multiplication table—at the centre of his focus. Such fundamental concepts as generators, subgroups, and isomorphisms appear in this early part of the dissertation.

In line with the title of his work, Sushkevich placed a great emphasis on operations on sets, and in the third chapter of his dissertation he turned to the notion of a unary operation on a finite set. For a finite collection of n distinct elements $\xi_1, \xi_2, \ldots, \xi_n$, and a unary operation $\mathscr{D}$ acting upon them, Sushkevich denoted by $\mathscr{D}\xi_i$ in a very natural way the result of applying $\mathscr{D}$ to the element ξ_i. His notation for the overall result of applying $\mathscr{D}$ to each element of the set in turn was essentially the familiar two-row notation for transformations:

$$\begin{pmatrix} \xi_1 & \xi_2 & \cdots & \xi_n \\ \mathscr{D}\xi_1 & \mathscr{D}\xi_2 & \cdots & \mathscr{D}\xi_n \end{pmatrix}.$$

Since the $\mathscr{D}\xi_i$ are not necessarily distinct, Sushkevich referred to this as a *substitution with repeating symbols*. In his wider work, it became, more simply, a *generalised substitution*. Later on in the dissertation, Sushkevich hinted at an

[7] It should be noted that Sushkevich also gave attention in his publications (in particular, [222]) to non-associative generalised groups, the forerunners of quasigroups: see [184]. Indeed, his interest in generalised groups was far from being confined to semigroups and extended even to objects with ternary operations, including the notion of 'heap' that we will meet in Sect. 4. The later chapters of his monograph [228] deal with a range of different notions of generalised group.

extension of Cayley's theorem: that any generalised group may be embedded in some collection of generalised substitutions. In the dissertation he offered a proof only for finite generalised groups with one-sided cancellation, but later extended this to any finite generalised group in a published paper of 1926 [220], wherein he began by recalling the proof of this result in the group case: given a finite group $\mathfrak{G} = \{A_1, A_2, \ldots, A_n\}$, we associate a permutation

$$\overline{A_i} = \begin{pmatrix} A_1 & A_2 & \cdots & A_n \\ A_1A_i & A_2A_i & \cdots & A_nA_i \end{pmatrix}$$

with each A_i, and then verify that there is an isomorphism between the A_i and the $\overline{A_i}$. In the semigroup case, however, Sushkevich noted a problem, namely, that we may have $\overline{A_i} = \overline{A_j}$ for some i and j, and so the representation will not be injective. In his dissertation, he had resolved this issue by introducing one-sided cancellation, but the more general case required a small modification to the above $\overline{A_i}$ by the introduction of a 'separation symbol' $\mathcal{E}$, tantamount to adjoining an identity to a generalised group $\mathfrak{G}$. This introduction of $\mathcal{E}$ enables us to associate a transformation

$$\overline{A_i} = \begin{pmatrix} A_1 & A_2 & \cdots & A_n & \mathcal{E} \\ A_1A_i & A_2A_i & \cdots & A_nA_i & A_i \end{pmatrix}$$

with each $A_i \in \mathfrak{G}$. The effect of the presence of $\mathcal{E}$ is to 'separate' the $\overline{A_i}$ and make them all distinct, thereby providing a faithful representation of the generalised group $\mathfrak{G}$ by means of generalised substitutions: the extended right regular representation that appears in modern textbooks on semigroup theory [116, p. 7]. The significance of this theorem for his wider theory was not lost on Sushkevich:

> it reduces the study of our abstract groups to the study of the concrete case of generalised substitution groups, which is in many respects easier. [220, p. 372]

In his more abstract works on semigroups, Sushkevich nevertheless lived up to his comments here by illustrating his results with concrete examples in terms of substitutions. It was also via substitutions, perhaps the most natural bridging language, that Sushkevich was eventually able to draw connections between his investigations in this direction and those of others, such as Henrich Brandt and Alfred Loewy.[8]

By the late 1920s, Sushkevich's work on generalised groups was starting to reach a wider readership—not just those in the USSR, but also mathematical readers in Europe more generally, thanks to his habit of publishing in a range of journals in Western European languages. In 1928, he attended the International Congress of Mathematicians that took place that year in Bologna. The paper that he presented [223] appears in the congress proceedings and provides a rather pleasing diagrammatic approach to generalised substitutions, inspired by the

[8] In [31] and [150], respectively. For comments on their works, see [107, §6.2].

standard decomposition of permutations into cycles. Sushkevich took as an example the generalised substitution

$$A = \begin{pmatrix} 0 & 1 & 2 & 3 & 4 & 5 & 6 & 7 & 8 & 9 & \alpha & \beta & \gamma & \delta & \varepsilon \\ 5 & \alpha & 7 & 3 & 3 & 6 & 4 & \beta & 1 & 1 & 2 & \alpha & 5 & 4 & 3 \end{pmatrix},$$

which he depicted by means of the following diagram:

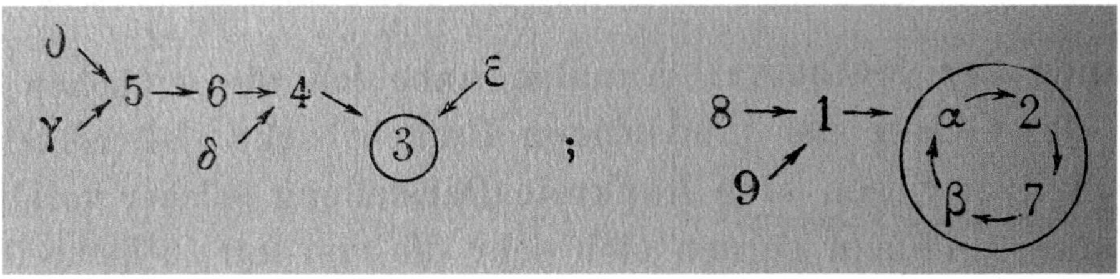

As Sushkevich noted, unlike the pictures that we might draw for ordinary permutations, such depictions of generalised substitutions may have 'tails' as well as 'cycles'. He went on to explore the use of such diagrams in the study of powers of substitutions, building on some comments that he had included in his dissertation on what we would now term monogenic subsemigroups.

Moving into the 1930s, we see that Sushkevich's published research began to consist largely of extensions of the work that we will discuss in Sect. 3 and also that he began to expand his research into generalised groups in other directions, such as using matrices to represent them [224]. Having mostly studied finite generalised groups, he also attempted to extend some of his results to the infinite case [225], in which connection he found it necessary to extend his notion of substitution to one on a countably infinite number of elements, possibly having been inspired by Alfréd Haar's similar extension for permutations [83]. Here, Sushkevich classified the different types of substitutions that may arise and used them to categorise different types of infinite generalised groups satisfying weakened versions of postulates from Frobenius' definition of a group. He would return to such substitutions in two papers of 1940, in which he explored different means of decomposing them into products of substitutions of a more amenable form [229].

The most complete presentation of Sushkevich's work on generalised groups is his 1937 monograph [228], which essentially summarises all of the algebraic investigations that he had carried out in and after his doctoral dissertation. In the years following the Second World War, Sushkevich's mathematical papers would largely dry up, replaced by writings on the history of mathematics, as we have already noted. Thus, in the years in which semigroup theory was starting to take off in the West, Sushkevich had nothing more to say on the subject.

Sushkevich's legacy in semigroup theory is a rather mixed one. From the early 1920s, he was advocating what he saw as the very natural study of collections of generalised substitutions and published at least some of his work in journals outside the USSR (for example, [221, 222]). The main thing that was picked up, however, was his rather more abstract result of 1928 on minimal ideals of semigroups, to which we will turn in Sect. 3. Sushkevich was certainly introducing generalised substitutions into his teaching at the beginning of the 1930s [227], and the uptake, though limited, of his work elsewhere in the Soviet Union may have been a factor in the strong predilection for studying semigroups of different types of transformations that we see in the later Soviet semigroup school of the 1950s [107, ch. 9] (we will see an example of this in Sect. 4). But it is perhaps in his very advocacy of the study of generalised substitutions that we find the reasons for the minimal impact of Sushkevich's work on the development of semigroup theory: as central as transformations of a set are to the foundations of semigroup theory, there is only so far you can go with them in their purely concrete and more arbitrary form.

3 The Rees Theorem

We have already alluded to another strand of Sushkevich's work, concerning minimal ideals, that was rather better known than his work on transformations—indeed, Sushkevich's main article on this other topic [221] would later be described by Clifford and Preston as 'a paper of fundamental importance' [42, vol. 1, p. ix] and by Preston alone, probably in ignorance of the earlier aspects of Sushkevich's work, as 'the first substantial paper on semigroups' [192, p. 19]. Although, as we have seen, Sushkevich's approach to semigroups arguably had an initial 'group-theoretic' flavour to it, his work on minimal ideals stands as a precursor to one of those early strands of semigroup theory that sought to mimic results for rings, in this case the Wedderburn–Artin theorem, which characterises a semisimple ring as a product of matrix rings over division algebras [43, §5.7]. Although Sushkevich's formulation did little to emphasise the connection with rings, it would later be subsumed within the analogous result for semigroups, the Rees theorem, proved by David Rees in 1940 and concerning the structure of completely 0-simple semigroups [193].

The relevant paper by Sushkevich appeared in *Mathematische Annalen* in 1928, in which he described its contents as an attempt 'to construct an abstract theory of finite groups, whose operation is not uniquely invertible' [221, p. 30]. At the start of the paper, he acknowledged that the multiplicative part of the hypercomplex number system studied by J. H. M. Wedderburn [247] would serve as an example of such a 'group'—but his knowledge of Wedderburn's work seems to have come only after he had completed the paper.[9] As in his work on transformations, Sushkevich's inspiration was broadly group-theoretic, as we see from his use of the term 'unique invertibility', lifted from Frobenius' definition of a finite group (Definition 1).[10]

[9] Sushkevich thanked Emmy Noether for providing him with the reference [221, p. 30, n. 1]; at that time, she served as an unofficial editor of *Mathematische Annalen* [241].

[10] In the original German of Frobenius, the cancellation law had appeared as 'das Gesetz der eindeutigen Umkehrbarkeit'; elsewhere, Sushkevich rendered this rather awkwardly into English as 'the rule of uniform reversibility' [222].

The main object of Sushkevich's investigations was a finite generalised group $\mathfrak{G}$ in which the left-hand side of the 'law of unique invertibility' holds; because of a quirk of the way in which he labelled left and right, this corresponds to the *right* cancellation condition. Thus, Sushkevich was studying finite right cancellative semigroups. His first observation regarding these was that, for any elements P, Q of such a semigroup $\mathfrak{A}$, the equation $XP = Q$ always has a solution X; this is the left-hand side of a condition that he elsewhere termed the 'law of unrestricted invertibility' (see, for example, [225]), but which we would now call left simplicity. Sushkevich adopted the term *left group* for any right cancellative semigroup; his usage thus tallies in the finite case with the modern sense of this term (for a semigroup that is both right cancellative and left simple). For what follows, it is also worth noting explicitly that Sushkevich's left groups are left simple in the sense of having no proper left ideals; the explicit adaptation of the property of simplicity from rings to semigroups would later be made by Rees.

Having set down his definition of left groups, Sushkevich moved next to the exploration of their basic properties. He noted, for example, that in a left group all idempotents E, which he termed the *principal elements*, are right identities for the whole semigroup. He also noted that every element has a left identity, to which it is said to 'belong'. Taking the principal elements of a left group $\mathfrak{A}$ to be $E_1, E_2, \ldots, E_s$, he showed that

$$\mathfrak{A} = \bigcup_{\kappa=1}^{s} E_s \mathfrak{A},$$

where the $\mathfrak{C}_\kappa := E_\kappa \mathfrak{A}$ are disjoint, mutually isomorphic ordinary groups. In addition, $\mathfrak{F} = \{E_1, E_2, \ldots, E_s\}$ forms a semigroup, the *left principal group* of $\mathfrak{A}$, under the multiplication $E_\kappa E_\lambda = E_\kappa$. Thus, Sushkevich had proved that any finite left group $\mathfrak{A}$ is the disjoint union $\bigcup_{\kappa \in \mathfrak{F}} \mathfrak{C}_\kappa$ of isomorphic copies of a group, indexed by a left zero semigroup $\mathfrak{F}$. Putting this into slightly different language and noting that this can in fact be extended to the infinite case, we have that any left group is isomorphic to the direct product of a group and a left zero semigroup [42, vol. 1, Theorem 1.27]. Only the proof of the finite case appears in Sushkevich's 1928 paper; the infinite extension appears in his 1937 monograph [228, §43].[11] Sushkevich's group-theoretic approach to semigroups is made plain here: he described the structure of his objects of interest in terms of groups.

Sushkevich's study of left groups (and of *right* groups, which he defined analogously) was preparatory for the real goal of his paper: to investigate the minimal ideals of a semigroup. For an arbitrary finite semigroup $\mathfrak{G}$, he turned his

[11] The result was also proved independently by A. H. Clifford in a paper of 1933 that was motivated by a modification of the group axioms [38]; Clifford's term for these semigroups was 'multiple groups'—an appropriate term, since they are unions of groups. Clifford would later turn to the study of unions of groups more generally [40]. On the range of independent characterisations of right and left groups that have appeared in the literature over the decades, see [110, Appendix C].

attention to the subsets $\mathfrak{G}P$ for elements P of $\mathfrak{G}$. Among these, he focused on one of smallest size and denoted it by $\mathfrak{A}$. Applying modern terminology, it is clear that this is a minimal left ideal; Sushkevich demonstrated, moreover, that it is a left group and that all minimal left ideals are isomorphic to this. Denoting the minimal left ideals of $\mathfrak{G}$ by $\mathfrak{A}_1, \mathfrak{A}_2, \ldots, \mathfrak{A}_r$, Sushkevich was able to use what he had shown previously to write each $\mathfrak{A}_\kappa$ in the form

$$\mathfrak{A}_\kappa = \bigcup_{\lambda=1}^{s} \mathfrak{C}_{\kappa\lambda},$$

where the $\mathfrak{C}_{\kappa\lambda}$ are disjoint, mutually isomorphic groups. Applying similar reasoning to the minimal *right* ideals of $\mathfrak{G}$, Sushkevich showed that these are right groups $\mathfrak{B}_\lambda$ that can be written in the form

$$\mathfrak{B}_\lambda = \bigcup_{\kappa=1}^{r} \mathfrak{C}_{\kappa\lambda},$$

where in fact $\mathfrak{C}_{\kappa\lambda} = \mathfrak{A}_\kappa \cap \mathfrak{B}_\lambda$.

By analogy with a definition of Wedderburn, Sushkevich next took the *kernel* $\mathfrak{K}$ of $\mathfrak{G}$ to be the particular (minimal) ideal of $\mathfrak{G}$ that is defined by

$$\mathfrak{K} = \bigcup_{\kappa=1}^{r} \mathfrak{A}_\kappa = \bigcup_{\lambda=1}^{s} \mathfrak{B}_\lambda = \bigcup_{\kappa=1}^{r} \bigcup_{\lambda=1}^{s} \mathfrak{C}_{\kappa\lambda}.$$

Using the machinery that he had developed up to this point, he showed that the kernel $\mathfrak{K}$ of a finite semigroup $\mathfrak{G}$ is completely determined as an ideal of $\mathfrak{G}$ by the following:

(i) The structure of the abstract group $\mathfrak{C}$ which is isomorphic to the $\mathfrak{C}_{\kappa\lambda}$.
(ii) The numbers r and s.
(iii) The $(r-1)(s-1)$ products $E_{11}E_{\kappa\lambda}$ ($\kappa = 2, \ldots, r$; $\lambda = 2, \ldots, s$), where $E_{\kappa\lambda}$ denotes the identity of $\mathfrak{C}_{\kappa\lambda}$.

He noted also that the above three conditions may be chosen arbitrarily in order to construct a kernel that no longer appears as an ideal of another semigroup—this is then of course a (finite) simple semigroup. Although it is by no means obvious from the notation used, Sushkevich's characterisation of kernels is essentially a finite version of the Rees theorem.[12] Harking back to his original motivation in studying semigroups, Sushkevich closed his discussion by presenting an example of such a semigroup, constructed in terms of transformations of a set [221, pp. 49–50].

[12] From the point of view of characterising finite simple semigroups, Clifford and Preston described Sushkevich's criteria as not being 'in a readily usable form' [42, vol. 1, p. 208].

As we have already indicated, a full characterisation of completely 0-simple semigroups, in the form of a semigroup analogue of the Wedderburn–Artin theorem, was eventually provided by David Rees in 1940. Having studied in Cambridge during the late 1930s, Rees had attended the lectures given there by Philip Hall on universal algebra [107, p. 152], which influenced his own approach to the subject and in particular sparked a short-lived interest in semigroups which would shape much of the subsequent theory.

Rees' 1940 paper, entitled simply 'On semi-groups' [193] (where the word was used in its modern sense, modulo a hyphen), generalises the results obtained by Sushkevich on his kernels, but does so in a much more ring-theoretic spirit. In order to develop a purely multiplicative version of the theory of semisimple rings, Rees was obliged to define the necessary fundamental notions, arguably doing it much more clearly than Sushkevich has done. First among these was the notion of an ideal, in terms of which Rees presented the construction that is now termed the *Rees quotient*. Using this and perhaps working in a theory-building mode akin to what Hall had done in his Cambridge lectures, Rees immediately developed semigroup versions of the second and third isomorphism theorems and Zassenhaus' lemma [193, Lemmas 1.322, 1.321 and 1.323]. While laying the groundwork for what was to come, Rees excluded the zero ideal as a 'proper' ideal (if it was present), and so we see that his definition of a 'simple' semigroup as one with no 'proper' ideals (and with the zero semigroup of order 2 explicitly excluded) covers both the modern notions of simple and 0-simple semigroups. At the end of the first part of his paper, Rees concluded that Sushkevich's kernels are precisely finite simple semigroups without zero, together with the zero semigroup.

It is in the second part of this 1940 paper that Rees turned to the more substantial problem of characterising completely 0-simple semigroups.[13] To this end, he began with the following definition [193, p. 393], for which we recall that a *primitive* idempotent is a non-zero idempotent f such that, for all other non-zero idempotents e, $ef = fe = e$ implies that $e = f$:

Definition 2 A semigroup S is *completely (0-)simple* if:

R1. S is (0-)simple.

R2. If $x \in S$, then there exist idempotents $e, f \in S$ such that $xe = x$ and $fx = x$, or, equivalently, x belongs to at least one set of the form fSe, where e and f are idempotents.

R3. There is a primitive idempotent f under every non-primitive idempotent e in the ordering where $i < j$ whenever $i = ij = ji$.

On the face of it, this does not appear to be the same as the modern definition of these semigroups, and yet we note first of all Rees' own observation that R3 may be replaced by the condition that every idempotent be primitive, and we record also an

[13] In the interests of clarity for the present-day reader, we adopt the modern terminology for these semigroups in what follows.

additional result proved by Rees in a short note of the following year: that condition R2 is in fact redundant [194, Lemma 2]. We therefore have here the familiar notion of a completely 0-simple semigroup. Note that at this stage Rees made no explicit mention of descending chain conditions on ideals: primitive idempotents were his main tool here, perhaps because of their usefulness in ring theory.[14]

The parallels between Rees' work and the Wedderburn–Artin theorem start to become clearer when we observe that the manner in which Rees sought to understand completely 0-simple semigroups was through semigroups of matrices. With this end in mind, he gave the following definition [193, §2.9]:

Definition 3 Let L and M be non-empty sets and let G be a group; the group with 0 adjoined is denoted G^0. A matrix $A = (x_{ij})$, where $x_{ij} \in G^0$, for $i \in L$ and $j \in M$, is called an (L, M)*-matrix over* G^0. The (L, M)-matrix over G^0 with just one non-zero entry $x \in G^0$ in the (i, j) position is denoted by $(x)_{ij}$.

Let S be the semigroup consisting of (L, M)-matrices $(x)_{ij}$ over G^0, together with the zero matrix O. Multiplication in S is defined by

$$O(x)_{ij} = (x)_{ij}O = OO = O;$$
$$(x)_{ij}(y)_{kl} = (x)_{ij}P(y)_{kl} = (xp_{jk}y)_{il},$$

where $P = (p_{qr})$ is an (M, L)-matrix over G^0. Then S is called the *matrix semigroup over* G^0 *with defining matrix* P.

If each row and each column of P contains a non-zero entry, then S is called the *regular matrix semigroup over* G^0 *with defining matrix* P.

It is an easy exercise to show that this is equivalent to the 'triples' definition of a Rees matrix semigroup that is in common use today and which was first given by Clifford [40].[15]

The present-day reader will not be surprised to learn that Rees went on to employ his regular matrix semigroups in proving the following result [193, Theorems 2.92 and 2.93]:

Theorem 1 *A regular matrix semigroup over a group with zero is completely* 0-*simple. Conversely, a completely* 0-*simple semigroup is isomorphic to a regular matrix semigroup over a group with zero.*

[14] An early application of chain conditions in the semigroup setting was that of Štefan Schwarz in 1943 [210]. The earliest explicit use of the descending chain condition specifically in connection with completely 0-simple semigroups was probably due to Clifford [41], who had previously employed chain conditions in his heavily ring-theory-inspired work on factorisation in semigroups [39]; see [107, ch. 4].

[15] A similar such structure had also been exhibited by Heinrich Brandt, though in a rather different setting, in 1926 [31]. On Brandt's 'groupoids', which become completely 0-simple semigroups when we adjoin a zero element, see [102] or [107, §6.2].

This result, the Rees theorem [116, Theorem 3.2.3], has justifiably been hailed as semigroup theory's first substantial structure theorem. As we see here quite clearly, it was guided by analogy with results from elsewhere in mathematics—an effective strategy in the early development of any new theory.

4 Inverse Semigroups

We turn finally in this survey to the emergence of what is probably one of the most-studied classes of semigroups—certainly one that has connections with a great range of other areas of mathematics—namely, inverse semigroups. These are also a prime example of the duplication of mathematical efforts that occurred between East and West during the years of the Cold War, inverse semigroups having been developed independently by V. V. Wagner in the USSR in 1952 and by G. B. Preston in the UK the year after. We have seen briefly how semigroup-theoretic questions connected with transformations arose as an extension of ideas from group theory and how other semigroup-related ideas emerged as restrictions of the corresponding notions and results in ring theory. Although the closeness of inverse semigroups to groups has often been emphasised, the inspiration for their definition came from a rather different direction: differential geometry.

As has been related in a number of sources (see the appendix), the story of inverse semigroups can be traced back to the 'Erlanger Programm' advocated by Felix Klein in the later years of the nineteenth century [87]. This was the point of view whereby any given geometry was to be regarded as the theory of invariants of a particular group. The shortcomings of this view were soon recognised, however—certainly in regard to differential geometry, where the corresponding group of automorphisms is trivial. Thus, in the early part of the twentieth century, investigations began into a possible extension of the 'Erlanger Programm' in terms of generalisations of groups. It was with such an extension in view that Oswald Veblen and J. H. C. Whitehead introduced the notion of a 'pseudogroup' in their 1932 monograph on differential geometry [236]. This was a collection of homeomorphisms between open sets of a topological space that was closed under the taking of 'local' inverses and also under the (partial) compositions of mappings, whereby we compose two functions only if the image of the first coincides with the domain of the second. Efforts to devise an appropriate abstract version of such an object soon followed, and it was these investigations with regard specifically to algebraic (as opposed to topological) properties, in conjunction seemingly with a general distaste for partially defined operations, that eventually gave rise to the notion of an inverse semigroup.

Perhaps because of the relative trickiness of defining such natural objects as subsystems, homomorphisms, congruences, etc. in the case of partially defined operations, the first point to be addressed by mathematicians seeking to abstract the properties of a pseudogroup was the completion of the operation to something defined everywhere.[16] Early efforts in this direction saw the definition of the composition of two partial transformations extended to the case when the image of the first has nonempty intersection with the domain of the second [209], while another contribution to the theory composed them if the image of the first is contained in the domain of the second [74]. In each of these cases, however, the composition remains only partially defined. A mental block was arguably at play here: in order to complete the operation we must acknowledge the possibility that the intersection of the image and domain in question may be empty and allow for this eventuality by introducing the 'empty transformation' ε which plays the role of the zero in the symmetric inverse monoid $\mathcal{I}_X$. This latter step was eventually supplied by the Russian geometer Viktor Vladimirovich Wagner at the beginning of the 1950s.

A differential geometer by training, Wagner cultivated an interest in the algebraic foundations of the discipline, and it was precisely this combination of expertise that led him to the notion of an inverse semigroup, prompted by his involvement in the preparation of a Russian translation of Veblen and Whitehead's monograph [237]. In developing algebraic unpinnings for geometrical concepts, Wagner had found the language of binary relations to be particularly useful. With the realisation here that the composition of (partial) functions is merely a special case of the composition of binary relations and that the latter may be regarded as subsets of a Cartesian product, the difficulty that had previously attended the introduction of the empty transformation was removed, since it appeared more natural to introduce ε in the context of sets.

Having noted their usefulness for geometry in earlier writings, Wagner turned to the study of partial transformations for their own sake in a short note of 1951, in which he considered bijections between open sets of a topological space and applied these to the study of 'coordinate structures' in differential geometry [242]. In connection with the latter, he introduced particular systems with a ternary operation that would go on to be closely linked to his development of inverse semigroups, which he began to unfold in two further short notes of 1952, one concerned very closely with partial bijections (that is, one-one transformations between subsets of a given set) and the other to a rather more abstract framing.

It was in one of these 1952 notes [243] that Wagner first recorded explicitly the observation made above: that a partial transformation α on a set X may be regarded as the binary relation

$$\{(x, y) \in \operatorname{dom}\alpha \times \operatorname{im}\alpha : x\alpha = y\}$$

on X, and therefore that the composition of two such transformations α and β appears as a special case of the composition of binary relations. As such, the (left-to-right) composition $\alpha\beta$ is then defined on the domain

[16] Not everyone insisted on working with fully defined operations, however: Charles Ehresmann, for example, pursued the study of groupoids from the starting point of pseudogroups [57].

$$\operatorname{dom} \alpha\beta = (\operatorname{im} \alpha \cap \operatorname{dom} \beta)\alpha^{-1}.$$

Taken at face value, this first 1952 note is phrased entirely in concrete terms as an investigation of the collection $\mathfrak{B}(A \times A)$ of all binary relations on a given set A, which Wagner showed to be a semigroup under composition of binary relations. Exploring further, he noted the presence in $\mathfrak{B}(A \times A)$ of a natural ordering that is compatible with composition and defined a unary operation $^{-1}$ on arbitrary binary relations ρ by the condition that $x\,\rho^{-1}\,y$ if and only if $y\,\rho\,x$. Inside $\mathfrak{B}(A \times A)$, he labelled the collection of all partial bijections by $\mathfrak{M}(A \times A)$ and demonstrated that this has (in later terminology) the structure of an inverse semigroup. Although this was ostensibly a paper about binary relations and seemingly phrased throughout in entirely concrete terms, Wagner never once appealed to the properties of binary relations: having first developed the formal algebraic properties of his elements of interest, he applied only these. The fully abstract formulation would be presented at greater length in a paper the following year, to which we will turn in a moment.

Following closely upon his study of what are essentially concrete inverse semigroups, Wagner formally introduced the abstract version of these—'generalised groups' in his terminology, using the label in a much more restrictive sense than Sushkevich had before him—in his other short note of 1952. To the eyes of the modern semigroup theorist, this note is particularly striking, since it not only contains a fully formed definition of an inverse semigroup and a description of its natural partial order, but also statements and brief proofs of the two key theorems relating to them [244, Theorems 5 and 6]:

Theorem 2 *Every semigroup of partial bijections of some set that is closed under composition and the taking of inverses forms a generalised group with respect to composition of partial transformations and every generalised group may be represented in this form.*

This very short paper is a taster for the much longer piece that came the following year [245], in which Wagner pulled together some of the themes of the preceding papers: binary relations, partial transformations, ternary operations, and generalised groups.[17]

The use of ternary operations, alien perhaps to much other algebraic thinking, had arisen for Wagner in the context of axiomatising coordinate atlases in an earlier paper [242],[18] but the associated ideas translated very naturally to his new setting of binary relations, this time in connection with binary relations between two different sets A and B. Consider the collection $\mathfrak{B}(A \times B)$ of all binary relations between elements of A and B, that is, all subsets of the Cartesian product $A \times B$. The composition of binary relations that Wagner employed above cannot of course be applied here if $A \neq B$, but, making use of his previously defined unary operation, he was able now to define a *ternary* operation for relations $\rho_1, \rho_2, \rho_3 \in \mathfrak{B}(A \times B)$:

[17] For a translation of this 1953 paper and some of the associated shorter pieces, see [113].

[18] See the discussion in [202, p. 194].

$$[\rho_1, \rho_2, \rho_3] = \rho_3 \circ \rho_2^{-1} \circ \rho_1,$$

where $\circ$ is the ordinary composition of binary relations.[19] It is an easy observation that $[\rho_1, \rho_2, \rho_3] \in \mathfrak{B}(A \times B)$, so $\mathfrak{B}(A \times B)$ is closed under this 'triple multiplication'. Wagner's next steps were to characterise the structure of such a collection abstractly, to investigate its special subsystems (in particular, those consisting of partial bijections), and to draw a connection with his generalised groups.

The first property of 'triple multiplication' that Wagner identified was a version of associativity that was later termed *pseudo-associativity* by another author [24]. Moving to an abstract setting, let K be a set upon which we define a ternary operation, denoted by $[k_1\ k_2\ k_3]$ for $k_1, k_2, k_3 \in K$. Pseudo-associativity is then expressed formally as

$$[[k_1\ k_2\ k_3]\ k_4\ k_5] = [k_1\ [k_4\ k_3\ k_2]\ k_5] = [k_1\ k_2\ [k_3\ k_4\ k_5]],$$

for all $k_1, k_2, k_3, k_4, k_5 \in K$. In fact, Wagner was not the first person to study systems with ternary operations in an algebraic setting [113, §3.3]; the above property of pseudo-associativity had previously appeared in a 1929 paper by Reinhold Baer [22], where it had been used in conjunction with the further property

$$[k_1\ k_2\ k_2] = [k_2\ k_2\ k_1] = k_1,$$

for all $k_1, k_2 \in K$, to define what he had called a *Schar*. Through a slightly tortuous etymology, this would eventually become known in English as a *heap*, hence Wagner's name for a system in which only pseudo-associativity holds: a *semiheap*.[20] Thus, $\mathfrak{B}(A \times B)$ forms a semiheap under triple multiplication of binary relations.

For Wagner, semiheaps went hand-in-hand with semigroups in a very natural way: starting from the study of binary relations from a set A to a set B, the semiheap structure becomes that of a semigroup when we set $A = B$. Much of Wagner's long 1953 paper is concerned with establishing the details of this connection between semiheaps and semigroups. In particular, he linked his generalised groups with so-called *generalised heaps*: a semiheap K forms a generalised heap if the following additional conditions hold for all $k, k_1, k_2 \in K$:

$$[[k\ k_1\ k_1]\ k_2\ k_2] = [[k\ k_2\ k_2]\ k_1\ k_1];$$

$$[k_1\ k_1\ [k_2\ k_2\ k]] = [k_2\ k_2\ [k_1\ k_1\ k]];$$

$$[k\ k\ k] = k.$$

---

[19] In fact, Wagner used a very slightly different notation here, but we employ the above for ease of typesetting.

[20] For the derivation of the name and an indication of other terms that have been used, see [107, p. 264, n. 32].

One of the tools that Wagner used to bridge the theories of semiheaps and semigroups was the notion of a semigroup with involution $^{-1}$: a unary operation $^{-1}$ such that $(g^{-1})^{-1} = g$ and $(g_1 g_2)^{-1} = g_2^{-1} g_1^{-1}$ for any elements g_1, g_2 of the semigroup. By analogy with the above triple multiplication of binary relations, such a semigroup S may be regarded as a semiheap, denoted K_S, by defining the ternary operation $[g_1\ g_2\ g_3] = g_1 g_2^{-1} g_3$ on S. If S is a monoid with identity e, then we may return from K_S to S by defining

$$g_1 g_2 = [g_1\ e\ g_2], \qquad g^{-1} = [e\ g\ e],$$

for all $g, g_1, g_2 \in K_S$. It may not always be possible, however, to construct a semigroup with involution from an arbitrary semiheap. In order to do so, we must have access to at least one *biunitary element*: an element b of a semiheap K with the property that $[k\ b\ b] = [b\ b\ k] = k$, for all $k \in K$. Then, if we have such a b, we may replace e by b in the conditions above in order to construct a semigroup K_b with involution from a semiheap K with a biunitary element b. The latter element will then be idempotent in K_b, and any two such semigroups constructed with respect to different biunitary elements will be isomorphic. The problem of needing the presence of a biunitary element in order to perform this construction is mitigated somewhat by Wagner's proof that every semiheap may be embedded in a semiheap with a biunitary element [245, Theorem 2.10].

After setting out the basics of the theories of binary relations and semiheaps, Wagner next revisited the ideas that he had outlined in one of his short papers of 1952, namely, those relating to the notion of generalised groups, which found a natural place in his 1953 paper as precisely those semigroups with involution whose idempotents commute. Here the neat symmetry of Wagner's definitions begins to assert itself, as he proved, for example, that if a generalised group is regarded as a semiheap in the manner outlined above, then it is in fact a generalised heap [245, Theorem 3.10], and similarly that a generalised heap with a biunitary element gives rise to a generalised group [245, Theorem 3.11].

In his final section, Wagner brought his paper to completion by returning to the study of binary relations and, in particular, partial bijections. As preparation for what follows, Wagner observed, for instance, that $\mathfrak{B}(A \times A)$ is a semigroup with involution, the latter being the unary operation that he had previously defined on binary relations. Denoting by $\mathfrak{K}(A \times B)$ the semiheap of all partial bijections from A to B, Wagner proved that this is in fact a generalised heap, and moreover that $\mathfrak{K}(A \times A)$, previously his $\mathfrak{M}(A \times A)$, is a generalised group [245, Theorems 5.1 and 5.2]. As in the short paper of 1952, Wagner showed that any generalised group may be embedded in some $\mathfrak{K}(A \times A)$, but now supplemented the result with the corresponding theorem concerning generalised heaps: that every generalised heap may be embedded in some $\mathfrak{K}(A \times B)$ [245, Theorem 5.19]. Although his work on generalised heaps was little known, at least in the West, during the subsequent decades, Wagner appears to have viewed the extensive interconnections between this and the theory of generalised groups, all situated in relation to the more concrete study of binary relations and partial bijections, not to mention the further applications to differential geometry, as serving to justify the whole.

Differential geometry and binary relations were, however, not at the front of the mind Gordon Preston, the British mathematician to whom credit must also go for the introduction (and the name) of inverse semigroups. He certainly had contact the geometry, as his 1953 Oxford DPhil was initially supervised by J. H. C. Whitehead, but when the latter left for the United States, Preston became a student of the algebraist E. C. Thompson, under whose supervision he eventually produced a dissertation in universal algebra [187]. In parallel with his main topic (the development of an appropriate notion of ideal in certain types of algebra), Preston also cultivated an interest in the burgeoning theory of semigroups, seemingly because of acquaintances, principally David Rees, that he had made while serving at Bletchley Park during the Second World War [192, p. 17]. The result was a curious extra chapter on inverse semigroups at the end of his DPhil dissertation, largely unconnected to the rest of its content, and completed in ignorance of Wagner's investigations. In the first instance, the dissertation chapter spawned three papers of 1953, all published in *Journal of the London Mathematical Society* [188–190]. Preston's approach to inverse semigroups had a considerably more group-theoretic flavour than Wagner's work.

By the beginning of the 1950s, Preston had certainly read Rees' 1940 paper on semigroups and, like Rees, had been influenced by Philip Hall's Cambridge lectures on universal algebra—in his case via notes obtained from Sandy Green [192, p. 19]. Preston had also read a rather different, and very short, paper of Rees' of 1947, concerning partial transformations [195]. Perhaps motivated by recent interest in semigroup embeddings [106], since he illustrated his results by applying them to an embedding problem, Rees had considered the partial transformations of a set, composing them only when the image of the first and the domain of the second have nonempty intersection. Any collection Σ of partial transformations that is closed both under this composition and also under the taking of local inverses, he termed a *regular set of partial transformations*. Defining a particular congruence $\sim$ on Σ, he showed that Σ factored by $\sim$ is a group; his $\sim$ was in fact what we would now call the minimum group congruence [116, p. 159]. As Preston would later note [192, p. 22], allowing for the lack of the empty transformation, Rees was effectively considering an inverse subsemigroup of a symmetric inverse monoid. These ideas piqued Preston's interest, and he set out to axiomatise the structures touched upon by Rees. Although he never attained an axiomatisation with which he was satisfied [192, p. 23], the search for one nevertheless led Preston to the notion of an inverse semigroup.

In his dissertation, Preston introduced the following definition:

Definition 4 Let S be a semigroup with 0. For $a \in S$, let E_a denote the (possibly empty) set of idempotents e for which $ea \neq 0$. We call S a *mapping semigroup* if it satisfies the following conditions:

M1. If $a \in S$, then there is at least one element $e \in S$ for which $ea = a$ and such that the equation $sx = e$ has a solution $x \in S$ (it follows that $e \in E_a$).

M2. If e, f are idempotents in S, then $ef = fe$.
M3. If $a \in S$ and $E_a \neq \emptyset$, then there is at least one $e \in E_a$ such that $ef = f$ and $f^2 = f$ together imply that either $f = 0$ or $f = e$.

Furthermore, a mapping semigroup S is termed a *complete mapping semigroup* if it satisfies the following additional condition:

M4. If $a, b \in S$ have the property that, for all $x, y \in S$, $xay = 0$ if and only if $xby = 0$, then $a = b$.

Condition M2 is of course familiar from the modern theory of inverse semigroups, and it is an easy observation that M1 asserts that every element of S has a generalised inverse; M3 in fact turns out to be redundant and was dropped by Preston in his subsequent writings. Preston's mapping semigroups are therefore inverse semigroups. His set of axioms M1–M4 was intended as a complete axiomatisation of the semigroup of all partial bijections of a set, but it suffers from the slight glitch that an arbitrary inverse semigroup with 0 need not satisfy M4. According to Preston's later recollections [192, pp. 23–24], it was questions from his DPhil examiners, Rees and Whitehead, that enabled him to polish this set of axioms, and it was the refined versions that soon appeared in print in the first of Preston's three introductory papers on inverse semigroups.

In this first paper [188], Preston eschewed his earlier terminology of 'mapping semigroup' in favour of 'inverse semigroup', for he now placed the emphasis on the presence of generalised inverses for every element: an inverse semigroup is defined by his conditions M1 and M2, but in a sequence of lemmas following his initial definition, Preston brought out the existence of inverses more explicitly. His goal in this first paper was the development of certain notions for inverse semigroups that naturally mimicked existing ideas elsewhere in abstract algebra. For example, he proved that every congruence on an inverse semigroup has a kernel which is a normal subsemigroup [188, Theorem 1], where 'kernel' and 'normal' are defined in appropriate senses. He also provided a version of the Jordan–Hölder–Schreier theorem for a series of kernels of what we would now term idempotent-separating morphisms, based upon results already established by others for general algebras [75].

Continuing the theme of providing analogues for existing abstract algebraic ideas, Preston's second paper [189] develops an ideal theory for inverse semigroups. Some of his results here arise as special cases of aspects of his DPhil dissertation. Elsewhere in the article, he sought to embed his work more thoroughly in the steadily growing semigroup-theoretic literature by making links with the papers of Rees [193, 194]. Preston proved, for example, that the union of all minimal right ideals in an inverse semigroup S is equal to the union of the principal ideals SeS, where e ranges over the primitive idempotents of S. But it is perhaps the third of Preston's papers of 1954, concerning representations of inverse semigroups [190], that is the most interesting for us in the present context. Here the theory-building continues, with Preston showing that any inverse semigroup may be embedded in some symmetric inverse monoid. This is the only of Preston's three papers to deal explicitly with partial bijections, and so, like other authors before him, he was required to determine how best to compose these. Although his composition does amount to the same as Wagner's, his approach to it was subtly different in that he did not explicitly employ the empty transformation, and that his transformations were not *a priori* partial bijections: for bijections α, β, mapping from a set A to a set B, whenever $\operatorname{im}\alpha \cap \operatorname{dom}\beta = \emptyset$, Preston took an abstract symbol 0 with the properties that $0^2 = 0$ and $0\alpha = 0 = \alpha 0$ for any transformation α and *defined* the composition $\alpha\beta$ to be equal to 0. Any collection of partial bijections that is closed under this composition (regardless of whether or not it was necessary to introduce 0) he termed simply a *semigroup of (1-1) mappings*. If such a semigroup is also closed under taking inverses, then he applied the additional label *complete*. At no point, however, were the domains and images of Preston's bijections explicitly assumed to be subsets of some universal set. His was thus arguably a purely algebraic and formal solution of the problem of the empty transformation, in contrast to Wagner's more concrete approach via binary relations. Preston's ultimate goal was, however, the same, namely, the proof of the following theorem [190, Theorem 1]:

Theorem 3 *A semigroup admits a faithful representation as a complete semigroup of (1-1) mappings if and only if it is an inverse semigroup.*

In the remaining pages of his paper, Preston once again picked up that tool which Rees had shown to be so useful in the study of the structure of semigroups: their idempotent elements. Preston sought now to represent inverse semigroups by one-one mappings of their idempotents.[21] Here, however, the corresponding theorems did not work quite so nicely as when arbitrary partial bijections are employed: in the first instance, Preston could not guarantee that his representation was faithful [190, Theorem 2]. It was only in the case of inverse semigroups with primitive idempotents that Preston was able to ensure that his embedding was injective and only then by placing additional restrictions on his semigroups. We omit the details, except to note that condition M4 of Preston's dissertation once again came in useful [190, Theorem 4].

The reason why we declared Preston's third paper of 1954 to be particularly interesting for our purposes is that as well as providing the representation result that is the keystone of the elementary theory of inverse semigroups, it also marks the point at which Preston finally became aware of Wagner's prior work: at the end of the paper, Preston thanked B. H. Neumann for drawing his attention to a paper by E. S. Lyapin [152], which cited Wagner in turn. As Preston later recorded [192, p. 25], he therefore set out to learn all that he could about Wagner's work, though not without difficulties of language. Over the following decades, the wider theory of inverse semigroups continued to develop in both East and West, still with the occasional duplication of results [107, ch. 10].

[21] This is a technique that would reappear some years later in the form of the *Munn representation* of an inverse semigroups—see [140, pp. 141–142] and [108, §4].

5 Later Developments

In the present chapter, we have by necessity given only the briefest snapshot of three aspects of the early development of classical semigroup theory. There is much more that we might say about each of these topics, but we hope that what we have included here gives sufficient illustration of the themes that we have identified.

The study of semigroups of transformations has of course remained at the heart of semigroup theory, and we note in particular that under the quiet influence of Sushkevich's early papers, it remained a strong theme in the later Soviet schools of semigroup theory [107, ch. 9]—we caught a glimpse of this in Wagner's study of partial bijections in Sect. 4. For a historical survey of the study of semigroups of transformations more generally, see [219].

In briefly outlining the origins of the Rees theorem, we have given just one example of the way in which a ring-theoretic influence was felt in the early development of semigroup theory. We can see other examples in the study of factorisation properties in semigroups and in efforts to embed semigroups in groups, by analogy with the embedding of integral domains in fields, both of which were active areas of research during the 1930s.[22] But in this connection it is probably the Rees theorem, in conjunction with parallel results obtained by A. H. Clifford in 1941 on completely regular semigroups,[23] that has exerted the strongest influence by providing a model for later semigroup structure theorems—on this last point, see the survey article [167].

Turning next to inverse semigroups, we can hardly understate the place that the study of such semigroups has occupied and continues to occupy, within the broader theory of semigroups. Once introduced, an elaborate theory grew up around them, developed by Wagner, Preston, and their followers. The history of these semigroups has consequently also seen much study—see the various sources listed in the appendix.

Of course there are very many other strands to semigroup theory. These include other topics that might be deemed 'classical', such as Green's relations; the histories of some such branches of the theory have been dealt with to varying degrees in [107]. Regarding the study of semigroups from other perspectives, a number of articles by K. H. Hofmann and J. D. Lawson have treated the theory of topological semigroups (see the appendix), while a few brief comments have been made on the development of the study of finite semigroups in [160]. Some early steps have recently been taken into the study of the history of combinatorial semigroup theory in [178, 179]. But the histories of many other branches of the theory are ripe for study. Any such investigations are likely to yield rich results and insight not only into the development of semigroup-theoretic ideas, but also into the working practices of mathematicians, drawn in many cases directly from the people involved and often

[22] On the former topic, see [107, ch. 4] and on the latter either [107, ch. 4] or [106], with reference to the correction that has recently been made in [180].

[23] In [40]; on this paper, see [107, §6.6].

taking place against the backdrop of the changing geopolitics of the second half of the twentieth century.

Acknowledgments I am grateful to Melanija Mitrović and Mahouton Norbert Hounkonnou for their invitation to contribute a historical chapter to this volume, which has enabled me to revisit some interesting topics after several years away from them.

Appendix: Sources on the History of Semigroup Theory

Sources by Topic

Combinatorial semigroup theory [27, 178, 179, 181]
Embeddings [106]
Finite semigroups [160]
General histories [51, 102, 107, 118, 128]
Inverse semigroups [104, 108, 113, 140, 192, 206, 207]
Leningrad [154]
Semigroup Forum [98, 172]
Smolenice conference 1968 [173]
Topological semigroups [91–96, 138, 139]
Transformations [219]
USSR [69]

Biographies of Contributors to Semigroup Theory

S. I. Adyan [25]
H. Brandt [4, 65]
A. H. Clifford [11, 13, 99, 169, 191]
P. Dubreil [52, 134, 148]
M.-L. Dubreil-Jacotin [146, 147]
J. B. Fountain [79]
L. M. Gluskin [26, 155, 204, 205]
J. A. Green [8, 16, 18, 44, 49, 58, 127]
P. A. Grillet [144]
H.-J. Hoehnke [46, 47, 165]
K. H. Hofmann [141, 142]
J. M. Howie [76, 174, 200, 216]
G. Lallement [3]
J. Lambek [23]
J. D. Lawson [97]
L. Lesieur [34]
M. M. Lesokhin [132]
A. E. Liber [60]
E. S. Lyapin [1, 9, 33, 73, 77, 105, 125, 157, 186, 199, 246]
A. I. Maltsev [2, 10, 29, 30, 35, 48, 68, 78, 126, 133, 137, 158, 159, 177, 199, 201, 217]
W. D. Munn [36, 53, 54, 64, 88, 89, 117, 119, 196]
K. S. S. Nambooripad [168]
M. Petrich [232]
G. B. Preston [21, 84, 115, 192]
V. Pták [17, 62, 63, 208, 234, 235, 239]
L. Rédei [6, 7, 162, 164, 183, 218, 248]
D. Rees [12, 143, 211–215]
J. Rhodes [161]
B. M. Schein [32]
M.-P. Schützenberger [14, 15, 135, 136, 185, 249]
Š. Schwarz [37, 55, 82, 120–123, 129, 170, 173, 198, 253]

L. N. Shevrin [238]
E. G. Shutov [203]
O. Steinfeld [163]
A. K. Sushkevich [5, 50, 67, 70–72, 103, 131, 156, 166, 251, 252]
D. Tamari [19]
T. Tamura [20, 86]
N. N. Vorobev [130]
V. V. Wagner [56, 59, 113, 149, 151, 199, 201, 202, 233]
A. D. Wallace [100, 101]
M. Ward [28, 145]

References

1. Aizenshtat, A.Ya., Schein, B.M.: Evgeniy Sergeyevich Lyapin — in memoriam. Aequationes Math. **73**, 1–9 (2007)
2. Aleksandrov, P.S., Ershov, Yu.L., Kargapolov, M.I., Kuzmin, E.N., Smirnov, D.M., Taimanov, A.D., Shirshov, A.I.: Anatolii Ivanovich Maltsev: obituary. Uspekhi matematicheskikh nauk **23**(3), 159–170 (in Russian); English translation: Russ. Math. Surv. **23**(3), 157–168 (1968)
3. Almeida, J., Perrin, D.: Obituary: Gérard Lallement (1935–2006). Semigroup Forum **78**, 379–383 (2009)
4. Anon: Heinrich Brandt. Jahresber. Dtsch. Math.-Ver. **57**, 8 (1955)
5. Anon: Anton Kazimirovich Sushkevich: obituary. Uspekhi matemticheskikh nauk **17**(2), 165 (1962). In Russian
6. Anon: Academician László Rédei is 70 years old, A Magyar Tudományos Akadémia Matematikai és Fizikai Tudományok Osztályának Közleményei **20**(3–4), 197–201 (1972). In Hungarian
7. Anon: László Rédei. Acta Sci. Math. (Szeged) **43**, 3–4 (1981)
8. Anon: Citation for James Alexander Green (Senior Berwick Prize 1984). Bull. Lond. Math. Soc. **16**, 654–656 (1984)
9. Anon: Evgenii Sergeevich Lyapin, on his 70th birthday. In: Gluskin, L.M., Zhitomirskii, G.I., Lyapin, E.S., Spivak, M.A., Fortunatov, V.A. (eds.) Theory of Semigroups and Its Applications, p. 3. Saratov State University, Saratov (1984)
10. Anon: Academic Anatoly Ivanovich Maltsev (on the 80th anniversary of his birth). Sibirskii matematicheskii zhurnal **30**(6), 3–6 (1989). In Russian
11. Anon: Obituary: Alfred Hoblitzelle Clifford. Semigroup Forum **46**, 137 (1993)
12. Anon: Citation for David Rees, FRS (Pólya prize, 1993). Bull. Lond. Math. Soc. **26**, 413 (1994)
13. Anon: Articles by A. H. Clifford. Semigroup Forum **52**, 3–6; Erratum, ibid. **87**, 494 (1996)
14. Anon: Interview: Marcel-Paul Schützenberger: the miracles of Darwinism. Origins Design **17**(2) (1996)
15. Anon: Nécrologie: Marcel-Paul Schützenberger 1920–1996. Gazette des mathématiciens **71**, 35–36 (1997)
16. Anon: An interview with J. A. Green. Bull. Int. Center Math. **4**, 5–7 (1998)
17. Anon: List of publications of Vlastimil Pták, 1995–1999. Linear Algebra Appl. **310**, 23–24 (2000)
18. Anon: Prizewinners. James Alexander Green: De Morgan Medal 2001. Bull. Lond. Math. Soc. **34**, 633–634 (2002)
19. Anon: Dov Tamari, 1911–2006. Focus: Newsletter of the MAA, October, 8 (2007)
20. Anon: Obituary: Takayuki Tamura (1919–2009). Semigroup Forum **79**, 1 (2009)
21. Anon: Vale Gordon Preston, Monash University, 22nd April 2015 (2015). https://www.monash.edu/news/articles/vale-gordon-preston. Accessed 27 Jan 2024
22. Baer, R.: Zur Einführung des Scharbegriffs. J. für reine und angewandte Mathematik **160**, 199–207 (1929)

23. Barr, M.: An appreciation of Jim Lambek at McGill: a talk given on the occasion of Jim Lambek's 75th birthday. Theory Appl. Categories **6** (The Lambek Festschrift), 2–4 (1999)
24. Behanzin, L.: Quelques considérations sur la théorie des demi-amas, Séminaire Dubreil. Algèbre et théorie des nombres **12**(1), exp. no. 3, 1–18 (1958–1959)
25. Beklemishev, L.D., Lysenok, I.G., Mal'tsev, A.A., Novikov, S.P., Pentus, M.R., Razborov, A.A., Semenov, A.L., Uspenskii, V.A.: Sergeĭ Ivanovich Adyan (on the occasion of his 75th birthday). Uspekhi matematicheskikh nauk **61**(3(369)), 179–191 (in Russian); English translation: Russian Math. Surv. **61**(3), 575–588 (2006)
26. Belousov, V.D., Berman, S.D., Lyapin, E.S., Mikhalyov, A.V., Novikov, B.V., Plotkin, B.I., Shevrin, L.N., Skornyakov, L.A.: Lazar Matveevich Gluskin (obituary), Uspekhi matematicheskikh nauk **42**(4), 113–114 (in Russian); English translation: Russian Math. Surv. **42**(4), 139–140 (1987)
27. Berstel, J., Perrin, D.: The origins of combinatorics on words. Eur. J. Comb. **28**, 996–1022 (2007)
28. Bohnenblust, F., Badger, R., DuBridge, L.A.: Morgan Ward, 1901–1963. California Institute of Technology, Pasadena, CA (1963)
29. Bokut, L.A.: Memories of Anatolii Ivanovich Maltsev. Byull. Sib. Mat. Ob. Novosibirsk, 22–25 (1989). In Russian
30. Bokut, L.A.: Anatolii Illarionovich Shirshov (1921–1981) and Anatolii Ivanovich Maltsev (1909–1967) in my life. In: Anatolii Illarionovich Shirshov — From a Cohort of Great Scientists, Istoriko-kraevedcheskii muzei MU "Kulturno-dosugovii tsentr" administratsii g. Aleiska, pp. 27–37 (2003). In Russian
31. Brandt, H.: Über eine Verallgemeinerung des Gruppenbegriffes. Math. Ann. **96**, 360–366 (1926)
32. Breen, M., Molchanov, V.A., Trokhimenko, V.S.: Boris M. Schein's 70th birthday. Aequationes Math. **82**, 1–30 (2011)
33. Budyko, M.I., Wagner, V.V., Vulikh, B.Z., Gluskin, L.M., Evseev, A.E., Faddeev, D.K., Shevrin, L.N.: Evgenii Sergeevich Lyapin (on his sixtieth birthday). Uspekhi matematicheskikh nauk **30**(3), 187–194 (in Russian); English translation: Russian Math. Surv. **30**(3), 139–147 (1975)
34. Cauchon, G.: Léonce Lesieur (1914–2002). Gazette des Mathématiciens **93**, 119–120 (2002)
35. Cheremisin, A.: Scientist and man. Rabochii Krai **27**(18969) (1984). 27th November (in Russian)
36. Churchhouse, B.: Douglas Munn. Lond. Math. Soc. Newsl. **381**, 5 (2009)
37. Čižmár, J.: The origins of modern algebra in Slovakia (Š. Schwarz, M. Kolibiar, J. Jakubík), in Bečvář, J., Fuchs, E. (eds.). Mathematics Throughout the Ages II, Dějiny Matematiky/History of Mathematics, vol. 16, pp. 251–262. Prometheus, Prague (2001). In Slovak
38. Clifford, A.H.: A system arising from a weakened set of group postulates. Ann. Math. (2) **34**, 865–871 (1933)
39. Clifford, A.H.: Arithmetic and ideal theory of commutative semigroups. Ann. Math. (2) **39**, 594–610 (1938)
40. Clifford, A.H.: Semigroups admitting relative inverses.Ann. Math. (2) **42**, 1037–1049 (1941)
41. Clifford, A.H.: Semigroups without nilpotent ideals. Am. J. Math. **71**, 834–844 (1949)
42. Clifford, A.H., Preston, G.B.: The Algebraic Theory of Semigroups, 2 vols., Mathematical Surveys, no. 7. American Mathematical Society, Providence (1961–67)
43. Cohn, P.M.: Basic Algebra: Groups, Rings and Fields. Springer, Berlin (2003)
44. Collins, M.: Sandy Green: Mathematician who worked at Bletchley before becoming a leading figure in the discipline of representation theory. The Independent, Sunday 20th July (2014)
45. Corry, L.: Modern Algebra and the Rise of Mathematical Structures, 2nd edn. Birkhäuser, Basel (2004)
46. Denecke, K.: To the memory of Hans-Jürgen Hoehnke (1925–2007). Diss. Math. – General Algebra Appl. **28**(1), 5–9 (2008)

47. Denecke, K.: Hans-Jürgen Hoehnke. Scientiae Mathematicae Japonicae **68**(2), 177–181 (2008)
48. Dimitrić, R.: Anatoly Ivanovich Maltsev. Math. Intell. **14**(2), 26–30 (1992)
49. Donkin, S., Erdmann, K.: James Alexander Green, 26 February 1926–7 April 2014. Biogr. Mem. Fellows R. Soc. **67**, 173–190 (2019)
50. Dubovitskiy, F.I.: And a lot lived through.... Chernovolovskaya Gazeta **6**(810) (2007). In Russian
51. Dubreil, P.: Apparition et premiers développements de la théorie des demi-groupes en France. Cahiers du séminaire d'histoire des mathématiques **2**, 59–65 (1981)
52. Dubreil, P.: Souvenirs d'un boursier rockefeller 1929–1931. Cahiers du séminaire d'histoire des mathématiques **4**, 61–73 (1983)
53. Duncan, L.: Obituary: Professor Douglas Munn. Stirling Observer, 12th Nov (2008)
54. Duncan, L.: Professor Walter Douglas Munn: an appreciation. The Herald 12th Nov. (2008)
55. Dvurečenskij, A.: Academician Štefan Schwarz (1914–1996). Math. Slovaca **46**(4), 433–434 (1996)
56. Efimov, N.V., Liber, A.E., Lyapin, E.S., Rashevskii, P.K.: Viktor Vladimirovich Wagner (on his seventieth birthday). Uspekhi matematicheskikh nauk **34**(4), 227–229 (in Russian); English translation: Russian Math. Surv. **34**(4), 209–212 (1979)
57. Ehresmann, Ch.: Gattungen von lokalen Strukturen. Jahresbericht der Deutschen Mathematiker-Vereinigung **60**, 49–77 (1957)
58. Erdmann, K.: Obituary: James Alexander (Sandy) Green. Lond. Math. Soc. Newsl. **437**, 29 (2014)
59. Ermakov, Yu.I., Laptev, B.L., Liber, A.E., Norden, A.P., Shirokov, A.P.: In memory of Viktor Vladimirovich Wagner. Izvestiya vysshikh uchebnykh zavedenii. Matematika **10**, 85–88 (1981). In Russian
60. Ermakov, Yu.I., Laptev, B.L., Lyapin, E.S., Norden, A.P., Ostianu, N.M., Shirokov, A.P.: Aleksandr Evgenevich Liber (on his 70th birthday). In: Buchko, I.Yu. (ed.). Differential Geometry: Structures in Manifolds and Their Applications, no. 8, pp. 3–9. Izdat. Saratov University (1985). In Russian
61. Ferreirós Domínguez, J.: Labyrinth of Thought: A History of Set Theory and Its Role in Modern Mathematics, 2nd rev. edn. Birkhäuser, Basel (2007)
62. Fiedler, M.: Vlastimil Pták (8 November 1925–9 May 1999). Linear Algebra Appl. **310**, 21–22 (2000)
63. Fiedler, M., Müller, V.: Professor Vlastimil Pták died. Czech. Math. J. **50**(125), 445–447; Math. Bohemica **125**(3), 371–373 (2000)
64. Fountain, J.: The work of Douglas Munn and its legacy. Semigroup Forum **81**(1), 2–25 (2010); erratum: ibid. **82**(1), 197 (2011)
65. Fritzsche, R., Hoehnke, H.-J.: Heinrich Brandt: 1886–1986. Wissenschaftliche Beiträge 47, Martin-Luther-Universität, Halle-Wittenberg, Halle (1986)
66. Frobenius, F.G.: Über endliche Gruppen, Sitzungsber. Akad. Wiss. Berlin, 81–112 (1895); also: Gesammelte Abhandlungen. Serre, J.-P. (ed.), pp. 632–663. Band II. Springer, Berlin (1968)
67. Gaiduk, Yu.M.: Anton Kazimirovich Sushkevich. Istoriko-matematichna zbirka **3**, 3–6 (1962). In Ukrainian
68. Gainov, A.T., Goncharov, S.S., Ershov, Yu.L., Zakharov, D.A., Kuzmin, E.N., Maksimova, L.L., Merzlyakov, Yu.L., Smirnov, D.M., Taimanov, A.D., Kharchenko, V.K., Khukhro, E.I.: On the eightieth birthday of outstanding Soviet mathematician Academic A. I. Maltsev. Algebra i logika **28**(6), 615–618 (1989). In Russian
69. Gluskin, L.M.: Theory of semigroups. In Shtokalo, I.Z., Bogolyubov, A.N. (eds.) (1966). History of National Mathematics, 4 vols., Akademiya nauk SSSR/Akademiya nauk USSR, Naukova Dumka, Kiev, 1966–1970, vol. 3, pp. 321–332 (1968). In Russian
70. Gluskin, L.M., Lyapin, E.S.: Anton Kazimirovich Sushkevich (on his seventieth birthday). Uspekhi matemticheskikh nauk **14**, 255–260 (1959). In Russian

71. Gluskin, L.M., Schein, B.M.: The theory of operations as the general theory of groups (Anton Suškevič, dissertation, Voronezh, 1922): an historical review. Semigroup Forum **4**, 367–371 (1972)
72. Gluskin, L.M., Zhmud, E.M., Fedoseev, M.V.: In memory of a scholar and a teacher (on the 10th anniversary of the death of Anton Kazimirovich Sushkevich). Matematika v shkole **1**, 82 (1972). In Russian
73. Gluskin, L.M., Evseev, E.E., Kolesnikova, T.L., Krivenko, V.M., Faddeev, D.K., Shevrin, L.N.: Evgenii Sergeevich Lyapin (on his seventieth birthday). Uspekhi matematicheskikh nauk **40**(2), 211–212 (In Russian); English translation: Russian Math. Surv. **40**(2), 251–253 (1985)
74. Gołąb, S.: Über den Begriff der 'Pseudogruppe von Transformationen'. Mathematische Annalen **116**, 768–780 (1939)
75. Goldie, A.W.: The Jordan–Hölder Theorem for general abstract algebras. Proc. Lond. Math. Soc. (2) **52**, 107–131 (1950)
76. Gomes, G.M.S., Ruškuc, N.: John Macintosh Howie: work and legacy. Semigroup Forum **89**, 2–19 (2014)
77. Gordeev, N.L., Verner, A.L., Evseev, A.E., Kublanovskii, S.I., Ponizovskii, J.S., Yakovlev, A.V.: Evgenii Sergeevich Lyapin: obituary. Uspekhi matematicheskikh nauk **60**(2), 143–144 (in Russian); English translation: Russian Math. Surv. **60**(2), 335–336 (2005)
78. Goryaeva, E. (ed.): Academic Maltsev: man, scientist, pedagogue. Yuniversitetskaya zhizn (Novosibirsk) **32**(175), 21st October (1986). In Russian
79. Gould, V.A.R., Lawson, M.V.: A tribute to John B. Fountain. Semigroup Forum **99**, 1–8 (2019)
80. Gray, J.: Worlds Out of Nothing: A Course in the History of Geometry in the 19th Century. Spinger, London (2007)
81. Gray, J.: A History of Abstract Algebra: From Algebraic Equations to Modern Algebra, Springer, Cham (2018)
82. Grošek, O., Satko, L., Schein, B.: Eightieth birthday of Professor Štefan Schwarz. Semigroup Forum **49**, 1–5 (1994)
83. Haar, A.: Über unendliche kommutative Gruppen. Math. Z. **33**(1), 129–159 (1931)
84. Hall, T.E.: G. B. Preston: his work so far, in [85], pp. 2–15 (1991)
85. Hall, T.E., Jones, P.R., Meakin, J.C.: (eds.): Monash Conference on Semigroup Theory, Melbourne 1990. World Scientific, River Edge (1991)
86. Hamilton, H.B., Nordahl, T.E.: Tribute for Takayuki Tamura on his 90th birthday. Semigroup Forum **79**(1), 2–14 (2009)
87. Hawkins, T.: The Erlanger Programm of Felix Klein: reflections on its place in the history of mathematics. Hist. Math. **11**(4), 442–470 (1984)
88. Hickey, J.: Professor Douglas Munn. The Scotsman, 12th Nov (2008)
89. Hickey, J.: Douglas Munn. Lond. Math. Soc. Newsl. **380** (April), 10 (2009)
90. Hilbert, D.: Grundlagen der Geometrie. Teubner, Leipzig (1899)
91. Hofmann, K.H.: Topological semigroups: history, theory, applications. Jahresber. Dtsch. Math.-Ver. **78**, 9–59 (1976)
92. Hofmann, K.H.: Semigroups in the 19th century? A historical note. In: Theory of Semigroups—Theorie der Halbgruppen. Proceedings of the International Conference Held in Greifswald, November 12–16, 1984, pp. 44–55. Mathematische Gesellschaft der DDR, Berlin (1985)
93. Hofmann, K.H.: Zur Geschichte des Halbgruppenbegriffs. Hist. Math. **19**, 40–59 (1992)
94. Hofmann, K.H.: Semigroups and Hilbert's fifth problem. Math. Slovaca **44**(3), 365–377 (1994)
95. Hofmann, K.H.: From a topological theory of semigroups to a geometric one. Semigroup Forum **50**, 123–134 (1995)
96. Hofmann, K.H.: A history of topological and analytical semigroups: a personal view. Semigroup Forum **61**, 1–25 (2000)

97. Hofmann, K.H.: Jimmie D. Lawson on the occasion of his 75th birthday. Semigroup Forum **97**, 1–6 (2018)
98. Hofmann, K.H., Lawson, J.D.: Throwing a glance at 100 volumes of a journal and at half a century of its existence. Semigroup Forum **100**(1), 2–8 (2020)
99. Hofmann, K.H., Mislove, M.W. (eds.): Semigroup theory and its applications: proceedings of the 1994 conference commemorating the work of Alfred H. Clifford. LMS Lecture Note Series, no. 231. Cambridge University Press, Cambridge (1996)
100. Hofmann, K.H., Koch, R.J., Mostert, P.S.: Alexander Doniphan Wallace on his 68th birthday. Semigroup Forum **7**, 10–31 (1974)
101. Hofmann, K.H., Koch, R.J., Mostert, P.S.: Alexander Doniphan Wallace in memoriam. Semigroup Forum **34**, 1–4 (1986)
102. Hollings, C.: The early development of the algebraic theory of semigroups. Arch. Hist. Exact Sci. **63**(5), 497–536 (2009)
103. Hollings, C.: Anton Kazimirovich Suschkewitsch (1889–1961). BSHM Bull. J. Brit. Soc. History Math. **24**(3), 172–179 (2009)
104. Hollings, C.: The Ehresmann–Schein–Nambooripad Theorem and its successors. Euro. J. Pure Appl. Math. **5**(4), 414–450 (2012)
105. Hollings, C.: The case of Evgenii Sergeevich Lyapin. Math. Today **48**(4), 184–186 (2012)
106. Hollings, C.: Embedding semigroups in groups: not as simple as it might seem. Arch. Hist. Exact Sci. **68**(5), 641–692 (2014)
107. Hollings, C.: Mathematics Across the Iron Curtain: A History of the Algebraic Theory of Semigroups. American Mathematical Society, Providence (2014)
108. Hollings, C.: Three approaches to inverse semigroups. Euro. J. Pure Appl. Math. **8**(3), 294–323 (2015)
109. Hollings, C.D.: Scientific communication across the Iron Curtain. SpringerBriefs in History of Science and Technology, Springer, Berlin (2016)
110. Hollings, C.D.: "Nobody could possibly misunderstand what a group is": a study in early twentieth-century group axiomatics. Arch. Hist. Exact Sci. **71**(5), 409–481 (2017)
111. Hollings, C.D.: The need for closure. Math. Gaz. **103**(557), 248–256 (2019)
112. Hollings, C.D.: Mathematical study in Berlin 1906–1911: The notebooks of A. K. Sushkevich. To appear in The Mathematical Intelligencer **47**(2)
113. Hollings, C.D., Lawson, M.V. (eds.): Wagner's Theory of Generalised Heaps. Springer, Cham (2017)
114. Howie, J.M.: An Introduction to Semigroup Theory. Academic Press, London (1976); updated edition: [116]
115. Howie, J.M.: Gordon Bamford Preston. Semigroup Forum **51**, 269–271 (1995)
116. Howie, J.M.: Fundamentals of semigroup theory. LMS Monographs, New Series, no. 12. Clarendon Press, Oxford (1995); updated edition of [114]
117. Howie, J.M.: Tribute: Walter Douglas Munn. Semigroup Forum **59**, 1–7 (1999)
118. Howie, J.M.: Semigroups, past, present and future. In: Hemakul, W. (ed.) Proceedings of the International Conference on Algebra and Its Applications, pp. 6–20. Department of Mathematics, Chulalongkorn University, Bangkok (2002)
119. Howie, J.M.: Professor Walter Douglas Munn. The Herald, Glasgow (2008)
120. Jakubík, J., Kolibiar, M.: Sixtieth anniversary of the birthday of Academician Štefan Schwarz, Czech. Math. J. **24**(99)(2), 331–340; Slovak versions: Časopis pro pěstování matematiky **99**, 200–213; Matematicky časopis. Slovenskej Akadémie Vied **24**(2), 99–111 (1974)
121. Jakubík, J., Kolibiar, M.: Seventieth anniversary of birthday of Academician Štefan Schwarz. Czech. Math. J. **34**(109), 490–498; Slovak version: Mathematica Slovaca **34**(2), 239–246 (1984)
122. Jakubík, J., Kolibiar, M.: Eighty years of Professor Štefan Schwarz. Math. Slovaca **44**(2), i–x (1994)
123. Jakubík, J., Kolibiarová, B., Kolibiar, M.: On the seventieth anniversary of academic Štefan Schwarz. Časopis pro pěstování matematiky **109**(3), 329–334 (1984). In Slovak

124. Katz, V.J., Parshall, K.H.: Taming the Unknown: A History of Algebra from Antiquity to the Early Twentieth Century. Princeton University Press, Princeton (2014)
125. Khait, N.M.: Pride of our city. Istoriya Peterburga **1**(23), 11–17 (2005). In Russian
126. Khalezov, E.A.: On the 75th year of Academic A. I. Maltsev. Sibirskii matematicheskii zhurnal **25**, 1–2 (1984). In Russian
127. Khammash, A.A.: Sandy Green As I Knew Him: personal impressions to memorialize Professor J. A. Green as a teacher and a friend. Not. Am. Math. Soc. **63**(6), 657–658 (2016)
128. Knauer, U.: Zur Entwicklung der algebraischen Theorie der Halbgruppen. Simon Stevin **54**, 165–177 (1980)
129. Kolibiar, M.: On the fiftieth anniversary of academic Štefan Schwarz. Matematicko-fyzikálny časopis. Slovenskej Akadémie Vied **14**(2), 150–157 (1964). In Slovak
130. Korbut, A.A., Yanovskaya, E.B.: In memoriam: N. N. Vorob'ev (1925–1995). Games Econ. Behav. **12**, 283–285 (1996)
131. Korzhova, O.V., Grigorovich, D.S.: The views of A. K. Sushkevich on the use of tests in the teaching of mathematics in high schools. In Belousova, L.I. (ed.) Research Students as a Factor in Improving the Training of Future Teachers: Scientific Papers, pp. 127–130. Kharkiv National Pedagogical University, Kharkiv (2010). In Ukrainian
132. Kublanovsky, S.I.: M. M. Lesokhin (1933–1998). Semigroup Forum **59**, 159–166 (1999)
133. Kurosh, A.G.: Anatoly Ivanovich Maltsev (on his fiftieth birthday). Uspekhi matematicheskikh nauk **14**(6), 203–211 (1959). In Russian
134. Lallement, G.: Paul Dubreil (1904–1994) in memoriam. Semigroup Forum **50**, 1–7 (1995)
135. Lallement, G., Perrin, D.: Marcel-Paul Schützenberger (1920–1996). Semigroup Forum **55**, 135–151 (1997)
136. Lallement, G., Simon, I.: List of publications of Marcel-Paul Schützenberger. Theor. Comput. Sci. **204**, 3–9 (1998)
137. Lavrov, I.A.: On the 100th birthday of Academic A. I. Maltsev. In: Seriya "Estestvennye, obshchestvennye nauki". Ivanovskii gosudarstvennyi universitet, pp. 139–147 (2009). In Russian
138. Lawson, J.D.: Historical links to a Lie theory of semigroups. Seminar Sophus Lie **2**, 263–278 (1992)
139. Lawson, J.D.: The earliest semigroup paper? Semigroup Forum **52**, 55–60 (1996)
140. Lawson, M.V.: Inverse Semigroups: The Theory of Partial Symmetries. World Scientific, Singapore (1998)
141. Lawson, J.D.: An interview with Karl H. Hofmann on the occasion of his seventieth birthday. Semigroup Forum **65**, 317–328 (2002)
142. Lawson, J.D.: A tribute to Karl H. Hofmann on the occasion of his 75th birthday. Semigroup Forum **77**, 2–4 (2008)
143. Lawson, M.V., O'Carroll, L., Rees, S.: David Rees 1918–2013. Semigroup Forum **88**, 513–522 (2014)
144. Leech, J.: Pierre Antoine Grillet at 75: a tribute. Semigroup Forum **94**, 189–193 (2017)
145. Lehmer, D.H.: The mathematical work of Morgan Ward. Math. Comput. **61**(203) (Special issue dedicated to Derrick Henry Lehmer), 307–311 (1993)
146. Leray, J.: Marie–Louise Dubreil: 7 juillet 1905–19 octobre 1972. Annuaire des Anciens Élèves de l'École Normale Supérieure; English translation: https://mathshistory.st-andrews.ac.uk/Extras/Dubreil-Jacotin/ (1974)
147. Lesieur, L.: Marie–Louise Dubreil–Jacotin, 1905–1972. Semigroup Forum **6**, 1–2 (1973); English translation: https://mathshistory.st-andrews.ac.uk/Extras/Dubreil-Jacotin/
148. Lesieur, L.: Paul Dubreil (1904–1994). Gazette des Mathématiciens **60**, 74–75 (1994)
149. Liber, A.E., Penzov, Yu.E., Rashevskii, P.K.: Viktor Vladimirovich Wagner (on his fiftieth birthday). Uspekhi matematicheskikh nauk **13**, 221–227 (1958). In Russian
150. Loewy, A.: Über abstrakt definierte Transmutationssysteme oder Mischgruppen. J. für reine und angewandte Mathematik **157**, 239–254 (1927)
151. Losik, M.V., Rozen, V.V. (eds.): Viktor Vladimirovich Wagner, on the 100th anniversary of his birth. In: Proceedings of Saratov University, Series Mathematics, Mechanics and Informatics, vol. 8. Saratov State University, Saratov (2008). In Russian

152. Lyapin, E.S.: Associative systems of all partial transformations. Dokl. Akad. Nauk SSSR **88**, 13–15 (1953); Errata: ibid. **92**, 692 (1953). In Russian
153. Lyapin, E.S.: Semigroups, Gosudarstvennoe Izdatelstvo, Fiziko-Matematicheskoi Literatury, Moscow (1960). In Russian; English translation: Translations of Mathematical Monographs, vol. 3, American Mathematical Society, Providence (1963)
154. Lyapin, E.S., Zybina, L.D.: The study of semigroups in the algebra cathedra of the A. I. Herzen Leningrad State Pedagogical Institute from 1946 to 1967. Uchenye zapiski Leningradskogo gosudarstvennogo pedagogicheskogo instituta **404**, 16–58 (1971). In Russian
155. Lyapin, E.S., Zhitomirskii, G.I., Kolesnikova, O.V.: Lazar Matveevich Gluskin (on his 60th birthday). In Wagner, V.V., Gluskin, L.M., Zhitomirskii, G.I., Lyapin, E.S., Fortunatov, V.A. (eds.) Theory of Semigroups and Its Applications: Polyadic Semigroups, Semigroups of Transformations, pp. 3–10. Saratov State University, Saratov 1983. In Russian
156. Lyubich, Yu.I., Zhmud, E.M.: Anton Kazimirovich Sushkevich. Kharkov State University Newspaper April (1989). In Russian
157. Makaridina, V.A., Mogilyanskaya, E.M.: Evgeniy Sergeevich Lyapin, 1914–2005: reflections by two pupils on their teacher. Semigroup Forum **77**, 143–151 (2008)
158. Malcev, I.A.: Anatolii Ivanovich Malcev (on the centenary of his birth). Uspekhi matematicheskikh nauk **65**(5), 197–203. In Russian; English translation: Russian Math. Surv. **65**(5), 991–997 (2010)
159. Marchuk, G.I., Kulikov, L.Ya., Kargapolov, M.I., Taimanov, A.D., Plotkin, B.I., Kemkhadze, Sh.S., Andrunakievich, V.A., Ershov, Yu.L., Yanovskii, R.G., Shirshov, A.I., Morozov, V.V., Smirnov, S.V.: Memories of A. I. Maltsev (from the opening ceremony of the Tenth All-Union Algebra Colloquium, dedicated to the memory of Academic A. I. Maltsev, Novosibirsk, 20–26 September 1967). In Collection Dedicated to the Memory of A. I. Maltsev, “Nauka”, Sibirskoe otdelenie, Novosibirsk (1973). In Russian
160. Margolis, S.W.: The q-theory of finite semigroups: history and mathematics (2015). arXiv:1409.2308
161. Margolis, S.W., Steinberg, B.: Introduction, International conference on semigroups and groups in honor of the 65th birthday of Prof. John Rhodes. Int. J. Algebra Comput. **14**, iii–xii (2004)
162. Márki, L.: A tribute to L. Rédei. Semigroup Forum **32**, 1–21 (1985)
163. Márki, L.: Ottó Steinfeld 1924–1990. Semigroup Forum **43**, 127–134 (1991)
164. Márki, L., Steinfeld, O., Szép, J. Short review of the work of László Rédei. Studia Scientiarum Mathematicarum Hungarica **16**(1–2), 3–14 (1981)
165. Márki, L., Pöschel, R., Vogel, H.-J.: Hans-Jürgen Hoehnke. Semigroup Forum **52**, 112–118 (1996)
166. Maznitsa, A.: “Abstract world” in reality. Zerkalo Nedeli **33**(202), 15th–21st August (1998) In Russian
167. Meakin, J.: The Rees construction in regular semigroups, in Semigroups (Szeged, 1981). Colloquia Mathematica Societatis János Bolyai 39, pp. 115–155. North-Holland, Amsterdam (1985)
168. Meakin, J., Rajan, A.R.: Tribute to K.S.S. Nambooripad. Semigroup Forum **91**, 299–304 (2015)
169. Miller, D.D.: A. H. Clifford: the first sixty-five years. Semigroup Forum **7**, 4–9 (1974)
170. Mišík, L.: Academician Štefan Schwarz awarded the 1980 National Prize of the Slovak Socialist Republic. Czech. Math. J. **31**(106), 338–339 (1981)
171. Mitrović, M.: Semilattices of Archimedean Semigroups. University of Niš, Serbia (2003)
172. Mostert, P.S.: The founding of the semigroup forum on the 50th anniversary of its formation. Semigroup Forum **100**(1), 9–15 (2020)
173. Mostert, P.S.: A front row seat: Štefan Schwarz and the Smolenice meeting. Semigroup Forum **100**, 16–21 (2020)
174. Munn, W.D.: John Mackintosh Howie: an appreciation. Semigroup Forum **73**, 1–9 (2006)
175. Neumann, P.M.: What groups were: A study of the development of the axiomatics of group theory. Bull. Aust. Math. Soc. **60**, 285–301 (1999)

176. Neumann, P.M.: The Mathematical Writings of Évariste Galois. European Mathematical Society, Zürich (2011)
177. Nikolskii, S.M.: Excerpts from a memoir on A. I. Maltsev. Uspekhi matematicheskikh nauk **27**(4), 223–230 (1972). In Russian; English translation: Russian Math. Surv. **27**(4), 179–187
178. Nyberg-Brodda, C.-F.: The word problem for one-relation monoids: a survey. Semigroup Forum **103**, 297–355 (2021); correction, ibid. **105**, 834–838 (2022)
179. Nyberg-Brodda, C.-F. (transl.): S. I. Adian's papers on the undecidability of algorithmic problems in group theory or the Adian–Rabin Theorem. arXiv:2208.08560v2 (2022)
180. Nyberg-Brodda, C.-F. (transl.): Six Ukrainian articles on semigroup theory by A. K. Sushkevich. arXiv:2302.02996v1 (2023)
181. Nyberg-Brodda, C.-F.: G. S. Tseytin's seven-relation semigroup with undecidable word problem. arXiv:2401.11757 (2024)
182. Paalman-de Miranda, A.B.: Topological Semigroups. Mathematisch Centrum, Amsterdam (1964)
183. Pálfy, P.P., Szép, J. The work of László Rédei in group theory. Matematikai lapok **33**(4), 243–254 (1982). In Hungarian
184. Pflugfelder, H.O.: Historical notes on loop theory. Comment. Math. Univ. Carol. **41**(2), 359–370 (2000)
185. Pin, J.-É.: Marcel-Paul Schützenberger (1920–1996). Int. J. Algebra Comput. **9**(3–4), 227–239 (1999)
186. Ponizovskii, J.S.: Evgeniĭ Sergejevich Lyapin on his 80th birthday. Semigroup Forum **49**, 271–274 (1994)
187. Preston, G.B.: Some Problems in the Theory of Ideals. DPhil thesis. University of Oxford, Oxford (1953)
188. Preston, G.B.: Inverse semi-groups. J. Lond. Math. Soc. **29**, 396–403 (1954)
189. Preston, G.B.: Inverse semi-groups with minimal right ideals. J. Lond. Math. Soc. **29**, 404–411 (1954)
190. Preston, G.B.: Representations of inverse semi-groups. J. Lond. Math. Soc. **29**, 411–419 (1954)
191. Preston, G.B.: A. H. Clifford: an appreciation of his work on the occasion of his sixty-fifth birthday. Semigroup Forum **7**, 32–57 (1974)
192. Preston, G.B.: Personal reminiscences of the early history of semigroups, in [85], pp. 16–30 (1991)
193. Rees, D.: On semi-groups. Proc. Camb. Philos. Soc. **36**, 387–400 (1940)
194. Rees, D.: Note on semi-groups. Proc. Camb. Philos. Soc. **37**, 434–435 (1941)
195. Rees, D.: On the group of a set of partial transformations. J. Lond. Math. Soc. **22**, 281–284 (1947)
196. Reilly, N.R.: Obituary: Walter Douglas Munn 1929–2008. Semigroup Forum **78**, 1–6 (2009)
197. Rhodes, J.: Algebraic theory of semigroups, Séminaire Dubreil–Pisot. Algèbre et théorie des nombres **23**(2), exp. no. DG10, DG1–DG9 (1969–1970)
198. Riečan, B.: Štefan Schwarz (1914–1996). Czech. Math. J. **47**(122)(2), 375–382 (1997)
199. Riguet, J.: Travaux récents de Malčev, Vagner, Liapin, Séminaire Châtelet–Dubreil; partie complémentaire: demi-groupes **7**(1953–1954), exp. no. 18, 9pp. (1953)
200. Robertson, E.: John Howie. Lond. Math. Soc. Newsl. **412**(March), 11 (2012)
201. Rosenfeld, B.A.: Reminiscences of Soviet mathematicians. In: Zdravkovska, S., Duren, P.L. (eds.) Golden Years of Moscow Mathematics, pp. 75–100. AMS/LMS, Providence (2007)
202. Schein, B.M.: Obituary: Victor Vladimirovich Vagner (1908–1981). Semigroup Forum **23**, 189–200 (1981)
203. Schein, B.M.: Obituary: Eduard Grigorievich Shutov. Semigroup Forum **25**, 387–394 (1982)
204. Schein, B.M.: Obituary: L. M. Gluskin (1922–1985). Semigroup Forum **32**, 221–231 (1985)
205. Schein, B.M.: L. M. Gluskin in memoriam. Aequationes Math. **31**, 1–6 (1986)
206. Schein, B.M.: Prehistory of the theory of inverse semigroups. In: Koch, R.J., Hildebrandt, J.A. (eds.) Proceedings of the 1986 LSU Semigroup Conference: Kochfest 60, pp. 72–76. Louisiana State University, Baton Rouge (1986)

207. Schein, B.M.: Book review: 'Inverse semigroups: the theory of partial symmetries' by Mark V. Lawson. Semigroup Forum **65**, 149–158 (2002)
208. Schneider, H.: Introduction to a group of articles in tribute to Vlastimil Pták. Linear Algebra Appl. **310**, 1–4 (2000)
209. Schouten, J.A., Haantjes, J.: On the theory of the geometric object. Proc. Lond. Math. Soc. **42**, 356–376 (1937)
210. Schwarz, Š.: Theory of semigroups. Sborník prác Prírodovedeckej Fakulty Slovenskej Univerzity v Bratislave **6**, 1–64 (1943). In Slovak
211. Sharp, R.: David Rees (1918–2013). Math. Today **49**(5), 197 (2013)
212. Sharp, R.: David Rees. Lond. Math. Soc. Newsl. **429**, 33–34 (2013)
213. Sharp, R.: Obituary: David Rees. The Guardian, Thursday 29th August (2013)
214. Sharp, R.Y.: David Rees, 29 May 1918–16 August 2013. Biogr. Mem. Fellows R. Soc. **61**, 379–401 (2015)
215. Sharp, R.Y.: David Rees, FRS 1918–2013. Bull. Lond. Math. Soc. **48**(3), 557–576 (2016)
216. Shaw, A.: Obituary: Professor John Howie, academic who helped reform Scottish education. The Scotsman, Monday 23rd January (2012)
217. Sinai, Y. (ed.): Russian Mathematicians in the 20th Century. World Scientific, Singapore (2003)
218. Steinfeld, O., Szép, J.: László Rédei is 70 years old. Matematikai Lapok **21**, 189–197 (1970). In Hungarian
219. Sullivan, R.P.: Transformation semigroups: past, present and future. In: Smith, P., Giraldes, E., Martins, P. (eds.) Proceedings of the International Conference on Semigroups (Braga, Portugal, 18–23 June 1999), pp. 191–243. World Scientific, Singapore (2000)
220. Suschkewitsch [Sushkevich], A.K.: Über die Darstellung der eindeutig nicht umkehrbaren Gruppen mittels der verallgemeinerten Substitutionen. Matematicheskii sbornik **33**, 371–374 (1926)
221. Suschkewitsch [Sushkevich], A.K.: Über die endlichen Gruppen ohne das Gesetz der eindeutigen Umkehrbarkeit. Math. Ann. **99**, 30–50 (1928)
222. Suschkewitsch [Sushkevich], A.K.: On a generalization of the associative law. Trans. Am. Math. Soc. **31**, 204–214 (1929)
223. Suschkewitsch [Sushkevich], A.K.: Untersuchungen über verallgemeinerte Substitutionen. In Zanichelli, N. (ed.) Atti del congresso internazionale dei matematici, Bologna, 3–10 Set. 1928, vol. II, Bologna, pp. 147–157 (1930)
224. Suschkewitsch [Sushkevich], A.K.: Über die Matrizendarstellung der verallgemeinerten Gruppen. Zapiski Kharkovskogo matematicheskogo obshchestva (4) **6**, 27–38 (1933)
225. Suschkewitsch [Sushkevich], A.K.: Über einen merkwürdigen Typus der verallgemeinerten unendlichen Gruppen. Zapiski Kharkovskogo matematicheskogo obshchestva (4) **9**, 39–44 (1934)
226. Sushkevich, A.K.: The theory of operations as the general theory of groups, dissertation, Voronezh (1922). In Russian
227. Sushkevich, A.K.: Lectures on the theory of finite groups, steklograph edition, Dnepropetrovsk (1931). In Russian
228. Sushkevich, A.K.: Theory of generalised groups. DNTVU, Kharkov–Kiev (1937). In Russian
229. Sushkevich, A.K.: Investigations on infinite substitutions. In: Schmidt, O.Yu. (ed.) Collection dedicated to D. A. Grave, Moscow, pp. 245–253; also published in Zapiski nauchno-issledovatelskogo instituta matematiki i mekhaniki i Kharkovskogo matematicheskogo obshchestva **18**, 27–37 (1940), both in Russian
230. Sushkevich, A.K.: Materials for the history of algebra in Russia in the 19th and the beginning of the 20th centuries. Istoriko-matematicheskie issledovanie **4**, 237–451 (1951). In Russian
231. Sushkevich, A.K.: Dissertations in mathematics at Kharkov University during the years 1805–1917. Zapiski matematicheskovo otdeleniya fiziko-matematicheskovo fakulteta KhGU i Kharkovskogo matematicheskovo obshchestva **24**, 91–115 (1956). In Russian
232. Szendrei, M.B., Silva, P.V.: A tribute to Mario Petrich. Semigroup Forum **104**, 1–9 (2022)

233. Vasil'ev, A.M., Efimov, N.V., Konstrikin, A.I., Liber, A.E., Lopshits, A.M., Lyapin, E.S., Rashevskiĭ, P.K.: Viktor Vladimirovich Wagner (obituary). Uspekhi matematicheskikh nauk **37**(2(224)), 171–173 (1982); Russian translation: Russian Math. Surv. **37**(2), 193–195
234. Vavřín, Z.: Miroslav Fiedler and Vlastimil Pták: life and work. Linear Algebra Appl. **223/224**, 3–29 (1995)
235. Vavřín, Z.: Seventy years of Professor Vlastimil Pták (biography and interview). Czech. Math. J. **46**(2), 337–349; Math. Bohem. **121**(3), 315–327 (1996)
236. Veblen, O., Whitehead, J.H.C.: The Foundations of Differential Geometry. Cambridge Tract No. 29. Cambridge University Press, Cambridge (1932)
237. Veblen, O., Whitehead, J.H.C.: The Foundations of Differential Geometry, with Appendix by V. V. Wagner. Izdatelstvo, Inostrannoy Literatury, Moscow (1949). Russian
238. Volkov, M.V.: Lev Naumovich Shevrin: fifty years in the service of mathematics. Semigroup Forum **76**, 185–191 (2008)
239. Vrbová, P.: Sixty years of Professor Vlastimil Pták. Czech. Math. J. **35**(4), 662–670 (1985)
240. van der Waerden, B.L.: Moderne Algebra, 2 vols. Springer, Berlin (1930–31)
241. van der Waerden, B.L.: Nachruf auf Emmy Noether. Math. Ann. **111**, 469–476 (1935)
242. Wagner, V.V.: Ternary algebraic operations in the theory of coordinate structures. Dokl. Akad. Nauk SSSR **81**, 981–984 (1951). In Russian
243. Wagner, V.V.: On the theory of partial transformations. Dokl. Akad. Nauk SSSR **84**, 653–656 (1952). In Russian
244. Wagner, V.V.: Generalised groups. Dokl. Akad. Nauk SSSR **84**, 1119–1122 (1952). In Russian
245. Wagner, V.V.: Theory of generalised heaps and generalised groups. Matematicheskii sbornik **32**, 545–632 (1953). In Russian
246. Wagner, V.V., Gluskin, L.M., Aizenshtat, A.Ya.: Evgenii Sergeevich Lyapin (on his fiftieth birthday). Uspekhi matematicheskikh nauk **20**(1(121)), 244–245 (1965). In Russian; English translation: Russian Math. Surv. **20**(1), 175–176
247. Wedderburn, J.H.M.: On hypercomplex numbers. Proc. Lond. Math. Soc. (2) **6**, 77–118 (1907)
248. Wiegandt, R.: Rédei—personal recollections with historical background. Matematikai lapok **8/9**(3–4) (1998/1999), 74–90 (1998). In Hungarian
249. Wilf, H.: Marcel-Paul Schützenberger (1920–1996). Electr. J. Combinat. **3**(1) (1996)
250. Wussing, H.: The Genesis of the Abstract Group Concept: A Contribution to the History of the Origin of Abstract Group Theory. MIT Press, Cambridge (1984)
251. Zhmud, E.M., Dakhiya, S.A.: Anton Kazimirovich Sushkevich (on the centenary of his birth). In: Yubilei nauki, Academy of Sciences, UkrSSR/G. M. Dobova Research Centre for Scientifico-Technical Potential and History of Science, pp. 23–29. Naukova Dumka, Kiev (1990). In Russian
252. Zinov'eva, N.S., Savochkina, T.I.: The contribution of A. K. Sushkevich in the theory semigroups. In: Belousova, L.I. (ed.) Research Students as a Factor in Improving the Training of Future Teachers: Scientific Papers, pp. 117–120. Kharkiv National Pedagogical University, Kharkiv (2010). In Ukrainian
253. Znám, Š., Katriňák, T.: Fifteen questions for academic Štefan Schwarz. Pokroky matematiky, fyziky a astronomie **24**(5), 245–253 (1979). In Slovak

Introduction to Inverse Semigroups

Mark V. Lawson

Abstract This is an account of the theory of inverse semigroups, assuming only that the reader knows the basics of semigroup theory.

Keywords Pseudogroup of transformations · Inverse semigroup

1 Introduction: A Little History

Inverse semigroups[1] were introduced in the 1950s by Wagner [116] in the Soviet Union[2], by Preston [99] in the UK[3] and by Ehresmann [21] in France[4] as algebraic analogues of pseudogroups of transformations. Inverse semigroups can be viewed as generalizations of groups. Whereas group theory is based on the notion of a symmetry—that is, a structure-preserving bijection—inverse semigroup theory is based on the notion of a partial symmetry—that is, a structure-preserving partial

[1] The early development of inverse semigroup theory occurred at a time of acute tensions between East and West. See the book by Christopher Hollings [36] for an in-depth analysis.

[2] The book [37] contains more information on Wagner and his approach to mathematics. See also Wagner's obituary by Boris Schein [106].

[3] The connection between Preston's introduction of what he termed 'inverse semigroups' and the theory of pseudogroups of transformations is the result of a conversation I had with Preston in Australia in the late 1980s.

[4] It is hard to single out a single paper of Ehresmann as the starting point for his approach to inverse semigroups—which was via a class of ordered groupoids—since he stayed faithful to his interest in pseudogroups as the way in which local structures were defined. In any event, Ehresmann's papers in this area reflect what might be called 'an evolving corpus of work'. See [22].

M. V. Lawson (✉)
Department of Mathematics and the Maxwell Institute for Mathematical Sciences, Heriot-Watt University, Edinburgh, Scotland
e-mail: m.v.lawson@hw.ac.uk

M. Mitrović, M. N. Hounkonnou (eds.), *Algebra Without Borders*, Mathematics in Mind, https://doi.org/10.1007/978-3-031-86477-3_2

bijection. The passage from bijections to partial bijections introduces a host of new algebraic phenomena.

In writing this chapter, I have assumed the reader is familiar with the basics of semigroup theory such as could be gleaned from the first few chapters of Howie [38].[5] Here are a few things you should know and which I take for granted. The multiplication in a semigroup will usually be denoted by $\cdot$ or by concatenation. If S is any semigroup, then S^0 is the semigroup S with an adjoined zero, and S^1 is the semigroup S with an adjoined identity. If A and B are any subsets of a semigroup S, then AB is the set ab of all products where $a \in A$ and $b \in B$. The study of congruences in semigroups is unavoidable since there are not, in general, the analogues of normal subgroups as in groups or ideals as in rings. If $\theta \colon S \to T$ is a homomorphism between semigroups, then $\ker(\theta)$ is the congruence that θ induces on S. On the other hand, every congruence ρ on a semigroup S has an associated *natural homomorphism* from S to S/ρ which maps s to $\rho(s)$. If both S and T are monoids, then a *monoid homomorphism* maps the identity of S to the identity of T; if S and T are both semigroups with zero, then a homomorphism of such semigroups maps the zero to the zero. If the semigroups S and T are isomorphic, then we write $S \cong T$. A subset $I \subseteq S$ is said to be a *right ideal* if $IS \subseteq I$ and a *left ideal* if $SI \subseteq I$. It is said to be an *ideal* if $SI \subseteq I$ and $IS \subseteq I$. If $a \in S$, then we write a instead of $\{a\}$. The right ideal aS^1 is called the *principal right ideal generated by* a, whereas the left ideal S^1a is called the *principal left ideal generated by* a. We call S^1aS^1 the *principal ideal generated by* a. Our use of S^1 is just a trick to ensure that a belongs to the ideal. Define $a \mathscr{R} b$ if $aS^1 = bS^1$ and $a \mathscr{L} b$ if $S^1a = S^1b$. Recall that $\mathscr{L} \circ \mathscr{R} = \mathscr{R} \circ \mathscr{L}$. Put $\mathscr{D} = \mathscr{L} \circ \mathscr{R}$, another equivalence relation. Put $\mathscr{H} = \mathscr{L} \cap \mathscr{R}$. Define $a \mathscr{J} b$ if $S^1aS^1 = S^1bS^1$. These are the familiar *Green's relations*. As usual, if $\mathscr{G}$ is one of Green's relations, then the $\mathscr{G}$-class that contains the element a is denoted by G_a. Although ideals are useful in semigroup theory, the connection between ideals and congruences is weaker for semigroups than it is for rings. If ρ is a congruence on a semigroup with zero S, then the set $I = \rho(0)$ is an ideal of S; however, examples show that the congruence is not determined by this ideal. Nevertheless, ideals can be used to construct some congruences on semigroups. Let I be an ideal in the semigroup S. Define a relation ρ_I on S by

$$(s,t) \in \rho_I \Leftrightarrow \text{either } s,t \in I \text{ or } s = t.$$

Then ρ_I is a congruence. The quotient semigroup S/ρ_I is isomorphic to the set $(S \setminus I) \cup \{0\}$ (we may assume that $0 \notin S \setminus I$) equipped with the following product: if $s,t \in S \setminus I$, then their product is st if $st \in S \setminus I$; all other products are defined to be 0. Such quotients are called *Rees quotients*.

There are currently two books entirely devoted to inverse semigroup theory: Petrich's [98] and mine [52]. Petrich's book is pretty comprehensive up to 1984

[5] On page 128, Howie quotes Clifford and Preston as saying that inverse semigroups were the most promising class of semigroups for future study. This prediction has turned out to be correct.

and is still a useful reference. Its only drawback is the poor index which makes finding particular topics a bit of a chore. My book is less ambitious; its goal is to motivate the study of inverse semigroups by concentrating on concrete examples and was completed in 1998.

In writing this chapter, I have drawn mainly upon my own book [52] and some notes I wrote [64] for a workshop at the University of Ottawa in 2010, but I have taken the opportunity to radically rethink what I wrote there in the light of subsequent research. Any results below that are stated but not proved should be treated as exercises.

2 Motivation: Pseudogroups of Transformations

Good mathematics has to be based on good examples. For example, finite semigroup theory grew out of the theory of finite-state automata since a finite semigroup can be associated with every finite-state automaton. Similarly, inverse semigroup theory grew out of the theory of pseudogroups of transformations, which goes back to the works of Sophus Lie and Henri Cartan. If you have ever taken an introductory course in differential geometry, then you have been exposed to pseudogroups of transformations even if that phrase was never once uttered. So, there I shall begin. However, for those of you who have not taken such a course, I shall quickly move on to the collection of all partial bijections on a set, jettisoning any topology or differential structure, and this will take us to the core motivation for studying inverse semigroups.

Let X be a nice space (for example, X should be Hausdorff but this will play no role in what follows). Let $\mathbb{R}^n$ be the usual space of real n-vectors. We shall coordinatize X using $\mathbb{R}^n$ but we shall only do so *locally* not *globally*. Thus, we assume that for each point $x \in X$ there exists an open set $x \in U$ and a homeomorphism $\phi \colon U \to V$, called a *chart*, where V is an open subset of $\mathbb{R}^n$. A chart gives one way of coordinatizing elements of U by means of elements of V. However, it is quite possible that x lies in another chart. In other words, there could well exist another open subset of X; let us say U', which also contains x and for which there is a homeomorphism $\psi \colon U' \to V'$ where V' is also an open subset of $\mathbb{R}^n$. This map gives another way of coordinatizing elements of U' by means of elements of V'. However, for any element in $y \in U \cap U'$, there are now two possible co-ordinate representations: $\phi(y)$ and $\psi(y)$. How are they to be related? This depends on the nature of the map $\psi\phi^{-1} \colon \phi(U \cap U') \to \psi(V \cap V')$, called the *change of co-ordinates map*, which connects them. This map is a homeomorphism from the open subset $\phi(U \cap U')$ of $\mathbb{R}^n$ to the open subset $\psi(V \cap V')$ of $\mathbb{R}^n$. If the map $\psi\phi^{-1}$ is always assumed to be *smooth*, then we say that we have defined on X the structure of a *differential manifold* [107].

We now dispense with X and concentrate instead on the set $\Gamma(\mathbb{R}^n)$ of all smooth maps between the open subsets of $\mathbb{R}^n$. This is just the set of all possible changes of co-ordinates and is our first example of a pseudogroup. Usually, it is the set X

which is the centre of attention, but we have shifted focus to the set of allowable co-ordinate transformations; if we change the nature of those, then we change the nature of the structure we are defining on X.

More generally, let X be any topological space with topology τ. Denote the set of all homeomorphisms between open subsets of X by $\mathcal{I}(X, \tau)$. If τ is the usual topology on $\mathbb{R}^n$, then $\Gamma(\mathbb{R}^n)$ is a subset of $\mathcal{I}(\mathbb{R}^n, \tau)$ closed under certain operations. The sets $\mathcal{I}(X, \tau)$ are also examples of pseudogroups. The terminology 'pseudogroup' comes from the fact that, just like groups of transformations, the sets $\mathcal{I}(X, \tau)$ are closed under the composition of partial functions—do not forget the empty function!—and are also closed under 'inverses' where we interpret inverses as inverses of partial functions. The set $\Gamma(\mathbb{R}^n)$ is likewise closed under composition of partial functions and said inverses. There is a further property that is needed to qualify as a pseudogroup. It is this. If $\{\phi_i : i \in I\}$ is any subset of $\mathcal{I}(X, \tau)$ such that $\bigcup_{i \in I} \phi_i$ is a partial bijection, then, in fact, $\bigcup_{i \in I} \phi_i \in \mathcal{I}(X, \tau)$. This is a completeness property. We shall call semigroups of the form $\mathcal{I}(X, \tau)$ *full pseudogroups of transformations*.

For modern applications of pseudogroups, see [17, 81].

To make our lives easier, let us now take the non-empty set X with the discrete topology. We denote by $\mathcal{I}(X)$ the set of all bijections between subsets of X. If $f : A \to B$, where $A, B \subseteq X$, then we call A the *domain (of definition)* of f and B the *range* of f. We write $\text{dom}(f) = A$ and $\text{ran}(f) = B$. The set $\mathcal{I}(X)$ is closed under the composition of partial functions and so is a semigroup. It is, in fact, a monoid since the identity function defined on X belongs to $\mathcal{I}(X)$. By analogy with the symmetric group, it is called the *symmetric inverse monoid*. What we mean by the term 'inverse monoid' will be explained in the next section. Clearly, the structure of the semigroup $\mathcal{I}(X)$ depends only on the cardinality of X. Thus, if X is a finite set with n elements, it makes sense to write $\mathcal{I}_n$ instead of $\mathcal{I}(X)$.

There are a number of structures we could look at in the semigroups $\mathcal{I}(X)$ but we begin with the following which turns out to be the right place to start. Let $f \in \mathcal{I}(X)$. Consider the following two equations in $\mathcal{I}(X)$:

$$f = fgf \text{ and } g = gfg.$$

These have the unique solution $g = f^{-1}$. This result is the basis for the definition of an inverse semigroup in the next section.

3 Basic Inverse Semigroup Theory

A semigroup S is said to be *inverse* if for each $s \in S$ there exists a unique element, denoted by s^{-1}, such that the following two equations are satisfied:

$$s = ss^{-1}s \text{ and } s^{-1} = s^{-1}ss^{-1}.$$

The semigroups $\mathcal{I}(X)$ really are inverse monoids. If $f \in \mathcal{I}(X)$, then f^{-1} is the unique solution of the equations $f = fgf$ and $g = gfg$ where g is the unknown.

Observe that if s^{-1} is the inverse of s, then s is the inverse of s^{-1}. We have therefore proved the following.

Lemma 1 *Let S be an inverse semigroup and let $s \in S$. Then $(s^{-1})^{-1} = s$.*

The elements $s^{-1}s$ and ss^{-1} are idempotents, where an *idempotent* in a semigroup is an element e such that $e^2 = e$. For example, $s^{-1}s$ is an idempotent because

$$(s^{-1}s)^2 = (s^{-1}s)(s^{-1}s) = (s^{-1}ss^{-1})s = s^{-1}s.$$

If e is an idempotent in an inverse semigroup S, then $e = ee = eee$. We have therefore proved the following.

Lemma 2 *Let S be an inverse semigroup and let $e \in S$ be an idempotent. Then $e^{-1} = e$.*

It follows that every idempotent in an inverse semigroup is of the form $s^{-1}s$ for some elements s or, equivalently, of the form ss^{-1}, for some element s.

Idempotents play an important role in inverse semigroup theory. For that reason we introduce some special notation. If S is an inverse semigroup, we denote its set of idempotents by $\mathsf{E}(S)$. Two special idempotents are the *identity* element, if it exists, usually denoted by 1, and the *zero* element, if it exists, usually denoted by 0. An inverse semigroup with identity is called an *inverse monoid* and an inverse semigroup with zero is called an *inverse semigroup with zero*.[6] It is important to be aware right at the beginning that even if S is an inverse monoid, it is not true in general that $s^{-1}s = 1 = ss^{-1}$.

The idempotents in $\mathcal{I}(X)$ are precisely the identity functions defined on the subsets of X. Thus, if $A \subseteq X$, then a typical idempotent is 1_A which is the identity function on the set A. The identity is 1_X and the zero is $1_\varnothing$.

The difference between monoids and semigroups is sometimes viewed as trivial. It is not. For example, a commutative unital C^*-algebra is constructed from a *compact* space, whereas a commutative non-unital C^*-algebra is constructed from a *locally compact* space.

If S is any monoid, then we can look at its *group of units* which we denote by $\mathsf{U}(S)$. If e is any idempotent in a semigroup S, then eSe is a monoid. I call these *local monoids* though you will find the misleading term 'local submonoid' in the literature. The group of units of eSe is denoted by H_e.

The units of $\mathcal{I}(X)$ are the bijective functions and these form a group $\mathcal{S}(X)$, called the *symmetric group* on X.

---

[6] There is no one term in English for a 'semigroup with zero'.

Lemma 3 *Let S be an inverse semigroup and let e be an idempotent in S. Then eSe is an inverse monoid.*

Proof It is clear that eSe is a monoid. We prove that it is inverse. Let $a \in eSe$. Then $a = aa^{-1}a$ and $a^{-1} = a^{-1}aa^{-1}$ in S. You can check that $ea^{-1}e$ is also an inverse of a since $ea = a = ae$. So, by uniqueness, we have that $a^{-1} = ea^{-1}e$. We have therefore proved that if $a \in eSe$, then $a^{-1} \in eSe$. It follows that eSe is an inverse monoid. □

The idempotents $s^{-1}s$ and ss^{-1} are so important that we have some special notation for them:

$$\mathbf{d}(s) = s^{-1}s \text{ and } \mathbf{r}(s) = ss^{-1}.$$

Observe that $a\mathbf{d}(a) = a$ and $\mathbf{r}(a)a = a$. If the inverse semigroup has a zero, then $a = 0$ if and only if $\mathbf{d}(a) = 0$ (respectively, $\mathbf{r}(a) = 0$.) You can easily check that in an inverse semigroup $a \mathscr{R} b$ if and only if $\mathbf{r}(a) = \mathbf{r}(b)$ and $a \mathscr{L} b$ if and only if $\mathbf{d}(a) = \mathbf{d}(b)$. For this reason, the explicit use of Green's relations in inverse semigroups is not very common.

An *inverse subsemigroup* of an inverse semigroup is a subsemigroup that is also closed under inverses. If S is an inverse subsemigroup of T and $\mathsf{E}(S) = \mathsf{E}(T)$, we say that S is a *wide*[7] inverse subsemigroup of T.

The inverse semigroup $\Gamma(\mathbb{R}^n)$ is, in fact, a wide inverse submonoid of $\mathcal{I}(\mathbb{R}^n, \tau)$ where τ is the usual topology on $\mathbb{R}^n$.

It is now time for our first example. We can, in fact, express this as a lemma.

Lemma 4

(i) *A group is an inverse semigroup having a unique idempotent.*
(ii) *An inverse semigroup having a unique idempotent is a group.*

Proof

(1) Let G be a group. Then for any element g in G we have that $g^{-1}g = 1 = gg^{-1}$. Multiply the first equation on the left by g to get $gg^{-1}g = g$ and multiply the second equation on the left by g^{-1} to get $g^{-1}gg^{-1} = g^{-1}$. On the other hand, suppose that $gxg = g$ and $xgx = x$. Multiply the first equation by g^{-1} on the left and by g^{-1} on the right. This gives us $x = g^{-1}$ which also satisfies the second equation. It follows that every group is an inverse semigroup. Let g be an idempotent in G. Then $gg = g$. Multiply the left-hand side of this equation by g^{-1} to get $g = 1$, and 1 is clearly an idempotent. Thus the only idempotent in a group is the identity.

[7] You will often see the term 'full' used to mean 'wide'. I prefer the term 'wide' because, as we shall see, it fits in well with groupoid theory, whereas 'full' has categorical connotations which are not what we want.

(2) Let S be an inverse semigroup with exactly one idempotent. Call this idempotent e. For any element $a \in S$, we already know that $a^{-1}a$ and aa^{-1} are idempotents. It follows that $a^{-1}a = e = aa^{-1}$ for any element $a \in S$. On the other hand, e is the identity. To see why, observe that $ea = (aa^{-1})a = a$. Similarly, $ae = a$. We have therefore proved that S is a group with identity e.

□

The most important fact about inverse semigroups, and one that is not at all obvious from the definition, is that if e and f are any idempotents in an inverse semigroup S, then $ef = fe$. We say that the *idempotents commute*. For the benefit of ring theorists, observe that we are not saying that the idempotents are central. Inverse semigroups in which the idempotents are central are called *Clifford semigroups*.

Proposition 1 *Idempotents commute in an inverse semigroup.*

Proof This is an example of a proof that is elementary but not easy. Let e and f be idempotents in the inverse semigroup S. Put $x = (ef)^{-1}$. We prove first that fxe is an idempotent:

$$(fxe)^2 = (fxe)(fxe) = f(x(ef)x)e = fxe.$$

Next, we claim that fxe is the inverse of ef:

$$ef(fxe)ef = ef^2xe^2f = efxef = ef$$

and

$$(fxe)ef(fxe) = fxe^2f^2xe = (fxe)^2 = fxe,$$

by what we proved above. It follows that

$$(ef)^{-1} = fxe.$$

We have therefore proved that $(ef)^{-1}$ is an idempotent. Now, we apply Lemma 1 and Lemma 2 to deduce that ef is an idempotent. We have therefore proved that in an inverse semigroup the product of any two idempotents is itself an idempotent. It follows that both ef and fe are idempotents. Next, we show that fe is the inverse of ef:

$$(ef)(fe)(ef) = ef^2e^2f = efef = ef$$

and

$$(fe)(ef)(fe) = fe^2f^2e = fefe = fe.$$

We have therefore proved that $(ef)^{-1} = fe$. But ef is an idempotent and so, by Lemma 2, it is its own inverse. We have therefore proved that $ef = fe$, as required. □

In the symmetric inverse monoid, the product of the idempotents 1_A and 1_B is the idempotent $1_{A\cap B}$. Because the intersection operation on two sets is commutative, we see why the idempotents commute in this special case.

Because we know that idempotents commute, we can now prove the following:

Lemma 5 *Let S be an inverse semigroup. Then $(st)^{-1} = t^{-1}s^{-1}$.*

Proof We prove that $t^{-1}s^{-1}$ is the inverse of st. We have that

$$st(t^{-1}s^{-1})st = s(tt^{-1})(s^{-1}s)t = s(s^{-1}s)(tt^{-1})t = st$$

where we have used the fact that idempotents in an inverse semigroup commute. Similarly,

$$t^{-1}s^{-1}(st)t^{-1}s^{-1} = t^{-1}s^{-1}.$$

By virtue of the fact that inverses are unique, we deduce that $(st)^{-1} = t^{-1}s^{-1}$. □

We also see that the analogue of conjugation holds. We shall use the term 'conjugation' below.

Lemma 6 *If e is an idempotent of an inverse semigroup and s is any element of that inverse semigroup, then ses^{-1} is an idempotent.*

Proof This follows by a direct calculation:

$$(ses^{-1})^2 = ses^{-1}ses^{-1} = se^2(s^{-1}s)s^{-1} = ses^{-1},$$

where we have again used the fact that idempotents commute. □

Let e and f be idempotents in an inverse semigroup. Then we may define a relation on them by $e \leq f$ precisely when $e = ef$. Because idempotents commute, it is not hard to show that this defines a partial order on the set of idempotents. We can now characterize Clifford semigroups.

Lemma 7 *Let S be an inverse semigroup. Then S is a Clifford semigroup if and only if $\mathbf{d}(s) = \mathbf{r}(s)$ for every $s \in S$.*

Proof We use the partial order we have defined on the idempotents of an inverse semigroup. Let S be a Clifford semigroup and let $s \in S$. Since the idempotents are central $s = s(s^{-1}s) = (s^{-1}s)s$. Multiply the last equality on the right by s^{-1} to deduce that $\mathbf{r}(s) \leq \mathbf{d}(s)$. It now follows by symmetry that $\mathbf{d}(s) = \mathbf{r}(s)$.

Suppose now that $\mathbf{d}(s) = \mathbf{r}(s)$ for all $s \in S$. Let e be any idempotent and s arbitrary. We shall prove that $se = es$. By assumption, $\mathbf{d}(es) = \mathbf{r}(es)$. That is $(es)^{-1}es = es(es)^{-1}$. We now use Lemmas 5 and 2 and the fact that idempotents commute to get $s^{-1}es = ss^{-1}e$. Because of our assumption, $ss^{-1} = s^{-1}s$. Thus $s^{-1}es = s^{-1}se$. Multiplying on the left by s, and using the fact that idempotents commute and our assumption, gives $es = se$, as required. □

An inverse semigroup S is said to be a *union of groups* if $S = \bigcup_{e \in \mathsf{E}(S)} H_e$. By Lemma 7, we may now deduce the following

Lemma 8 *Let S be an inverse semigroup. Then S is a Clifford semigroup if and only if it is a union of groups.*

Proof Let S be a Clifford semigroup and let $s \in S$. Put $e = \mathbf{d}(s) = \mathbf{r}(s)$. Then $s \in eSe$. But, in fact, s belongs to the group of units of eSe which is H_e. Thus $s \in H_e$ and so we have proved that every Clifford semigroup is a union of groups.

Conversely, suppose that S is an inverse semigroup which is a union of groups. Then for each $s \in S$, we have that $s \in H_e$ for some idempotent e. We have that $\mathbf{d}(s) = e = \mathbf{r}(s)$. It follows that S Clifford. □

I have included Lemma 8 because group theorists sometimes tend to see inverse semigroups as unions of groups. These are, in fact, a very special class of inverse semigroups. Clifford semigroups can be described explicitly as 'strong semilattices of groups' or, what amounts to the same thing, as presheaves of groups over meet semilattices. This applies to, in particular, abelian inverse semigroups. For strong semilattices of groups, see [38, Chapter IV, Section 2].

We can obtain another characterization of Clifford semigroups which is analogous to part of [31, part (c) of Theorem 3.2] but at the price of an extra assumption on the semilattice of idempotents. We say that a meet semilattice is *0-disjunctive* if for all $0 \neq f < e$ there exists $0 \neq g \leq e$ such that $fg = 0$. Thus, semilattices which are 0-disjunctive have a weak notion of complement. A non-zero element a of an inverse semigroup is said to be an *infinitesimal* if $a^2 = 0$.

Lemma 9 *Let S be an inverse semigroup, the semilattice of idempotents of which is 0-disjunctive. Then S is a Clifford semigroup if and only if S contains no infinitesimals.*

Proof Suppose that S is a Clifford semigroup. Let a be an infinitesimal. Then $a^2 = 0$ and thus $aa = 0$. It follows that $\mathbf{d}(a)\mathbf{r}(a) = 0$. But, in a Clifford semigroup $\mathbf{d}(a) = \mathbf{r}(a)$ by Lemma 7. It follows that $\mathbf{d}(a) = 0$ and so $a = 0$ which is a contradiction. Thus there are no infinitesimals.

To prove the converse, suppose that there are no infinitesimals. Let $a \in S$ be arbitrary and non-zero. We shall prove that $\mathbf{d}(a) = \mathbf{r}(a)$. Suppose not. Put $e = \mathbf{d}(a)\mathbf{r}(a)$. If $e = 0$, then $a^{-1}aaa^{-1} = 0$ and so $a^2 = 0$. But this means that a is an infinitesimal. It follows that $e \neq 0$. There are now two cases. Suppose, first, that $e = \mathbf{d}(a)$. Then $\mathbf{d}(a) < \mathbf{r}(a)$. By assumption, there exists $0 < f \leq \mathbf{r}(a)$ such that $f\mathbf{d}(a) = 0$. Put $b = fa$. If $b = 0$, then it is easy to see that $f = 0$, which is a contradiction. It follows that $b \neq 0$. You can easily check that b is an infinitesimal. This is a contradiction. We can now deal with the second case where $e < \mathbf{d}(a)$. By assumption, there is a non-zero idempotent f such that $ef = 0$ and $f \leq \mathbf{d}(a)$. Put $b = af$. If $b = 0$, then $(af)^{-1}af = 0$ which implies that $f = 0$. It follows that

$b \neq 0$. However, $b^2 = afaf = a\mathbf{d}(a)f\mathbf{r}(a)af = aefaf = 0$. But this contradicts the assumption that there are no infinitesimals. It follows that $\mathbf{d}(a) = \mathbf{r}(a)$ for all $a \in S$. Thus by Lemma 7, we have shown that S is a Clifford semigroup. □

Although the idempotents in an inverse semigroup are not in general central, we do have the following result, which tells us how idempotents 'pass through' non-idempotent elements.

Lemma 10 *Let S be an inverse semigroup in which e is an idempotent and s is any element.*

(i) *$es = sf$ for some idempotent f.*
(ii) *$se = gs$ for some idempotent g.*

Proof We prove only (1) since the proof of (2) is similar. We have that $es = ess^{-1}s = e(ss^{-1})s$. Now we use the fact that idempotents commute to get $es = (ss^{-1})es$. But this is equal to $s(s^{-1}es)$ and now we use Lemma 6 to deduce that $f = s^{-1}es$ is an idempotent. We have therefore proved that $es = sf$. □

We have described one extreme example of inverse semigroup. We now describe another. By a *meet semilattice*, we mean a partially ordered set $(P, \leq)$ with the property that every pair of elements $a, b \in P$ has a *greatest lower bound*, denoted by $a \wedge b$. Thus, $a \wedge b \leq a, b$ and if $c \leq a, b$, then $c \leq a \wedge b$. We call $a \wedge b$ the *meet* of a and b.[8] The proofs of the following are immediate from the definitions.

Lemma 11 *Let $(P, \leq)$ be a meet semilattice and let $a, b, c \in P$. Then*

(i) $a \wedge (b \wedge c) = (a \wedge b) \wedge c$.
(ii) $a \wedge b = b \wedge a$.
(iii) $a \wedge a = a$.

A semigroup is called a *band* if every element is an idempotent. Observe that we have proved in Lemma 11 that if $(P, \leq)$ is a meet semilattice, then $(P, \wedge)$ is a commutative band.

Lemma 12 *Let S be a commutative band. Define $a \leq b$ by $a = ab$. Then $(S, \leq)$ is a partially ordered set and, in fact, a meet semilattice in which $a \wedge b = ab$.*

Proof We first check that $\leq$ is a partial order. By assumption, each element of S is an idempotent. It follows that $a \leq a$. Suppose that $a \leq b$ and $b \leq a$. Then $a = ab$ and $b = ba$. But the semigroup is commutative and so $a = b$. Finally, suppose that $a \leq b$ and $b \leq c$. Then $a = ab$ and $b = bc$. It follows that $a = abc = ac$. Thus $a \leq c$. We have verified that $(S, \leq)$ is a partially ordered set. We now show that $a \wedge b = ab$. Observe first that $(ab)a = a^2b = ab$, using commutativity and the fact that every element is idempotent. Also, $(ab)b = ab^2 = ab$. We have shown that $ab \leq a, b$. Suppose, now, that $c \leq a, b$. Then $c = ca = cb$, and so $c(ab) = cb = c$. This shows that $c \leq ab$ and thus proves that $ab = a \wedge b$. □

[8] We shall usually denote the meet by $\cdot$ or concatenation.

As a result of Lemmas 11 and 12, we can think about meet semilattices in two equivalent ways: either as partially ordered sets in which each pair of elements has a greatest lower bound or as commutative bands.

Observe that if S is any inverse semigroup, then, since the idempotents commute, $\mathsf{E}(S)$ is a subsemigroup of S. It follows that $\mathsf{E}(S)$ is always a commutative band and so it becomes a meet semilattice when we define $e \wedge f = ef$. For this reason, it is usual to refer to $\mathsf{E}(S)$ as the *semilattice of idempotents* of S.

We now have the following companion to Lemma 4.

Lemma 13

(i) *A meet semilattice is an inverse semigroup with respect to the meet operation in which every element is an idempotent.*
(ii) *An inverse semigroup in which every element is an idempotent is a meet semilattice.*

Proof The proof of (1) is immediate by Lemma 11. (2) Suppose that S is an inverse semigroup in which every element is an idempotent. Then $S = \mathsf{E}(S)$. The result now follows by Lemma 12, since $\mathsf{E}(S)$ is a commutative band. □

Our next result is often useful in showing that a semigroup is or is not inverse. We need a definition first. A semigroup S is said to be *regular* if for each $a \in S$ there exists an element b such that $a = aba$ and $b = bab$. The element b is called *an inverse* of a. Observe that both ab and ba are idempotents. In showing that a semigroup is regular, it is enough to check that for each element a there is an element x such that $a = axa$. The reason is that $b = xax$ is then an inverse of a. Inverse semigroups are the regular semigroups in which every element has a unique inverse.

Proposition 2 *A regular semigroup is inverse if and only if its idempotents commute.*

Proof Let S be a regular semigroup in which the idempotents commute and let u and v be inverses of x. Then

$$u = uxu = u(xvx)u = (ux)(vx)u,$$

where both ux and vx are idempotents. Thus, since idempotents commute, we have that

$$u = (vx)(ux)u = vxu = (vxv)xu = v(xv)(xu).$$

Again, xv and xu are idempotents and so

$$u = v(xu)(xv) = v(xux)v = vxv = v.$$

Hence, $u = v$. The converse follows by Proposition 1. □

As an example of Proposition 2 in action, observe that the full transformation monoid $\mathcal{T}(X)$, of all functions from X to itself where X has at least two elements, is regular but not inverse. We show first that it is regular. Let $f \in \mathcal{I}(X)$. There are two cases. Suppose first that f is surjective; for each $y \in X$ choose $x_y \in f^{-1}(y)$. Define $g: X \to X$ by $g(y) = x_y$. That is, g maps an element to one of its pre-images under f. We calculate fgf. Let $x \in X$ and $y = f(x)$. Then

$$(fgf)(x) = (fg)(f(x)) = f(g(y)) = f(x_y) = y.$$

It follows that $f = fgf$. Suppose now that f is not surjective. Choose any element $x_0 \in X$. For each element y in the image of f choose $x_y \in f^{-1}(y)$. Now define $g: X \to X$ as follows: if y is in the image of f, then define $g(y) = x_y$, and if y is not in the image of f, then define $g(y) = x_0$. A similar calculation to the one above shows that $f = fgf$. We have therefore shown that the full transformation monoid is regular. Choose two distinct elements from X that we call x_1 and x_2. Define two functions $c_1, c_2: X \to X$ where c_1 is the constant function to x_1 and c_2 is the constant function to x_2. The functions c_1 and c_2 are distinct and both are idempotents. But $c_1c_2 = c_1$, whereas $c_2c_1 = c_2$. We have shown that the idempotents do not commute. Thus $\mathcal{T}(X)$ is regular but not inverse.

We now consider homomorphisms between inverse semigroups. Our first result is actually slightly more general.

Lemma 14 *Let S be an inverse semigroup and let T be any semigroup. Suppose that $\theta: S \to T$ is a semigroup homomorphism. Then $\theta(S)$, the image of S, is an inverse semigroup.*

Proof We shall use Proposition 2 to prove that $\theta(S)$ is inverse. We show first that it is regular. Let $\theta(s) \in \theta(S)$. Then, since $s \in S$ inverse, there is an element $s^{-1} \in S$ such that $s = ss^{-1}s$ and $s^{-1} = s^{-1}ss^{-1}$. It follows that $\theta(s) = \theta(s)\theta(s^{-1})\theta(s)$ and $\theta(s^{-1}) = \theta(s^{-1})\theta(s)\theta(s^{-1})$. This shows that $\theta(S)$ is regular.

We now prove that the idempotents in $\theta(S)$ commute. To begin with, we have to show that idempotents in the image come from idempotents in the domain.[9] Let $\theta(s)$ be an idempotent in $\theta(S)$. We shall prove that there is an idempotent e in S such that $\theta(e) = \theta(s)$. Consider the element $e = ss^{-2}s = (ss^{-1})(s^{-1}s)$. It is the product of two idempotents in S and so it is an idempotent. In particular, e is an idempotent in S. We calculate $\theta(e)$. We have that

$$\theta(e) = \theta(ss^{-2}s) = \theta(s)\theta(s^{-2})\theta(s).$$

But $\theta(s) = \theta(s)^2 = \theta(s^2)$. We therefore have that

$$\theta(e) = \theta(s^2s^{-2}s^2) = \theta(s^2) = \theta(s)^2 = \theta(s).$$

[9] This is an instance of what is often termed *Lallement's lemma*.

Now, let $\theta(s)$ and $\theta(t)$ be idempotents in $\theta(S)$. By our result above, there are idempotents e and f in S such that $\theta(e) = \theta(s)$ and $\theta(f) = \theta(t)$. It follows that $\theta(ef) = \theta(s)\theta(t)$. But idempotents in inverse semigroups commute and so $ef = fe$. We have therefore shown that $\theta(s)\theta(t) = \theta(t)\theta(s)$. Thus, the idempotents in $\theta(S)$ commute. We have therefore proved that the image of θ is an inverse semigroup by Proposition 2. □

Homomorphisms of inverse semigroups are just semigroup homomorphisms. *Isomorphisms of inverse semigroups* are just semigroup isomorphisms.

Lemma 15 *Let $\theta: S \to T$ be a homomorphism between inverse semigroups. Then:*

(i) $\theta(s^{-1}) = \theta(s)^{-1}$ *for all $s \in S$.*
(ii) *If e is an idempotent, then $\theta(e)$ is an idempotent.*
(iii) *If $\theta(s)$ is an idempotent, then there is an idempotent e in S such that $\theta(s) = \theta(e)$.*
(iv) *$\theta(S)$ is an inverse subsemigroup of T.*
(v) *If U is an inverse subsemigroup of T, then $\theta^{-1}(U)$ is an inverse subsemigroup of S.*

Proof

(1) Clearly, $\theta(s)\theta(s^{-1})\theta(s) = \theta(s)$ and $\theta(s^{-1})\theta(s)\theta(s^{-1}) = \theta(s^{-1})$. Thus, by uniqueness of inverses, we have that $\theta(s^{-1}) = \theta(s)^{-1}$.
(2) $\theta(e)^2 = \theta(e)\theta(e) = \theta(e^2) = \theta(e)$.
(3) If $\theta(s)^2 = \theta(s)$, then $\theta(s^{-1}s) = \theta(s^{-1})\theta(s) = \theta(s)^{-1}\theta(s) = \theta(s)^2 = \theta(s)$, where we have used the fact that every idempotent in an inverse semigroup is its own inverse.
(4) This is immediate by Lemma 14.
(5) Straightforward.

□

Let $\theta: S \to T$ be a homomorphism of inverse semigroups. Then the restriction map, $(\theta \mid \mathsf{E}(S))\colon \mathsf{E}(S) \to \mathsf{E}(T)$, is well-defined since homomorphisms map idempotents to idempotents by part (2) of Lemma 15 The homomorphism θ is said to be *idempotent-separating* if $(\theta \mid \mathsf{E}(S))$ is actually injective. We say that a congruence is idempotent-separating if its associated natural homomorphism is idempotent-separating. We say that S is an *idempotent-separating cover* of T if θ is both surjective and idempotent-separating.

It remains to be shown that inverse semigroups really do encode partial bijections. Given an inverse semigroup S our goal is therefore to construct an injective homomorphism $\lambda: S \to \mathcal{I}(X)$ for some set X. We shall do just this by generalizing the familiar Cayley's Theorem from group theory. As a starting point, we follow the group theory and take as our underlying set X the set S itself. We now have to associate an element of $\mathcal{I}(S)$ with each element $a \in S$. This is delivered by the following lemma.

Lemma 16 *Let S be an inverse semigroup and let $a \in S$. Then $\lambda_a \colon \mathbf{d}(a)S \to \mathbf{r}(a)S$ defined by by $\lambda_a(x) = ax$ is a well-defined partial bijection.*

Proof This is well-defined because $aS = aa^{-1}S$ as the following set inclusions show

$$aS = aa^{-1}aS \subseteq aa^{-1}S \subseteq aS.$$

Also $\lambda_{a^{-1}} \colon \mathbf{r}(a)S \to \mathbf{d}(a)S$, $\lambda_{a^{-1}}\lambda_a$ is the identity on $\mathbf{d}(a)S$, and $\lambda_a\lambda_{a^{-1}}$ is the identity on $\mathbf{r}(a)S$. Thus λ_a is a bijection and $\lambda_a^{-1} = \lambda_{a^{-1}}$. □

We now define a function $\lambda \colon S \to \mathcal{I}(S)$ by $\lambda(a) = \lambda_a$. The following theorem completes the circle of ideas begun in Sect. 2 and justifies the claim that inverse semigroups are the abstract theory of partial bijections.

Theorem 1 (Wagner-Preston Representation Theorem) *Every inverse semigroup can be embedded in a symmetric inverse monoid.*

Proof We prove first that λ is a semigroup homomorphism. To that end, we prove that $\lambda_a\lambda_b = \lambda_{ab}$. If e and f are any idempotents of an inverse semigroup S, then

$$eS \cap fS = efS.$$

Thus

$$\mathrm{dom}(\lambda_a) \cap \mathrm{im}(\lambda_b) = a^{-1}aS \cap bb^{-1}S = a^{-1}abb^{-1}S.$$

Hence,

$$\mathrm{dom}(\lambda_a\lambda_b) = \lambda_b^{-1}(a^{-1}abb^{-1}S) = b^{-1}a^{-1}aS = b^{-1}a^{-1}abS$$

where we use the following subset inclusions:

$$b^{-1}a^{-1}aS = b^{-1}bb^{-1}a^{-1}aS = b^{-1}a^{-1}abb^{-1}S \subseteq b^{-1}a^{-1}abS \subseteq b^{-1}a^{-1}aS.$$

Thus $\mathrm{dom}(\lambda_a\lambda_b) = \mathrm{dom}(\lambda_{ab})$. It is immediate from the definitions that $\lambda_a\lambda_b$ and λ_{ab} have the same effect on elements, and so λ is a homomorphism. It remains to prove that λ is injective. Suppose that $\lambda_a = \lambda_b$. Then $a = ba^{-1}a$ and $b = ab^{-1}b$. Observe that $ab^{-1}b = (ba^{-1}a)b^{-1}b = b(b^{-1}b)a^{-1}a = ba^{-1}a = a$, where we have used the fact that idempotents commute. It follows that $a = b$. □

We conclude this section by stating a deeper result that we shall not prove. Let S be a *finite* semigroup. Then for each $s \in S$ there exists some power s^n which is an idempotent. Let $\theta \colon S \to T$ be a homomorphism between finite semigroups.

Suppose that $\theta(s)$ is an idempotent in T. Then for some natural number $n \geq 1$, the element s^n is also an idempotent. But $\theta(s^n) = \theta(s)^n = \theta(s)$. Thus, for finite semigroups, if $\theta(s)$ is an idempotent, then there is an idempotent e such that $\theta(e) = \theta(s)$. If S is an inverse semigroup, then every subsemigroup (notice: I did not say *inverse* subsemigroup on purpose) has commuting idempotents. We say that a semigroup T *divides* a semigroup S if there is a subsemigroup S' of S such that T is a homomorphic image of S'. Suppose now that S is a finite inverse semigroup. Then every subsemigroup S' of S is a finite semigroup that has commuting idempotents. Thus every homomorphic image of S' is a finite semigroup with commuting idempotents. We have therefore proved the following:

Lemma 17 *Every semigroup that divides a finite inverse semigroup has commuting idempotents.*

The following theorem was proved by Chris Ash [3] and is deep. It is the converse of the above lemma.

Theorem 2 (Ash) *Every finite semigroup with commuting idempotents divides a finite inverse semigroup.*

This is a beautiful result and a stunning piece of mathematics. It uses finite combinatorics, in the guise of Ramsey theory, as part of the proof. See [50] for the connections between combinatorics and semigroup theory.

4 The Natural Partial Order

Partial bijections can be compared with each other: we can say that one partial bijection is the restriction of another. This leads to a partial order on the set of partial bijections of a set. It might be thought that this partial order has to be imposed on an inverse semigroup but, remarkably, this partial order can be defined in purely algebraic terms. It is therefore called the natural partial order. It will follow that every inverse semigroup is, in fact, a partially ordered semigroup with respect to this order.

On an inverse semigroup, define $s \leq t$ iff $s = ts^{-1}s$. This looks one-sided because the idempotent appears on the right-hand side. It is here that we invoke Lemma 10.

Lemma 18 *In an inverse semigroup, the following are equivalent:*

(i) $s \leq t$.
(ii) $s = te$ *for some idempotent* e.
(iii) $s = ft$ *for some idempotent* f.
(iv) $s = ss^{-1}t$.

Proof (1) $\Rightarrow$(2). This is immediate.

(2) $\Rightarrow$(3). This is immediate by Lemma 10.

(3) $\Rightarrow$(4). Suppose that $s = ft$. Then $fs = s$ and so $fss^{-1} = ss^{-1}$. It follows that $s = ss^{-1}t$.

(4) $\Rightarrow$(1). Suppose that $s = ss^{-1}t$. By Lemma 10, we know that $s = ti$ for some idempotent i. Observe that $si = s$ and so $s^{-1}si = s^{-1}s$. It readily follows that $s = ts^{-1}s$ giving $s \leq t$. □

Let S be a semigroup equipped with a partial order $\leq$. We say that S is a *partially ordered semigroup* if $a \leq b$ and $c \leq d$ imply that $ac \leq bd$. We may now establish the main properties of the relation we have defined.

Lemma 19 *Let S be an inverse semigroup.*

(i) *The relation $\leq$ is a partial order.*
(ii) *If $s \leq t$, then $s^{-1} \leq t^{-1}$.*
(iii) *The semigroup S is partially ordered with respect to $\leq$.*
(iv) *If e and f are idempotents, then $e \leq f$ if and only if $e = ef = fe$.*
(v) *$s \leq e$, where e is an idempotent, implies that s is an idempotent.*
(vi) *Let T be an inverse semigroup and let $\theta\colon S \to T$ be a homomorphism. Then $a \leq b$ in S implies that $\theta(a) \leq \theta(b)$ in T.*

Proof

(1) Observe that $a \leq a$ since $a = a\mathbf{d}(a)$. Suppose that $a \leq b$ and $b \leq a$. Then $a = b\mathbf{d}(a)$ and $b = a\mathbf{d}(b)$. Then $b\mathbf{d}(a) = b$ from which it follows that $a = b$. Suppose that $a \leq b$ and $b \leq c$. Using Lemma 18, it is easy to show that $a \leq c$.
(2) This is immediate using the definition and Lemma 18.
(3) This is immediate using the definition, Lemmas 18 and 10.
(4) Straightforward.
(5) This is immediate from the definition and the fact that the product of idempotents is an idempotent.
(6) This is immediate from the fact that the natural partial order is algebraically defined, Lemma 18, and the fact that idempotents are mapped to idempotents by homomorphisms.

□

Part (1) of Lemma 19 leads us to dub $\leq$ the *natural partial order* on S. If a partial order is studied in relation to an inverse semigroup, then it is always this one.

Part (2) of Lemma 19 needs to be highlighted since readers familiar with lattice-ordered groups might have been expecting something different.

Part (4) of Lemma 19 tells us that when the natural partial order is restricted to the semilattice of idempotents we get back the usual ordering on the idempotents that we have already defined.

The natural partial order in the symmetric inverse monoids $\mathcal{I}(X)$ is precisely the usual restriction order on partial functions. Observe that by the Wagner-Preston representation theorem Theorem 1, we have that $a \leq b$ if and only if $\lambda_a \subseteq \lambda_b$.

The natural partial order is used to define a class of inverse monoids. We say that an inverse monoid S is *factorizable* if for each element $s \in S$ there exists a unit g such that $s \leq g$,

The symmetric inverse monoids $\mathcal{I}(X)$ are factorizable if X is finite. Let $f \in \mathcal{I}(X)$. Then not only do $\mathrm{dom}(f)$ and $\mathrm{ran}(f)$ have the same cardinality but so too do the sets $X \setminus \mathrm{dom}(f)$ and $X \setminus \mathrm{ran}(f)$. Choose any bijection g from $X \setminus \mathrm{dom}(f)$ to $X \setminus \mathrm{ran}(f)$. Then $f \cup g$ is a bijection and so is an element of the group of units of $\mathcal{I}(X)$. However, there are many choices for g and so there are many units that extend f. It is precisely this lack of uniqueness in the unit that implies that we cannot reduce inverse monoid theory to group theory.

Our next result tells us that the partial order encodes how far from being a group an inverse semigroup is.

Lemma 20 *An inverse semigroup is a group if and only if the natural partial order is the equality relation.*

Proof Let S be an inverse semigroup in which the natural partial order is equality. If e and f are any idempotents, then $ef \leq e, f$ and so $e = f$. It follows that there is exactly one idempotent. We deduce that S is a group by Lemma 4. The proof of the converse is immediate. □

The natural partial order is very important in studying inverse semigroups. For this reason, it is appropriate here to introduce some terminology and notation from the theory of partially ordered sets.

In any partially ordered set $(X, \leq)$, a subset $Y \subseteq X$ is said to be an *order ideal* if $x \leq y \in Y$ implies that $x \in Y$. More generally, if Y is any subset of X, then define

$$Y^{\downarrow} = \{x \in X : x \leq y \text{ for some } y \in Y\}.$$

This is the *order ideal generated by* Y. If $y \in X$, then we denote $\{y\}^{\downarrow}$ by $y^{\downarrow}$ and call it the *principal order ideal generated by* y. If Y is any subset of X, define

$$Y^{\uparrow} = \{x \in X : x \geq y \text{ for some } y \in Y\}.$$

If $Y = \{y\}$, we denote $\{y\}^{\uparrow}$ by $y^{\uparrow}$. If P and Q are partially ordered sets, then a function $\theta : X \to Y$ is said to be *isotone* if $x \leq y$ in X implies that $\theta(x) \leq \theta(y)$ in Y. An *order isomorphism* between two partially ordered sets is a bijective isotone function whose inverse is also isotone.

Part (5) of Lemma 19 tells us that the semilattice of idempotents of an inverse semigroup S is an order ideal in S with respect to the natural partial order.

Part (6) of Lemma 19 tells us that homomorphisms between inverse semigroups are isotone.

Idempotents and non-idempotents are closely related in an inverse semigroup.

Lemma 21 *Let S be an inverse semigroup. Then there is an order isomorphism between the set $a^{\downarrow}$ and the set $\mathbf{d}(a)^{\downarrow}$ (and, likewise, with the set $\mathbf{r}(a)^{\downarrow}$).*

Proof Define a map from $a^{\downarrow}$ to $\mathbf{d}(a)^{\downarrow}$ by $b \mapsto \mathbf{d}(b)$. This is well-defined since if $b \leq a$, then $\mathbf{d}(b) \leq \mathbf{d}(a)$. It is isotone since if $b_2 \leq b_1 \leq a$, then $\mathbf{d}(b_2) \leq \mathbf{d}(b_1) \leq \mathbf{d}(a)$. From the definition of the natural partial order, it is immediate that this map is injective. We define a map from $\mathbf{d}(a)^{\downarrow}$ to $a^{\downarrow}$ by $e \mapsto ae$. It is routine to check that this is well-defined and isotone. These two maps are mutually inverse. It follows that the partially ordered sets are order-isomorphic. □

Looking below an idempotent we see only idempotents, but what happens if we look up? The answer is that we do not necessarily see only idempotents. The symmetric inverse monoid is an example. This leads us to the following definition. An inverse semigroup S is said to be *E-unitary* if $e \leq s$, where e is an idempotent, implies that s is an idempotent. An inverse semigroup with zero S is said to be *E^*-unitary*[10] if $0 \neq e \leq s$ where e is an idempotent implies that s is an idempotent.

Lemma 22 *Let S be an inverse semigroup with zero. Then it is E-unitary if and only if it is a meet semilattice.*

Proof Suppose that S is E-unitary. If $a \in S$, then $0 \leq a$. But 0 is an idempotent, and so a is an idempotent. The proof of the converse is immediate. □

The above lemma explains why we have made two definitions above, depending on whether the inverse semigroup does not or does have a zero. This bifurcation between inverse semigroup without zero and inverse semigroup with zero permeates the subject.

There was a time when the study of E-unitary inverse semigroups was the centre of attention. The two papers by Don McAlister [85, 86] are probably the most significant in that they describe the structure of E-unitary inverse semigroups in terms of simpler building blocks and relate them to arbitrary inverse semigroups. The theory of E^*-unitary inverse semigroups has also been pursued. There are both interesting examples of such inverse semigroups and the analogue of McAlister's theory can be developed in the case of the so-called strongly E^*-unitary inverse semigroups. See [12, 53, 54], for example.

The following lemma tells us that in an inverse semigroup, there is a relationship between two elements that have a common upper bound.

Lemma 23 *Let S be an inverse semigroup and suppose that $a, b \leq c$. Then $a^{-1}b$ and ab^{-1} are idempotents.*

Proof By part (2) of Lemma 19, we have that $a^{-1}, b^{-1} \leq c^{-1}$. By part (3) of Lemma 19, we have that $a^{-1}b \leq c^{-1}c$ and $ab^{-1} \leq cc^{-1}$. It follows by part (5) of Lemma 19 that both $a^{-1}b$ and ab^{-1} are idempotents. □

Lemma 23 leads us to the following definition. Let $a, b \in S$, an inverse semigroup. Define $a \sim b$ if and only if $a^{-1}b, ab^{-1} \in \mathsf{E}(S)$. This is called the *compatibility relation*. If $a \sim b$ and if their least upper bound exists, we denote it by

[10] The term 0-*E*- *unitary* is also used.

$a \vee b$ and call it the *join* of a and b. A subset of an inverse semigroup is said to be *compatible* if the elements are pairwise compatible. For example, for each element $a \in S$ in an inverse semigroup, the set $a^{\downarrow}$ is compatible.

In an inverse semigroup with zero, there is a refinement of the compatibility relation which is important. In such a semigroup, a pair of idempotents e and f is said to be *orthogonal*, denoted by $e \perp f$, if and only if $ef = 0$. We define an arbitrary pair of elements a and b to be *orthogonal*, denoted by $a \perp b$, precisely when $\mathbf{d}(a) \perp \mathbf{d}(b)$ and $\mathbf{r}(a) \perp \mathbf{r}(b)$ You can easily check that $a \perp b$ precisely when $a^{-1}b = 0 = ab^{-1}$. We call $\perp$ the *orthogonality relation*. Observe that $a \perp b$ implies that $a \sim b$. If an orthogonal subset has a least upper bound, then it is said to have an *orthogonal join*.

Let $f, g \in \mathcal{I}(Y)$. Then $f \sim g$ if and only if $f \cup g$ is a partial bijection, and $f \perp g$ if and only if the domain of f is disjoint from the domain of g and the range of f is disjoint from the range of g.

Lemma 24 *Let S be an inverse semigroup with zero. If $a \perp b$ and $c \in S$, then $ac \perp bc$ and $ca \perp cb$.*

Proof We prove that $ac \perp bc$; the proof of the other case is similar. It is routine to check that $\mathbf{d}(ac)\mathbf{d}(bc) = 0$. The result now follows by symmetry. □

The compatibility relation is reflexive and symmetric but not, in general, transitive, as the symmetric inverse monoid shows. However, we do have an exact criterion for when the compatibility relation is transitive.

Proposition 3 *The compatibility relation is transitive if and only if the semigroup is E-unitary.*

Proof Suppose that $\sim$ is transitive. Let $e \leq s$, where e is an idempotent. Then se^{-1} is an idempotent because $e = se = se^{-1}$, and $s^{-1}e$ is an idempotent because $s^{-1}e \leq s^{-1}s$. Thus $s \sim e$. Clearly $e \sim s^{-1}s$, and so, by our assumption that the compatibility relation is transitive, we have that $s \sim s^{-1}s$. But $s(s^{-1}s)^{-1} = s$, so that s is an idempotent.

Conversely, suppose that S is E-unitary and that $s \sim t$ and $t \sim u$. Clearly $(s^{-1}t)(t^{-1}u)$ is an idempotent and

$$(s^{-1}t)(t^{-1}u) = s^{-1}(tt^{-1})u \leq s^{-1}u.$$

But S is E-unitary and so $s^{-1}u$ is an idempotent. Similarly, su^{-1} is an idempotent. Hence, $s \sim u$. □

There is a connection between compatible elements and certain kinds of meets. We shall examine meets in greater generality later on in this section.

Lemma 25 *Let S be an inverse semigroup.*

(i) *$s \sim t$ if and only if $s \wedge t$ exists and $\mathbf{d}(s \wedge t) = \mathbf{d}(s) \wedge \mathbf{d}(t)$ and $\mathbf{r}(s \wedge t) = \mathbf{r}(s) \wedge \mathbf{r}(t)$.*

(ii) *If* $s \sim t$, *then*

$$s \wedge t = t\mathbf{d}(s) = s\mathbf{d}(t) = \mathbf{r}(s)t = \mathbf{r}(t)s.$$

Proof

(1) We prove first that st^{-1} is an idempotent if and only if the greatest lower bound $s \wedge t$ of s and t exists and $(s \wedge t)^{-1}(s \wedge t) = s^{-1}st^{-1}t$. The full result then follows by the dual argument. Suppose that st^{-1} is an idempotent. Put $z = st^{-1}t$. Then $z \leq s$ and $z \leq t$, since st^{-1} is an idempotent. Let $w \leq s, t$. Then $w^{-1}w \leq t^{-1}t$ and so $w \leq st^{-1}t = z$. Hence, $z = s \wedge t$. Also

$$z^{-1}z = (st^{-1}t)^{-1}(st^{-1}t) = t^{-1}ts^{-1}st^{-1}t = s^{-1}st^{-1}t.$$

Conversely, suppose that $s \wedge t$ exists and $(s \wedge t)^{-1}(s \wedge t) = s^{-1}st^{-1}t$. Put $z = s \wedge t$. Then $z = sz^{-1}z$ and $z = tz^{-1}z$. Thus $sz^{-1}z = tz^{-1}z$, and so $st^{-1}t = ts^{-1}s$. Hence, $st^{-1} = ts^{-1}st^{-1}$, which is an idempotent.

(2) We shall prove that $s \wedge t = t\mathbf{d}(s)$ since the other equalities follow by symmetry. Observe that $t\mathbf{d}(s) = ts^{-1}s \leq s, t$ since $ts^{-1} = (st^{-1})^{-1}$ is an idempotent. Suppose that $x \leq s, t$. Then $x = xx^{-1}x \leq ts^{-1}s$. It follows that $s \wedge t = t\mathbf{d}(s)$.

□

The following result is useful since it enables us to deduce that two elements are equal from apparently weaker conditions.

Lemma 26 *Let S be an inverse semigroup. If $a \sim b$ and $\mathbf{d}(a) = \mathbf{d}(b)$ (respectively, $\mathbf{r}(a) = \mathbf{r}(b)$), then $a = b$.*

Proof Suppose that $a \sim b$; then by Lemma 25 the meet $a \wedge b$ exists and $\mathbf{d}(a \wedge b) = \mathbf{d}(a)\mathbf{d}(b)$. By assumption, $\mathbf{d}(a) = \mathbf{d}(b)$. Thus, $a \wedge b \leq a$ and $\mathbf{d}(a \wedge b) = \mathbf{d}(a)$. It follows that $a \wedge b = a$. Similarly, $a \wedge b = b$. We have therefore proved that $a = b$. □

Inverse semigroups generalize groups, but we can also construct groups from inverse semigroups. The idea is this. Groups are abstract versions of groups of bijections, and bijections can be constructed by glueing together compatible sets of partial bijections. Thus, we could construct groups from inverse semigroups by glueing together suitable compatible subsets. We show first how to construct groups from arbitrary inverse semigroups. The motivation for how to do this comes from Lemma 20, which tells us that groups are those inverse semigroups in which the natural partial order is equality, and part (6) of Lemma 19, which tells us that homomorphisms between inverse semigroups are isotone. These two results lead us to make the following definition. On an inverse semigroup S, with $s, t \in S$, define the relation σ by $s \,\sigma\, t$ if and only if there is an element u such that $u \leq s, t$.

Theorem 3 *Let S be an inverse semigroup.*

(i) *σ is a congruence on S.*

(ii) *S/σ is a group.*
(iii) *If ρ is any congruence on S such that S/ρ is a group, then $\sigma \subseteq \rho$.*

Proof

(1) We begin by showing that σ is an equivalence relation. Reflexivity and symmetry are immediate. To prove transitivity, let $(a, b), (b, c) \in \sigma$. Then there exist elements $u, v \in S$ such that $u \leq a, b$ and $v \leq b, c$. Thus $u, v \leq b$. The set $b^{\downarrow}$ is a compatible subset and so $u \wedge v$ exists by Lemma 25. But $u \wedge v \leq a, c$ and so $(a, c) \in \sigma$. The fact that σ is a congruence follows from the fact that the natural partial order is compatible with the multiplication.
(2) Clearly, all idempotents are contained in a single σ-class (possibly with non-idempotent elements). Consequently, S/σ is an inverse semigroup with a single idempotent. Thus S/σ is a group by Lemma 20.
(3) Let ρ be any congruence such that S/ρ is a group. Let $(a, b) \in \sigma$. Then $z \leq a, b$, for some z, by definition. Hence, $\rho(z) \leq \rho(a), \rho(b)$ since homomorphisms between inverse semigroups are isotone. But S/ρ is a group and so its natural partial order is equality. Hence, $\rho(a) = \rho(b)$.

□

In the light of Theorem 3, it is natural to call the congruence σ the *minimum group congruence* and the group S/σ the *maximum group image* of S.

A homomorphism $\theta \colon S \to T$ is said to be *idempotent-pure* if $\theta(a)$ an idempotent implies that a is an idempotent.

Lemma 27 *Let S be an inverse semigroup.*

(i) $\sim \subseteq \sigma$.
(ii) *The congruence ρ is idempotent-pure if and only if $\rho \subseteq \sim$.*

Proof

(1) Suppose that $a \sim b$. Then $a \wedge b$ exists by Lemma 25. It follows that $a \,\sigma\, b$.
(2) Suppose that ρ is idempotent-pure and that $a \,\rho\, b$. Then $ab^{-1} \,\rho\, bb^{-1}$. But ρ is idempotent-pure and so ab^{-1} is an idempotent. Similarly, $a^{-1}b$ is an idempotent. We have proved that $a \sim b$. Conversely, suppose that ρ is a congruence such that $\rho \subseteq \sim$ and $a \,\rho\, e$, where e is an idempotent. Observe that $a^{-1} \,\rho\, e$. Thus $aa^{-1} \,\rho\, e$. It follows that $aa^{-1} \,\rho\, a$. Thus $(aa^{-1})a$ is an idempotent and so a is an idempotent.

□

Inverse semigroups and their homomorphisms form a category (of structures). The category of groups and their homomorphisms is a subcategory. The properties of the minimum group congruence lead naturally to the following result on the category of inverse semigroups.

Proposition 4 *The category of groups is a reflective subcategory of the category of inverse semigroups.*

Proof Let S be an inverse semigroup and let $\sigma^{\natural}\colon S \to S/\sigma$ be the associated natural homomorphism. Let $\theta\colon S \to G$ be a homomorphism to a group G. Then $\ker(\theta)$ is a group congruence on S and so $\sigma \subseteq \ker(\theta)$ by Theorem 3. Thus by standard semigroup theory, there is a unique homomorphism θ^* from S/σ to G such that $\theta = \theta^*\sigma^{\natural}$. □

It follows by standard category theory, such as [79, Chapter IV, Section 3], that there is a functor from the category of inverse semigroups to the category of groups which takes each inverse semigroup S to S/σ and if $\theta\colon S \to T$ is a homomorphism of inverse semigroups, then the function $\psi\colon S/\sigma \to T/\sigma$ defined by $\psi(\sigma(s)) = \sigma(\theta(s))$ is the corresponding group homomorphism (this can be checked directly).

There is another characterization of E-unitary inverse semigroups which is interesting in this context.

Lemma 28 *Let S be an inverse semigroup. Then the following conditions are equivalent:*

(i) *S is E-unitary.*
(ii) $\sim = \sigma$.
(iii) *σ is idempotent-pure.*
(iv) *$\sigma(e) = \mathsf{E}(S)$ for any idempotent e.*

Proof (1) $\Rightarrow$(2). By part (1) of Lemma 27, the compatibility relation is contained in σ. Let $(a, b) \in \sigma$. Then $z \leq a, b$ for some z. It follows that $z^{-1}z \leq a^{-1}b$ and $zz^{-1} \leq ab^{-1}$. But S is E-unitary and so $a^{-1}b$ and ab^{-1} are both idempotents. Hence, $a \sim b$.

(2) $\Rightarrow$(3). By part (2) of Lemma 27, a congruence is idempotent-pure precisely when it is contained in the compatibility relation.

(3) $\Rightarrow$ (4). This is immediate from the definition of an idempotent-pure congruence.

(4) $\Rightarrow$ (1) Suppose that $e \leq a$ where e is an idempotent. Then $(e, a) \in \sigma$. But by our assumption, the element a is an idempotent. □

For inverse semigroups with zero the minimum group congruence is not very interesting since the group degenerates to the trivial group. The following example shows one concrete way to deal with this issue.

Let G be a group and let S be the inverse monoid of all isomorphisms between the subgroups of G. The semilattice of idempotents is isomorphic to the partially ordered set of subgroups of G. The trivial group is a subgroup of every group. It follows that the isomorphism from the trivial subgroup to itself is the zero of S. Suppose now that G is infinite. Consider the inverse subsemigroup $\Omega(G)$ of S which consists of all the isomorphisms between the subgroups of G of finite index. The group $\mathrm{Comm}(G) = \Omega(G)/\sigma$ is called the *abstract commensurator* of G [7]. See also [91]. The elements of $\mathrm{Comm}(G)$ are 'hidden symmetries' to use the terminology of Farb and Weinberger [24].

The above example shows that we may form groups from inverse semigroups by looking at the 'large' elements of the inverse semigroup (and so excluding the zero).

Here is another example. Let S be an inverse semigroup. A non-zero idempotent e of an inverse semigroup S is said to be *essential* if $ef \neq 0$ for all non-zero idempotents f of S. An element s is said to be *essential* if both $s^{-1}s$ and ss^{-1} are essential. Essential elements were first defined in [9] and applied to construct groups in [56, 57]. Denote by S^e the set of all essential elements of S. We regard the elements of S^e as being 'large' elements of S. We call S^e the *essential part of S*.

Lemma 29 *Let S be an inverse semigroup. Then, if non-empty, S^e is an inverse subsemigroup of S.*

Proof It is clear that S^e is closed under inverse. Let $a, b \in S^e$. We prove that $\mathbf{d}(ab)$ is essential; the proof that $\mathbf{r}(ab)$ is essential is similar. Observe, first, that $b^{-1}a^{-1}ab \neq 0$ since both bb^{-1} and $a^{-1}a$ are essential. Let e be a non-zero idempotent. We calculate $\mathbf{d}(ab)e$. Observe that $beb^{-1} \neq 0$ and is an idempotent, since it is the conjugate of an idempotent. Thus $a^{-1}abeb^{-1} \neq 0$. It follows that $a^{-1}abe \neq 0$ and so $b^{-1}a^{-1}abe \neq 0$. □

We expect the groups S^e/σ to be interesting, which indeed they are if we choose S carefully. See [57, 71, 74] for applications of the essential part of an inverse semigroup in constructing groups.

We have seen that there is a precondition that must be satisfied in order that a pair of elements have a join. The same is not true for meets. If every pair of elements of an inverse semigroup has meets, we say that it is a *meet-semigroup*. Such semigroups were first studied by Leech [76]. There is an alternative way to characterize inverse meet-semigroups which is often useful.

Lemma 30 *Let S be an inverse semigroup.*

(i) *S has all binary meets if and only if for each element $a \in S$ there is an idempotent, denoted by $\phi(a)$, such that $\phi(a) \leq a$ and if $e \leq a$, where e is an idempotent, then $e \leq \phi(a)$. Thus, $\phi(a)$ is the largest idempotent below a.*
(ii) *The map $\phi \colon S \to \mathsf{E}(S)$, defined in part (1) above, is an order-preserving idempotent function with $\mathsf{E}(S)$ as its fixed-point set such that $\phi(ae) = \phi(a)e$ and $\phi(ea) = e\phi(a)$ for all $e \in \mathsf{E}(S)$.*

Proof

(1) Suppose first that S has all binary meets. For each $a \in S$ define $\phi(a) = a \wedge \mathbf{d}(a)$. Clearly, $\phi(a)$ is an idempotent (because it is beneath an idempotent). Let e be an idempotent such that $e \leq a$. Then $e \leq \mathbf{d}(a)$ and so $e \leq \phi(a)$. We have therefore proved that if all binary meet exist, then the function ϕ exists. Conversely, suppose that such a function ϕ exists. Let $a, b \in S$ and consider the element $\phi(ab^{-1})b$. Clearly, $\phi(ab^{-1})b \leq b$. But, by definition, $\phi(ab^{-1}) \leq ab^{-1}$ and so $\phi(ab^{-1})b \leq ab^{-1}b \leq a$. Let $c \leq a, b$. Then $cc^{-1} \leq ab^{-1}$ and so $cc^{-1} \leq \phi(ab^{-1})$. Now $c \leq b$ and so $c = (cc^{-1})c \leq \phi(ab^{-1})b$. It follows that $a \wedge b = \phi(ab^{-1})b$.
(2) It is easy to prove that ϕ is order-preserving. We prove that $\phi(ae) = \phi(a)e$ for all $e \in \mathsf{E}(S)$. By definition $\phi(a) \leq a$. Thus $\phi(a)e \leq ae$. But $\phi(a)e$ is an

idempotent. It follows that $\phi(a)e \leq \phi(ae)$. We have that $\phi(ae) \leq ae \leq a$. But $\phi(ae)$ is an idempotent. It follows that $\phi(ae) \leq \phi(a)$ and so $\phi(ae)e \leq \phi(a)e$. But $\phi(ae) \leq ae$, and so $\phi(ae)e = \phi(ae)$. Thus $\phi(ae) \leq \phi(a)e$. We have therefore proved that $\phi(ae) = \phi(a)e$.

□

The function ϕ is called a *fixed-point operator.*

We can deduce right away that the symmetric inverse monoids are meet-monoids because if $f \in \mathcal{I}(X)$, then we can define the idempotent 1_A where A is the set of all points that f fixes (and if f does not fix any points, then $A = \varnothing$.)

The E^*-unitary semigroups also enjoy a property that is more significant than it looks. The following was first noted by Lenz [77].

Proposition 5 *An E^*-unitary inverse semigroup has meets of all pairs of elements.*

Proof Let s and t be any pair of elements. Suppose that there exists a non-zero element u such that $u \leq s, t$. Then $uu^{-1} \leq st^{-1}$ and uu^{-1} is a non-zero idempotent. Thus st^{-1} is an idempotent. Similarly $s^{-1}t$ is an idempotent. It follows that $s \wedge t$ exists by Lemma 25. If the only element below s and t is 0, then $s \wedge t = 0$. □

To conclude this section, we state another deep result about finite inverse monoids. We need a definition first. An inverse monoid S is said to be *F-inverse* if every σ-class contains a greatest element.

Lemma 31 *Every F-inverse monoid is E-unitary.*

Proof Let S be an F-inverse monoid. Suppose that $e \leq a$. Then $\sigma(e) = \sigma(a)$. Each σ-class contains a maximum element. Since we are working in a monoid, $\sigma(e) = \sigma(1)$. Let the maximum element in $\sigma(1)$ be x. Then $1 \leq x$. It follows that $1 = x1 = x$. Thus the maximum element of $\sigma(1)$ is 1 itself. Thus $a \leq 1$ and so a is an idempotent. □

You can find out a lot more about F-inverse monoids, here [6]. In [5, Theorem 2.7], the following deep theorem is proved and the background to it explained.

Theorem 4 *Every finite inverse monoid has a finite F-inverse monoid cover.*

The implications of this theorem for finite inverse monoid theory are explained in [51], but it is enough to say that this is another example of a beautiful piece of mathematics and is related to Ash's deep result Theorem 2.

5 Inverse Semigroups as Non-commutative Lattices

It is only a slight exaggeration to say that in the four decades after inverse semigroups were introduced the main focus of researchers was on the purely algebraic properties of inverse semigroups.[11] Subsequently, the properties of the natural partial order have come to the fore. In this section, we shall study inverse semigroups with respect to this natural partial order. From this point of view, inverse semigroups can themselves be regarded as 'non-commutative meet semilattices'. However, we shall find it fruitful to regard certain inverse semigroups as being 'non-commutative lattices' of various kinds, always with the proviso that, because of Lemma 23, the join will not always be defined. Meets of idempotents will always usually be denoted by concatenation. Specifically, we shall take as our points of departure the various classes of lattice (each with top 1 and bottom 0): a *frames* are the complete infinitely distributive lattices; *distributive lattices* have all binary meets and binary joins with binary meets distributing over binary joins and vice versa; *Boolean algebras* are those distributive lattices in which each element x has a *complement* $\bar{x}$ such that $x\bar{x} = 0$ and $x \vee \bar{x} = 1$. Observe that homomorphisms of distributive lattices automatically preserve complements. Boolean algebras are particularly important so it will be useful to have a way of defining them which involves only products and complements. For the standard axioms for a Boolean algebra see [62, Chapter 2]. The following lemma contains some axioms due to Frink [27]. If you have trouble proving that this really is a Boolean algebra, see [94].

Lemma 32 *Consider the following structure $(B, \cdot, a \mapsto \bar{a}, 0)$ where $(B, \cdot)$ is a commutative band and $ab = a$ if and only if $a\bar{b} = 0$. Define $a + b = \overline{(\bar{a} \cdot \bar{b})}$. Then $(B, \cdot, +, 0, \bar{0})$ is a Boolean algebra.*

We call these *Frink's axioms for a Boolean algebra.*

We begin with frames. The lattice of open sets of a topological space is an example of a frame The theory of frames can be viewed as the theory of topological spaces in which the open sets, and not the points, are taken as primary. As well as being an interesting theory in its own right [39] with important applications, it is also a key ingredient in topos theory [80]. Johnstone discusses the origins of frame theory in his book on this subject [39, Chapter II]. One sentence is significant for the goals of this section. He writes on page 76:

> It was Ehresmann ... and his student Bénabou ... who first took the decisive step in regarding complete Heyting algebras as 'generalized topological spaces'.

However, Johnstone does not say *why* Ehresmann was led to his frame-theoretic viewpoint of topological spaces. In fact, it was Ehresmann's paper [21], which we have cited above as being one of the origins of inverse semigroup theory, which led to the theory of frames. Ehresmann was interested in pseudogroups of transformations. Among those are the full transformation pseudogroups $\mathcal{I}(X, \tau)$ of homeomorphisms between the open subsets of τ. The idempotents of such pseudogroups are the identity functions defined on the open subsets of τ. Of course, these

[11] This is a tendency rather than an absolute. For example, the papers [104] and [105] deal with the natural partial order.

form frames. More generally, we define a *pseudogroup* to be an inverse semigroup in which all compatible subsets have joins and in which multiplication distributes over such joins. The semilattice of idempotents of a pseudogroup is a frame and so we can regard pseudogroups as being non-commutative generalizations of frames. Using this language, we can say that Schein [105] proved that associated with every inverse semigroup is a universal pseudogroup. Pseudogroups themselves are the subject of [103].

We can continue in this vein, but consider instead only finite joins. We say that an inverse monoid is *distributive* if it has all finite joins and multiplication distributes over such joins. The semilattice of idempotents of a distributive inverse monoid is a distributive lattice. We can therefore regard distributive inverse monoids as being non-commutative distributive lattices. Distributive inverse monoids were first studied in [43].

An inverse monoid is said to be *Boolean* if it is distributive and its semilattice of idempotents is in fact a Boolean algebra. We can therefore regard Boolean inverse monoids as being non-commutative Boolean algebras. Boolean inverse monoids have shown themselves to be particularly interesting.

The relationships between these various classes of inverse semigroup are described in [67]. Boolean inverse monoids were introduced in [60], and the fact that associated with every inverse semigroup is a universal Boolean inverse semigroup is proved in [63]. Associated with every Boolean inverse monoid is its *type monoid*: this is always an abelian refinement monoid [48]. The type monoid is then studied in great detail in [118]. The following table summarizes this whole approach to inverse semigroups:

Commutative	**Non − commutative**
Meet semilattice	Inverse semigroup
Frame	Pseudogroup
Distributive lattice	Distributive inverse monoid
Boolean algebra	Boolean inverse monoid

In the remainder of this section, we shall be particularly interested in Boolean inverse monoids, but we shall prove some results in greater generality. Our first results hold in any inverse semigroup

Lemma 33 *Let S be an inverse semigroup.*

(i) *Here, we shall need the fact that the inverse semigroup has a zero. Suppose that $a, b \leq c$. If $\mathbf{d}(a) \perp \mathbf{d}(b)$, then $\mathbf{r}(a) \perp \mathbf{r}(b)$.*
(ii) *If $a \wedge b$ exists, then $c(a \wedge b) = ca \wedge cb$ and $(a \wedge b)c = ac \wedge bc$.*

Proof

(1) We are given that $a = c\mathbf{d}(a)$ and $b = c\mathbf{d}(b)$. Using these equations and the fact that idempotents commute, it is easy to check that $\mathbf{r}(a)\mathbf{r}(b) = 0$.

(2) We are given that $a \wedge b$ exists. We prove that $ca \wedge cb$ exists and that $(a \wedge b)c = ac \wedge bc$. The proof of the other case is similar. Observe that $c(a \wedge b) \leq ca, cb$. Now let $y \leq ca, cb$. Then $c^{-1}y \leq c^{-1}ca, c^{-1}cb$ and so $c^{-1}y \leq a, b$. It follows that $c^{-1}y \leq a \wedge b$. Thus $cc^{-1}y \leq c(a \wedge b)$. But $y \leq ca$ and so $y = ca\mathbf{d}(y)$. It follows that $cc^{-1}y = y$. Thus $y \leq ca, cb$ implies that $y \leq c(a \wedge b)$. It follows that $c(a \wedge b) = ca \wedge cb$

□

We now specialize to distributive inverse monoids. Result (2) below is what remains of one of the distributive laws.

Lemma 34 *Let S be a distributive inverse monoid.*

(i) *Suppose that $a \sim b$. Then $\mathbf{d}(a \vee b) = \mathbf{d}(a) \vee \mathbf{d}(b)$ and $\mathbf{r}(a \vee b) = \mathbf{r}(a) \vee \mathbf{r}(b)$*
(ii) *Suppose that both $a \vee b$ and $c \wedge (a \vee b)$ exist. Then both $c \wedge a$ and $c \wedge b$ exist, the join $(c \wedge a) \vee (c \wedge b)$ exists and $c \wedge (a \vee b) = (c \wedge a) \vee (c \wedge b)$.*

Proof

(1) We prove the first result; the proof of the second is similar. We calculate $(a \vee b)^{-1}(a \vee b)$. We first use the fact that $a \mapsto a^{-1}$ is an order isomorphism. It follows that $(a \vee b)^{-1}(a \vee b) = (a^{-1} \vee b^{-1})(a \vee b)$. Now multiply out to get

$$a^{-1}a \vee a^{-1}b \vee b^{-1}a \vee b^{-1}b.$$

But $a \sim b$ and so both $a^{-1}b$ and $b^{-1}a$ are idempotents. Moreover, $a^{-1}b \leq a^{-1}a$ and $b^{-1}a \leq b^{-1}b$. The result now follows.

(2) Let $x \leq c, a$. Then $x \leq c \wedge (a \vee b)$. It follows that $x\mathbf{d}(a) \leq (c \wedge (a \vee b))\mathbf{d}(a)$. But $x \leq a$ implies that $x\mathbf{d}(a) = x$. It follows that $x \leq (c \wedge (a \vee b))\mathbf{d}(a)$. On the other hand, $(c \wedge (a \vee b))\mathbf{d}(a) \leq c, a$. It follows that

$$c \wedge a = (c \wedge (a \vee b))\mathbf{d}(a).$$

Similarly,

$$c \wedge b = (c \wedge (a \vee b))\mathbf{d}(b).$$

Observe that since $c \wedge a, c \wedge b \leq a \vee b$ it follows that $(c \wedge a) \sim (c \wedge b)$. Thus $(c \wedge a) \vee (c \wedge b)$ exists. Observe that $(c \wedge a) \vee (c \wedge b) \leq c \wedge a, c \wedge b$. It follows that $(c \wedge a) \vee (c \wedge b) \leq c, a \vee b$. Thus,

$$(c \wedge a) \vee (c \wedge b) \leq c \wedge (a \vee b).$$

Let $x \leq a \vee b, c$. Then $x = (a \vee b)\mathbf{d}(x)$. From $x \leq a \vee b$ we get that $x\mathbf{d}(a) \leq (a \vee b)\mathbf{d}(a)$. Now, $(a \vee b)\mathbf{d}(a) = a$. It follows that $x\mathbf{d}(a) \leq a$. But $x \leq c$ and so $x\mathbf{d}(a) \leq c\mathbf{d}(a)$. Thus we have proved that $x\mathbf{d}(a) \leq a, c\mathbf{d}(a)$. We now apply Lemma 33 and deduce that $(a \wedge c)\mathbf{d}(a) = a \wedge c\mathbf{d}(c)$. Thus $x\mathbf{d}(a) \leq a \wedge c\mathbf{d}(c)$.

Similarly, $x\mathbf{d}(b) \leq c\mathbf{d}(b) \wedge b$. It follows that

$$x\mathbf{d}(a) \vee x\mathbf{d}(b) \leq (c\mathbf{d}(a) \wedge a) \vee (c\mathbf{d}(b) \wedge b)$$

and so

$$x\mathbf{d}(a \vee b) \leq (c\mathbf{d}(a) \wedge a) \vee (c\mathbf{d}(b) \wedge b).$$

By the above, we deduce that

$$x \leq (c\mathbf{d}(a) \wedge a) \vee (c\mathbf{d}(b) \wedge b).$$

But $(c\mathbf{d}(a) \wedge a) \vee (c\mathbf{d}(b) \wedge b) \leq (c \wedge a) \vee (c \wedge b)$ and the result follows.

□

By induction, the second result above can be generalized to any number of joins.

We now specialize to Boolean inverse monoids. Let S be a Boolean inverse monoid. If $y \leq x$, define

$$x \setminus y = x\overline{\mathbf{d}(y)}.$$

Lemma 35 *Let S be a Boolean inverse monoid.*

(i) $\mathbf{d}(x \setminus y) = \mathbf{d}(x)\overline{\mathbf{d}(y)}$.
(ii) *If* $y \leq x$, *then* $y \perp (x \setminus y)$ *and* $x = y \vee (x \setminus y)$.
(iii) *Suppose that* $a \leq x$ *is such that* $a \perp y$ *and* $x = y \vee a$, *then* $a = x \setminus y$.
(iv) $\mathbf{r}(x \setminus y) = \mathbf{r}(x)\overline{\mathbf{r}(y)}$.

Proof

(1) Straightforward from the definition.
(2) By definition $(x \setminus y) \leq x$ and from part (1), we have that $\mathbf{d}(x \setminus y) = \mathbf{d}(x)\overline{\mathbf{d}(y)}$. It follows that $\mathbf{d}(y) \perp \mathbf{d}(x \setminus y)$. We now apply Lemma 33, to deduce that $y \perp (x \setminus y)$. Clearly, $y \vee (x \setminus y) \leq x$. But $\mathbf{d}(y \vee (x \setminus y)) = \mathbf{d}(y) \vee \mathbf{d}(x)\overline{\mathbf{d}(y)}) = \mathbf{d}(x)$. It follows that $x = y \vee (x \setminus y)$.
(3) Observe that $\mathbf{d}(x) = \mathbf{d}(y) \vee \mathbf{d}(a)$ and $\mathbf{d}(a) \perp \mathbf{d}(y)$. It follows that $\mathbf{d}(a) = \mathbf{d}(x)\overline{\mathbf{d}(y)}$. But $a = x\mathbf{d}(a) = x\overline{\mathbf{d}(y)} = x \setminus y$.
(4) This follows by part (3) above.

□

A *morphism* between distributive inverse monoids is a homomorphism of monoids with zero that preserves binary joins.

In working with Boolean inverse monoids, it is often much easier to calculate with orthogonal joins than arbitrary ones. The following result provides circumstances under which we lose nothing by doing this.

Lemma 36 *Let S be an inverse semigroup.*

(i) *We make the following assumptions about S: the semilattice of idempotents of S forms a Boolean algebra under the natural partial order; all finite orthogonal joins exist; multiplication distributes over finite orthogonal joins. Then S is a Boolean inverse monoid.*

(ii) *Let S and T be Boolean inverse monoids. Let $\theta \colon S \to T$ be a monoid homomorphism of monoid with zero that preserves orthogonal joins. Then θ preserves all binary compatible joins.*

Proof

(1) We have to prove that all binary joins exist and that multiplication distributes over such joins. Let $e, f \in \mathsf{E}(S)$. Then using the distributive law, we have that $e\bar{f} \vee f = e \vee f$. But $e\bar{f}$ and f are orthogonal. It follows that $a(e \vee f) = a(e\bar{f} \vee f) = ae\bar{f} \vee af$. Observe that $(ae\bar{f} \vee af)\mathbf{d}(ae) = ae$. It follows that $ae, af \leq ae\bar{f} \vee af$. On the other hand, if $ae, af \leq x$, then $ae\bar{f} \vee af \leq x$. We have therefore proved that

$$a(e \vee f) = ae \vee af$$

where e and f are any idempotents. Suppose, now, that $a \sim b$. Then $a \wedge b$ exists by Lemma 25. We have that

$$(a \setminus a \wedge b) \vee b = a\overline{\mathbf{d}(a)\mathbf{d}(b)} \vee b$$

since, in this case, $\mathbf{d}(a \wedge b) = \mathbf{d}(a)\mathbf{d}(b)$ by Lemma 25. But $\overline{\mathbf{d}(a)\mathbf{d}(b)} = \overline{\mathbf{d}(a)} \vee \overline{\mathbf{d}(b)}$. We now use our result above to deduce that

$$(a \setminus a \wedge b) \vee b = a\overline{\mathbf{d}(b)} \vee b,$$

an orthogonal join. Observe that $(a\overline{\mathbf{d}(b)} \vee b)\mathbf{d}(a) = a$, where we use the fact that $b\mathbf{d}(a) = a\mathbf{d}(b)$ by Lemma 25, our result above, and a little Boolean algebra. We therefore have that $a, b \leq a\overline{\mathbf{d}(b)} \vee b$. Suppose that $a, b \leq x$. Then it is routine to check that $a\overline{\mathbf{d}(b)} \vee b \leq x$. We have therefore proved that $a \vee b$ exists and is equal to the orthogonal join $a\overline{\mathbf{d}(b)} \vee b$.

It remains to show that multiplication distributes over binary joins. Suppose that $a \sim b$. It can be proved directly that $ca \sim cb$. We prove that

$$c(a \vee b) = ca \vee cb.$$

We have that

$$c(a \vee b) = c(a\overline{\mathbf{d}(b)} \vee b) = ca\overline{\mathbf{d}(b)} \vee cb.$$

Thus

$$ca \leq ca\overline{\mathbf{d}(b)} \vee cb.$$

Now,

$$(ca\overline{\mathbf{d}(b)} \vee cb)\mathbf{d}(ca) = ca\overline{\mathbf{d}(b)} \vee cb\mathbf{d}(ca).$$

We now use the fact that $ca \sim cb$ and Lemma 25 to get

$$ca\overline{\mathbf{d}(b)} \vee cb\mathbf{d}(ca) = ca\overline{\mathbf{d}(b)} \vee ca\mathbf{d}(cb) \leq ca.$$

It follows that

$$ca = (ca\overline{\mathbf{d}(b)} \vee cb)\mathbf{d}(ca).$$

We deduce that $ca, cb \leq ca\overline{\mathbf{d}(b)} \vee cb$. Suppose that $ca, cb \leq x$. Then it is routine to check that $ca\overline{\mathbf{d}(b)} \vee cb \leq x$. Whence, $ca \vee cb = ca\overline{\mathbf{d}(b)} \vee cb$ and so $c(a \vee b) = ca \vee cb$.

(2) Suppose that $a \sim b$. Then $a \vee b = a\overline{\mathbf{d}(b)} \vee b$, which is an orthogonal join. By assumption,

$$\theta(a \vee b) = \theta(a\overline{\mathbf{d}(b)}) \vee \theta(b) = \theta(a)\theta(\overline{\mathbf{d}(b)})) \vee \theta(b).$$

The result now follows since $\theta(\bar{e}) = \overline{\theta(e)}$ where e is any idempotent.

□

For the remainder of this section, we shall focus on Boolean inverse monoids. As we shall prove in Theorem 5, they arise naturally as soon as we try to map inverse monoids to the multiplicative monoids of rings, such as in representation theory. Accordingly, we begin with some results about idempotents in rings. Recall that all our rings will be unital. If e and f are idempotents in a ring R, then we say they are *orthogonal* if $ef = 0$ denoted by $e \perp f$. A finite set of idempotents is *orthogonal* if each distinct pair of elements is orthogonal. A sum of a finite number of orthogonal idempotents will be called an *orthogonal sum*. More generally, a sum of orthogonal elements taken from an inverse submonoid will be called an *orthogonal sum*.

Lemma 37 *Let R be a unital ring.*

(i) *If e is an idempotent in R, then $1 - e$ is an idempotent in R; in addition, the idempotents e and $1 - e$ are orthogonal.*
(ii) *If e and f are orthogonal idempotents, then $e + f$ is an idempotent; this can be generalized to any finite sets of orthogonal idempotents.*
(iii) *If e and f are orthogonal idempotents, then*

$$1 - (e + f) = (1 - e)(1 - f).$$

(iv) *If e and f are commuting idempotents, then*

$$e + f - ef = f + e(1 - f),$$

an orthogonal sum.

(v) *If* e *and* f *are commuting idempotents, then*

$$e \circ f = e + f - ef$$

is an idempotent.

(vi) *If* e *and* f *are commuting idempotents, then*

$$(1 - e)(1 - f) = 1 - e \circ f.$$

(vii) *If* e, f, g *is a set of commuting idempotents, then* e *commutes with* $f \circ g$ *and* g *commutes with* $e \circ f$ *and*

$$(e \circ f) \circ g = e \circ (f \circ g).$$

(viii) *If* $e_1, \ldots, e_m$ *are orthogonal idempotents, then*

$$1 - \left(\sum_{i=1}^{m} e_i\right) = \prod_{i=1}^{m}(1 - e_i).$$

(ix) *Let* $e_1, \ldots, e_m$ *be a set of commuting idempotents. Define*

$$[e_1, \ldots, e_m] = (\ldots((e_1 \circ e_2) \circ e_3)\ldots) \circ e_m.$$

Then

$$\prod_{i=1}^{m}(1 - e_i) = 1 - [e_1, \ldots, e_m].$$

(x) *Let* $e_1, \ldots, e_m$ *be a set of commuting idempotents. Then*

$$[e_1, \ldots, e_m]$$

is defined to be

$$e_1(1 - e_2)\ldots(1 - e_m) + e_2(1 - e_3)\ldots(1 - e_m) + \ldots + e_{m-1}(1 - e_m) + e_m,$$

a sum of orthogonal idempotents.

Proof The proofs of (1)–(7) are straightforward.

(8) By part (3) and induction.

(9) By part (6) and induction.

(10) By part (4) and induction. □

We begin with a lemma which is a special case of the result we want to prove.

Lemma 38 *Let $E \subseteq R$, where R is a unital ring. Suppose that $0, 1 \in E$ and E is a commutative band under the induced multiplication from R. Then there is a subset $E \subseteq B \subseteq R$ such that B is a Boolean algebra with meet given by the product in the ring.*

Proof If E is any semilattice in R, define

$$E' = \{e(1 - e_1) \ldots (1 - e_m) \colon e, e_1, \ldots, e_m \in E\} \cup E.$$

Then it is easy to check that this is a commutative band. If E is any semilattice in R, define $E^{\perp}$ to be the set of all sums of finite sets of orthogonal elements of E. Then it is easy to check that this is a commutative band. Put $B = (E')^{\perp}$. This is a commutative band. To prove that B is a Boolean algebra, we use Lemma 32. We claim that if $e \in B$, then $1 - e \in B$. An element of B is an orthogonal join of idempotents that belong to E'. Observe that if $e_1, \ldots, e_m$ are any orthogonal idempotents of B, then $1 - \left(\sum_{i=1}^{m} e_i\right) = \prod_{i=1}^{m}(1 - e_i)$ by part (8) of Lemma 37 where each $e_i \in E'$. If $e \in E$, then $1 - e \in E'$. Thus we need only concentrate on idempotents of the form

$$e(1 - f_1) \ldots (1 - f_n) \in E'$$

where $f_1, \ldots, f_n \in E$. Then

$$1 - e(1 - f_1) \ldots (1 - f_n) = (1 - e) + e[f_1, \ldots, f_n],$$

using part (9) of Lemma 37, which is a sum of orthogonal idempotents by part (10) of Lemma 37. We have therefore proved that B is closed under taking complements. Let $e, f \in B$ and suppose that $ef = e$. Then $e(1 - f) = e - ef = e - e = 0$. On the other hand, suppose that $e(1 - f) = 0$. Then $e = ef$. We have that B satisfies Frink's axioms and so B is a Boolean algebra. □

To help us prove our main result, it is useful to have the following two lemmas. The first says that to check if an inverse monoid is Boolean it is enough to prove that we have orthogonal joins.

Lemma 39 *Let $S \subseteq R$, where R is a unital ring. Suppose that $0, 1 \in S$ and that S is an inverse monoid under the induced multiplication from R. In addition, we suppose that S is closed under sums of pairs of orthogonal elements, and if $e \in S$ is an idempotent, then $1 - e \in S$. Then S is a Boolean inverse monoid.*

Proof It is immediate that $\mathsf{E}(S)$ satisfies Frink's axioms and so is a Boolean algebra. Let e and f be orthogonal idempotents. Then $e + f = e \vee f$; to see why observe that $(e + f)e = e$ and $(e + f)f = f$. Thus $e, f \leq e + f$. On the other hand, suppose that $e, f \leq i$. Then $(e + f)i = e + f$. Thus $e + f \leq i$. It follows that $e + f = e \vee f$. Let $a, b \in S$ be arbitrary orthogonal elements. They are, in particular, compatible.

We prove that $a+b = a \vee b$ and the result then follows by Lemma 36. Observe that

$$(a + b)\mathbf{d}(a) = a\mathbf{d}(a) + b\mathbf{d}(b) = a\mathbf{d}(a) + a\mathbf{d}(b) = a\mathbf{d}(a) = a$$

where we have used Lemma 25. Thus $a \leq a+b$. Similarly $b \leq a+b$. Suppose that $a, b \leq x$. Then $a + b = x\mathbf{d}(a) + x\mathbf{d}(b) = x(\mathbf{d}(a) \vee \mathbf{d}(b))$. Thus $a + b \leq x$. We have proved that $a + b = a \vee b$ if a and b are orthogonal. □

Our second result deals with orthogonal sums.

Lemma 40 *Let $S \subseteq R$, where R is a unital ring. Suppose that $0, 1 \in S$ and that S is an inverse monoid under the induced multiplication from R. Define $S^{\perp}$ to be the sums of all finite orthogonal sets of S. Then $S^{\perp}$ is an inverse monoid and its set of idempotents is $\mathsf{E}(S)^{\perp}$.*

Proof Let $a, b \in S^{\perp}$. Then the fact that $ab \in S^{\perp}$ follows by Lemma 24. Observe that if $a = a_1 + \ldots + a_n$, an orthogonal sum of elements of S, then defining $a^{-1} = a_1^{-1} + \ldots + a_n^{-1}$ we get that $a = aa^{-1}a$. Thus $S^{\perp}$ is a regular semigroup where, for example, $aa^{-1} = a_1a_1^{-1} + \ldots + a_na_n^{-1}$. Suppose that $a = a_1 + \ldots + a_n$ is an idempotent. Then $a^2 = a$ and so $a^2a_1^{-1} = a_1a_1^{-1}$. It follows that $a_1^2a_1^{-1} = a_1a_1^{-1}$. We deduce that a_1 is an idempotent. In a similar way, we may deduce that each of $a_1, \ldots, a_n$ is an idempotent and so a is a sum of orthogonal idempotents. We have therefore provided that $\mathsf{E}(S^{\perp}) = \mathsf{E}(S)^{\perp}$. It now follows that $S^{\perp}$ is an inverse monoid since it is regular and the idempotents commute. □

We can now prove the more general result which shows us that Boolean inverse monoids arise naturally.

Theorem 5 *Let $S \subseteq R$, where R is a unital ring. Suppose that $0, 1 \in S$ and S is an inverse monoid under the induced multiplication from R. Then there is a subset $S \subseteq T \subseteq R$ such that T is a Boolean inverse monoid.*

Proof Put $E = \mathsf{E}(S)$. Define E' as in the proof of Lemma 38. It is easy to check that E' is closed under conjugation by elements of S. Put $S' = SE'$. Then S' is a regular monoid and the set of idempotents of S' is precisely the set E'. It follows by Proposition 1 that S' is also an inverse monoid. Define T to be the sums of finite orthogonal subsets of S'. Then T is an inverse monoid whose set of idempotents is precisely the set $(\mathsf{E}(S)')^{\perp}$, using the notation of the proof of Lemma 38. This is proved using Lemma 40. It now follows by Lemma 39 that T is a Boolean inverse monoid. □

6 The Underlying Groupoid

The product we have defined on the symmetric inverse monoid $\mathcal{I}(X)$ is not the only one, nor perhaps even the most obvious one, that we might define. Given partial bijections f and g, we could define fg only when the domain of f is equal to the range of g. When we do this, we are regarding f and g as being functions rather than as partial functions. With respect to this 'restricted product', $\mathcal{I}(X)$ becomes a groupoid. What we have done for the special case of the symmetric inverse monoids, we can also do for arbitrary inverse semigroups, but first we review the basics of groupoid theory we shall need.

Categories are usually regarded as categories of structures with their morphisms. They can, however, also be regarded as algebraic structures in their own right, no different from groups, rings and fields except that the binary operation is only partially defined. We shall define categories from this purely algebraic point of view. Let C be a set equipped with a partial binary operation which we shall denote by $\cdot$ or by concatenation. If $x, y \in C$ and the product $x \cdot y$ is defined, we write $\exists x \cdot y$. An element $e \in C$ is called an *identity* if $\exists e \cdot x$ implies $e \cdot x = x$, and $\exists x \cdot e$ implies $x \cdot e = x$. The set of identities of C is denoted C_o, where the subscript 'o' stands for 'object'. The pair $(C, \cdot)$ is said to be a *category* if the following axioms hold:

(C1) $x \cdot (y \cdot z)$ exists if and only if $(x \cdot y) \cdot z$ exists, in which case they are equal.
(C2) $x \cdot (y \cdot z)$ exists if and only if $x \cdot y$ and $y \cdot z$ exist.
(C3) For each $x \in C$ there exist identities e and f such that $\exists x \cdot e$ and $\exists f \cdot x$.

From axiom (C3), it follows that the identities e and f are uniquely determined by x. We write $e = \mathbf{d}(x)$ and $f = \mathbf{r}(x)$, where $\mathbf{d}(x)$ is the *domain* identity and $\mathbf{r}(x)$ is the *range* identity.[12] Observe that $\exists x \cdot y$ if and only if $\mathbf{d}(x) = \mathbf{r}(y)$. The elements of a category are called *arrows*. We say that the arrow x *starts at* $\mathbf{d}(x)$ and *ends at* $\mathbf{r}(x)$. If C is a category and e and f identities in C, then we put

$$\hom(e, f) = \{x \in C \colon \mathbf{d}(x) = f \text{ and } \mathbf{r}(x) = e\},$$

the set of *arrows from f to e*. Subsets of C of the form $\hom(e, f)$ are called *hom-sets*. We also put $\operatorname{end}(e) = \hom(e, e)$, the *local monoid at e*. We define *subcategories* in the obvious way. Viewed in this light, a category is a monoid with many identities since the categories with exactly one identity are precisely the monoids.

We shall not need arbitrary categories but those that generalize groups. A category C is said to be a *groupoid* if for each $x \in C$ there is an element x^{-1} such that $x^{-1}x = \mathbf{d}(x)$ and $xx^{-1} = \mathbf{r}(x)$. The element x^{-1} is unique with these properties. In the case of groupoids, the local monoids are, in fact, *local groups*. It can happen that a groupoid consists entirely of its local units; in this case, we say that the groupoid is a *union of groups*. A groupoid with exactly one identity is a group. Two elements x and y of a groupoid are said to be *connected* if there is an arrow starting at $\mathbf{d}(x)$ and ending at $\mathbf{d}(y)$. This defines an equivalence relation on the groupoid whose equivalence classes are called the *connected components*[13] of

---

[12] As you will see, there is no contradiction with the notation we introduced earlier.

[13] The use of the word 'connected' here is unfortunate. It has nothing to do with topology. The groupoids here are discrete.

the groupoid. A groupoid with one connected component is said to be *connected*. Every groupoid can be written as a disjoint union of connected groupoids. We say that a groupoid is *principal* if for any identities e and f there is at most one arrow from e to f.

Lemma 41 *Let G be a groupoid. Then G is principal if and only if all its local groups are trivial.*

Proof If a groupoid is principal, it is clear that the local groups are trivial. Suppose, now, that the local groups are trivial. We prove that the groupoid is principal. Suppose that x and y are arrows that start at e and end at f. Then $y^{-1}x$ begins and ends at e. This means that, under our assumption, $y^{-1}x = e$. We deduce that $x = y$. □

We define *subgroupoids* in the obvious way.

Example 1 Let X be a set. The set $X \times X$ becomes a groupoid when we define $\mathbf{d}(x, y) = (y, y)$, $\mathbf{r}(x, y) = (x, x)$ and $(x, y)^{-1} = (y, x)$; define a partial product by $(x, y)(y, z) = (x, z)$. Now, let $\sim$ be an equivalence relation on X. Define G to consist of those ordered pairs (x, y) where $x \sim y$. It is easy to check that G is a subgroupoid of $X \times X$. It is, in fact, a principal groupoid. If G is an arbitrary principal groupoid, then it defines an equivalence relation on the set $X = G_o$. In fact, principal groupoids and equivalence relations are different ways of defining the same thing.

At this point, category theorists should look away since we shall convert a category into a semigroup. If C is a category as we have defined it above, then we can convert it into a semigroup with zero by adjoining a zero and defining all undefined products to be zero. We denote the semigroup with zero that arises by C^0.

Now we can return to inverse semigroups. Let S be an arbitrary inverse semigroup. Define the *restricted product*[14] of two elements s and t in S to be $s \cdot t = st$ if $s^{-1}s = tt^{-1}$ and undefined otherwise. The following result simply tells us that what we expect to happen actually does happen.

Proposition 6 *Every inverse semigroup S is a groupoid with respect to its restricted product.*

Proof We begin by showing that all idempotents of S are identities of $(S, \cdot)$. Let $e \in S$ be an idempotent and suppose that $e \cdot x$ is defined. Then $e = xx^{-1}$ and $e \cdot x = ex$. But $ex = (xx^{-1})x = x$. Similarly, if $x \cdot e$ is defined, then it is equal to x. We now check that the axioms (C1), (C2) and (C3) hold.

Axiom (C1) holds: suppose that $x \cdot (y \cdot z)$ is defined. Then

$$x^{-1}x = (y \cdot z)(y \cdot z)^{-1} \text{ and } y^{-1}y = zz^{-1}.$$

[14] Sometimes referred to as the *trace product*.

But

$$(y \cdot z)(y \cdot z)^{-1} = yzz^{-1}y^{-1} = yy^{-1}.$$

Hence, $x^{-1}x = yy^{-1}$, and so $x \cdot y$ is defined. Also $(xy)^{-1}(xy) = y^{-1}y = zz^{-1}$. Thus $(x \cdot y) \cdot z$ is defined. It is clear that $x \cdot (y \cdot z)$ is equal to $(x \cdot y) \cdot z$. A similar argument shows that if $(x \cdot y) \cdot z$ exists, then $x \cdot (y \cdot z)$ exists and they are equal.

Axiom (C2) holds: suppose that $x \cdot y$ and $y \cdot z$ are defined. We show that $x \cdot (y \cdot z)$ is defined. We have that $x^{-1}x = yy^{-1}$ and $y^{-1}y = zz^{-1}$. Now

$$(yz)(yz)^{-1} = y(zz^{-1})y^{-1} = y(y^{-1}y)y^{-1} = yy^{-1} = x^{-1}x.$$

Thus $x \cdot (y \cdot z)$ is defined. The proof of the converse is straightforward.

Axiom (C3) holds: for each element x we have that $x \cdot (x^{-1}x)$ is defined, and we have seen that idempotents of S are identities. Thus we put $\mathbf{d}(x) = x^{-1}x$. Similarly, we put $xx^{-1} = \mathbf{r}(x)$. It is now clear that $(S, \cdot)$ is a category. The fact that it is a groupoid is immediate. □

We call $(S, \cdot)$ the *(underlying) groupoid* of S. We can use the underlying groupoid to reveal something of the structure of inverse semigroups.

We can interpret Lemma 8 in terms of the structure of the underlying groupoid. An inverse semigroup is a Clifford semigroup if and only if its underlying groupoid is a union of groups.

In the light of Proposition 6, it is now natural to picture an element a of an inverse semigroup as follows:

$$\mathbf{d}(a) \xrightarrow{a} \mathbf{r}(a).$$

where we now call $\mathbf{d}(a)$ the *domain idempotent* of a and $\mathbf{r}(a)$ is the *range idempotent of* a. The underlying groupoid structure more or less does away with the need to deal directly with Green's relations. If e and f are idempotents, we write $e \mathscr{D} f$ to mean that there exists an element a such that $\mathbf{d}(a) = e$ and $\mathbf{r}(a) = f$. We write $a \mathscr{D} b$ to mean that $\mathbf{d}(a) \mathscr{D} \mathbf{d}(b)$. The relation $\mathscr{D}$ really is Green's relation $\mathscr{D}$; by the same token $a \mathscr{L} b$ if and only if $\mathbf{d}(a) = \mathbf{d}(b)$ and $a \mathscr{R} b$ if and only if $\mathbf{r}(a) = \mathbf{r}(b)$. We define $a \mathscr{H} b$ if and only if $a \mathscr{L} b$ and $a \mathscr{R} b$; in other words, a and b belong to the same *hom-set*. An inverse semigroup is said to be *bisimple* if its underlying groupoid consists of one connected component. An inverse semigroup with zero is said to be 0-*bisimple* if its underlying groupoid consists of two connected components.

In passing from an inverse semigroup to its underlying groupoid, we do lose some information. As the following result shows, the information that is lost is encoded by the natural partial order.

Lemma 42 *Let S be an inverse semigroup. Then for any $s, t \in S$ there exist elements $s' \leq s$ and $t' \leq t$ such that $st = s' \cdot t'$ where the product on the right is the restricted product.*

Proof Put $e = \mathbf{d}(s)\mathbf{r}(t)$ and define $s' = se$ and $t' = et$. Observe that $\mathbf{d}(s') = e$ and $\mathbf{r}(t') = e$ and that $st = s't'$. □

It is possible to formalize the idea of a groupoid equipped with a partial order in such a way that the original semigroup product can be recaptured. This is the approach that Ehresmann took to studying inverse semigroups. See [52, Chapter] for exactly how this is done.[15]

We shall now deal with the missing $\mathscr{J}$-relation. First, we need a slight extension of Lemma 42.

Lemma 43 *Let S be an inverse semigroup. Then $abc = a' \cdot b' \cdot c'$, which is a restricted product where $a' \leq a$, $b' \leq b$ and $c' \leq c$.*

Proof Because of associativity, it does not matter how we bracket. We write $abc = (ab)c$. Put $d = ab$. Then $dc = d' \cdot c'$. But $d' = ab\mathbf{r}(c)$. Thus $(a(b\mathbf{r}(c))) \cdot c'$. Now write $a(b\mathbf{r}(c)) = a' \cdot b'$ where $b' \leq b\mathbf{r}(c) \leq b$. We have therefore written $abc = a' \cdot b' \cdot c'$ using the fact that the multiplication in a groupoid is associative. □

We can now describe the $\mathscr{J}$-relation on inverse semigroups.

Lemma 44 *Let S be an inverse semigroup. Then $SaS \subseteq SbS$ if and only if $a \mathscr{D} b' \leq b$ for some element $b' \in S$.*

Proof Suppose, first, that $SaS \subseteq SbS$. Then $a = xby$ for some $x, y \in S$. By Lemma 43, we may write $a = x' \cdot b' \cdot y'$. You can check that $a \mathscr{D} b'$. But $b' \leq b$. We have therefore proved one direction. To prove the converse, suppose that $a \mathscr{D} b' \leq b$ for some element $b' \in S$. Let $\mathbf{d}(a) \xrightarrow{x} \mathbf{d}(b')$. Put $y = b' \cdot x \cdot a^{-1}$. Then $a = y^{-1} \cdot b' \cdot x$. Similarly, $b' = y \cdot a \cdot x^{-1}$ It follows that $SaS = Sb'S$. Clearly, $Sb'S \subseteq SbS$. We have therefore shown that $SaS \subseteq SbS$. □

We can use the theory we have developed to generalize Lemma 4.

Lemma 45 *Let S be an inverse semigroup with zero. Then S is isomorphic to a groupoid with a zero adjoined if and only if the natural partial order is equality on the set $S \setminus \{0\}$.*

Proof We prove one direction only. Let S be an inverse semigroup with zero such that the natural partial order is equality on the set $S \setminus \{0\}$. Let G be the underlying groupoid of S with the component $\{0\}$ removed. There is an obvious bijection between S and G^0. We prove that this is a homomorphism. Let $a, b \in S$ be any non-zero elements. Then $ab = a' \cdot b'$ where $a' \leq a$ and $b' \leq b$ by Lemma 42. As

[15] This groupoid, which is discrete, is quite different from the one that Paterson constructs [95].

a result of our assumption on the natural partial order, either $a' = a$ and $b' = b$, in which case the product is a groupoid product, or at least one of a' and b' is equal to zero—in which case $ab = 0$. Thus a product in S is either equal to zero or a restricted product but not both. □

In addition to the underlying groupoid, we may sometimes be able to associate another, smaller, groupoid to an inverse semigroup with zero. Let S be an inverse semigroup with zero. An element $s \in S$ is said to be an *atom* if $t \leq s$ implies that $t = 0$ or $t = s$.

Lemma 46 *Let S be an inverse semigroup.*

(i) *If x is an atom, then x^{-1} is an atom.*
(ii) *If x is an atom, then both $\mathbf{d}(x)$ and $\mathbf{r}(x)$ are atoms*
(iii) *Suppose that $\mathbf{d}(x)$ is an atom and then x is an atom; similarly, if $\mathbf{r}(x)$ is an atom, then x is an atom.*
(iv) *If x and y are atoms and the restricted product $x \cdot y$ is defined, then $x \cdot y$ is an atom.*
(v) *If x and y are distinct compatible atoms, then $x \perp y$.*

Proof

(1) Immediate from the properties of the inverse.
(2) We prove that $\mathbf{d}(x)$ is an atom; the proof that $\mathbf{r}(x)$ is an atom is similar. Suppose that $e \leq \mathbf{d}(x)$. Then $xe \leq x$. It follows that either $xe = 0$ or $xe = x$. Suppose, first, that $xe = 0$. Then $e\mathbf{d}(x) = 0$ and so $e = 0$. Alternatively, if $xe = x$, then $e\mathbf{d}(x) = \mathbf{d}(x)$ and so $e = \mathbf{d}(x)$. This shows that $\mathbf{d}(x)$ is an atom.
(3) We prove that if $\mathbf{d}(x)$ is an atom, then x is an atom; the proof of the other statement is analogous. Suppose that $y \leq x$. Then $\mathbf{d}(y) \leq \mathbf{d}(x)$. Since $\mathbf{d}(x)$ is an atom either $\mathbf{d}(y) = 0$ or $\mathbf{d}(y) = \mathbf{d}(x)$. It follows that either $y = 0$ or $y = x$. We have therefore proved that x is an atom.
(4) We are given that x and y are atoms and that $x \cdot y$ exists. Observe first that $x \cdot y \neq 0$. Let $z \leq xy$. Then $z = x(y\mathbf{d}(z))$. Now, $y\mathbf{d}(z) = 0$ or $y\mathbf{d}(z) = y$, since y is an atom. If the former, then $z = 0$, and if the latter, then $z = xy$. We have therefore proved that $x \cdot y$ is an atom.
(5) Let x be an atom. Then by part (2) above it follows that $\mathbf{d}(x)$ is an atom. Similarly, $\mathbf{d}(y)$ is an atom. If the product $\mathbf{d}(x)\mathbf{d}(y)$ is non-zero, then in fact $\mathbf{d}(x) = \mathbf{d}(y)$. But $x \sim y$ and so $x = y$ by Lemma 26, which contradicts our assumption that x and y are distinct. It follows that $\mathbf{d}(x) \perp \mathbf{d}(y)$. A similar argument shows that $\mathbf{r}(x) \perp \mathbf{r}(y)$ from which it follows that $x \perp y$.

□

It follows by Lemma 46 that the set of atoms of S, if non-empty, forms a groupoid, which we shall call the *atomic groupoid* of S and denote by $\mathsf{A}(S)$.

The finite symmetric inverse monoid $\mathcal{I}(X)$ has an interesting atomic groupoid. It consists of those partial bijections the domains of which contain exactly one element of X. This groupoid is isomorphic to the groupoid $X \times X$ defined in Example 1.

We shall describe all finite Boolean inverse monoids in terms of groupoids. The reader will recall that the finite Boolean algebras are isomorphic to the powerset Boolean algebras defined on the finite set of atoms. We shall replace finite sets by finite groupoids. We need some definitions first. If A and B are subsets of a category C, then AB is the set of all products ab where $a \in A$, $b \in B$ and $\mathbf{d}(a) = \mathbf{r}(b)$. If A is a subset of a groupoid, then A^{-1} is the set of all a^{-1} where $a \in A$. We first show how to construct finite Boolean inverse monoids from finite groupoids. If G is a groupoid, then a subset $A \subseteq G$ is said to be a *local bisection* if both AA^{-1} and $A^{-1}A$ consist entirely of identities. You can check that a subset $A \subseteq G$ is a local bisection if $a, b \in A$ and $\mathbf{d}(a) = \mathbf{d}(b)$ (respectively, $\mathbf{r}(a) = \mathbf{r}(b)$), then $a = b$,

Proposition 7 *Let G be a finite groupoid. Then $\mathsf{K}(G)$, the set of all local bisections of G under subset multiplication, is a finite Boolean inverse monoid, the set of atoms of which forms a groupoid isomorphic to G.*

Proof You can check that the product of two local bisections is a local bisection. If A is a local bisection, then so is $A^{-1} = \{a^{-1} \colon a \in A\}$ and $A = AA^{-1}A$. The idempotents are just the subsets of G_o and the product of two idempotents is just the intersection of these two sets. It follows that $\mathsf{K}(G)$ is an inverse semigroup since it is a regular semigroup with commuting idempotents. It is a monoid with identity G_o and has a zero $\varnothing$. It has a Boolean algebra of idempotents. Observe that $A \leq B$ if and only if $A \subseteq B$. You can check that $A \sim B$ if and only if $A \cup B$ is a local bisection. It is now easy to check that $\mathsf{K}(G)$ is a Boolean inverse monoid. The atoms are the singleton sets $\{g\}$ and form a groupoid isomorphic to G. □

We shall now go in the opposite direction. Our first result is an immediate consequence of finiteness.

Lemma 47 *Let S be a finite Boolean inverse monoid. Then each non-zero element is above an atom.*

We now connect elements with the atoms beneath them. Let $a \in S$. Define $\theta(a) = a^{\downarrow} \cap \mathsf{A}(S)$. Observe that $\theta(0) = \varnothing$.

Lemma 48 *Let S be a Boolean inverse semigroup. For each $a \in S$, the set $\theta(a)$ is a local bisection of the groupoid $\mathsf{A}(S)$.*

Proof If $a = 0$, then $\theta(a) = \varnothing$. If $a \neq 0$, then it is above at least one atom by Lemma 47 and so is non-empty. Let $x, y \in \theta(a)$ such that $\mathbf{d}(x) = \mathbf{d}(y)$. Then $x = y$. Dually, if $\mathbf{r}(x) = \mathbf{r}(y)$, then $x = y$. □

If S is a finite Boolean inverse monoid, then by Lemma 47 we may define a function $\theta \colon S \to \mathsf{K}(\mathsf{A}(S))$ by $\theta(0) = \varnothing$ and $\theta(a) = a^{\downarrow} \cap \mathsf{A}(S)$. It is quite rare that we can say anything about the structure of finite semigroups belonging to some class. Thus the following result is a pleasant surprise.

Theorem 6 (The Structure of Finite Boolean Inverse Monoids) *Let S be a finite Boolean inverse monoid. Then S is isomorphic to the Boolean inverse monoid $\mathsf{K}(\mathsf{A}(S))$.*

Proof It remains to show that θ (as defined above) is an isomorphism of semigroups. First, θ is a homomorphism. Let x be an atom such that $x \leq ab$. Then $x = a(b\mathbf{d}(x))$. Thus by Lemma 42, we may write $x = a' \cdot b'$ where $a' \leq a$ and $b' \leq b$. It is easy to check that a' and b' are themselves atoms. We have therefore proved that $\theta(ab) \subseteq \theta(a)\theta(b)$. Conversely, let $x \in \theta(a)$ and $y \in \theta(b)$ such that the restricted product $x \cdot y$ is defined. Then $x \cdot y = xy \leq ab$. But the restricted product of atoms is an atom and so we have proved the first claim.

It remains to prove that θ is a bijection. We show first that $a = \bigvee \theta(a)$. Put $b = \bigvee \theta(a)$. Then, clearly, $b \leq a$. Suppose that $b \neq a$. It is here that we use the Boolean structure. Then $a \setminus b \neq 0$. It follows by Lemma 47 that $a \setminus b$ is above an atom x. But then $x \leq a$ and so $x \leq b$ (by definition). Thus $x \leq b, a \setminus b$ which implies that $x = 0$. But atoms are non-zero. It follows that $a = \bigvee \theta(a)$ and so θ is an injection.

Now let $A \in \mathsf{K}(\mathsf{A}(S))$. We do not lose any generality by assuming that it is non-empty. Then A is a set of compatible elements. Put $a = \bigvee A$. Clearly, $A \subseteq \theta(a)$. Let x be an atom such that $x \leq a$. Then $x = \bigvee_{a \in A}(x \wedge a)$ by part (2) of Lemma 34. Remembering that both x and a are atoms, it follows that $x = a$ for some $a \in A$. We have therefore proved that $\theta(a) = A$. □

We now have a complete description of the finite Boolean inverse monoids: as a result of Proposition 7 and Theorem 6, they are precisely the inverse monoids of the form $\mathsf{K}(G)$ where G is a finite groupoid. We can apply the theory we have developed to the representation theory of *arbitrary* finite inverse monoids, although a more elementary account can be found in [112, Chapter 9]. We prove first that every finite inverse monoid S can be embedded into the Boolean inverse monoid constructed from its underlying groupoid.

Lemma 49 *Let S be a finite inverse monoid with underlying groupoid G.*

(i) *Let $a \in S$. Then $a^{\downarrow}$ is a local bisection of G.*
(ii) *We have that $(ab)^{\downarrow} = a^{\downarrow}b^{\downarrow}$.*

Proof

(1) Let $x, y \leq a$ such that $\mathbf{d}(x) = \mathbf{d}(y)$. Then it is immediate that $x = y$. Similar reasoning shows that if $x, y \leq a$ such that $\mathbf{r}(x) = \mathbf{r}(y)$, then $x = y$. We have therefore shown that the set $a^{\downarrow}$ is a local bisection of G.
(2) Suppose that $x \leq a$ and $y \leq b$. Then $xy \leq ab$. On the other hand, if $c \leq ab$, then we can write $c = (\mathbf{r}(c)a)(b\mathbf{d}(c))$ and so $x \in a^{\downarrow}b^{\downarrow}$.

□

Let S be an arbitrary inverse monoid with underlying groupoid G. An element a with the property that $a^{\downarrow} = \{a\}$, a singleton set, will be said to be *at the bottom*. Define $\beta\colon S \to \mathsf{K}(G)$ by $\beta(a) = a^{\downarrow}$. Then by Lemma 49, β is an injective homomorphism of inverse semigroups. The elements in $1^{\downarrow}$ are precisely the idempotents, which are the identities of G; the set of identities of G is the monoid identity of $\mathsf{K}(G)$. Thus the homomorphism is a monoid homomorphism. The above result, on its own, does not take us very far because we already know that every finite inverse monoid can be embedded in a finite Boolean inverse monoid. The key point of this embedding is that the map $\beta \colon S \to \mathsf{K}(G)$ has a special property described by the following proposition.

Proposition 8 *Let S be a finite inverse monoid with underlying groupoid G and let $\alpha \colon S \to T$ be any monoid homomorphism to a Boolean inverse monoid T. Then there is a unique morphism of Boolean inverse monoids $\gamma \colon \mathsf{K}(G) \to T$ such that $\gamma\beta = \alpha$.*

Proof Our first step is to define the function $\gamma \colon \mathsf{K}(G) \to T$. Define $\gamma(\varnothing) = 0$. If a is any element of S, then $\{a\}$ is an element of $\mathsf{K}(G)$. Let the set of elements strictly below a in S be $\{a_1, \ldots, a_m\}$. This set could well be empty if a is at the bottom—this will not cause us any problems. Since $a_1, \ldots, a_m \leq a$, these elements are pairwise compatible. It follows that $\alpha(a_1), \ldots, \alpha(a_m)$ is a compatible subset of T. Thus the join $\alpha(a_1) \vee \ldots \vee \alpha(a_m)$ exists; if the set of elements strictly below a is empty, then this join is just 0. It is clearly less than $\alpha(a)$ and so we may form the element $\alpha(a) \setminus (\alpha(a_1) \vee \ldots \vee \alpha(a_m))$. On the basis of the above, define

$$\gamma(\{a\}) = \alpha(a) \setminus (\alpha(a_1) \vee \ldots \vee \alpha(a_m)).$$

If A is a local bisection which is neither empty nor a singleton set, define

$$\gamma(A) = \bigvee_{a \in A} \gamma(\{a\}).$$

This makes sense, for if $a, b \in A$, then $\{a\}, \{b\} \subseteq A$ and so $\{a\} \sim \{b\}$ in $\mathsf{K}(G)$. This completes the definition of γ.

We show that γ preserves binary joins. If $A \sim B$ in $\mathsf{K}(G)$, then $A \vee B = A \cup B$. From the definition of γ, it is clear that γ preserves binary joins. It maps the empty set to the zero by definition.

We prove that $\gamma(a^{\downarrow}) = \alpha(a)$. This proves that we have a monoid homomorphism and that $\gamma\beta = \alpha$. Define the *height* of a in S to be the length of a chain of maximum length from a. Those elements with height zero are precisely those which are at the bottom. If a has height zero, then

$$\gamma(a^{\downarrow}) = \alpha(a).$$

We assume that we have proved that $\gamma(b^{\downarrow}) = \alpha(b)$ for all elements b of height at most n. Let a be an element of height $n+1$. Let $b_1, \ldots, b_m$ be all the elements immediately below a. Then, by the induction hypothesis, we have that

$$\gamma(b_i^{\downarrow}) = \alpha(b_i).$$

Let the elements strictly less than a be $a_1, \ldots, a_m$. These include the elements $b_1, \ldots, b_m$, for example, so in general each element a_i is beneath one of the b_j.

It follows that we can write

$$\alpha(a) \setminus (\alpha(a_1) \vee \ldots \vee \alpha(a_m))$$

as

$$\alpha(a) \setminus (\alpha(b_1 e_1) \vee \ldots \vee \alpha(b_n e_n))$$

where the idempotents $\alpha(e_j)$ gather together by means of a join all the idempotents that arise from showing that $a_i \leq b_j$ for various i. By definition

$$\gamma(a^{\downarrow}) = \alpha(a) \setminus (\alpha(a_1) \vee \ldots \vee \alpha(a_m)) \vee \gamma(\{a_1\}) \vee \ldots \vee \gamma(\{a_m\})$$

but we can write

$$\gamma(a^{\downarrow}) = \gamma(\{a\}) \vee \gamma(\{a_1, \ldots, a_m\}^{\downarrow}).$$

But

$$\{a_1, \ldots, a_m\}^{\downarrow} = b_1^{\downarrow} \cup \ldots \cup b_n^{\downarrow}$$

and so

$$\gamma(\{a_1, \ldots, a_m\}^{\downarrow}) = \gamma(b_1^{\downarrow}) \vee \ldots \vee \gamma(b_n^{\downarrow}).$$

Thus

$$\gamma(\{a_1, \ldots, a_m\}^{\downarrow}) = \alpha(b_1) \vee \ldots \vee \alpha(b_n)$$

using the induction hypothesis. Thus

$$\gamma(a^{\downarrow}) = \alpha(a) \setminus (\alpha(a_1) \vee \ldots \vee \alpha(a_m)) \vee \alpha(b_1) \vee \ldots \vee \alpha(b_n.)$$

By our argument above

$$\gamma(a^{\downarrow}) = \alpha(a) \setminus (\alpha(b_1 e_1) \vee \ldots \vee \alpha(b_n e_n)) \vee \alpha(b_1) \vee \ldots \vee \alpha(b_n).$$

This is equal to

$$\alpha(a) \setminus (\alpha(b_1 e_1) \vee \ldots \vee \alpha(b_n e_n)) \vee (\alpha(b_1 e_1) \vee \ldots \vee \alpha(b_n e_n))) \vee \alpha(b_1) \vee \ldots \vee \alpha(b_n)$$

which is just

$$\alpha(a) \vee \alpha(b_1) \vee \ldots \vee \alpha(b_n)$$

which is equal to $\alpha(a)$.

We now prove uniqueness. Suppose that $\gamma' \colon \mathsf{K}(S) \rightarrow T$ is a morphism of Boolean inverse monoids such that $\gamma'\beta = \alpha$. We show that $\gamma' = \gamma$. It is immediate from our assumption on γ' that for all elements of height zero we have that $\gamma(a) = \gamma'(a)$. So, let $a \in S$ be any element which does not have height zero. Let $\{a_1, \ldots, a_m\}$ be the set of all elements strictly less than a. Then

$$\{a_1, \ldots, a_m\} = a_1^{\downarrow} \cup \ldots \cup a_m^{\downarrow}.$$

It follows that

$$\gamma'(\{a_1, \ldots, a_m\}) = \gamma'(a_1^{\downarrow}) \vee \ldots \vee \gamma'(a_m^{\downarrow}).$$

But this is just equal to

$$\gamma'(\beta(a_1)) \vee \ldots \vee \gamma'(\beta(a_m))$$

which is equal to

$$\gamma(\beta(a_1)) \cup \ldots \cup \gamma(\beta(a_m))$$

by assumption. We have therefore proved that

$$\gamma'(\{a_1, \ldots, a_m\}) = \gamma(\{a_1, \ldots, a_m\}).$$

Observe that $\{a\} = a^{\downarrow} \setminus \{a_1, \ldots, a_m\}$. Thus

$$\gamma'(\{a\}) = \gamma'(a^{\downarrow}) \setminus \gamma'(\{a_1, \ldots, a_m\}).$$

But

$$\gamma'(a^{\downarrow}) = \gamma(a^{\downarrow})$$

by assumption, and

$$\gamma'(\{a_1, \ldots, a_m\}) = \gamma(\{a_1, \ldots, a_m\})$$

by what we proved above. It follows that $\gamma'(\{a\}) = \gamma(\{a\})$ for all elements $a \in S$. Now, let A be any non-empty local bisection. Then, since γ' is a morphism of Boolean inverse monoids, we have that $\gamma'(A) = \bigvee_{a \in A} \gamma'(\{a\})$. It follows that $\gamma' = \gamma$. □

We now apply our results to the study of the representation theory of finite inverse monoids. The starting point is to say what we mean by the representation theory of groupoids. Let G be a finite groupoid. Then we get a semigroup with zero G^0 by adjoining a zero; in fact, this is a special kind of inverse semigroup by Lemma 45.

We can therefore consider homomorphisms $\theta \colon G^0 \to R$ to the multiplicative monoid of the unital ring R. Each identity $e \in G$ gives rise to an idempotent $\theta(e)$ in the ring R. If e and f are distinct identities of the groupoid G, then $\theta(e)$ and $\theta(f)$ are orthogonal. By assumption, the groupoid G is finite. Put $f = \sum_{e \in G_o} \theta(e)$. Then f is an idempotent in R. Consider the ring $fRf = \{a \in R \colon faf = a\}$. This has identity f. Let $g \in G$ be an arbitrary element of G. Let e, e' be the identities such that $a = e'ae$. Then $\theta(a) = \theta(e')\theta(a)\theta(e)$. But for any identity e, we have that $\theta(e) = f\theta(e)f$. It follows that $\theta(a) \in fRf$. We therefore define a *representation* of a finite groupoid G in a ring R to be a semigroup with zero homomorphism $\theta \colon G^0 \to R$ such that $1 = \sum_{e \in G_o} \theta(e)$.

We shall also need the following definition. Let S be a Boolean inverse monoid. A monoid homomorphism $\phi \colon S \to R$ to the multiplicative monoid of the ring R which maps zero to zero is said to be *additive* if $\phi(a \vee b) = \phi(a) + \phi(b)$, whenever $a \perp b$. We refer the reader to part (2) of Lemma 36 for the rationale for this definition.

The following theorem was proved in a different way in [112, Theorem 9.3].

Theorem 7 (Representation Theory of Finite Inverse Monoids) *Let S be a finite inverse monoid with underlying groupoid G and let R be a unital ring. Then there is a bijective correspondence between the set of representations of S in R and the set of representations of the finite groupoid G in R.*

Proof Let $\theta \colon S \to R$ be a monoid homomorphism to the multiplicative monoid of the ring R. The image $\theta(S)$ is an inverse submonoid of the multiplicative monoid of the ring R by Lemma 14 and so $\theta(S)^0$, the inverse monoid $\theta(S)$ with the zero of the ring R adjoined, is an inverse monoid with zero. Thus by Theorem 5, there is a Boolean inverse monoid T such that $\theta(S)^0 \subseteq T \subseteq R$. By Proposition 8, there is therefore a unique morphism of Boolean inverse monoids $\phi \colon \mathsf{K}(G) \to T$ such that $\phi\beta = \theta$. We want to regard ϕ as a map from $\mathsf{K}(G)$ to the ring R. This is an additive homomorphism. We have proved that every homomorphism $\theta \colon S \to R$ gives rise to an additive homomorphism $\phi \colon \mathsf{K}(G) \to R$. On the other hand, given an additive homomorphism $\phi \colon \mathsf{K}(G) \to R$, we can construct a monoid homomorphism $\phi\beta \colon S \to R$. This leads to a bijective correspondence between representations of S in R and additive homomorphisms of $\mathsf{K}(G)$ in R.

We described all finite Boolean inverse monoids in Theorem 6. In what follows, therefore, we may assume that S is a finite Boolean inverse monoid with atomic groupoid G. Let $\theta \colon S \to R$ be an additive homomorphism of S. Then, by restriction, we get a semigroup homomorphism θ' from G^0 to the ring R. Since S is a finite Boolean inverse monoid, the identity of S is an orthogonal join of the atomic idempotents. It follows that the sum of the idempotents $\theta'(e)$, where $e \in G_o$, is equal to the identity of R. Thus, we have defined a representation of the groupoid G. We now go in the opposite direction. Let S be a finite Boolean inverse monoid with atomic groupoid G and suppose that there a representation $\theta' \colon G^0 \to R$. We shall now define a homomorphism θ of S that extends θ'. Let $a \in S$. Define $\theta \colon S \to R$ by $\theta(a) = \theta'(a_1) + \ldots + \theta'(a_m)$ where $a_1, \ldots, a_m \leq a$ are all the atoms below a; we can assume this is an orthogonal set by part (5) of Lemma 46 This is an additive homomorphism of S. We therefore have a bijective correspondence between the additive homomorphisms of a Boolean inverse monoid and the representations of its atomic groupoid.

If we put our two results together, then we have established a bijective correspondence between representations of an inverse monoid S and the representations of its underlying groupoid G. □

7 Fundamental Inverse Semigroups

In Section 3, we introduced the Clifford semigroups. These are the inverse semigroups in which every element is central. Living inside every inverse semigroup is a Clifford semigroup. For every inverse semigroup S, define $\mathsf{Z}(\mathsf{E}(S))$, the *centralizer of the idempotents*, to be set of all elements of S which commute with every idempotent. Then $\mathsf{Z}(\mathsf{E}(S))$ is a wide inverse subsemigroup of S which is Clifford. If $\mathsf{Z}(\mathsf{E}(S)) = \mathsf{E}(S)$, we say the inverse semigroup is *fundamental*. Fundamental inverse semigroups are important. In this section, we shall study them in more detail.

To do this, we shall need the following. Define the relation μ on an arbitrary inverse semigroup by

$$(s,t) \in \mu \Leftrightarrow (\forall e \in \mathsf{E}(S))(ses^{-1} = tet^{-1}).$$

It is routine to check that this is a congruence.

Lemma 50 *Let S be an inverse semigroup.*

(i) *If $(s,t) \in \mu$, then $\mathbf{r}(s) = \mathbf{r}(t)$ and $\mathbf{d}(s) = \mathbf{d}(t)$. It follows that $\mu \subseteq \mathscr{H}$.*
(ii) *In the definition, we may restrict to those idempotents in $\mathbf{d}(s)^{\downarrow}$*
(iii) *If $(e,f) \in \mu$, where e and f are idempotents, then $e = f$.*

Proof

(1) If $(s,t) \in \mu$, then $\mathbf{r}(s) \leq \mathbf{r}(t)$. Symmetry now delivers the answer. To prove the second claim, observe that if $(s,t) \in \mu$, then $(s^{-1}, t^{-1}) \in \mu$. We can now use the first claim to prove the second claim. It follows that $\mu \subseteq \mathscr{H}$.
(2) We have used all idempotents in the definition of μ but we may restrict to those idempotents in $\mathbf{d}(s)^{\downarrow}$ simply by multiplying by $\mathbf{d}(s)$.
(3) Immediate.

□

By part (3) of Lemma 50, it follows that μ is an *idempotent-separating* congruence. In fact, we have the following.

Lemma 51 *μ is the largest idempotent-separating congruence on S.*

Proof Let ρ be any idempotent-separating congruence on S and let $(s, t) \in \rho$. Let e be any idempotent. Then $(ses^{-1}, tet^{-1}) \in \rho$ but ρ is idempotent-separating and so $ses^{-1} = tet^{-1}$. It follows that $(s, t) \in \mu$. Thus we have shown that $\rho \subseteq \mu$. □

We can now explain the connection between the congruence μ and fundamental inverse semigroups.

Lemma 52 *Let S be an inverse semigroup. Then S is fundamental if and only if μ is the equality relation*

Proof Suppose that S is fundamental. We prove that μ is the equality relation. Let $(s, t) \in \mu$. Then $(st^{-1}, tt^{-1}) \in \mu$. Let e be any idempotent. Then $(st^{-1})e(st^{-1})^{-1} = tt^{-1}ett^{-1}$. It follows that

$$st^{-1}e = e\mathbf{r}(t)st^{-1} = e\mathbf{r}(s)st^{-1} = est^{-1}.$$

We have therefore proved that st^{-1} is central. By assumption it must be an idempotent. It follows that $st^{-1} = tt^{-1}$. Thus $st^{-1}t = t$ but $\mathbf{d}(t) = \mathbf{d}(s)$ and so $s = t$. To prove the converse, suppose that μ is the equality relation. Let s commute with every idempotent. Then $(s, ss^{-1}) \in \mu$. Thus, by assumption, $s = ss^{-1}$ and so s is an idempotent. □

We can easily construct fundamental inverse semigroups.

Lemma 53 *Let S be an inverse semigroup. Then S/μ is fundamental.*

Proof Suppose that $\mu(s)$ and $\mu(t)$ are μ-related in S/μ. Every idempotent in S/μ is of the form $\mu(e)$ where $e \in E(S)$. Thus

$$\mu(s)\mu(e)\mu(s)^{-1} = \mu(t)\mu(e)\mu(t)^{-1}$$

so that $\mu(ses^{-1}) = \mu(tet^{-1})$. But both ses^{-1} and tet^{-1} are idempotents, so that $ses^{-1} = tet^{-1}$ for every $e \in E(S)$. Thus $(s, t) \in \mu$. □

The symmetric inverse monoid is constructed from an arbitrary set. We now show how to construct an inverse semigroup from a meet semilattice. Let $(E, \leq)$ be a meet semilattice, and denote by T_E be the set of all order isomorphisms between the principal order ideals of E. Clearly, T_E is a subset of $\mathcal{I}(E)$. In fact we have the following.

Lemma 54 *The set T_E is an inverse subsemigroup of $\mathcal{I}(E)$ whose semilattice of idempotents is isomorphic to E.*

The semigroup T_E is called the *Munn semigroup* of the semilattice E. We can now construct inverse semigroups having specific semilattices of idempotents.

Theorem 8 (Munn Representation Theorem) *Let S be an inverse semigroup. Then there is an idempotent-separating homomorphism $\delta \colon S \to T_{\mathsf{E}(S)}$ whose image is a wide inverse subsemigroup of $T_{\mathsf{E}(S)}$. The kernel of δ is μ.*

Proof For each $s \in S$ define the function

$$\delta_s \colon \mathbf{d}(s)^{\downarrow} \to \mathbf{r}(s)^{\downarrow}$$

by $\delta_s(e) = ses^{-1}$. We first show that δ_s is well-defined. Let $e \leq s^{-1}s$. Then $ss^{-1}\delta_s(e) = \delta_s(e)$, and so $\delta_s(e) \leq ss^{-1}$. To show that δ_s is isotone, let $e \leq f \in (s^{-1}s)^{\downarrow}$. Then

$$\delta_s(e)\delta_s(f) = ses^{-1}sfs^{-1} = sefs^{-1} = \delta_s(e).$$

Thus $\delta_s(e) \leq \delta_s(f)$. Consider now the function $\delta_{s^{-1}} \colon (ss^{-1})^{\downarrow} \to (s^{-1}s)^{\downarrow}$. This is isotone by the argument above. For each $e \in (s^{-1}s)^{\downarrow}$, we have that

$$\delta_{s^{-1}}(\delta_s(e)) = \delta_{s^{-1}}(ses^{-1}) = s^{-1}ses^{-1}s = e.$$

Similarly, $\delta_s(\delta_{s^{-1}}(f)) = f$ for each $f \in (ss^{-1})^{\downarrow}$. Thus δ_s and $\delta_{s^{-1}}$ are mutually inverse, and so δ_s is an order isomorphism.

Define $\delta \colon S \to T_{\mathsf{E}(S)}$ by $\delta(s) = \delta_s$. To show that δ is a homomorphism, we begin by calculating $\mathrm{dom}(\delta_s\delta_t)$ for any $s, t \in S$. We have that

$$\mathrm{dom}(\delta_s\delta_t) = \delta_t^{-1}((s^{-1}s)^{\downarrow} \cap (tt^{-1})^{\downarrow}) = \delta_t^{-1}((s^{-1}stt^{-1})^{\downarrow}).$$

But $\delta_t^{-1} = \delta_{t^{-1}}$ and so

$$\mathrm{dom}(\delta_s\delta_t) = ((st)^{-1}st)^{\downarrow} = \mathrm{dom}(\delta_{st}).$$

If $e \in \mathrm{dom}\,\delta_{st}$, then

$$\delta_{st}(e) = (st)e(st)^{-1} = s(tet^{-1})s^{-1} = \delta_s(\delta_t(e)).$$

Hence, $\delta_s\delta_t = \delta_{st}$.

To show that δ is idempotent-separating, suppose that $\delta(e) = \delta(f)$ where e and f are idempotents of S. Then $\mathrm{dom}\,\delta(e) = \mathrm{dom}\,\delta(f)$. Thus $e = f$.

The image of δ is a wide inverse subsemigroup of $T_{\mathsf{E}(S)}$ because every idempotent in $T_{\mathsf{E}(S)}$ is of the form $1_{[e]}$ for some $e \in E(S)$, and $\delta_e = 1_{[e]}$.

Suppose that $\delta(s) = \delta(t)$. Then $(s, t) \in \mathscr{H}$. It is now immediate from the definition that, precisely, $(s, t) \in \mu$. □

The Munn representation should be contrasted with the Wagner-Preston representation: the Wagner-Preston was injective, whereas the Munn representation has a non-trivial kernel. Fundamental inverse semigroups arise in the following way.

Theorem 9 *Let S be an inverse semigroup. Then S is fundamental if and only if S is isomorphic to a wide inverse subsemigroup of the Munn semigroup* $T_{\mathsf{E}(S)}$.

Proof Let S be a fundamental inverse semigroup. By Theorem 8, there is a homomorphism $\delta\colon S \to T_{\mathsf{E}(S)}$ such that $\ker(\delta) = \mu$. By assumption, μ is the equality congruence by Lemma 52, and so δ is an injective homomorphism. Thus S is isomorphic to its image in $T_{\mathsf{E}(S)}$, which is a wide inverse subsemigroup.

Conversely, let S be a wide inverse subsemigroup of a Munn semigroup T_E. Clearly, we can assume that $E = \mathsf{E}(S)$. We calculate the maximum idempotent-separating congruence of S. Let $\alpha, \beta \in S$ and suppose that $(\alpha, \beta) \in \mu$ in S. Then $\mathrm{dom}(\alpha) = \mathrm{dom}(\beta)$. Let $e \in \mathrm{dom}(\alpha)$. Then $1_{[e]} \in S$, since S is a wide inverse subsemigroup of $T_{\mathsf{E}(S)}$. By assumption $\alpha 1_{[e]}\alpha^{-1} = \beta 1_{[e]}\beta^{-1}$. It is easy to check that $1_{[\alpha(e)]} = \alpha 1_{[e]}\alpha^{-1}$ and $1_{[\beta(e)]} = \beta 1_{[e]}\beta^{-1}$. Thus $\alpha(e) = \beta(e)$. Hence, $\alpha = \beta$, and so S is fundamental. □

The following is a special case of an argument due to Wagner. Let (X, τ) be a T_0-space. We prove that $\mathcal{I}(X, \tau)$ is fundamental. Let $f \in \mathcal{I}(X, \tau)$ be a non-idempotent. Then there is an element $x \in X$ such that $f(x) \neq x$. Since X is T_0 there is an open set U such that either $f(x) \in U$ and $x \notin U$ or $f(x) \notin U$ and $x \in U$. In either event, the elements $f 1_U$ and $1_U f$ are not equal, where 1_u is an idempotent. It follows that $\mathcal{I}(X, \tau)$ is fundamental.

As an example of fundamental inverse semigroups, we can easily construct the fundamental finite Boolean inverse monoids.

Proposition 9 *A finite Boolean inverse monoid is fundamental if and only its groupoid of atoms is principal.*

Proof Let S be fundamental. We shall prove that the groupoid of atoms is principal by proving that the local groups are trivial. Suppose that $e \xrightarrow{a} e$ where e is an atom and an idempotent. We shall prove that a is an idempotent. Then a is an atom. Let f be any idempotent. Then $fa \leq a$. It follows that $fa = 0$ or $fa = a$. Suppose that $fa = 0$. Then $fe = 0$ and so $af = 0$. Thus $fa = af$. Suppose now that $fa = a$. Then $fe = e = fe$ and so $af = a$. It follows again that $fa = af$. We have therefore proved that a commutes with every idempotent, but under our assumption that the inverse semigroup is fundamental, we deduce that $a = e$.

Conversely, suppose that $e \xrightarrow{a} e$, where e is an atomic idempotent, implies that $a = e$. We shall prove that our semigroup is fundamental. Let a commute with all idempotents. We prove that a is an idempotent. We can write $a = \bigvee_{i=1}^{m} a_i$ where the a_i are atoms. We prove that $\mathbf{d}(a_i) = \mathbf{r}(a_i)$ for all i from which the result follows. Since $\mathbf{r}(a_j)a = a\mathbf{r}(a_j)$, by assumption, we have that $a_j = \bigvee_{i=1}^{m} a_i\mathbf{r}(a_j)$. But $a_i\mathbf{r}(a_j) \leq a_j$. Thus either $a_i\mathbf{r}(a_j) = 0$ or $a_i\mathbf{r}(a_j) = a_j$ since a_j is an atom. But $a_i\mathbf{r}(a_j) \leq a_i$. It follows that $a_i\mathbf{r}(a_j) = a_i$ and so $a_i = a_j$. Thus $a_j\mathbf{r}(a_j) = a_j$. Hence, $\mathbf{d}(a_j) \leq \mathbf{r}(a_j)$ and so $\mathbf{d}(a_j) = \mathbf{r}(a_j)$, since both $\mathbf{d}(a_j)$ and $\mathbf{r}(a_j)$ are atoms. By assumption a_j is an idempotent. It follows that a is an idempotent. □

The above result can be used to obtain a more explicit description of the finite fundamental Boolean inverse monoids.

Theorem 10 *Let S be a finite Boolean inverse monoid. Then it is fundamental if and only if S is isomorphic to a finite direct products $\mathscr{I}_{n_1} \times \ldots \times \mathscr{I}_{n_r}$.*

Proof It can be checked that the product of two fundamental inverse semigroups is itself a fundamental inverse semigroup, and that the product of two Boolean inverse monoids is again a Boolean inverse monoid. So, one direction is easy to prove. Let, now, S be a fundamental finite Boolean inverse monoid. Then by Proposition 9, S is isomorphic to a Boolean inverse monoid of the form $\mathsf{K}(G)$ where G is a principal groupoid. Now, $G = \bigcup_{i=1}^{i=m} H_i$ is a finite disjoint union of connected principal groupoids. It can be checked that $\mathsf{K}(G) \cong \mathsf{K}(H_1) \times \ldots \times \mathsf{K}(H_m)$. Let H be any finite connected principal groupoid with set of identities X. Then $\mathsf{K}(H) \cong \mathcal{I}(X)$. The result now follows. □

The above result can be specialized to characterize the finite symmetric inverse monoids. Let S be a Boolean inverse monoid. A semigroup ideal $I \subseteq S$ is said to be an *additive ideal* if $a, b \in I$ and $a \sim b$ implies that $a \vee b \in I$. Clearly, both $\{0\}$ and S itself are additive ideals. If these are the only ones we say that S is 0-*simplifying*. A Boolean inverse monoid is said to be *simple* if it is both 0-simplifying and fundamental.

We now have the following theorem which can be derived from Theorem 10 (or see [61]).

Theorem 11 *The finite simple Boolean inverse monoids are precisely the finite symmetric inverse monoids.*

The above theorem suggests that the groups of units of simple Boolean inverse monoids should be regarded as generalizations of finite symmetric inverse monoids.

8 Congruence-Free Inverse Semigroups with Zero

An inverse semigroup is said to be *congruence-free* if its only congruences are equality and the universal congruence. In this section, we shall characterize those inverse semigroups with zero which are congruence-free.[16] We begin by ruling out the existence of various kinds of congruence. An inverse semigroup with zero S is said to be 0-*simple* if the only ideals are $\{0\}$ and S. The following characterization uses Lemma 44 where we describe the $\mathscr{J}$-relation in terms of the $\mathscr{D}$-relation and the natural partial order.

Lemma 55 *Let S be an inverse semigroup with zero. Then S is* 0*-simple if and only if for any two non-zero idempotents e and f in S there exists an idempotent i such that $e \,\mathcal{D}\, i \leq f$ and an idempotent j such that $f \,\mathcal{D}\, j \leq e$.*

Proof Suppose first that S is 0-simple. Let e and f be any two non-zero idempotents. Observe that $e \in SeS$ and $f \in SfS$. So, both SeS and SfS are not equal to

[16] Douglas Munn once remarked to me that this was one of the few instances where the theory for inverse semigroups with zero was easier than it was for the one without.

$\{0\}$. It follows that $S = SeS = SfS$. Thus $e \mathcal{J} f$. We now use Lemma 44 to deduce the result where we have used the fact that the idempotents form an order-ideal in an inverse semigroup. We now prove the converse. Let $I \neq \{0\}$ be any non-zero ideal of S. Suppose that $I \neq S$. Let $a \in S \setminus I$. Observe that $\mathbf{d}(a) \in S \setminus I$ because if $\mathbf{d}(a) \in I$, then $a = a\mathbf{d}(a) \in I$, since I is an ideal. Let $b \in I$ be any non-zero element. Then $\mathbf{d}(b) \in I$ since I is an ideal. By assumption, $\mathbf{d}(a)$ and $\mathbf{d}(b)$ are nonzero idempotents. It follows by the assumption and Lemma 44 that $S\mathbf{d}(a)S = S\mathbf{d}(b)S$. This implies that $\mathbf{d}(a) \in I$ and so $a \in I$, which is a contradiction. It follows that, in fact, $I = S$. □

Let S be any inverse semigroup with zero. Define

$$(s,t) \in \xi_S \Leftrightarrow (\forall a, b \in S)(asb = 0 \Leftrightarrow atb = 0).$$

It is left to the reader to check that this really is a congruence. We shall denote it by ξ when the semigroup it is defined on is clear. In the case where S is a meet-semilattice, the above definition simplifies somewhat. Let E be a meet-semilattice with zero, the operation of which is denoted by concatenation. Then $(e, f) \in \xi$ if and only $(\forall g \in E)(eg = 0$ if and only if $fg = 0)$. A congruence ρ is said to be 0-*restricted* if the ρ-class containing 0 is just 0. We now have the following characterization of the congruence ξ.

Lemma 56 *Let S be an inverse semigroup with zero. The congruence ξ is the maximum 0-restricted congruence on S.*

Proof Observe, first, that ξ is 0-restricted. Suppose that $(s, 0) \in \xi$. By definition, for all $a, b \in S$ we have that $asb = 0$ if and only of $a0b = 0$. However, if we put $a = ss^{-1}$ and $b = s^{-1}s$, then we deduce that $s = 0$. Now, let ρ be any 0-restricted congruence on S and let $s \,\rho\, t$. Suppose that $asb = 0$. Then $asb \,\rho\, atb$. Since ρ is 0-restricted, we have that $atb = 0$. Thus $asb = 0$ implies that $atb = 0$. By symmetry, we deduce that $a \,\xi\, b$. Thus $\rho \subseteq \xi$, as required. □

Our next result shows, amongst other things, the relationship between μ and ξ.

Lemma 57 *Let S be an inverse semigroup with zero.*

(i) $\mu \subseteq \xi$.
(ii) *The congruence ξ restricted to $\mathsf{E}(S)$ is precisely $\xi_{\mathsf{E}(S)}$.*

Proof

(1) Let $s \,\mu\, t$. We prove that $s \,\xi\, t$. Suppose that $asb = 0$ for $a, b \in S$. We shall prove that $atb = 0$. However, $asb \,\mu\, atb$. By definition, for all idempotents e, we have that $(asb)e(asb)^{-1} = (atb)e(atb)^{-1}$. Choose $e = b^{-1}b$. It follows that $\mathbf{r}(asb) = \mathbf{r}(atb)$. We are given that $asb = 0$ and so $\mathbf{r}(asb) = 0$. It follows that $\mathbf{r}(atb) = 0$ and so $atb = 0$. By symmetry, this shows that $s \,\xi\, t$.
(2) Let e and f be idempotents. To say that $(e, f) \in \xi$ means that for all $a, b \in S$ we have that $aeb = 0$ if and only if $afb = 0$. It is clear that $(e, f) \in \xi_{\mathsf{E}(S)}$. Suppose, now, that $(e, f) \in \xi_{\mathsf{E}(S)}$. We prove that $(e, f) \in \xi$ in S. Let

$aeb = 0$. Then $a^{-1}aebb^{-1} = 0$. Thus $a^{-1}abb^{-1}e = 0$ and so, by assumption, $a^{-1}abb^{-1}f = 0$. Hence, $a^{-1}afbb^{-1} = 0$ and so $afb = 0$. The reverse direction is proved similarly.

□

We now link ξ being the equality relation on the meet-semilattice E with the property of E being 0-disjunctive which we introduced just before Lemma 9.

Lemma 58 *Let E be a meet semilattice with zero. Then ξ is the equality relation on E if and only if E is 0-disjunctive.*

Proof Suppose first that ξ is the equality relation on E. We prove that E is 0-disjunctive. Suppose that $0 < f < e$ and assume that there does not exist any $0 \neq g \leq e$ such that $fg = 0$. If $ei = 0$, then clearly $fi = 0$. Suppose that $fi = 0$. Then $fi = fei = 0$ and so $f(ei) = 0$. Clearly, $ei \leq e$ and so by our assumption above, we must have $ei = 0$. It follows that $(e, f) \in \xi$. But this implies that $e = f$, which is a contradiction.

We assume that E is 0-disjunctive and prove that ξ is the equality relation. Suppose that $(e, f) \in \xi$, where e and f are both non-zero. Then $e \, \xi \, ef$ and so $ef \neq 0$ since ξ is 0-restricted. Suppose that $ef \neq e$. Then $0 < ef < e$. Then, by assumption, there exists $0 \neq g \leq e$ such that $(ef)g = 0$. But, clearly, $g = eg \neq 0$. However, $e \, \xi \, ef$ and so we have a contradiction. It follows that $ef = e$. Similarly, $ef = f$ and so $e = f$, as required. □

We are nearly at our goal.

Lemma 59 *Let S be an inverse semigroup with zero. Then ξ is the equality relation if and only if $\mathsf{E}(S)$ is 0-disjunctive and S is fundamental.*

Proof Suppose first that ξ is the equality relation. Then $\mathcal{E}(S)$ is 0-disjunctive by Lemmas 57 and 58, and it is fundamental by Lemma 57. To prove the converse, suppose that $\mathsf{E}(S)$ is 0-disjunctive and S is fundamental. Then by Lemmas 57 and 58, ξ restricted to $\mathsf{E}(S)$ is the equality relation and so ξ is idempotent-separating. It follows by Lemma 51 that $\xi \subseteq \mu$. But S is fundamental and so μ is the equality relation thus ξ is the equality relation. □

We may now state the characterization of congruence-free inverse semigroups with zero.

Theorem 12 (Congruence-Free Inverse Semigroups with Zero) *An inverse semigroup with zero S is congruence-free if and only if S is fundamental, 0-simple, and $\mathsf{E}(S)$ is 0-disjunctive.*

Proof Suppose that S is congruence-free. Then μ is equality, there are no non-trivial ideals and ξ is equality. Thus S is fundamental, 0-simple, and $\mathsf{E}(S)$ is 0-disjunctive by Lemma 58

To prove the converse, suppose that S is fundamental, 0-simple, and $\mathsf{E}(S)$ is 0-disjunctive. Let ρ be a congruence on S which is not the universal relation. Then $\rho(0)$ is an ideal which is not S. Thus this ideal must be equal to $\{0\}$. It follows that

ρ is a 0-restricted congruence and so $\rho \subseteq \xi$ by Lemma 56. Now we use the fact that the semigroup is fundamental together with Lemma 59, to deduce that ξ is the equality congruence and so ρ is the equality congruence. □

9 Further Reading and Examples

My interest in inverse semigroups developed as a result of John Fountain's undergraduate course in semigroups at the University of York, UK. It was deepened by attending the seminar at Oxford organized by Peter Cameron, Wilfrid Hodges, Angus Macintyre, and Peter Neumann, in which partial automorphisms and the work of Fraïssé played a central role. The connections between Fraïssé's work and inverse semigroups were clarified in [115] though based on the work of Benda to be found [8]. Partial automorphisms also play a significant role in Fraïssé's approach to the study of relational structures [26]. The work of Ehresmann on pseudogroups [22] convinced me that inverse semigroups really were worth studying. I was, however, bugged by the question of whether there were 'natural' examples of inverse semigroups. But what makes an area of mathematics 'natural'? I can think of at least two answers to this question: the existence of deep problems and the proliferation of good examples. When I embarked on writing my book [52], I was primarily motivated by finding good examples. The work of Kellendonk [45–47] and Girard [30], as mediated by Peter Hines [35], provided such good examples. Since that time, it has become clear that partial bijections, and therefore inverse semigroups, arise naturally in many different parts of mathematics. There are also deep questions such as whether every finite inverse monoid has a finite F-inverse cover [5] or what can be said about finite semigroups with commuting idempotents [3]. In this section, I shall concentrate on sources of good examples, since these look like promising growth points for the theory.

If there is one area of mathematics which makes essential use of inverse semigroups, it is in the theory of C^*-algebras. There are so many books and papers that are relevant here that I will only mention a few to get you started. The connection between inverse semigroups and C^*-algebras was first explored in [49] and [102]. A rather more recent book on this topic is [95] which will repay reading. There are many, many papers on this topic. A good place to start is [23] and [25] but it is worth checking out [100, page 45] which is on the special case of Cantor minimal systems. Let me, here, give one example of the theory of C^*-algebras leading to new semigroup theory. A monoid is said to be *finitely aligned* if the intersection of any two principal right ideals is a finitely generated right ideal. This concept was first defined within the theory of C^*-algebras in [101] and made extensive use of in [108, 109]. However, it seems to have been first defined within semigroup theory in [32], quite early on, but was not picked up by workers in inverse semigroup theory; this is but one illustration of the importance of mathematicians in different fields talking to one another. See also the more recent [13]. It would be most naturally applied in [14].

The connection between inverse semigroups and C^*-algebras has come to be mediated by what are now termed *Steinberg algebras*; see [111, 113]. Even where no direct connection between inverse semigroups and C^*-algebras appears to exist, the theory of C^*-algebras has sometimes led to new developments in the theory of inverse semigroups. A good example is Steinberg's notion of strong Morita equivalence of inverse semigroups motivated by the corresponding theory for C^*-algebras [110].

The connections between inverse semigroups and quasi-crystals were first spelt out by Kellendonk [45–47]. The inverse semigroup perspective on the topological groupoids that Kellendonk constructs was first clarified by Lenz [77] and then developed in [73]. Aspects of this connection were described in [67]. This led to what we might term 'non-commutative Stone duality' which connects, in particular, Boolean inverse monoids with a class of étale groupoids. For a survey of non-commutative Stone duality, see [66].

I once asked the late Pieter Hofstra, an expert in topos theory, why I kept seeing inverse semigroups everyone. I do not know whether Pieter was being polite, but he explained that it was because étendues were everywhere, where an étendue is a particular kind of topos. This seems to me a particularly fruitful area of research. We refer the reader to [28, 29].

If the theory of inverse semigroups is to be more fully developed, then there will be a need to study invariants associated with inverse semigroups. Motivated by the Banach-Tarski paradox [16] and the theory of K_0-groups [31], the type monoid was introduced in [48, 117] and its theory developed by Wehrung [118]. Type monoids are commutative refinement monoids and provide the first example of invariants of inverse semigroups. Another source of examples of invariants of inverse semigroups are cohomology theories. These have barely been developed; see the work of Lausch [75] and, particularly, Loganathan [78]. An application of the extension theory of inverse semigroups in the case of Von Neumann algebras can be found in [20]. MV-algebras form a natural class of invariants for the class of factorizable Boolean inverse monoids S in which $S/\mathscr{J}$ is a lattice. MV-algebras are another generalization of Boolean algebras but it was proved in [68] that all countable MV-algebras arise in this way and this was generalized to all MV-algebras by Wehrung [118] using very different methods. Thus every MV-algebra arises from a particular kind of Boolean inverse monoid. The theory of MV-algebras goes back to Tarski's book [114], which is also noteworthy in that partial automorphisms of structures are singled out in [114, Theorem 11.6]. MV-algebras are studied in their own right in [15, 88, 89] but their applications to the study of certain kinds of Boolean inverse monoids has so far not been developed.

I have indicated that certain kinds of groups can be constructed from inverse semigroups. See [82, 83] for how groups arise as so-called 'topological full groups' of certain kinds of groupoids; under non-commutative Stone duality, this means that the groups arise as the groups of units of certain kinds of Boolean inverse monoids. The Thompson groups form an important class of such groups. They, too, can be constructed from inverse monoids [42, 71, 74] generalizing some initial work in [9]. Groups with a different flavour arise from what Dehornoy calls 'geometry monoids'.

This is explained in his book [18] but the explicit connection with inverse monoids is made in [55].

If I were to single out one class of inverse semigroups that justifies the field it would be the polycyclic inverse monoids introduced by Nivat and Perrot [93]. They were generalized by Perrot [96, 97]. As I pointed out in [58], Perrot's thesis contains the first definition of self-similar group actions [92] and was motivated entirely by inverse semigroup theory. My paper was subsequently generalized by Alistair Wallis [117]; see also [69, 70]. The polycyclic inverse monoids are intimately related to the classical Thompson groups [56, 57, 65]. The representation theory (by means of partial bijections) of the polycyclic inverse monoids is the subject of [40, 41, 59] and was motivated by Bratteli and Jorgensen [11] and by Kawamura [44], which on the face of it have nothing to do with the polycyclic inverse monoids. The polycyclic inverse monoids arise from the free monoid. Analogous inverse monoids arise from free categories [4, 42]. There is a natural connection between such inverse semigroups and Leavitt path algebras [1]. This is part of the general topic of studying algebras of various kinds associated with inverse semigroups. This goes back to pioneering work of Douglas Munn [90] and it reaches its apotheosis in the applications of inverse semigroups to the theory of C^*-algebras mentioned above. This is a subject in its own right, but we mention in passing [19] and [72], which are relevant.

To conclude, I have not said anything about the classical theory of inverse semigroups. Here, I would simply highlight the papers [2, 84, 87] with the suggestion that there is more to do here. I have also said nothing about free inverse semigroups. These are introduced in my book [52, Chapter 6] but a much more recent paper on their structure is [34]. In passing, I would like to mention that free inverse monoids can also be constructed using the doubly pointed pattern classes of Kellendonk applied to the graph of the free group regarded as a 'dendritic tiling' [46]. In fact, it was this example which convinced me that Kellendonk was on to something. The study of free inverse monoids leads naturally to the study of presentations of inverse semigroups. A recent paper on this topic is [33].

Acknowledgments I would like to thank Victoria Gould, Bernard Rybolowicz, and Aidan Sims for reading the first version of this chapter and giving me good notes.

References

1. Abrams, G., Ara, P., Molino, M.S.: Leavitt Path Algebras. Springer, Berlin (2017)
2. Ash, Ch.: Uniform labelled semilattices. J. Austral. Math. Soc. (Ser. A) **28**, 385–397 (1979)
3. Ash, Ch.J.: Finite semigroups with commuting idempotents. J. Austral. Math. Soc. (Ser. A) **43**, 81–90 (1987)
4. Ash, Ch.J., Hall, T.E.: Inverse semigroups on graphs. Semigroup Forum **11**, 140–145 (1975)
5. Auinger, K., Bitterlich, J., Otto, M.: Solution to the Henckell-Rhodes problem: finite F-inverse covers do exist. arXiv:2208.03273v3

6. Auinger, K., Kudryavtseva, G., Szenderei, M.B.: F-inverse monoids as algebraic structures in enriched signature. Indiana Univ. Math. J. **70**, 2107–2131 (2021)
7. Bartholdi, L., Bogopolski, O.: On abstract commensurators of groups. arXiv:0902.4542v1
8. Benda, M.: Modeloids I. Trans. Am. Math. Soc. **250**, 47–90 (1979)
9. Birget, J.-C.: The groups of Richard Thompson and complexity. Inter. J. Alg. and Comput. **14**, 569–626 (2004)
10. Boothby, W.M.: An Introduction to Differentiable Manifolds and Riemnannian Geometry, 2nd edn. Academic Press, Cambridge (1986)
11. Bratteli, O., Jorgensen, P.E.T.: Iterated Function Systems and Permutation Representations of the Cuntz Algebra. AMS, Amsterdam (1999)
12. Bulman-Fleming, S., Fountain, J., Gould, V.: Inverse semigroups with zero: covers and their structure. J. Austral. Math. Soc. (Ser. A) **67**, 15–30 (1999)
13. Carson, S., Gould, V.: Right ideal Howson semigroups. Semigroup Forum **102**, 62–85 (2021)
14. Cherubini, A.. Petrich, M.: The inverse hull of right cancellative semigroups. J. Alg. **111**, 74–113 (1987)
15. Cignoli, R., D'Ottaviano, I.M.L., Mundici, D.: Algebraic Foundations of Many-Valued Reasoning. Trends in Logic, vol. 7. Kluwer, Dordrecht (2000)
16. Ceccherini-Silberstein, T., Grigorchuk, R., de la Harpe, P.: Amenability and paradoxical decompositions for pseudogroups and for discrete metric spaces. Proc. Steklov Inst. Math. **224**, 57–97 (1999)
17. Cornulier, Y.: Commensurating actions for groups of piecewise continuous transformations. arXiv:1803.08572
18. Dehornoy, P.: Braids and Self-Distributivity. Brikhäuser, Basel (2000)
19. Donsig, A.P., Milan, D.: Joins and covers in inverse semigroups and tight C^*-algebras. Bull. Austral. Math. Soc. **90**, 121–133 (2014)
20. Donsig, A.P., Fuller, A.H., Pitts, D.R.: Von Neumann algebras and extensions of inverse semigroups. Proc. Edinb. Math. Soc. **60**, 57–97 (2017)
21. Ehresmann, Ch.: Gattungen von lokalen Strukturen. Jahresber. Dtsch. Math.-Ver. **60**, 49–77 (1957)
22. Ehresmann, Ch.: Oeuvres complètes et commentées, Ehresmann, A.C. (Ed.) Supplements to Cahiers de topologie et géométrie différentielle, Amiens (1980–83)
23. Exel, R.: Inverse semigroups and combinatorial C^*-algebras. Bull. Braz. Math. Soc. (N.S.) **39**, 191–313 (2008)
24. Farb, B., Weinberger, S.: Hidden symmetry and arithmetic manifolds. arXiv:math/0405160
25. Farthing, C., Muhly, P.S., Yeend, T.: Higher-rank graph C^*-algebras: an inverse semigroup and groupoid approach. Semigroup Forum **71**, 159–187 (2005)
26. Fraïssé, R.: Theory of Relations. North-Holland, Amsterdam (1986)
27. Frink, O.: Representation of Boolean algebras. Bull. Amer. Math. Soc. **47**, 775–776 (1941)
28. Funk, J., Hofstra, P.: Toposes for semigroups: an invitation. Semigroup Forum **103**, 715–776 (2021)
29. Funk, J., Hofstra, P.: Pseudogroups and their torsors. Semigroup Forum **104**, 281–319 (2022)
30. Girard, J.-Y.: The geometry of interaction III: accommodating the additives. In: Girard, J.-Y., Lafont, Y., Regnier, L. (Eds.) Advances in Linear Logic. Cambridge University Press, Cambridge (1995)
31. Goodearl, K.R.: Von Neumann Regular Rings. Pitman, New Jersey (1979)
32. Gould, V.: Coherent monoids. J. Austral. Math. Soc. **53**, 166–182 (1992)
33. Gray, R.D.: Undecidability of the word problem for one-relator inverse monoids via right-angled Artin subgroups of one-relator groups. Invent. Math. **219**, 987–1008 (2020)
34. Gray, R.D., Steinberg, B.: Free inverse monoids are not FP_2. C. R. Maths. **359**, 1047–1057 (2021)
35. Hines, P.M.: The algebraic theory of self-similarity and its applications, PhD Thesis, University of Wales (1997)
36. Hollings, Ch.: Scientific Communication Across the Iron Curtain. Springer, Berlin (2016)
37. Hollings, Ch., Lawson, M.V.: Wagner's Theory of Generalised Heaps. Springer, Berlin (2017)

38. Howie, J.M.: An Introduction to Semigroup Theory. Academic, Cambridge (1976)
39. Johnstone, P.T.: Stone Spaces. CUP (1986)
40. Jones, D.G.: Polycyclic monoids and their generalisations, PhD Thesis, Heriot-Watt University (2011)
41. Jones, D.G., Lawson, M.V.: Strong representation of the polycyclic inverse monoids: cycles and atoms. Period. Math. Hung. **64**, 53–87 (2012)
42. Jones, D.G., Lawson, M.V.: Graph inverse semigroups: their characterization and completion. J. Algebra **409**, 444–473 (2014)
43. Kaarli, K., Márki, L.: A characterization of the inverse monoid of bi-congruences of certain algebras. Inter. J. Alg. Comput. **19**, 791–808 (2009)
44. Kawamura, K.: Polynomial endomorphisms of the Cuntz algebras arising from permutations I. General theory. Lett. Math. Phys. **71**, 149–158 (2005)
45. Kellendonk, J.: Noncommutative geometry of tilings and gap labelling. Rev. Math. Phys. **7**, 1133–1180 (1995)
46. Kellendonk, J.: The local structure of tilings and their integer group of coinvariants. Comm. Math. Phys **187**, 115–157 (1997)
47. Kellendonk, J.: Topological equivalence of tilings. J. Math. Phys. **38**, 1823–1842 (1997)
48. Kudryavtseva, G., Lawson, M.V., Lenz, D.H., Resende, P.: Invariant means on Boolean inverse monoids. Semigroup Forum **92**, 77–101 (2016)
49. Kumjian, A.: On localizations and simple C^*-algebras. Pac. J. Math. **112**, 141–192 (1984)
50. Lallement, G.: Semigroups and Combinatorial Applications. Wiley, Hoboken (1981)
51. Lawson, M.V.: Almost factorisable inverse semigroups. Glasg. Math. J. **36**, 97–111 (1994)
52. Lawson, M.V.: Inverse Semigroups: The Theory of Partial Symmetries. World Scientific, Singapore (1998)
53. Lawson, M.V.: The structure of 0-E-unitary inverse semigroups I: the monoid case. Proc. Edinb. Math. Soc. **42**, 497–520 (1999)
54. Lawson, M.V.: E^*-unitary inverse semigroups. In: Gomes, G.M.S., Pin, J.-E., Silva, P.V. (Eds.) Semigroups, Algorithms, Automata and Languages, pp. 195–214. World Scientific, Singapore (2002)
55. Lawson, M.V.: A correspondence between balanced varieties and inverse monoids. Inter. J. Alg. Comput. **16**, 887–924 (2006)
56. Lawson, M.V.: Orthogonal completions of the polycyclic monoids. Commun. Algebra **35**, 1651–1660 (2007)
57. Lawson, M.V.: The polycyclic monoids P_n and the Thompson groups $V_{n,1}$. Commun. Algebra **35**, 4068–4087 (2007)
58. Lawson, M.V.: A correspondence between a class of monoids and self-similar group actions I. Semigroup Forum **76**, 489–517 (2008)
59. Lawson, M.V.: Primitive partial permutation representations of the polycyclic monoids and branching function systems. Period. Math. Hung. **52**, 189–207 (2009)
60. Lawson, M.V.: A non-commutative generalization of Stone duality. J. Austral. Math. Soc. **88**, 385–404 (2010)
61. Lawson, M.V.: Non-commutative Stone duality: inverse semigroups, topological groupoids and C^*-algebras. Inter. J. Alg. Comput. **22**, 1250058 (2012)
62. Lawson, M.V.: A First Course in Logic. Routledge, Milton Park (2018)
63. Lawson, M.V.: The Booleanization of an inverse semigroup. Semigroup Forum **100**, 283–314 (2020)
64. Lawson, M.V.: Primer on inverse semigroups I. arXiv:2006:01628
65. Lawson, M.V.: The polycyclic inverse monoids and the Thompson groups revisited. In: Romeo, P.G., Rajan, A.R. (Eds.) Semigroups, Categories and Partial Algebras ICSAA 2019, vol. 345. Springer, Berlin (2021)
66. Lawson, M.V.: Non-commutative Stone duality. arXiv:2207.02688v2
67. Lawson, M.V., Lenz, D.H.: Pseudogroups and their étale groupoids. Adv. Math. **244**, 117–170 (2013)

68. Lawson, M.V., Scott, P.: AF inverse monoids and the structure of countable MV-algebras. J. Pure Appl. Algebra **221**, 45–74 (2017)
69. Lawson, M.V., Wallis, A.: A correspondence between a class of monoids and self-similar group actions II. Inter. J. Alg. Comput. **25**, 633–668 (2015)
70. Lawson, M.V., Wallis, A.: Levi categories and graphs of groups. Theory Appl. Categ. **32**, 780–802 (2017)
71. Lawson, M.V., Vdovina, A.: Higher dimensional generalizations of the Thompson groups. Adv. Math. **369**, 107191 (2020)
72. Lawson, M.V., Vdovina, A.: The universal Boolean inverse semigroup presented by the abstract Cuntz-Krieger relations. J. Noncommut. Geom. **15**, 279–304 (2021)
73. Lawson, M.V., Margolis, S.W., Steinberg, B.: The étale groupoid of an inverse semigroup as a groupoid of filters. J. Austral. Math. Soc. **94**, 234–256 (2014)
74. Lawson, M.V., Sims, A., Vdovina, A.: Higher dimensional generalizations of the Thompson groups via higher rank graphs. arXiv:2010.08960v3
75. Lausch, H.: Cohomology of inverse semigroups. J. Algebra **35**, 273–303 (1975)
76. Leech, J.: Inverse monoids with a natural semilattice ordering. Proc. Lond. Math. Soc. **70**, 146–182 (1995)
77. Lenz, D.H.: On an order-based construction of a topological groupoid from an inverse semigroup. Proc. Edinb. Math. Soc. **51**, 387–406 (2008)
78. Loganathan, M.: Cohomology of inverse semigroups. J. Algebra **70**, 375–393 (1981)
79. Mac Lane, S.: Categories for the Working Mathematician. Springer, Berlin (1971)
80. Mac Lane, S., Moerdijk, I.: Sheaves in Geometry and Logic. Springer, Berlin (1992)
81. Matte Bon, N.: Rigidity properties of full groups of pseudogroups over the Cantor set. arXiv:1801:10133
82. Matui, H.: Homology and topological full groups of étale groupoids on totally disconnected spaces. Proc. London Math. Soc. **104**(3), 27–56 (2012)
83. Matui, H.: Topological full groups of one-sided shifts of finite type. J. Reine Angew. Math. **705**, 35–84 (2015)
84. McAlister, D.B.: Inverse semigroups which are separated over a subsemigroup. Trans. Am. Math. Soc. **182**, 85–117 (1973)
85. McAlister, D.B.: Groups, semilattices and inverse semigroups. Trans. Am. Math. Soc. **192**, 227–244 (1974)
86. McAlister, D.B.: Groups, semilattices and inverse semigroups II. Trans. Amer. Math. Soc. **196**, 251–270 (1974)
87. Meakin, J.: The partially ordered set of $\mathscr{J}$-classes of a semigroup. J. London Math. Soc. **21**, 244–256 (1980)
88. Mundici, D.: Logic of infinite quantum systems. Int. J. Theor. Phys. **32**, 1941–1955 (1993)
89. Mundici, D.: MV-algebras, May 26th (2007). http://www.matematica.uns.edu.ar/IXCongresoMonteiro/Comunicaciones/Mundici_tutorial.pdf
90. Munn, W.D.: Inverse semigroup algebras. In: Group and Semigroup Rings. Johannesburg (1985). In: North-Holland Mathematics Studies, vol. 126. North-Holland, Amsterdam, (1986), pp. 197–223
91. Nekrashevych, V.: Virtual endomorphisms of groups. Algebra Discrete Math. Number **1**, 88–128 (2002)
92. Nekrashevych, V.: Self-Similar Groups. AMS, Providence (2005)
93. Nivat, M., Perrot, J.-F.: Une généralisation du monoïde bicyclique. C. R. Acad. Sci. Paris Sér. I Math. **A271**, 824–827 (1970)
94. Padmanabhan, R.: A first order proof of a theorem of Frink. Alg. Univ. **13**, 397–400 (1981)
95. Paterson, A.L.T.: Groupoids, Inverse Semigroups, and Their Operator Algebras. Progress in Mathematics, vol. 170. Birkhäuser, Boston (1998)
96. Perrot, J.-F.: Une famile de monoïdes inversifs 0-bisimples généralisant le monoïde bicyclique. Séminaire Dubreil. Algèbre **25**, 1–15 (1971–1972)
97. Perrot, J.-F.: Contribution a l'étude des monoïdes syntactiques et de certains groupes associés aux automates finis. L'Université Paris VI, Thése (1972)

98. Petrich, M.: Inverse Semigroups. Wiley, Hoboken (1984)
99. Preston, G.B.: Inverse semi-groups. J. Lond. Math. Soc. **29**, 396–403 (1954)
100. Putnam, I.F.: Cantor Minimal Systems. AMS, Amsterdam (2018)
101. Raeburn, I., Sims, A.: Product systems of graphs and the C^*-algebras of higher-rank graphs. J. Operator Th. **53**, 399–429 (2005)
102. Renault, J.: A Groupoid Approach to C^*-Algebras. Springer, Berlin (1980)
103. Resende, P.: Etale groupoids and their quantales. Adv. Math. **208**, 147–209 (2007)
104. Rinow, W.: Über die Vervollständigung induktiver Gruppoide. Math. Nachr. **33**, 199–222 (1963)
105. Schein, B.M.: Completions, translational hulls and ideal extensions of inverse semigroups. Czechoslov. Math. J. **23**, 575–610 (1973)
106. Schein, B.M.: Victor Vladimirovich Vagner (1908–1981). Semigroup Forum **23**, 189–200 (1981)
107. Sontz, S.B.: Principal Bundles. Springer, Berlin (2015)
108. Spielberg, J.: Groupoids and C^*-algebras for categories of paths. Trans. Am. Math. Soc. **366**, 5771–5819 (2014)
109. Spielberg, J.: Groupoids and C^*-algebras for left cancellative small categories. arXiv:1712:07720v2
110. Steinberg, B.: Strong Morita equivalence of inverse semigroups. Houston J. Math. **37**, 895–927 (2009)
111. Steinberg, B.: A groupoid approach to discrete inverse semigroup algebras. Adv. Math. **223**, 689–727 (2010)
112. Steinberg, B.: Representation Theory of Finite Monoids. Springer, Berlin (2016)
113. Steinberg, B., Szakács, N.: Simplicity of inverse semigroup and étale groupoid algebras. Adv. Math. **202**, 107611 (2022)
114. Tarski, A.: Cardinal Algebras. OUP, Oxford (1949)
115. Tiemans, L., Scott, D.S., Benzmüller, Ch., Benda, M.: Computer-supported exploration of a categorical axiomatization of modeloids. In: RAMiCS: Relational and Algebraic Models in Computer Science. LNTCS, vol. 12062, pp. 302–317. (2020)
116. Wagner, V.V.: Generalised groups. Dokl. Akad. Nauk SSSR **84**, 1119–1122 (1952). (In Russian)
117. Wallis, A.R.: Semigroup and category-theoretic approaches to partial symmetry, PhD Thesis, 2013, Heriot-Watt University, Edinburgh
118. Wehrung, F.: Refinement Monoids, Equidecomposability Types, and Boolean Inverse Semigroups. LNM, vol. 2188. Springer, Berlin (2017)

A Journey Through Constructive Inverse Semigroups with Apartness

Melanija Mitrović, Mahouton Norbert Hounkonnou, and Daniel Abraham Romano

> *Mathematics is that portion of our intellectual activity which transcends our biology and our environment [...] Mathematics, a creation of mind, is less arbitrary than biology or physics, creations of nature; the creatures we imagine inhabiting another world in another universe, with another biology and another physics, will develop a mathematics which in essence is the same as ours. In believing this we may be falling into a trap: Mathematics being a creation of our mind, it is, of course, difficult to imagine how mathematics could be otherwise without actually making it so, but perhaps we should not presume to predict the course of the mathematical activities of all possible types of intelligence. On the other hand, the pragmatic content of our belief in the transcendence of mathematics has nothing to do with alien forms of life. Rather it serves to give a direction to mathematical investigation, resulting from the insistence that mathematics be born of an inner necessity.* (Errett Bishop, **FCA**, 1967, [5])

Abstract An important goal of this chapter is to provide a clear, understandable picture of constructive inverse semigroups with apartness in the Errett Bishop's style of constructive mathematics, **BISH**. The theory of constructive semigroup with apartness is a new approach to semigroup theory (and not a new class of semigroups). It presents novelty within both mathematics: classical and constructive.

M. Mitrović (✉)
Center for Applied Mathematics of the Faculty of Mechanical Engineering Niš, CAM-FMEN, University of Niš, Niš, Serbia
e-mail: melanija.mitrovic@masfak.ni.ac.rs

M. N. Hounkonnou
International Chair in Mathematical Physics and Applications (ICMPA-UNESCO Chair), University of Abomey-Calavi, Cotonou, Benin
e-mail: norbert.hounkonnou@cipma.uac.bj

D. A. Romano
International Mathematical Virtual Institute, Banja Luka, Bosnia and Herzegovina

M. Mitrović, M. N. Hounkonnou (eds.), *Algebra Without Borders*, Mathematics in Mind, https://doi.org/10.1007/978-3-031-86477-3_3

The theory of semigroups with apartness was developed by M. Mitrović and co-authors: M. N. Hounkonnou, M. A. Baroni, S. Crvenković, D. A. Romano, and S. Silvestrov; see, for example, Crvenković et al. (Math. Logic Q. 59(6), 407–414, 2013; Semigroup Forum 92(3), 659–674, 2016), Mitrović et al. (Fundamenta Informaticae 184(3), 233–271, 2021; Constructive semigroups with apartness—a state of the art, 2024), Mitrović and Silvestrov ((Apartness) Isomorphism theorems for basic constructive algebraic structures with special emphasize on constructive semigroups with apartness—an overview, 2020), Mitrović et al. (J. Phys. Conf. Ser. 1194, 012076, 2019). In establishing our theory we were inspired by results obtained in interactive theorem proving the approach of formal verifications (more in Mitrović et al., Fundamenta Informaticae 184(3), 233–271, 2021). So how do we achieve all this? Simply by changing the logic with which we do our mathematics! Instead of using classical logic, we systematically use intuitionistic logic, and our work is based on the notion of apartness (between elements of the set and, later, between elements and its subsets). Based on the results presented in (Crvenković et al., Math. Logic Q. 59(6), 407–414, 2013; Crvenković et al., Semigroup Forum 92(3), 659–674, 2016), A. Cherubini and A. Frigeri introduced in 2019 constructive inverse semigroups with apartness (Cherubini and Frigeri, Semigroup Forum 98, 571–588, 2019). M. Mitrović and D. A. Romano contributed to the area too (Mitrović and Romano, Co-congruence on an inverse semigroup with apartness that separates idempotents; Romano, Acta Universitatis Apulensis 73, 101–138, 2023).

Keywords Set with apartness · Semigroup with apartness · Inverse semigroup with apartness · Inverse linked relations

1 Introduction

It is only recently, due to the work of M. Mitrović and co-authors, that constructive mathematics has paid attention to semigroup theory. The use of intuitionistic logic in constructive mathematics facilitates the extraction of computational content from constructive proofs, something almost impossible within classical mathematics. M. Mitrović and co-authors' approach to constructivism closely follows the paradigm of E. Bishop, whose informal system **BISH** is developed in his seminal book Foundations of Constructive Analysis published in 1967 [5]. Constructive algebra is a relatively old discipline developed among others by L. Kronecker, van der Waerden, and A. Heyting. For more information on the history, see [19, 31]. For more information on algebra in Bishop's style, see [18]. In the words of I. Yangui [35], constructive algebra "can be seen as an abstract version of computer algebra. In computer algebra, on the one hand, one attempts to construct efficient algorithms for solving concrete problems given in an algebraic formulation, where a problem is understood to be concrete if its hypotheses and conclusion have computational content. Constructive algebra, on the other hand, can be understood as a "preprocessing" step for computer algebra that leads to general algorithms, even if they are sometimes not efficient. In constructive algebra, one tries to give general algorithms for solving "virtually any" theorem of abstract algebra".

One of the main topics in constructive algebra is constructive algebraic structures with the relation of (tight) apartness, the second most important relation in constructive mathematics. The principal novelty in treating basic algebraic structures constructively is that (tight) apartness becomes a fundamental notion. For example, consider the reals: a basic ingredient of constructive real numbers is the apartness relation, the constructive version of the (classical) inequality on reals—two real numbers are apart if it can positively be decided that they are distinct from each other. In a line of conclusion, constructive algebra is quite more complicated than classical one in various aspects. For example, the carrier set of an algebraic structure as a rule does not carry a decidable equality relation. This difficulty is partly met by the introduction of a strong inequality relation, the abovementioned apartness relation on its carrier set. Roughly, the descriptive definition of a *structure with apartness* includes two main parts:

- the notion of a certain classical algebraic structure is straightforwardly adopted;
- a structure is equipped with an apartness with standard operations respecting that apartness.

The constructive theory of semigroups with an apartness relation is a new approach to semigroup theory and not a new class of semigroups. It presents a semigroup facet of a relatively well-established direction within constructive mathematics which, to the best of our knowledge, has not yet been considered within the classical semigroup community, as well as within the constructive one. Inspired by results obtained from the use of interactive theorem-provers and the general theory of formal verification of constructive proofs (more on that can be found in the work of M. Mitrović et al. [22]), a new constructive algebraic theory, the so-called theory of semigroups with apartness was developed by M. Mitrović and co-authors: M. N. Hounkonnou, M. A. Baroni, S. Crvenković, E. Darpö, D. A. Romano, and S. Silvestrov. See, for example, [12–14, 20, 22, 23, 25, 26].

Based on the results presented in [12, 13], A. Cherubini and A. Frigeri in 2019 made important steps forward introducing constructive inverse semigroups with apartness [11]. Later, M. Mitrović and D. A. Romano contributed to the area too [24, 28]. This chapter is a journey through some of the results obtained within the class of constructive semigroups with apartness with special accent put on its proper subclass of inverse semigroups with apartness.

The theory is partly inspired by the classical case, but it is distinguished from it in two significant aspects: it uses intuitionistic logic rather than classical throught; it is based on the notion of apartness (between elements of the set and, later, between elements and its subsets). Following A. Heyting [16], classical semigroup theory is seen, at least initially, as a guide that helps us to develop the theory of constructive semigroups with apartness. On the other hand, it cannot be said that the theory of constructive semigroups with apartness (including constructive inverse semigroups with apartness) aims at revising the whole classical framework in nature.

Surrounded with a couple of excellent chapters on classical inverse semigroup theory within this book, our chapter can be seen, among other things, as an excellent evidence of what is written at the beginning of this paragraph.

Within Sect. 2 the constructive view of the fundamental notions will be given. All presented material is broad rather than deep, and it is not intended to be comprehensive. It is heavily based on the treatments in other standards on constructive mathematics books such as, for example, [5, 7, 31]. The main novelty is in the selection and arrangement of material. Working within classical or intuitionistic logic, in order to analyze algebraic structures it is necessary to start with study on sets and ordered sets. Development of an appropriate (order) theory of constructive sets with apartness has been one of the main objectives of our work. Most of the selected results on constructive sets with apartness we have published to the date are given in Sect. 3. Development of the theory of constructive semigroup with apartness has been another important objective of our work. Selected results we have published to the date are given in Sect. 4. Results on constructive inverse semigroups with apartness are presented in Sects. 5 and 6.

Results presented in Sects. 3 and 4 are based on the ones published in [12, 13, 22, 23, 28], while those ones from Sects. 5 and 6 are based on [11, 24, 28], in addition, shortly, on the work in progress [21].

The present chapter is by no means an attempt to give a complete overview of our existing results to date.

2 Informal Foundations of Constructive Mathematics

Throughout this chapter constructive mathematics is understood as mathematics performed in the context of intuitionistic logic, that is, without the law of excluded middle, **LEM**. By constructive mathematics we mean Bishop-style mathematics, **BISH**. The principal features of the constructive mathematics which forms the framework within we work is indicated in the appendix of this chapter. A presentation of constructive mathematics given within the appendix is broad rather than deep: we first sketch the history that motivated the early workers in the field of constructive mathematics, we describe informal intuitionistic logic, and we give the briefest outline on the constructively valid arguments. As stated in [7], logic alone does not suffice as a foundation for constructive mathematics. Every approach to the foundations of mathematics begins by analyzing the fundamental notions that occur again and again in the actual mathematics and in the terms of which the mathematics can be understood. In what follows within this section the constructive view of the fundamental notions will be given. It is heavily based on the treatments in other standards on constructive mathematics books, such as, for example, [5, 7, 31]. The main novelty is in the selection and arrangement of material.

2.1 Primitive Notions

A set of positive numbers and constructions are two primitive notions, and, as such, they cannot be defined. The cornerstones for **BISH** include the notions sets, functions, and relations.

The set of natural numbers $\mathbb{N}$ is regarded as a basic set, and it is assumed that the positive numbers have the usual algebraic and order properties, including mathematical induction.

The notion of ***algorithm*** (algorithmic process, finite routine, rule, mechanical operation) is primitive. The idea that the notion of an algorithm is primitive has also been advanced by the Russian mathematicians V. A. Uspenskii and A. L. Semenov [33]: "The concept of algorithm like that of set of natural number is a such a fundamental concept that it cannot be explained through other concepts and should be regarded as [an]undefinable one" (source [19]).

2.2 Informal Constructive Set Theory

Constructive set theory provides a standard set theoretical framework for the development of constructive mathematics in the style of Errett Bishop and one of several such frameworks for constructive mathematics that have been considered. Following [1] there are just sets as in classical set theory. This means that mathematics in constructive set theory can look very much like ordinary ordinary classical mathematics. The advantage of this is that the ideas, conventions, and practice of the set theoretical presentation of ordinary mathematics can be used also in set theoretical development of constructive mathematics, provided that a suitable discipline is adhered to. In the first place only the methods of logical reasoning available in intuitionistic logic should be used.

Building on the set of natural numbers and using intuitionistic logic, Bishop's approach is followed in developing constructive mathematics in higher and higher levels of abstraction. In order to be able to do this, the notion of the set should be cleared up first. Contrary to the classical case, a set exists only when it is defined. To define a *set* S we must have two pieces of data:

- a description of how we construct the elements (members) of S using objects that have been, or could have been, constructed prior to S;
- a description of what it means for two elements of S to be equal.

The equality on a set S is a binary relation $=$ satisfying the usual properties of an equivalence relation. The issue of equality seems to get more attention in constructive mathematics than it does in classical mathematics. In the customary approach to set theory, one does not regard a set as "coming equipped" with a special equality relation of its own. There is the "universal" or "absolute" equality relation on the entire universe. In constructive mathematics, no such thing is assumed: we have only those equality relations which we can construct.

Remark 1 In some (constructive) mathematics' books can be found that the use of equivalence relations rather than intensional equality (that is, identity of description) in classical mathematics often goes unnoticed. For example, for the set $\mathbb{N}$ of all natural numbers as well as for the set $\mathbb{Z}$ of all integers the equality is the relation of *identity*, in which two elements of the set are regarded as equal if and only if they are identically described. However, the equality on the set $\mathbb{Q}$ of all rational numbers is not that of identity, but is the usual equivalence between rationals:

$$\frac{m}{n} = \frac{m'}{n'} \overset{\text{def}}{\Leftrightarrow} mn' = nm',$$

where $m, m' \in \mathbb{Z}$ and $n, n' \in \mathbb{N}$. $\diamond$

As usual, we write $x \in S$ to signify that x is an element of the set S and $x \notin S$ instead of $\neg(x \in S)$.

A *subset* $(X, =_X)$ of a set $(S, =_S)$ consists of elements drawn from S together with the equality relation induces on X by the given equality on S, that is,

$$x =_X y \overset{\text{def}}{\Leftrightarrow} x =_S y,$$

for $x, y \in X$. If X is a subset of S, then we write $X \subseteq S$.

A property P, which is applicable to the elements of a set S, determines a subset X of S denoted by $X = \{x \in S : P(x)\}$. Furthermore, we will be interested only in properties $P(x)$ which are *extensional* in the sense that for all $x_1, x_2 \in S$ with $x_1 = x_2$, $P(x_1)$ and $P(x_2)$ are equivalent. Informally, it means that "it does not depend on the particular description by which x is given to us".

The logical complement of a subset X of S is

$$\neg S = \{x \in X : x \notin S\}.$$

The empty subset of S is defined by

$$\emptyset_S = \neg S.$$

One of the main features of constructive mathematics is the so-called the splitting notion phenomena, in which the concepts that are equivalent in the presence of **LEM** need not be equivalent any more. In what follows during this and subsequent chapters, it will be pointed out that, for example, we distinguish nonempty and inhabited sets, several types of inequalities, two complements of a given set, etc.

A set $(S, =)$ is an *inhabited* set if we can construct an element of S, i.e., $\exists_x (x \in S)$. The distinction between the notions of a nonempty set and an inhabited set is a key in constructive set theories. While an inhabited set is nonempty, the converse does not hold in general. In what follows we will write set shortly by S instead $(S, =)$.

The notion of equality of different sets is not defined. The only way in which elements of two different sets can be regarded as equal is by requiring them to be subsets of a third set. Let X and Y be subsets of a set S.

- The relations $X \subseteq Y$, $X = Y$ are defined as follows:

$$X \subseteq Y \text{ if and only if } x \in X \Rightarrow x \in Y;$$
$$X = Y \text{ if and only if } X \subseteq Y \text{ and } Y \subseteq X.$$

- The subsets $X \cup Y$, $X \cap Y$, $X \setminus Y$ are defined by

$$X \cup Y = \{x : x \in X \vee x \in Y\};$$
$$X \cap Y = \{x : x \in X \wedge x \in Y\};$$
$$X \setminus Y = \{x : x \in X \vee x \notin Y\}.$$

The standard laws of the algebra of sets, such as commutative, associative, and distributive laws, follow.

Theorem 1 *Let X, Y, and Z be subsets of a set $(S, =)$. Then:*

(i) $X \cup Y = Y \cup X \quad X \cap Y = Y \cap X$.
(ii) $(X \cup Y) \cup Z = X \cup (Y \cup Z) \quad (X \cap Y) \cap Z = X \cap (Y \cap Z)$.
(iii) $X \cap (Y \cup Z) = (X \cap Y) \cup (X \cap Z) \quad X \cup (Y \cap Z) = (X \cup Y) \cap (X \cup Z)$.
(iv) $X \cap Y \subseteq X \quad X \subseteq X \cup Y$.
(v) $\neg(X \cup Y) = (\neg X) \cap (\neg Y)$.
(vi) $X \subseteq \neg\neg X$.
(vii) $(\neg X) \cup (\neg Y) \subseteq \neg(X \cap Y)$.

Remark 2 We note that neither in (vi) nor in (vii) we can prove the reverse inclusions constructively. ◇

Given two sets $(S, =_S)$ and $(T, =_T)$, it is permissible to construct the set $(T^S, =)$ of mappings between them. A *mapping* $f : S \to T$ is an algorithm which produces an element $f(x)$ of T when applied to an element x of S, which is extensional, that is,

$$\forall_{x,y \in S} (x =_S y \Rightarrow f(x) =_T f(y)).$$

The set S is called the domain of f, written *dom* f, and the set $f(S) = \{f(x) \in T : x \in S\}$, with equality inherited from T, is called the range of f, written *ran* f.

As usual, equality between two elements f and g from T^S is defined by

$$f =_{T^S} g \overset{\text{def}}{\Leftrightarrow} \forall_{x \in X} (f(x) =_T g(x)).$$

Mappings $f : S \to T$ and $g : T \to U$ can be composed, giving a mapping $g \circ f : S \to U$ defined by $(g \circ f)(x) = g(f(x))$, $x \in S$.

Lemma 1 *Composition of mappings is associative, i.e.,*

$$h \circ (g \circ f) = (h \circ g) \circ f$$

whenever the compositions are well defined.

Corollary 1 $(S^S, =; \circ)$ *is a semigroup.*

A mapping $f : S \to T$ is:

- *onto* or *surjection*: $\forall_{y \in T}\ \exists_{x \in S}\ (y =_T f(x))$, i.e., $ranf = T$;
- *one-one* or *injection*: $\forall_{x,y \in S}\ (f(x) =_T f(y) \Rightarrow x =_S y)$;
- *bijection* between S and T: one-one and onto.

The (Cartesian) product of two sets $(S, =_S)$ and $(T, =_T)$ is the set $(S \times T, =_{S\times T})$ defined by

$$S \times T \overset{\text{def}}{=} \{(x, y) : x \in S \wedge y \in T\}$$
$$((x, y) =_{S\times T} (u, v)) \overset{\text{def}}{\Leftrightarrow} (x =_S u \wedge y =_T v).$$

From now on, when the meaning is clear from the context, we write $=$ instead $=_S$.

A partial mapping $f : S \to T$ is a mapping from a subset of S into T. The subset of S consisting of all elements $x \in S$ for which $f(x)$ is defined is called the *domain of* f, denoted by $dom\ f$. The *rang* of f is the subset $ran\ f = f(dom\ f)$, i.e.,

$$ran\ f = \{y \in T : \exists_{x \in S}(y = f(x))\}.$$

For every subset X of S the identity partial mapping denoted by 1_X is a partial mapping from S to itself. *The image of a subset* X of S under f is the set

$$f(X) = \{f(x) : x \in X\}.$$

The *inverse image* of a subset Y of T under f is the set

$$f^{-1}(Y) = \{x \in dom\ f : f(x) \in Y\}.$$

A partial mapping $f : S \to T$ is said to be *a total* partial mapping on S if $dom\ f = S$.

A partial mapping $f : S \to T$ can be identified with its *graph*:

$$G_f = \{(x, y) \in S \times T : x \in dom\ f,\ y \in ran\ f,\ y = f(x)\},$$

a subset of the Cartesian product $S \times T$. We define two partial mappings $f, g : S \to T$ to be equal if their graphs are equal as sets. Thus,

$$f = g \overset{\text{def}}{\Leftrightarrow} dom\ f = dom\ g \text{ and } \forall_{x \in dom\ f}(f(x) = g(x)).$$

The composition of partial mappings $f : S \to T$ and $g : T \to U$ is the partial mapping $g \circ f : S \to U$, defined by

$$(g \circ f)(x) = g(f(x))$$

whenever the right side exists. A partial bijective mapping $f : S \to T$ has *inverse mapping*

$$f^{-1} : ran\ f \to dom\ f$$

defined by $f^{-1}(f(x)) = x$.

In what follows of particular interest will be the Cartesian product of a set S by itself, $S \times S$. An inhabited subset ρ of $S \times S$, or, equivalently, a property applicable to elements of $S \times S$, is called a *binary relation* on S. In general, there are many properties that binary relations may satisfy on a given set, like:

(R) reflexive: $(x, x) \in \rho$
(IR) irreflexive: $\neg((x, x) \in \rho)$
(S) symmetric : $(x, y) \in \rho \Rightarrow (y, x) \in \rho$
(AS) antisymmetric: $(x, y) \in \rho \wedge (y, x) \in \rho \Rightarrow x = y$
(T) transitive: $(x, y) \in \rho \wedge (y, z) \in \rho \Rightarrow (x, z) \in \rho$
(coT) co-transitive: $(x, y) \in \rho \Rightarrow (x, z) \in \rho \vee (z, y) \in \rho$.

Inherited from classical mathematics, **CLASS**, they "play game" under constructive rules.

In **CLASS** equivalence is the natural generalization of equality. A theory with equivalence involves the equivalence and functions and relations respecting this equivalence. In constructive mathematics the same works without difficulty.

Many sets come with binary relation called *inequality* satisfying certain properties and denoted by $\neq$, #, or $\not\preceq$. In general, more computational information is required to distinguish elements of a set S, then to show that elements are equal. Comparing with **CLASS**, the situation for inequality is more complicated. There are different types of inequalities (denial inequality, diversity, apartness, tight apartness—to mention a few), some of them completely independent, which only in **CLASS** are equal to one standard inequality [29]. So, in **CLASS** the study of equivalence relation suffices, but, in constructive mathematics inequality becomes a "basic notion in intuitionistic axiomatics".

An inequality relation is often denoted by $\neq$. A tuple $(S, =_S, \neq_S)$ is called a *set with inequality*. It has to be emphasized that $\neq$ is not, in general, the negation of $=$. The interpretation of the symbol $x \neq y$ depends on the context. In the literature on constructive mathematics, an *inequality relation* is often considered to be a binary relation that is *irreflexive* and *symmetric*.

Specific types of inequality relations include:

- the *denial inequality*: $x \neq y \overset{\text{def}}{\Leftrightarrow} \neg(x = y)$;
- a *tight inequality*: $\neg(x \neq y) \Rightarrow x = y$;

- an *apartness* relation: co-transitive inequality ((IR), (S), (coT));
- a *tight apartness*: co-transitive tight inequality.

Remark 3 In constructive mathematics apartness is more basic than equality. For example, apartness relation, often denoted by #, is a basic ingredient of constructive real numbers; it is a constructive version of the classical inequality on reals. Two real numbers are apart if it can positively be decided that they are distinct of each other. Usually in the formalization of reals restriction is made to a tight apartness, saying that the negation of apartness is the equality. In general apartness need not to be tight. For example, in the theory of constructive semigroups it is more convenient to work with non-tight apartness. As it is shown in [14], even starting with constructive semigroup with tight apartness related quotient semigroup need not be with tight apartness. ◇

3 Set with Apartness

There are many decisions a mathematician must make when deciding to replace classical logic with intuitionistic logic. Let us mention some of them: first of all, the choice of variant of constructive mathematics. Our choice was the Errett Bishop-style constructive mathematics, **BISH**.

Going through the literature there are several variants of what is considered to be a set with apartness—depending on the relations between equality and apartness defined on a set. Our choice—our starting structure—is a **set with apartness** $(S, =, \#)$ where:

- equality and apartness are basic notions,
- equality and apartness are independent of each other,
- apartness is not, in general, tight.

Such our choice was and is a novelty within constructive circles.

3.1 Basic Concepts and Important Examples

Let $(S, =)$ be an *inhabited* set. By an ***apartness*** on S we mean a binary relation # on S which satisfies the axioms of irreflexivity, symmetry, and co-transitivity:

(Ap1) $\quad \neg(x\#x)$
(Ap2) $\quad x\#y \;\Rightarrow\; y\#x,$
(Ap3) $\quad x\#z \;\Rightarrow\; \forall_y\,(x\#y \;\vee\; y\#z).$

The apartness on a set S is *tight* if

(Ap4) $\quad \neg(x\#y) \;\Rightarrow\; x = y.$

If apartness is tight, then $\neg(x\#y) \Leftrightarrow x = y$.

By extensionality, we have

(Ap5) $\quad x\#y \wedge y = z \Rightarrow x\#z,$

the equivalent form of which is

(Ap5') $\quad x\#y \wedge x = x' \wedge y = y' \Rightarrow x'\#y'.$

Given two sets with apartness $(S, =_S, \#_S)$ and $(T, =_T, \#_T)$, it is permissible to construct the set of mappings between them. An important property applicable to mapping f is that of strong extensionality. Namely, a mapping $f : S \to T$ is a *strongly extensional* mapping, or, for short, an *se-mapping*, if

$$\forall_{x,y\in S}\, (f(x)\#_T f(y) \Rightarrow x\#_S y).$$

Let us remember that strong extensionality of all mappings from $\mathbb{R}$ to $\mathbb{R}$ implies the Markov principle, **MP**.

An se-mapping f is:

- an *se-surjection* if it is surjective;
- an *se-injection* if it is injective;
- an *se-bijection* if it is bijective.

Furthermore, f is:

- *apartness injective*, shortly *a-injective*: $\forall_{x,y\in S}\, (x\#_S\, y \Rightarrow f(x)\#_T f(y))$;
- *apartness bijective*: a-injective, se-bijective.

Given a set X with apartness it is permissible to construct the set of all se-mappings of X into itself which inherits apartness from X.

Theorem 2 *Let $(X, =, \#)$ be a set with apartness. If $\mathcal{T}_S^{se}$ is the set of all se-mappings from X to X, then $(\mathcal{T}_S^{se}, =, \#)$ with*

$$f = g \Leftrightarrow \forall_{x\in X}\, (f(x) = g(x))$$
$$f\#g \Leftrightarrow \exists_{x\in X}\, (f(x)\#g(x))$$

is a set with apartness too.

3.1.1 Distinguishing Subsets

The presence of apartness implies the appearance of different types of substructures connected to it. Following Bridges and Vîţă [7], we define the relation $\bowtie$ between an element $x \in S$ and a subset Y of S by

$$x \bowtie Y \overset{\text{def}}{\Leftrightarrow} \forall_{y\in Y}(x\#y).$$

A subset Y of S has two natural complementary subsets: *the logical complement* of Y (we define at Sect. 2)

$$\neg Y \stackrel{\text{def}}{=} \{x \in S : x \notin Y\},$$

and *the apartness complement* or, shortly, the *a-complement* of Y

$$\sim Y \stackrel{\text{def}}{=} \{x \in S : x \bowtie Y\}.$$

If the apartness is not tight, we can find subsets Y with $\sim Y \subset \neg Y$ as in the following example.

Example 1 Let $S = \{a, b, c\}$ be a set with apartness defined by $\{(a, c), (c, a), (b, c), (c, b)\}$ and let $Y = \{a\}$. Then the a-complement $\sim Y = \{c\}$ is a proper subset of its logical complement $\neg Y = \{b, c\}$. ⋄

Remark 4 For a tight apartness, the two complements are constructive counterparts of the classical complement. In general, we have $\sim Y \subseteq \neg Y$. However, even for a tight apartness, the converse inclusion entails the Markov principle, **MP**. This result illustrates the abovementioned feature of constructive mathematics: classically equivalent notions could be no longer equivalent constructively. ⋄

Lemma 2 *If X and Y are subsets of S, then:*

(i) $X \subseteq \neg Y \Leftrightarrow Y \subseteq \neg X \Leftrightarrow X \cap Y = \emptyset$.
(ii) *If $\widetilde{x}$ is the a-complement of the singleton $\{x\}$, $x \in S$, and Y is a subset of S, then $x \in \sim Y$ if and only if $Y \subseteq \widetilde{x}$.*
(iii) $X \subseteq Y \Rightarrow \sim Y \subseteq \sim X$.
(iv) $\sim (X \cup Y) = (\sim X) \cap (\sim Y)$.
(v) $(\sim X) \cup (\sim Y) \subseteq \sim (X \cap Y)$.
(vi) $X \subseteq \sim\sim X$.
(vii) $\sim X = \sim\sim\sim X$.

Proof

(i) Follows immediately.
(ii) Follows by the definition of a-complement.
(iii) Let $x \in \sim Y$, or, by (ii), equivalently, $Y \subseteq \widetilde{x}$. Then, by the assumption, $X \subseteq \widetilde{x}$, i.e., $x \in \sim X$.
(iv) Follows by (ii).

$$\begin{aligned} x \in \sim (X \cup Y) &\Leftrightarrow X \cup Y \subseteq \widetilde{x} \\ &\Leftrightarrow X \subseteq X \cup Y \subseteq \widetilde{x} \wedge Y \subseteq X \cup Y \subseteq \widetilde{x} \\ &\Leftrightarrow X \subseteq \widetilde{x} \wedge Y \subseteq \widetilde{x} \\ &\Leftrightarrow x \in \sim X \wedge x \in \sim Y \\ &\Leftrightarrow x \in (\sim X) \cap (\sim Y). \end{aligned}$$

(v) Follows by (ii).

$$\begin{aligned} x \in (\sim X) \cup (\sim Y) &\Leftrightarrow x \in\sim X \vee x \in\sim Y \\ &\Leftrightarrow X \subseteq \widetilde{x} \vee Y \subseteq \widetilde{x} \\ &\Rightarrow X \cap Y \subseteq X \subseteq \widetilde{x} \vee X \cap Y \subseteq Y \subseteq \widetilde{x} \\ &\Rightarrow X \cap Y \subseteq \widetilde{x} \\ &\Leftrightarrow x \in\sim (X \cap Y). \end{aligned}$$

(vi) and (vii). See [7] (22–23 p).

□

The question which naturally arises here is: For which type of subset of a set with apartness do we have equality between its two complements? It turns out that the answer initiated a development of *order theory* for sets and semigroups with apartness.

The complements are used for the classification of subsets of a given set. A subset Y of S is:

- *a detachable* subset in S or, in short, a *d-subset* in S if

$$\forall_{x \in S} \, (x \in Y \vee x \in \neg Y),$$

- *a strongly detachable* subset of S, shortly *an sd-subset of* S, if

$$\forall_{x \in S} \, (x \in Y \vee x \in \sim Y),$$

- *a quasi-detachable* subset of S, shortly *a qd-subset of* S, if

$$\forall_{x \in S} \forall_{y \in Y} \, (x \in Y \vee x \# y).$$

A description of the relationships between those subsets of set with apartness, which, in turn, justifies the *constructive order theory* for sets and semigroups with apartness we develop, is given in the next theorem from our paper [22] (Theorem 3.1).

Theorem 3 *Let Y be a subset of S. Then:*

(i) *Any sd-subset is a qd-subset of S. The converse implication entails* ***LPO***.
(ii) *Any qd-subset Y of S satisfies $\sim Y = \neg Y$.*
(iii) *If any qd-subset is a d-subset, then* ***LPO*** *holds.*
(iv) *If any d-subset is a qd-subset, then* ***MP*** *holds.*
(v) *Any sd-subset is a d-subset of S. The converse implication entails* ***MP***.
(vi) *If any subset of a set with apartness S is a qd-subset, then* ***LPO*** *holds.*

Proof

(i) Let Y be an sd-subset of S. Then, applying the definition and logical axiom we have

$$\forall_{x\in S}\,(x\in Y\vee x\in{\sim}Y)\;\Leftrightarrow\;\forall_{x\in S}\,(x\in Y\;\vee\;\forall_{y\in Y}(x\#y))$$
$$\Rightarrow\;\forall_{x\in S}\,\forall_{y\in Y}\,(x\in Y\vee x\#y).$$

In order to prove the second part of this statement, we consider the real number set $\mathbb{R}$ with the usual (tight) apartness and the subset $Y=\widetilde{0}$. Then, for each real number x and for each $y\in Y$, it follows, from the co-transitivity of #, either $y\#x$ or $x\#0$, that is, either $x\in Y$ or $x\#y$. Consequently, Y is a qd-subset of $\mathbb{R}$. On the other hand, if Y is an sd-subset of $\mathbb{R}$, then for each $x\in\mathbb{R}$, either $x\in Y$ or $x\in\sim Y$. In the former case, $x\#0$, and in the latter, $x=0$; hence, **LPO** holds.

(ii) Let Y be a qd-subset, and let $a\in\neg Y$. By assumption we have

$$\forall_{x\in S}\,\forall_{y\in Y}\,(x\in Y\vee x\#y),$$

so substituting a for x, we get $\forall_{y\in Y}\,(a\in Y\vee a\#y)$, and since, by assumption, $\neg(a\in Y)$, it follows that $a\#y$ for all $y\in Y$. Hence, $a\in\sim Y$. See also [26].

(iii) Let S be the real number set $\mathbb{R}$ with the usual apartness #. As in the proof of (i), consider the qd-subset $\widetilde{0}$ of $\mathbb{R}$. If $\widetilde{0}$ is a d-subset of $\mathbb{R}$, then $x\in\widetilde{0}$ or $\neg(x\in\widetilde{0})$, for all real numbers x. In the latter case $\neg(x\#0)$, which is equivalent to $x=0$. Thus, we obtain the property $\forall_{x\in\mathbb{R}}\,(x\#0\vee x=0)$ which, in turn, is equivalent to **LPO**.

(iv) Consider a real number a with $\neg(a=0)$ and let S be the set $\{0,a\}$ endowed with the usual apartness of $\mathbb{R}$. For $Y=\{0\}$, since $0\in Y$ and $a\in\neg Y$, it follows that Y is a d-subset of S. On the other hand, if Y is a qd-subset of S, then $a\#0$. It follows that for any real number with $\neg(a=0)$, $a\#0$ which entails the Markov principle, **MP**.

(v) The first part follows immediately from (i), (ii), and the definition of d-subsets. The converse follows from (i) and (iv).

(vi) Consider again $\mathbb{R}$ with the usual apartness and define $Y=\{0\}$. If Y is a qd-subset of $\mathbb{R}$, then for all $x\in\mathbb{R}$ we have $x=0$ or $x\#0$; hence, **LPO** holds.

□

If the apartness is not tight, we can find subsets which are not qd-subsets, let alone sd-subsets. To show this, let us consider the set $S=\{a,b,c\}$ with the apartness defined in Example 1 and define $Y=\{a\}$. Then Y is not a qd-subset of S.

Remark 5 If we work with a tight apartness, although vacuously true in classical mathematics, the properties of detachability are not automatically satisfied in **BISH**. The Brouwerian examples from Theorem 3 motivate the use of qd-subsets. Once again, constructive mathematics brings to the light some notions which are invisible to the classical eye (here, the three notions of detachability). ◇

For all subsets Y of the set with apartness S for which two distinguished complements coincide, we will adopt the following notation:

$$\boldsymbol{Y^c} =\sim Y = \neg Y.$$

Lemma 3 *Let X and Y be subsets of a set with apartness S.*

(i) *If X is a qd-subset of S, then*

$$X \subseteq \neg Y \Leftrightarrow X \subseteq\sim Y.$$

(ii) *If X and Y are qd-subsets of S, then $X \cap Y$ is a qd-subset too.*

Proof

(i) Let $X \subseteq \neg Y$. Then, by Lemma 2(i),(iii) and by Theorem 3(ii) we have

$$\begin{aligned} X \subseteq \neg Y &\Leftrightarrow Y \subseteq \neg X =\sim X \\ &\Rightarrow \sim\sim X \subseteq\sim Y. \end{aligned}$$

By Lemma 2(vi) we finally have $X \subseteq\sim\sim X \subseteq\sim Y$.

Conversely, let $X \subseteq\sim Y$. Then, by Lemma 2(iii) and by Theorem 3(ii) it follows

$$X \subseteq\sim Y \Rightarrow\sim\sim Y \subseteq\sim X = \neg X.$$

Now Lemma 2(vi) gives us

$$Y \subseteq\sim\sim Y \subseteq \neg X,$$

which, by Lemma 2(i), implies $X \subseteq \neg Y$.

(ii) Let $s \in S$ and $x \in X \cap Y$. Then, by the assumption, we have $s \in X$ or $s\#x$, and, on the other hand, $s \in Y$ or $s\#x$. It follows that $s\#x$ or else $s \in X \cap Y$, that is, $X \cap Y$ is a qd-subset of S.

□

Remark 6 The condition (i) from Lemma 3 is Proposition 2.2 [14]. ◇

Sd-subsets play an important role in establishing the theory of constructive inverse semigroup with apartness. As evident from A. Cherubini and A. Frigeri's paper [11], the story begins with partial apartness bijections. A partial apartness bijection, or shortly, partial a-bijection $f : S \to T$ on S, is an *sd-partial a-bijection* if $dom\ f$ and $ran\ f$ are sd-subsets of S. We denote by I^S the set of all sd-partial a-bijections on S. Then we have the following important result from [11] for the development of constructive inverse semigroups with apartness.

Theorem 4 *Let $(S, =, \#)$ be a set with apartness. Then $(I^S, =, \#)$ is a set with apartness which equality is defined by*

$$f = g \overset{def}{\Leftrightarrow} dom\ f = dom\ g\ and\ \forall_{x \in dom\ f}(f(x) = g(x)),$$

while apartness is defined by

$$\begin{aligned} f\#g \overset{def}{\Leftrightarrow}\ & \exists_{x \in S}\big(x \in dom\ f \cap dom\ g^c\big) \vee \\ & \exists_{x \in S}\big(x \in dom\ f^c \cap dom\ g\big) \vee \\ & \exists_{x \in S}\big(x \in dom\ f \cap dom\ g \wedge f(x)\#g(x)\big). \end{aligned}$$

Proof Straightforward. □

Remark 7 In general apartness on I^S is neither tight nor decidable as evidenced in Example 1 from [11]. ⋄

3.2 Binary Relations

Let $(S \times S, =, \#)$ be an inhabited set with apartness. Recall an inhabited subset of $S \times S$, or, equivalently, a property applicable to the elements of $S \times S$ is called a *binary relation* on S. There are a few natural relations that may be defined on any set S: *the identity relation on* S or *diagonal* of S

$$\triangle = \{(x, x) : x \in S\},$$

the *co-diagonal* of S

$$\nabla = \{(x, y) \in S \times S : \neg(x = y)\}.$$

Let α be a relation on S. Then

$$(a, b) \bowtie \alpha \ \Leftrightarrow\ \forall_{(x,y) \in \alpha}\ ((a, b)\,\#\,(x, y)),$$

for any $(a, b) \in S \times S$. The apartness complement of α is the relation

$$\sim \alpha \ =\ \{(x, y) \in S \times S : (x, y) \bowtie \alpha\}.$$

In general, we have $\sim \alpha \subseteq \neg\alpha$, which is shown by the following example.

Example 2 Let $S = \{a, b, c\}$ be a set with apartness defined by $\{(a, c), (c, a), (b, c), (c, b)\}$. Let $\alpha = \{(a, c), (c, a)\}$ be a relation on S. Its a-complement of α

$$\sim \alpha = \{(a, a), (b.b), (c, c), (a, b), (b, a)\}$$

is a proper subset of its logical complement $\neg\alpha$. ⋄

Let α and β be relations on S. Then, the following operations can be defined:

- the *composition* or *product* of α and β:

$$\alpha \circ \beta = \{(x, z) \in S \times S : \exists_{y \in S} ((x, y) \in \alpha \wedge (y, z) \in \beta)\}$$

- the *co-composition* or *co-product* of α and β:

$$\alpha \star \beta = \{(x, z) \in S \times S : \forall_{y \in S} ((x, y) \in \alpha \vee (y, z) \in \beta)\}$$

- α is *associated* with β:

$$\alpha \hookleftarrow \beta \overset{\text{def}}{\Leftrightarrow} \forall_{x,y,z \in S} ((x, y) \in \alpha \wedge (y, z) \in \beta \Rightarrow (x, z) \in \alpha).$$

The relation α defined on a set with apartness S is:

- irreflexive if $\neg((x, x) \in \alpha))$
- strongly irreflexive if $(x, y) \in \alpha \Rightarrow x\#y$
- co-transitive if $(x, y) \in \alpha \Rightarrow \forall_{z \in S} ((x, z) \in \alpha \vee (z, y) \in \alpha)$
- *co-antisymmetric* if $\forall_{x,y \in S} (x\#y \Longrightarrow (x, y) \in \alpha \vee (x, y) \in \alpha^{-1})$.

An alternatively way of defining properties given above is:

- *irreflexive* if $\alpha \subset \nabla$
- *strongly irreflexive* if $\alpha \subseteq \#$
- *co-transitive* if $\alpha \subseteq \alpha \star \alpha$
- *co-antisymmetric* if $\# \subseteq (\alpha \cup \alpha^{-1})$.

Remark 8 It is easy to check that a strongly irreflexive relation is also irreflexive. For a tight apartness, the two notions of irreflexivity are classically equivalent but not so constructively. More precisely, if each irreflexive relation was strongly irreflexive, then **MP** would hold. ◇

A relation τ defined on a set with apartness S is a:

- *weak co-quasiorder* if it is irreflexive and co-transitive
- *co-quasiorder* if it is strongly irreflexive and co-transitive.

Remark 9 Even if the two classically (but not constructively) equivalent variants of a co-quasiorder are constructive counterparts of a quasiorder in the case of (a tight) apartness, the stronger variant, co-quasiorder, is, of course, the most appropriate for a constructive development of the theory of semigroups with apartness we develop, which is evident from our already published papers, for example, [22]. The weaker variant, that is, weak co-quasiorder, could be relevant in analysis. ◇

Example 3 Let $S = \{a, b, c\}$ be a set with its diagonal as the equality relation. If we denote by $K = \Delta_S \cup \{(a, b), (b, a)\}$, then we can define the apartness $\#$ on S to be $(S \times S) \setminus K$. Then $(S, =, \#)$ is a set with apartness. The relation $\alpha \subseteq S \times S$, defined by

$$\alpha = \{(c, a), (c, b)\},$$

is a co-quasiorder on S. ⋄

A co-quasiorder has the following important properties, ([22]—Proposition 2).

Proposition 1 *Let τ be a co-quasiorder on S. Then:*

(i) *τ is a qd-subset of $S \times S$.*
(ii) $\sim \tau = \neg \tau = \tau^c$.
(iii) *τ^c is a quasiorder on S.*

Proof

(i) Let $(x, y) \in S \times S$. Then, for all $(a, b) \in \tau$,

$$\begin{aligned} (a, x) \in \tau \vee (x, b) \in \tau &\Rightarrow (a, x) \in \tau \vee (x, y) \in \tau \vee (y, b) \in \tau \\ &\Rightarrow a\#x \vee (x, y) \in \tau \vee y\#b \\ &\Rightarrow (a, b)\#(x, y) \vee (x, y) \in \tau, \end{aligned}$$

that is, τ is a qd-subset.
(ii) It follows from (i) and Theorem 3(ii).
(iii) Let τ be a strongly irreflexive relation on S. For each $a \in S$, it can be easily proved that $(a, a)\#(x, y)$ for all $(x, y) \in \tau$. Thus, τ^c is reflexive.

If $(x, y), (y, z) \in \tau^c$, then, by the definition of $\tau^c = \sim \tau$, we have that $(x, y) \bowtie \tau$ and $(y, z) \bowtie \tau$. For an element $(a, b) \in \tau$, by co-transitivity of τ, we have $(a, x) \in \tau$ or $(x, y) \in \tau$ or $(y, z) \in \tau$ or $(z, b) \in \tau$. Thus, $(a, x) \in \tau$ or $(z, b) \in \tau$, which implies that $a\#x$ or $b\#z$, that is, $(x, z)\#(a, b)$. So, $(x, z) \bowtie \tau$ and $(x, z) \in \sim \tau = \tau^c$. Therefore, τ^c is transitive. □

Remark 10 As in Example 3 the a-complement of a relation can be a proper subset of its logical complement. If the relation in question is a co-quasiorder, then, by the above proposition, two complements coincide. ⋄

Starting from a qd-subset or an sd-subset, we are able to construct co-quasiorders.

Proposition 2 *Let A be a qd-subset of a set with apartness. Then the relation κ_A defined on S by*

$$(x, y) \in \kappa_A \overset{def}{\Leftrightarrow} x\#y \wedge x \in A$$

is a co-quasiorder on S.

Proof *Proof of co-transitivity.* Let $(x, y) \in \kappa_A$. Then:

$$
\begin{aligned}
(x, y) \in \kappa_A &\Leftrightarrow x\#y \wedge x \in A \\
&\Leftrightarrow x\#y \wedge x \in A \wedge x \in A \\
&\Rightarrow \forall_{z \in S}\, (x\#z \vee z\#y) \wedge \forall_{z \in S}\, (x\#z \vee z \in A) \wedge x \in A \\
&\Leftrightarrow \forall_{z \in S}\, (x\#z \vee (z\#y \wedge z \in A)) \wedge x \in A \\
&\Rightarrow (x\#z \vee (z\#y \wedge z \in A)) \wedge x \in A \\
&\Leftrightarrow (x\#z \wedge x \in A) \vee (z\#y \wedge z \in A \wedge x \in A) \\
&\Leftrightarrow (x, y) \in \kappa_A \vee (y, z) \in \kappa_A.
\end{aligned}
$$

So, κ_A is a co-quasiorder. □

Corollary 2 *Let T be an sd-subset of a set with apartness S. Then, the relation τ on S, defined by*

$$(a, b) \in \tau \overset{def}{\Leftrightarrow} a \in T^c \wedge b \in T,$$

is a co-quasiorder on S.

Proof Follows by Theorem 3 and Proposition 2. □

In general, to describe the relation we have to determine which ordered pairs belong to τ, that is, we have to determine $a\tau$ and τa, the left and the right τ-class of each element a from S. Left and right classes of a co-quasiorder given in Example 4 from [22] are sd-subsets of S. In general, we cannot prove that all left and/or right classes of a co-quasiorder are d-subsets or sd-subsets. More precisely, as it is shown in Proposition 3.2 from [22], we have the following important result, a consequence of Theorem 3.

Proposition 3 *Let τ be a co-quasiorder. Then:*

(i) *If $a\tau$ is a d-subset of S for any $a \in S$, then* **LPO** holds.
(ii) If $a\tau$ is an sd-subset of S for any $a \in S$, then **LPO** holds.

Proof

(i) Similar to the proof of Theorem 3(iii). It suffices to let τ be the usual apartness on the real number set and $a = 0$.
(ii) We can use the same example as above and apply Theorem 3(i).

□

The co-quasiorder is one of the main building blocks for the co-order theory of set with (non-tight) apartness that we develop. We can use them to define the following important types of relations:

- *co-equivalence*: a symmetric co-quasiorder,
- *co-order*: a co-antisymmetric co-quasiorder.

3.2.1 Co-equivalences

A quotient structure does not have, in general, a natural apartness relation. For most purposes, we overcome this problem using a co-equivalence instead of an equivalence. Existing properties of a co-equivalence κ guarantee its complement κ^c is an equivalence as well as the quotient set S/κ^c of that equivalence will inherit an apartness.

Theorem 5 *Let κ be a co-equivalence on S. Then*

(i) *The relation κ^c is an equivalence on S such that $\kappa \nleftarrow \kappa^c$.*
(ii) *$(S/\kappa^c, =, \#)$ is a set with apartness where*

$$a\kappa^c = b\kappa^c \Leftrightarrow (a, b) \bowtie \kappa$$

$$a\kappa^c \# b\kappa^c \Leftrightarrow (a, b) \in \kappa.$$

(iii) *The quotient mapping $\pi : S \to S/\kappa^c$, defined by $\pi(x) = x\kappa^c$, is an se-surjection.*

Proof

(i) By the Proposition 1(iii), κ^c is a quasiorder on S. As κ is symmetric, then

$$\begin{aligned}(x, y) \in \sim\kappa &\Leftrightarrow \forall_{(a,b)\in\kappa}\,((x, y)\#(a, b))\\ &\Rightarrow \forall_{(b,a)\in\kappa}\,((x, y)\#(b, a))\\ &\Rightarrow \forall_{(b,a)\in\kappa}\,(x\#b \ \vee \ y\#a)\\ &\Rightarrow \forall_{(a,b)\in\kappa}\,((y, x)\#(a, b))\\ &\Leftrightarrow (y, x) \in \sim\kappa = \kappa^c.\end{aligned}$$

Thus, κ^c is an equivalence.

Let $(x, y) \in \kappa$ and $(y, z) \in \kappa^c$. Thus, $(x, y) \in \kappa$ and $(y, z) \bowtie \kappa$. By the co-transitivity of κ we have $(x, z) \in \kappa$ or $(y, z) \in \kappa$. Thus, $(x, z) \in \kappa$, and $\kappa \nleftarrow \kappa^c$.

(ii) The strong irreflexivity of # is implied by its definition and by the strong irreflexivity of κ.

Let $a\kappa^c \# b\kappa^c$. Then $(a, b) \in \kappa$ implies that $(b, a) \in \kappa$, that is, $b\kappa^c \# a\kappa^c$.

Let $a\kappa^c \# b\kappa^c$ and $u\kappa^c \in S/\kappa^c$. Then $(a, b) \in \kappa$, and, by the co-transitivity of κ, we have $(a, u) \in \kappa$ or $(u, b) \in \kappa$. Finally, we have that $a\kappa^c \# u\kappa^c$ or $u\kappa^c \# b\kappa^c$, so the relation # is co-transitive.

(iii) Let $\pi(x)\#\pi(y)$, i.e., $x\kappa^c \# y\kappa^c$, which by (ii) means that $(x, y) \in \kappa$. Then, by the strong irreflexivity of κ, we have $x\#y$. So π is an se-mapping.

Let $a\kappa^c \in S/\kappa^c$ and $x \in a\kappa^c$. Then $(a, x) \in \kappa^c$, i.e., $a\kappa^c = x\kappa^c$, which implies that $a\kappa^c = x\kappa^c = \pi(x)$. Thus, π is an se-surjection.

□

Any se-mapping between two sets with apartness gives rise to a co-equivalence defined on domain set.

Lemma 4 *Let $f : S \to T$ be an se-mapping between sets with apartness S and T. The relation* coker f *defined on a set S*

$$\operatorname{coker} f \stackrel{def}{=} \{(x, y) \in S \times S : f(x)\# f(y)\}$$

is a co-equivalence on S.

Proof The strong irreflexivity of coker f is easy to prove: if $(x, y) \in \operatorname{coker} f$, then $f(x)\#f(y)$ and therefore $x\#y$. If $(x, y) \in \operatorname{coker} f$, then, by the symmetry of apartness in T, $f(y)\#f(x)$; so $(y, x) \in \operatorname{coker} f$. If $(x, y) \in \operatorname{coker} f$ and $z \in S$, i.e., $f(x)\#f(y)$ and $f(z) \in T$, then either $f(x)\#f(z)$ or $f(z)\#f(y)$; that is, either $(x, z) \in \operatorname{coker} f$ or $(z, y) \in \operatorname{coker} f$. Hence, coker f is a co-equivalence on S. □

The co-equivalence coker f is called the *co-kernel* of se-mapping f. Now, *the first apartness isomorphism theorem* for sets with apartness follows.

Theorem 6 *Let $f : S \to T$ be an se-mapping between sets with apartness. Then:*

(i) *Co-equivalence* coker f *is such that* $\operatorname{coker} f \nleftarrow \ker f$ *and* $\ker f \subseteq (\operatorname{coker} f)^c$.
(ii) *$(S/\ker f, =, \#)$ is a set with apartness, where*

$$a(\ker f) = b(\ker f) \Leftrightarrow (a, b) \in \ker f$$
$$a(\ker f)\#b(\ker f) \Leftrightarrow (a, b) \in \operatorname{coker} f.$$

(iii) *The mapping $\varphi : S/\ker f \to T$, defined by $\varphi(x(\ker f)) = f(x)$, is an a-injective se-injection such that $f = \varphi\pi$.*
(iv) *If f maps S onto T, then φ is an apartness bijection.*

Proof

(i) Let $(x, y) \in \operatorname{coker} f$ and $(y, z) \in$ ker f. Then $f(x)\#f(y)$ and $f(y) = f(z)$. Hence, $f(x)\#f(z)$, that is, $(x, z) \in \operatorname{coker} f$, and $\operatorname{coker} f \nleftarrow \ker f$.
Now let $(x, y) \in \ker f$, so $f(x) = f(y)$. If $(u, v) \in \operatorname{coker} f$, then, by the co-transitivity of coker f, it follows that $(u, x) \in \operatorname{coker} f$ or $(x, y) \in \operatorname{coker} f$ or $(y, v) \in \operatorname{coker} f$. Thus, either $(u, x) \in \operatorname{coker} f$ or $(y, v) \in \operatorname{coker} f$, and, by the strong irreflexivity of coker f, either $u\#x$ or $y\#v$; whence, we have $(x, y)\#(u, v)$. Thus, $(x, y) \bowtie \operatorname{coker} f$, or, equivalently, $(x, y) \in (\operatorname{coker} f)^c$.
(ii) This follows from the definition of # in $S/\ker f$ and (i).
(iii) Let us first prove that φ is well defined. Let $x(\ker f), y(\ker f) \in S/\ker f$ be such that $x(\ker f) = y(\ker f)$, that is, $(x, y) \in \ker f$. Then we have $f(x) = f(y)$, which, by the definition of φ, means that $\varphi(x(\ker f)) = \varphi(y(\ker f))$.
Now let $\varphi(x(\ker f)) = \varphi(y(\ker f))$; then $f(x) = f(y)$. Hence, $(x, y) \in \ker f$, which implies that $x(\ker f) = y(\ker f)$. Thus, φ is an injection.

Next let $\varphi(x(\ker f))\#\varphi(y(\ker f))$; then $f(x)\#f(y)$. Hence, $(x, y) \in \operatorname{coker} f$, which, by (iii), implies that $x(\ker f)\#y(\ker f)$. Thus, φ is an se-mapping.

Let $x(\ker f)\#y(\ker f)$, that is, by (iii), $(x, y) \in \operatorname{coker} f$. So we have $f(x)\#f(y)$, which, by the definition of φ means $\varphi(x(\ker f))\#\varphi(y(\ker f))$. Thus, φ is a-injective.

By the definition of composition of functions, Theorem 5, and the definition of φ, for each $x \in S$, we have

$$(\varphi\pi)(x) = \varphi(\pi(x)) = \varphi(x(\ker f)) = f(x).$$

(iv) Taking into account (iii), we have to prove only that φ is a surjection. Let $y \in T$. Then, as f is onto, there exists $x \in S$, such that $y = f(x)$. On the other hand $\pi(x) = x(\ker f)$. By (iv), we now have

$$y = f(x) = (\varphi\,\pi)(x) = \varphi(\pi(x)) = \varphi(x(\ker f)).$$

Thus, φ is a surjection.

□

4 Semigroups with Apartness

It is only recently that constructive mathematics has paid attention to semigroup theory. The constructive theory of semigroups with an apartness relation is a new approach to semigroup theory and not a new class of semigroups. It presents a semigroup facet of a relatively well-established direction within constructive mathematics which, to the best of our knowledge, has not yet been considered within the classical semigroup community, as well as within the constructive one. Inspired by results obtained from the use of interactive theorem-provers and the general theory of formal verification of constructive proofs (more on that can be found in the work of M. Mitrović et al. [22]), a new constructive algebraic theory, the so-called theory of semigroups with apartness, was developed by M. Mitrović and co-authors: M. N. Hounkonnou, M. A. Baroni, S. Crvenković, E. Darpö, D. A. Romano, and S. Silvestrov. See, for example, [12–14, 22, 23, 25, 26].

4.1 *Basic Concepts: Definitions, Examples, Se-embeddings*

A semigroup with apartness satisfies a number of extra conditions: firstly, the well-known axioms of apartness, and secondly, the semigroup operation has to be strongly extensional.

Given a set with apartness $(S, =, \#)$, the tuple $(S, =, \#, \cdot)$ is a *semigroup with apartness* if the binary operation $\cdot$ is associative

(A) $\quad \forall_{a,b,c \in S}\ [(a \cdot b) \cdot c = a \cdot (b \cdot c)]$

and strongly extensional

(S) $\quad \forall_{a,b,x,y \in S}\ (a \cdot x \# b \cdot y \Rightarrow (a \# b \vee x \# y)).$

As usual, we are going to write ab instead of $a \cdot b$.

An important example of a class of semigroup with apartness arises in the following way. For a given set with apartness X we can construct a semigroup with apartness $\mathcal{T}_X^{se}$ as given in the next theorem.

Theorem 7 *Let X be a set with apartness. Let S be the set of all se-functions from X to X with*

$$f = g \overset{def}{\Leftrightarrow} \forall_{x \in S}\ (f(x) = g(x))$$

and apartness

$$f \# g \overset{def}{\Leftrightarrow} \exists_{x \in A}\ (f(x) \# g(x)).$$

Then $(S, =, \#, \circ)$ is a semigroup with apartness with respect to the binary operation of composition of functions.

Proof By Theorem 2 $(S, =, \#)$ is a set with apartness.

Let $f, g \in S$ and suppose that $(f \circ g)(x) \# (f \circ g)(y)$ for some $x, y \in A$. Then, by the definition of the composition, $f(g(x)) \# f(g(y))$, and, as f is an se-mapping, we have $g(x) \# g(y)$. Finally, as g is an se-mapping as well, we have $x \# y$. Thus, $f \circ g$ is an se-mapping and $f \circ g \in S$. As in the classical case, composition of functions is associative, Lemma 1, so $(S, \circ)$ is a semigroup.

Let $f, g, h, w \in S$ and $f \circ h \# g \circ w$. Then, by the definition of apartness in S, there is an element $x \in A$ such that $(f \circ h)(x) \# (g \circ w)(x)$, i.e., $f(h(x)) \# g(w(x))$. Now we have

$$f(h(x)) \# f(w(x)) \vee f(w(x)) \# g(w(x)),$$

which, further, implies $h(x) \# w(x)$ (because f is an se-mapping) or $f \# g$ (by the definition of the apartness relation on S). Thus, $f \# g \vee h \# w$, that is, composition $\circ$ is an se-operation and $(S, =, \#, \circ)$ is a semigroup with apartness. □

Apartness from Theorem 7 does not have to be tight. The following example shows that we cannot even prove constructively that the apartness on every *finite* semigroup is tight.

Example 4 Let $X = \{a, b, c\}$ with the usual equality relation, that is, the diagonal Δ_X of $X \times X$. Let

$$K = \Delta_X \cup \{(b, c), (c, b)\},$$

and define an apartness # on X to be $(X \times X) \setminus K$. Then, by Theorem 7, $\mathcal{T}_X^{se}$ becomes a semigroup with apartness. Define mappings $f, g : X \to X$ by

$$f(a) = b,\ f(b) = b,\ f(c) = c,$$

$$g(a) = c,\ g(b) = b,\ g(c) = c.$$

In view of our definition of the apartness on X, there is no element $x \in X$ with $f(x)\#g(x)$; so, in particular, f and g are se-functions. However, if $f = g$, then $b = c$, which, by our definition of the equality on S, is not the case. Hence, the apartness on S is not tight. ◇

Let $(S, =_S, \#_S, \cdot)$ and $(T, =_T, \#_T, \cdot)$ be semigroups with apartness. A homomorphism $f\colon S \longrightarrow T$ is an *se-homomorphism* if it is an se-mapping. Several types of se-homomorphism have a specific name. A homomorphism $f : S \to T$ is:

- an *se-embedding* if it is an se-injection,
- an *apartness embedding*, or, shortly, *a-embedding* if it is an a-injective se-embedding,
- an *apartness isomorphism*, or, shortly, *a-isomorphism* if it is an apartness bijection.

As a consequence of Theorem 7, we can formulate *the constructive Cayley's theorem for semigroups with apartness* as follows.

Theorem 8 *Every semigroup with apartness se-embeds into the semigroup of all strongly extensional self-maps on a set.*

Proof Let $(S, =, \#, \cdot)$ be a semigroup with apartness. The semigroup S embeds into the monoid with apartness $S^1 = (S \cup \{1\}, =_1, \#_1, \cdot)$ with equality $=_1$ which consists of all pairs in $=$ and the pair $(1, 1)$ and with apartness $\#_1$ which consists of all pairs in # and the pairs $(a, 1)$, $(1, a)$ for each $a \in S$.

Let f_a be a left translation of S^1, i.e., $f_a(x) = ax$, for all $x \in S^1$. Then f_a is an se-mapping. Indeed, $f_a(x)\#_1 f_a(y)$ is equivalent to $ax\#_1 ay$. The strong extensionality of multiplication implies $x\#_1 y$.

Denote by T the set of all se-mapping from S^1 to S^1. As in **CLASS**, define a mapping $\varphi\colon S^1 \longrightarrow T$ letting

$$\varphi(a) = f_a,$$

for each $a \in S^1$. It is routine to verify that

$$\varphi(ab) = f_{ab} = f_a \circ f_b = \varphi(a)\varphi(b),$$

as well as

$$\varphi(a)\#_T\varphi(b) \Rightarrow a\#_1 b.$$

Thus, φ is an se-homomorphism. Also, $\varphi(a) =_T \varphi(b)$ iff $ax =_1 bx$ for all $x \in S^1$, and, for $x = 1$, we have $a =_1 b$. Therefore, φ is an embedding. □

A number of subsets of a semigroup enjoy special properties relative to the multiplication. A subset T of a semigroup S is:

- *completely isolated* if $ab \in T$ implies $a \in T$ or $b \in T$ for any $a, b \in S$,
- *left* (*right*) *convex* if $ab \in T$ implies $a \in T$ $(b \in T)$ for any $a, b \in S$,
- *convex* if $ab \in T$ implies both $a, b \in T$ for any $a, b \in S$,
- *subsemigroup* if for any $a, b \in T$ we have $ab \in T$,
- *ideal* if for any $a \in T$ and $s \in S$ we have $as, sa \in T$.

A subsemigroup T of S which is convex (resp. completely isolated) as a subset is called a *convex* (resp. *completely isolated*) *subsemigroup*. In an analogous way, we define a complex (completely isolated) ideal of S. We are also going to encounter qd-subsets or qd-subsemigroups which have some of the properties mentioned above: a quasi-detachable convex (respectively quasi-detachable completely isolated) subsemigroup of S is called, in short, an *qd-convex* (respectively qd-completely isolated) subsemigroup of S. Similarly, there are qd-convex and qd-completely isolated ideals of S.

Furthermore, a subset T of a semigroup S with apartness is:

- *co-subsemigroup* if it is a completely isolated qd-subset,
  - *strong co-subsemigroup* if it is a completely isolated sd-subset,
- *left* (*right*) *co-ideal* of S if it is left (right) convex qd-subset,
- *co-ideal* if it is a convex qd-subset,
  - *strong co-ideal* if it is a convex sd-subset.

Lemma 5 *Let S be a semigroup with apartness. The following conditions are true:*

(i) *If T is a co-subsemigroup of S, then, if inhabited, T^c is a convex subsemigroup of S.*
(ii) *Let I be a strong co-ideal of a semigroup with apartness S. If I^c is inhabited, then it is an ideal of S.*

Proof

(i) Let $xy \in T^c$ for some $x, y \in S$. Then $\neg(xy \in T)$, which, by the assumption and logic law, gives

$$\neg(x \in T \vee y \in T) \Leftrightarrow \neg(x \in T) \wedge \neg(y \in T),$$

that is $x, y \in T^c$. Thus, T^c is a convex subset. Let $x, y \in T^c$ be such that $xy \in T$. Then, by the assumption, we have $x \in T$ or $y \in T$, which is a contradiction. So, $\neg(xy \in T)$ which implies $xy \in T^c$.

(ii) Let $s \in S$ and $x \in I^c$ be such that $xs \in I$. Then, by the assumption we have $x, s \in I$, which is a contradiction. So we have $\neg(xs \in I)$, that is $xs \in I^c$. In a similar manner we can prove $sx \in I^c$. Thus, I^c is an ideal of S.

□

4.1.1 Co-congruences on a Semigroup with Apartness

A relation τ defined on a semigroup S with apartness is called:

- *left co-compatible*: $(zx, zy) \in \tau \Rightarrow (x, y) \in \tau$
- *right co-compatible*: $(xz, yz) \in \tau \Rightarrow (x, y) \in \tau$
- *co-compatible*: $(xz, yt) \in \tau \Rightarrow (x, y) \in \tau \vee (z, t) \in \tau$

for any $x, y, z, t \in S$.

The lemma which follows will be used without reference.

Lemma 6 *Let τ be a co-quasiorder on a semigroup with apartness S. Then, τ is co-compatible if and only if τ is a left and a right co-compatible.*

Proof Let τ be a co-compatible co-quasiorder on S and let $x, y, z \in S$. Then $(zx, zy) \in \tau$ implies $(x, y) \in \tau$ or $(z, z) \in \tau$. The latter is impossible because of strong irreflexivity of τ. Thus, $(x, y) \in \tau$, i.e., τ is left co-compatible.

Conversely, let τ be a left and a right co-compatible co-quasiorder on S. Let $x, y, z, t \in S$ be such that $(xz, yt) \in \tau$. By the co-transitivity of τ, it follows either $(xz, yz) \in \tau$ or $(yz, yt) \in \tau$. Now, by the assumption, we have $(x, y) \in \tau$ or $(z, t) \in \tau$, as required. □

A co-equivalence κ is a *co-congruence* if it is co-compatible.

Theorem 9 *Let S be a semigroup with apartness, and let κ be a co-congruence on S. Define*

$$a\kappa^c = b\kappa^c \Leftrightarrow (a, b) \bowtie \kappa,$$

$$a\kappa^c \# b\kappa^c \Leftrightarrow (a, b) \in \kappa,$$

$$a\kappa^c\, b\kappa^c = (ab)\kappa^c.$$

Then $(S/\kappa^c, =, \#, \cdot)$ is a semigroup with apartness. Moreover, the quotient mapping $\pi : S \to S/\kappa^c$, defined by $\pi(x) = x\kappa^c$, is an se-epimorphism.

Proof By Theorem 5, $(S/\kappa^c, =, \neq)$ is a set with apartness. The associativity of multiplication in S/κ^c follows from that of multiplication on S.

Let $a\kappa^c\, x\kappa^c \# b\kappa^c\, y\kappa^c$. Then $(ax)\kappa^c \# (by)\kappa^c$. By Theorem 5, we have that $(ax, by) \in \kappa$. But κ is a co-congruence, so either $(a, b) \in \kappa$ or $(x, y) \in \kappa$. Thus, by the definition of # in S/κ^c, either $a\kappa^c \# b\kappa^c$ or $x\kappa^c \# y\kappa^c$. So $(S/\kappa^c, =, \#, \cdot)$ is a semigroup with apartness. Using that fact and the definition of π, we have

$$\pi(xy) = (xy)\kappa^c = x\kappa^c\, y\kappa^c = \pi(x)\pi(y).$$

Hence, π is a homomorphism, and, by Theorem 5, π is an se-surjection. □

An important example of a co-congruence defined on a semigroup S is connected to an se-homomorphism between semigroups.

Lemma 7 *Let $f : S \to T$ be an se-homomorphism between semigroups with apartness. Then* coker f *is a co-congruence on* S.

Proof Taking into account Lemma 4 it is enough to prove that coker f is co-compatible with multiplication in S. Let $(ax, by) \in$ coker f, i.e., $f(ax) \neq f(by)$. Since f is a homomorphism, we have $f(a)f(x) \neq f(b)f(y)$. The strong extensionality of multiplication implies that either $f(a) \neq f(b)$ or $f(x) \neq f(y)$. Thus, either $(a, b) \in$ coker f or $(x, y) \in$ coker f, and therefore, coker f is a co-congruence on S. □

The *first apartness isomorphism theorem for semigroups with apartness* follows.

Theorem 10 *Let $f : S \to T$ be an se-homomorphism between semigroups with apartness. Then:*

(i) *Co-congruence* coker f *is such that* coker $f \nleftarrow\!\!\!\!\text{P}$ ker f *and* ker $f \subseteq (\text{coker } f)^c$.
(ii) $(S/\ker f, =, \#; \cdot)$ *is a semigroup with apartness, where*

$$a(\ker f) = b(\ker f) \Leftrightarrow (a, b) \in \ker f$$
$$a(\ker f)\#b(\ker f) \Leftrightarrow (a, b) \in \text{coker } f$$
$$a(\ker f)b(\ker f) = (ab)(\ker f).$$

(iii) *The mapping $\varphi : S/\ker f \to T$, defined by $\varphi(x(\ker f)) = f(x)$, is an apartness embedding such that $f = \varphi\pi$.*
(iv) *If f maps S onto T, then φ is an apartness isomorphism.*

Proof

(i) Follows immediately by Theorem 6(ii).
(ii) This follows by Theorems 6 and 9.
(iii) Using (ii) and the assumption that f is a homomorphism, we have

$$\begin{aligned}\varphi(x(ker\ f)\ y(ker\ f)) &= \varphi((xy)(ker\ f))\\ &= f(xy)\\ &= f(x)f(y)\\ &= \varphi(x(ker\ f))\,\varphi(y(ker\ f)).\end{aligned}$$

Now, by Theorem 6, φ is an apartness embedding.
(iv) This follows by Theorem 6 and (iii).

□

4.2 *Semigroups with Apartness with the Inhabited Set of Idempotents*

An element e of a semigroup S is *idempotent* if $e^2 = e$, and the set of all idempotents of a semigroup S is denoted by $E(S)$. An element a of a semigroup S is *apartness idempotent*, or shortly *a-idempotent* if $a^2\#a$, and the set of all a-idempotent is denoted by $A(S)$. An example of a semigroup with inhabited subsets of idempotents $E(S)$ and a-idempotents $A(S)$ follows.

Example 5 Let S be a semigroup given by

·	a	b	c	d	e
a	b	b	d	d	d
b	b	b	d	d	d
c	d	d	c	d	c
d	d	d	d	d	d
e	d	d	c	d	c

Let the equality on S be the diagonal $\triangle_S = \{(a, a), (b, b), (c, c), (d, d), (e, e)\}$. If we denote by $K = \triangle_S \cup \{(a, b), (b, a)\}$, then we can define an apartness # on S by $(S \times S) \setminus K$. It is easy to see that $E(S) = \{b, c, d\} \subset S$. On the other hand, $e^2 = c$, and, by the definition of apartness on S, $e^2 = c\#e$, so $A(S) = \{e\}$ is an inhabited set. $\diamond$

Some properties of sets $E(S)$ and $A(S)$ as well as their connections follow.

Proposition 4 *Let S be a semigroup with apartness and let* $e \in E(S)$. *The following conditions are true:*

(i) $\widetilde{e}$ *is a co-subsemigroup.*
(ii) $A(S)$ *is a qd-subset of S.*
(iii) $A(S) \subseteq \sim E(S) \subseteq \widetilde{e}$.
(iv) $E(S) \subseteq A(S)^c$.

Proof

(i) Let $x, y \in S$ such that $x \in \widetilde{e}$, i.e., $x\#e$. By the co-transitivity we have that $x\#y$ or $y\#e$, so we have $x\#y$ or $y \in \widetilde{e}$. Therefore, $\widetilde{e}$ is a qd-subset.
Let $x, y \in S$ and $xy \in \widetilde{e}$, i.e., $xy\#e$. Then, by the strong extensionality of multiplication, we have $x\#e$ or $y\#e$, that is, $x \in \widetilde{e}$ or $y \in \widetilde{e}$, and $\widetilde{e}$ is a completely isolated subset of S. Therefore, $\widetilde{e}$ is a co-subsemigroup.
(ii) Let $s \in S$ and $a \in A(S)$. Then $a^2\#a$, and, by co-transitivity, we have $a^2\#s^2$ or $s^2\#s$ or $s\#a$, which imply $a\#s$ or $s^2\#s$. Thus, $s \in A(S)$ or $a\#s$, so $A(S)$ is a qd-subset.
(iii) It is obvious that $A(S) \cap E(S) = \emptyset$. By Lemma 2(i) we have $A(S) \subseteq \neg E(S)$, which, further, by (ii) and Lemma 3(i), gives $A(S) \subseteq \sim E(S)$. Let $e \in E(S)$, i.e., $\{e\} \subseteq E(S)$. Then, by Lemma 2(iii), $\sim E(S) \subseteq \widetilde{e}$.

(iv) By (iii) we have $A(S) \subseteq\sim E(S)$, which, by Lemma 2(iii) and Theorem 3(ii), implies $\sim\sim E(S) \subseteq\sim A(S) = \neg A(S) = A(S)^c$. Finally, by Lemma 2(vi), $E(S) \subseteq A(S)^c$ follows.

□

If $f : S \to T$ an se-homomorphism between semigroup S with inhabited set of idempotents and a semigroup T, then a semigroup T is also with inhabited set of idempotents.

Lemma 8 *Let S be a semigroup with inhabited set of idempotents and T a semigroup. Let $f : S \to T$ be an se-homomorphism. If $e \in E(S)$, then:*

(i) $f(e) \in E(T)$.
(ii) *If f is an injection, then $f(\widetilde{e}) \subseteq \widetilde{f(e)}$.*

Proof

(i) $f(e) = f(ee) = f(e)f(e) = f(e)^2$, so $f(e) \in E(T)$.
(ii) Let $y \in f(\widetilde{e})$, i.e., $y = f(x)$ for some $x \in \widetilde{e}$. Then $x\#e$ and assumption imply $f(x)\#f(e)$. Thus, $y = f(x) \in \widetilde{f(e)}$.

□

If $f : S \to T$ an se-homomorphism between semigroup S with inhabited set of a-idempotents and a semigroup T, then we have a slightly different situation than that of Lemma 8.

Lemma 9 *Let S be a semigroup with apartness with inhabited set of a-idempotents $A(S)$. Let $f : S \to T$ be an se-homomorphism, then:*

(i) $\forall_{a\in S}\, (f(a) \in A(T) \Rightarrow a \in A(S))$.
(ii) *If f is a-injective, then $a \in A(S)$ implies $f(a) \in A(T)$.*

Proof

(i) Let $f(a) \in A(T)$, for some $a \in S$. Then $f(a)^2\#f(a)$ which implies $a^2\#a$. Thus, $a \in A(S)$.
(ii) Let $a \in A(S)$. Then $a^2\#a$, which, by a-injectivity of f, implies $f(a)^2\#f(a)$. Thus, $f(a) \in A(T)$.

□

4.2.1 Co-congruences on a Semigroup with Idempotents

As it is shown in Proposition 3, we cannot prove, in general, that left and/or right classes of a co-quasiorder τ defined on S are d-subsets or qd-subsets. However, any τ-class defined by an idempotent is a qd-subset; even more, it is a co-subsemigroup.

Lemma 10 *Let S be a semigroup with apartness with inhabited set of idempotents $E(S)$. Let $e \in E(S)$. If τ is a co-compatible co-quasiorder on S, then $e\tau$ and τe are co-subsemigroups.*

Proof Let $x, y \in S$ such that $y \in e\tau$. Then $(e, y) \in \tau$, which, by co-transitivity of τ, gives $(e, x) \in \tau$ or $(x, y) \in \tau$. So $x \in e\tau$ or $x \# y$. Thus, $e\tau$ is a qd-subset of S.

Let $x, y \in S$ such that $xy \in e\tau$. Then $(e, xy) \in \tau$, which, by co-compatibility of τ, gives $(e, x) \in \tau$ or $(e, y) \in \tau$. So $x \in e\tau$ or $y \in e\tau$, and $e\tau$ is a completely isolated subset. Therefore, $e\tau$ is a co-subsemigroup. □

For the co-congruence $\operatorname{coker} f$ and its class defined by an idempotent we have the following result.

Proposition 5 *Let $f : S \to T$ be an se-homomorphism between semigroups with apartness. Let $E(S)$ be an inhabited set. If $e \in E(S)$, then:*

(i) *$e(\operatorname{coker} f)$ is a co-subsemigroup.*
(ii) *$e(\operatorname{coker} f) \subseteq \widetilde{e}$.*
(iii) *If f is a-injective, then $e(\operatorname{coker} f) = \widetilde{e}$.*

Proof

(i) Follows by Lemma 7 and Lemma 10.
(ii) Let $x \in e(\operatorname{coker} f)$, i.e., $(e, x) \in \operatorname{coker} f$. By the strong irreflexivity of $\operatorname{coker} f$, $e \# x$ follows, i.e., $x \in \widetilde{e}$.
(iii) Let $x \in \widetilde{e}$. Then $e \# x$ which, by a-injectivity of f, implies $f(e) \# f(x)$, that is $x \in e(\operatorname{coker} f)$.

□

As a consequence of Proposition 5 and Lemma 8 we have the next corollary.

Corollary 3 *Let S be a semigroup with inhabited set of idempotents and T a semigroup. Let $f : S \to T$ be an se-homomorphism. If $e \in E(S)$, then:*

(i) *If f is a surjection, then $\widetilde{f(e)} \subseteq \underset{\sim}{f}(\widetilde{e})$.*
(ii) *If f is a bijection, then $f(\widetilde{e}) = \widetilde{f(e)}$.*

Proof

(i) Let $y \in \widetilde{f(e)}$. By the assumption there exists $x \in S$ such that $y = f(x)$ and $y = f(x) \# f(e)$. Thus, $x \# e$ which implies, by Proposition 5(ii), $x \in e(\operatorname{coker} f) \subseteq \widetilde{e}$. Hence, $y = f(x) \in f(\widetilde{e})$.
(ii) Follows by Lemma 8(ii) and (i).

□

We can prove a more general result.

Proposition 6 *Let κ be a co-congruence on a semigroup S with inhabited set of idempotents. Let $e \in E(S)$. Then:*

(i) *$e\kappa \subseteq \widetilde{e}$.*
(ii) *$e\kappa$ is a co-subsemigroup of S.*

Proof

(i) Let $x \in e\kappa$. Then $(x, e) \in \kappa$, which, by the strong irreflexivity of κ, implies $x \# e$. So $x \in \widetilde{e}$.

(ii) Let $x, y \in S$ be such that $xy \in e\kappa$. Then $(xy, e) \in \kappa$ which, by the co-compatibility of κ gives $(x, e) \in \kappa$ or $(y, e) \in \kappa$. Thus, $x \in e\kappa$ or $y \in e\kappa$, i.e., $e\kappa$ is a completely isolated subset of S.

Let $x, y \in S$ be such that $x \in e\kappa$. Then $(x, e) \in \kappa$ which, by the co-cotransitivity of κ, implies $(x, y) \in \kappa$ or $(y, e) \in \kappa$, and we have $x\#y$ or $y \in e\kappa$. Thus, $e\kappa$ is a qd-subset in S. So, $e\kappa$ is a co-subsemigroup of S.

□

4.2.2 Apartness Idempotent-separating Co-congruences on a Semigroup with Idempotents

A relation α defined on a semigroup with apartness S with inhabited set of idempotents is called *apartness idempotent-separating* or, shortly, *a-idempotent-separating* if

$$\forall_{e, f \in E(S)} (e\#f \Rightarrow (e, f) \in \alpha).$$

Lemma 11 *Let S be a semigroup with apartness with inhabited set of idempotents, and let $f : S \to T$ be a-injective homomorphism between semigroups S and T. The co-congruence* coker f *is an a-idempotent separating.*

Proof Let $g, h \in E(S)$ be such that $g\#h$. Then, by the assumption, $f(g)\#f(h)$, that is $(g, h) \in \text{coker } f$. □

5 Inverse Semigroups with Apartness

Based on the results presented in [12, 13], A. Cherubini and A. Frigeri in 2019 made important steps forward introducing constructive inverse semigroups with apartness [11]. Later, M. Mitrović and D. A. Romano contributed to the area too [24, 28].

5.1 Basic Notions and Examples

A semigroup with apartness $(S, =, \#; \cdot)$ is said to be *inverse* if:

(IS$_1$) for each $x \in S$ there exists a unique element, denoted by x^{-1}, such that the following two equations are satisfied: $x = xx^{-1}x$ and $x^{-1} = x^{-1}xx^{-1}$

(IS$_2$) $\forall_{x,y \in S} (x^{-1}\#y^{-1} \Rightarrow x\#y)$.

Inverse semigroups with apartness form a proper subclass of a class of semigroups with apartness. Semigroup S given in Example 5 is neither inverse nor regular. We have $aba = b$, $aca = ada = aea = d$, so $a \in S$ is not regular element.

Till the end of this chapter inverse semigroups will mean inverse semigroup with apartness.

In what follows we are going to list several basic results inherited from the classical inverse semigroup theory. Their proofs are analogous to the classical case—see, for example, [27] or [17].

Lemma 12 *Let S be an inverse semigroup and let* $x \in S$. *Then* $(x^{-1})^{-1} = x$.

Proof Relays on the fact that if x^{-1} is the inverse of x, then x is the unique inverse of x^{-1}. □

Starting from $x\#y$, by the previous lemma, we have $(x^{-1})^{-1} = x$ and $(y^{-1})^{-1} = y$. By extensionality $(x^{-1})^{-1}\#(y^{-1})^{-1}$ follows. Applying (IS_2) we obtain $x^{-1}\#y^{-1}$. Then, in the definition of an inverse semigroup with apartness, condition (IS_2) can be replaced by

(IS_3) $\forall_{x,y \in S}\ (x\#y \Leftrightarrow x^{-1}\#y^{-1})$.

Idempotents, like in classical inverse semigroup theory, play an important role.

Lemma 13 *Let S be an inverse semigroup with apartness and let* $e \in E(S)$ *Then* $e^{-1} = e$.

Proposition 7 *Idempotents commute in an inverse semigroup with apartness.*

Lemma 14 *Let S be an inverse semigroup with apartness. Then:*

(i) $\forall_{x \in S}\ (xx^{-1}, x^{-1}x \in E(S))$.
(ii) $\forall_{x \in S} \forall_{e \in E(S)}\ (xex^{-1}, x^{-1}ex \in E(S))$.

As a consequence of Proposition 7 and Lemma 14 we have the next result.

Lemma 15 *Let S be an inverse semigroup with apartness. Then* $(xy)^{-1} = y^{-1}x^{-1}$, *for any* $x, y \in S$.

Although the idempotents in an inverse semigroup are not in general central "we do have the following result, which tells us how idempotents 'pass through' non-idempotent elements". (See M. V. Lawson's chapter in this book.)

Lemma 16 *Let S be an inverse semigroup with apartness. Let* $s \in S$ *and* $e \in E(S)$. *Then:*

(i) $es = sf$ *for some* $f \in E(S)$.
(ii) $se = gs$ *for some* $g \in E(S)$.

A subset T of an inverse semigroup with apartness is an *inverse subset* if

$$\forall_{x \in S}\ (x^{-1} \in T \Rightarrow x \in T).$$

Some important examples of inverse subsets of an inverse semigroup with apartness follow.

Proposition 8 *Let S be an inverse semigroup with apartness. The following conditions are true:*

(i) *$A(S)$ is an inverse qd-subset of S.*
(ii) *$\sim E(S)$ is an inverse subset of S.*
(iii) *$\widetilde{e}$ is an inverse co-subsemigroup of S.*
(iv) *Any co-ideal K of S is an inverse subset of S.*

Proof

(i) By Proposition 4(ii) $A(S)$ is a qd-subset. Let $x^{-1} \in A(S)$. Then $(x^{-1})^2 \# x^{-1}$. Hence, $(x^{-1})^2 = x^{-1}xx^{-1}x^{-1}xx^{-1} = (x^{-1})^2x^2(x^{-1})^2$. On the other hand, $x^2 = x(x^{-1}x)(xx^{-1})x = x(xx^{-1})(x^{-1}x)x = x^2(x^{-1})^2x^2$. So $(x^2)^{-1} = (x^{-1})^2$ is the unique inverse for x^2. Now we have $(x^2)^{-1} \# x^{-1}$, which implies $x^2 \# x$, that is, $x \in A(S)$.
(ii) Let $x^{-1} \in \sim E(S)$. Then $x^{-1} \# e$, for any $e \in E(S)$. Then $x^{-1} \# e = e^{-1}$ which implies $x \# e$. So, $x \in \sim E(S)$ and $E(S)$ is an inverse subset.
(iii) By Proposition 4(i) $\widetilde{e}$ is a co-subsemigroup of S, for any $e \in E(S)$. Let $x \in S$ be such that $x^{-1} \in \widetilde{e}$, i.e., $x^{-1} \# e = e^{-1}$. So $x \# e$, which completes the proof.
(iv) Follows immediately by convexity of K.

□

An se-homomorphism $f : S \to T$ between inverse semigroups with apartness is an *I-se-homomorphism* if

$$f(x)^{-1} \# f(y)^{-1} \Rightarrow x^{-1} \# y^{-1},$$

for any $a, y \in S$.

We are going to use next result without special announcement.

Lemma 17 *Let $f : S \to T$ be an I-homomorphism between inverse semigroups with apartness. Then:*

(i) $\forall_{x \in S}\ (f(x)^{-1} = f(x^{-1}))$.
(ii) $\forall_{x,y \in S}\ (f(x)^{-1} \# f(y)^{-1} \Leftrightarrow f(x) \# f(y))$.

Proof

(i) Straightforward.
(ii) Based on (i) and strong extensionality of the operation of inversion in a semigroup T.

□

5.1.1 Symmetric Inverse Semigroup with Apartness on S

As a consequence of Theorem 4, an important example of an inverse semigroup with apartness follows.

Theorem 11 *Let $(S, =, \#)$ be a set with apartness. Let $(I^S, =, \#)$ be a set with apartness defined as in Theorem 4. Then $(I^S, =, \#; \circ)$ is an inverse semigroup with apartness where $\circ$ is the composition of two sd-partial a-bijections from I^S and with sd-partial a-bijection f^{-1} as the unique solution of the equations $f = f \circ g \circ f$ and $g = g \circ f \circ g$ for each $f \in I^S$.*

Proof See Theorem 2 and its proof from [11]. □

The semigroup $(I^S, =, \#, \circ)$ is called the *symmetric inverse semigroup with apartness on S*. Recall a subset T of a set (or semigroup) with apartness is called *decidable* if $x = y$ or $x\#y$ for each $x, y \in T$. *The constructive Wagner-Preston theorem* for inverse semigroups with apartness follows.

Theorem 12 *Let $(S, =, \#, \cdot)$ be an inverse semigroup with apartness with $E(S)$ decidable subset. Then S is se-embeddable in the symmetric inverse semigroup with apartness I^S on the set $(S, =, \#)$.*

Proof See Theorem 3 from [11] □

Theorem 12 is stated under the assumption that $E(S)$ is a decidable subset of S. In general, contrary to classical case, an inverse semigroup with apartness cannot be embedded into the symmetric inverse semigroup with apartness. Example 2 from [11] is an evidence of that. Furthermore, *the constructive Wagner-Preston theorem* is one more example that constructive version of a classical theorem can be obtained by strengthening the conditions and/or weakening the conclusion of the theorem.

5.2 Co-congruences on Inverse Semigroup with Apartness

We start this part with an example of a co-congruence defined on an inverse semigroup with apartness.

Theorem 13 *Let S be an semigroup with apartness and κ a relation defined on it by*

$$(a, b) \in \kappa \overset{def}{\Leftrightarrow} \exists_{x \in S}\, (a^{-1}xa\#b^{-1}xb).$$

Then κ is a co-congruence on S.

Proof *Strong irreflexivity.* Let $(a, b) \in \kappa$, $a, b \in S$. Then there exists $x \in S$ such that $a^{-1}xa\#b^{-1}xb$. By the definition of a semigroup with apartness we have $a^{-1}\#b^{-1}$ or $x\#x$ or $a\#b$. Thus, $a^{-1}\#b^{-1}$ or $a\#b$, so we have $a\#b$.

Symmetry. Obvious.

Co-transitivity. It follows by the co-transitivity of apartness. Namely, if $a, b \in S$, then there exists $x \in S$ such that $a^{-1}xa\#b^{-1}xb$. Now, by the co-transitivity of # we have $a^{-1}xa\#c^{-1}xc$ or $c^{-1}xc\#b^{-1}xb$ for any $c \in S$. Hence, $(a, c) \in \kappa$ or $(c, b) \in \kappa$. So, κ is a co-equivalence.

Co-compatibility. Let $a, b, u \in S$ be such that $(au, bu) \in \kappa$. Then there exists $x \in S$ such that $(au)^{-1}x(au)\#(bu)^{-1}x(bu)$. So we have $u^{-1}a^{-1}xau\#u^{-1}b^{-1}xbu$, which implies $a^{-1}xa\#b^{-1}xb$. Hence, $(a, b) \in \kappa$, and we prove the right co-compatibility of κ. In a similar manner we can also prove the left co-compatibility. Therefore, κ is a co-congruence on S. □

5.2.1 I-co-congruences

In the case of inverse semigroups with apartness, we have to add the co-compatibility with respect to the inversion. A co-congruence κ defined on an inverse semigroup with apartness S is an *I-co-congruence* if

$$\forall_{x,y\in S} \left((a^{-1}, b^{-1}) \in \kappa \Rightarrow (a, b) \in \kappa\right).$$

An important example of an I-co-congruence follows.

Lemma 18 *Let $f : S \to T$ be an se-homomorphism between inverse semigroups with apartness. Then* coker f *is an I-co-congruence on S.*

Proof By Lemma 7, coker f is a co-congruence on S. Let $(x^{-1}, y^{-1}) \in \operatorname{coker} f$. Then $f(x^{-1})\#f(y^{-1})$, which, by Lemma 17, implies $f(x)\#f(y)$. So $(x, y) \in \operatorname{coker} f$. □

Let us list some basic properties of I-co-congruences.

Proposition 9 *Let κ be a I-co-congruence on an inverse semigroup with apartness S; then:*

(i) $\forall_{a,b\in S} \left((a^{-1}, b^{-1}) \in \kappa \Leftrightarrow (a, b) \in \kappa\right)$.
(ii) $\forall_{a,b\in S} \left((a^{-1}, b^{-1}) \in \kappa^c \Leftrightarrow (a, b) \in \kappa^c\right)$.
(iii) $\forall_{a,b\in S} \left((a, b) \in \kappa^c \Rightarrow (aa^{-1}, bb^{-1}), (a^{-1}a, b^{-1}b), (ab, ba) \in \kappa^c\right)$.
(iv) $\forall_{a\in S}\forall_{e,f\in E(S)} \left((a^{-1}ea, a^{-1}fa) \in \kappa \Rightarrow (e, f) \in \kappa\right)$.

Proof

(i) Let $(a, b) \in \kappa$. That is, we have $((a^{-1})^{-1}, (b^{-1})^{-1}) \in \kappa$, which, by the assumption, implies $(a^{-1}, b^{-1}) \in \kappa$.
(ii) Let $(a, b) \in \kappa^c$ and $(x, y) \in \kappa$. By (i) it follows that $(x^{-1}, y^{-1}) \in \kappa$. Thus, $a\#x^{-1}$ or $b\#y^{-1}$, i.e., we have $(a^{-1})^{-1}\#x^{-1}$ or $(b^{-1})^{-1}\#y^{-1}$, which, further, implies $a^{-1}\#x$ or $b^{-1}\#y$. Therefore, $(a^{-1}, b^{-1})\#(x, y)$, i.e., $(a^{-1}, b^{-1}) \in \kappa^c$. Converse implication follows in the same way.
(iii) Let $x, y \in S$ be such that $(x, y) \in \kappa$. Then, by the co-transitivity of κ, $(x, ab) \in \kappa$ or $(ab, ba) \in \kappa$ or $(ba, y) \in \kappa$. If $(ab, ba) \in \kappa$, then, by the co-compatibility, $(a, b) \in \kappa$ follows, which is contradiction. Thus, $(x, ab) \in \kappa$ or $(ba, y) \in \kappa$, and by the strong irreflexivity of κ, it follows that $x\#ab$ or $y\#ba$, which further implies $(x, y)\#(ab, ba)$. Therefore, $(ab, ba) \in \kappa^c$.
(iv) Follows immediately by co-compatibility of κ.

□

Proposition 10 *Let κ be an I-co-congruence on an inverse semigroup S with apartness, and let $e \in E(S)$. Then $e\kappa$ is an inverse co-subsemigroup of S.*

Proof By Proposition 6(ii), $e\kappa$ is a co-subsemigroup of S. Let $x \in S$ be such that $x^{-1} \in e\kappa$. Then $(x^{-1}, e) \in \kappa$, which, by the definition of I-co-congruence, implies $(x, e) \in \kappa$, i.e., $x \in e\kappa$. □

Recall in Sect. 4 we define an a-idempotent separating relation defined on an semigroup with inhabited set with idempotents. The next theorem gives a way to establish an a-idempotent-separating I-co-congruence on an inverse semigroup with apartness.

Theorem 14 *Let S be an inverse semigroup. Relation η defined on S by*

$$(a, b) \in \eta \overset{def}{\Leftrightarrow} \exists_{e,f \in E(S)} \, (eaf\#ebf)$$

is an a-idempotent-separating I-co-congruence.

Proof *Strong irreflexivity*. It follows immediately from definition of η and strong extensionality of multiplication.

Symmetry. It is obvious.

Co-transitivity. Let $a, b, c \in S$ be such that $(a, c) \in \eta$. Then $eaf\#ecf$, which, by the co-transitivity of apartness, gives $eab\#ebf$ or $ebf\#ecf$. Thus, $(a, b) \in \eta$ or $(b, c) \in \eta$, i.e., η is co-transitive.

Co-compatibility. Let $a, b, u \in S$ be such that $(au, bu) \in \eta$. Then $eauf\#ebuf$, for some $e, f \in E(S)$. By Lemma 16, there exists $g \in E(S)$ such that $uf = gu$. Then $eauf = eagu$ and $ebuf = ebgu$, which imply $eauf = eagu\#ebuf = ebgu$. Hence, $eag\#ebg$, i.e., $(a, b) \in \eta$. In a similar manner it can be proved that from $(ua, ub) \in \eta$ it follows that $(a, b) \in \eta$. Therefore, η is a co-congruence.

I-co-congruence. Let $a, b \in S$ be such that $(a^{-1}, b^{-1}) \in \eta$. Then

$$ea^{-1}f\#eb^{-1}f \Leftrightarrow (fae)^{-1}\#(fbe)^{-1} \Leftrightarrow fae\#fbe \Leftrightarrow (a, b) \in \eta.$$

A-idempotent-separating. Let $e\#f$, $e, f \in E(S)$. By the co-transitivity of apartness it follows $e\#ef$ or $ef\#f$. If $e\#ef$, then $e = eee\#ef = eef = efe$, so $(e, f) \in \eta$. If $ef\#f$, then $f = fff\#eff = fef$ so $(e, f) \in \eta$. □

5.2.2 Apartness Isomorphism Theorems for Inverse Semigroups with Apartness

The next two theorems are consequences of the ones given in Sect. 4.

Theorem 15 *Let S be an inverse semigroup with apartness, and let κ be an I-co-congruence on S. Define*

$$a\kappa^c = b\kappa^c \Leftrightarrow (a, b) \bowtie \kappa,$$

$$a\kappa^c \# b\kappa^c \Leftrightarrow (a,b) \in \kappa,$$
$$a\kappa^c\, b\kappa^c = (ab)\kappa^c,$$
$$(a\kappa^c)^{-1} = (a^{-1})\kappa^c.$$

Then $(S/\kappa^c, =, \#, \cdot)$ *is an inverse semigroup with apartness. Moreover, the quotient mapping* $\pi : S \to S/\kappa^c$*, defined by* $\pi(x) = x\kappa^c$*, is an se-epimorphism.*

Proof By Theorem 9, $(S/\kappa^c, =, \#, \cdot)$ is a semigroup with apartness and the quotient mapping $\pi : S \to S/\kappa^c$, defined by $\pi(x) = x\kappa^c$, is an se-epimorphism.

Let $a\kappa^c \in S/\kappa^c, a \in S$. Then $a\kappa^c = (aa^{-1}a)\kappa^c = a\kappa^c(a^{-1})\kappa^c a\kappa^c$. On the other hand, $(a^{-1})\kappa^c = (a^{-1}aa^{-1})\kappa^c = (a^{-1})\kappa^c a\kappa^c(a^{-1})\kappa^c$. Thus, $(a^{-1})\kappa^c = (a\kappa^c)^{-1}$ is the unique inverse of $a\kappa$ in S/κ^c (as S is an inverse semigroup).

Let $a\kappa^c, b\kappa^c \in S/\kappa^c$, $a, b \in S$ be such that $(a\kappa^c)^{-1}\#(b\kappa^c)^{-1}$, that is, $(a^{-1})\kappa^c\#(b^{-1})\kappa^c$. Then, by the definition of apartness in the quotient semigroup with apartness, $(a^{-1}, b^{-1}) \in \kappa$, which, by the assumption, implies $(a, b) \in \kappa$. So we have $a\kappa^c\#b\kappa^c$. Therefore, $(S/\kappa^c, =, \#, \cdot)$ is an inverse semigroup. □

The *first apartness isomorphism theorem for inverse semigroups with apartness* follows.

Theorem 16 *Let* $f : S \to T$ *be an se-homomorphism between inverse semigroups with apartness. Then:*

(i) Coker f *is an I-co-congruence such that* coker $f \not\leftarrow$ ker f *and* ker $f \subseteq$ (coker $f)^c$.
(ii) $(S/\ker f, =, \#; \cdot)$ *is an inverse semigroup with apartness, where*

$$a(\ker f) = b(\ker f) \Leftrightarrow (a,b) \in \ker f$$
$$a(\ker f)\#b(\ker f) \Leftrightarrow (a,b) \in \operatorname{coker} f$$
$$a(\ker f)b(\ker f) = (ab)(\ker f)$$
$$(a(\ker f))^{-1} = a^{-1}(\ker f).$$

(iii) *The mapping* $\varphi : S/\ker f \to T$*, defined by* $\varphi(x(\ker f)) = f(x)$*, is an apartness embedding such that* $f = \varphi\pi$.
(iv) *If* f *maps* S *onto* T*, then* φ *is an apartness isomorphism.*

Proof By Theorem 10, $(S/\ker f, =, \#; \cdot)$ is a semigroup with apartness, and the mapping $\varphi : S/\ker f \to T$, defined by $\varphi(x(\ker f)) = f(x)$, is an apartness embedding such that $f = \varphi\pi$, and if f maps S onto T, then φ is an apartness isomorphism. Furthermore, coker f is a co-congruence such that coker $f \not\leftarrow$ ker f and ker $f \subseteq$ (coker $f)^c$. By Lemma 18, coker f is an I-co-congruence.

Now, we have to prove that the quotient semigroup $S/\ker f$ is inverse. We can prove that $a^{-1}(\ker f) = (a(\ker f))^{-1}$ is the unique inverse of $a(\ker f)$ in $S/\ker f$

in a similar way as in the proof of Theorem 15. Let $a^{-1}(\ker f)\#b^{-1}(\ker f)$. That is, $(a^{-1}, b^{-1}) \in \text{coker } f$, which, by Lemma 18, implies $(a, b) \in \text{coker } f$. Hence, $(S/\ker f, =, \#; \cdot)$ is an inverse semigroup. □

5.3 Natural Co-order

A co-quasiorder τ defined on a semigroup with apartness is a *co-order* if it is co-antisymmetric.

Let $\not\leq_l$ be a relation on an inverse semigroup with apartness S defined by

$$a \not\leq_1 b \overset{\text{def}}{\Leftrightarrow} a\#ba^{-1}a,$$

$a, b \in S$. In what follows the next lemma will be used without special announcement.

Lemma 19 *Let S be an inverse semigroup and let $a, b \in S$. Then the following conditions are equivalent:*

(i) $a\#ba^{-1}a$.
(ii) $aa^{-1}\#ba^{-1}$.

Proof

(i) $\Rightarrow$ (ii). Let $a, b \in S$ such that $a\#ba^{-1}a$. Then $aa^{-1}a\#ba^{-1}a$, which implies $aa^{-1}\#ba^{-1}$.
(ii) $\Rightarrow$ (i). Conversely, let $a, b \in S$ such that $aa^{-1}\#ba^{-1}$. Then $aa^{-1}\#ba^{-1}aa^{-1}$, which implies $a\#ba^{-1}a$.

□

Theorem 17 *If S is an inverse semigroup, then $\not\leq_1$ is a left co-compatible co-order on it.*

Proof *Strong irreflexivity.* Let $a, b \in S$ be such that $a \not\leq_1 b$, that is, $a\#ba^{-1}a$. From $aa^{-1}a\#ba^{-1}a$ it follows that $a\#b$.

Co-transitivity. Let $a, b, c \in S$ be such that $a \not\leq_1 c$, that is, $a\#ca^{-1}a$. By the co-transitivity of apartness we have $a\#ba^{-1}a$ or $ba^{-1}a\#ca^{-1}a$. If $a\#ba^{-1}a$, then, obviously, $a \not\leq_1 b$. Let $ba^{-1}a\#ca^{-1}a$. Then $b(a^{-1}a)(a^{-1}a)\#c(a^{-1}a)(a^{-1}a)$, and, by the co-transitivity of apartness, we have

$$b(a^{-1}a)(a^{-1}a)\#(cb^{-1}b)(a^{-1}a)(a^{-1}a) \text{ or } (cb^{-1}b)(a^{-1}a)(a^{-1}a)\#c(a^{-1}a)(a^{-1}a)$$

From $b(a^{-1}a)(a^{-1}a)\#(cb^{-1}b)(a^{-1}a)(a^{-1}a)$ it follows $b\#cb^{-1}b$, that is, $b \not\leq_1 c$. On the other hand, if $(cb^{-1}b)(a^{-1}a)(a^{-1}a)\#c(a^{-1}a)(a^{-1}a)$, it follows

$$(b^{-1}b)(a^{-1}a)(a^{-1}a)\#(a^{-1}a)(a^{-1}a).$$

So, $(a^{-1}a)(b^{-1}b)(a^{-1}a)\#(a^{-1}a)(a^{-1}a)$ which imply $(a^{-1}ab^{-1})(b(a^{-1}a))\#a^{-1}a$, i.e., we have

$$(b(a^{-1}a))^{-1}(b(a^{-1}a))\#a^{-1}a.$$

Thus, $(b(a^{-1}a))^{-1}\#a^{-1}$ or $b(a^{-1}a)\#a$. That means $a \not\leq_1 b$. Co-transitivity of $\not\leq_1$ is proved.

Co-antisymmetry. Let $a, b \in S$ be such that $a\#b$. Then, by the co-transitivity of apartness, $a\#ba^{-1}a$ or $ba^{-1}a\#b$, that is, $a \not\leq_1 b$ or $ba^{-1}a\#b$. If $ba^{-1}a\#b$, i.e., $b(b^{-1}b)(a^{-1}a)\#b$, then $b(b^{-1}b)(a^{-1}a)\#a(b^{-1}b)$ or $a(b^{-1}b)\#b$. Therefore, $a\#ba^{-1}a$ or $a(b^{-1}b)\#b$, i.e., $a \not\leq_1 b$ or $b \not\leq_1 a$.

Left co-compatibility. Let $a, b, u \in S$ be such that $ua \not\leq_1 ub$, i.e.,

$$(ub)(ua)^{-1}\#(ua)(ua)^{-1}.$$

Then we have $uba^{-1}u^{-1}\#uaa^{-1}u^{-1}$ which implies $aa^{-1}\#ba^{-1}$, and, by Lemma 19, $a \not\leq_1 b$ follows. □

Let $e \in E(S)$. Let us denote $\not\leq_1$-classes by

$$\langle e]_{\not\leq_1} = \{x \in S : x \not\leq e\} \text{ and } [e\rangle_{\not\leq_1} = \{x \in S : e \not\leq x\}.$$

As a consequence of Theorem 17 we have next

Corollary 4 *Let S be an inverse semigroup and let $e \in E(S)$. Then $\langle e]_{\not\leq_1}$ ($[e\rangle_{\not\leq_1}$) is a qd-subset and $\langle e]_{\not\leq_1} \subseteq \widetilde{e}$ ($[e\rangle_{\not\leq_1} \subseteq \widetilde{e}$).*

Proof Let $x, y \in S$ and $e \in E(S)$ be such that $y \in \langle e]_{\not\leq_1}$. Then $y \not\leq_1 e$, and, by co-transitivity of $\not\leq_1$, we have $y \not\leq_1 x$ or $x \not\leq_1 e$. Strong reflexivity of $\not\leq_1$ and definition of $\langle e]_{\not\leq_1}$ imply $x\#y$ or $x \in \langle e]_{\not\leq_1}$. Therefore, $\langle e]_{\not\leq_1}$ is a qd-subset.

Let $x \in \langle e]_{\not\leq_1}$. Then $x\#ex^{-1}x$, so $xx^{-1}x\#ex^{-1}x$. Thus, $x\#e$, i.e., $x \in \widetilde{e}$. □

In a similar manner we can define on S relation $\not\leq_2$ by

$$a \not\leq_2 b \overset{\text{def}}{\Leftrightarrow} a\#aa^{-1}b,$$

$a, b \in S$. The following results can be proved in a similar manner as done in the proofs of Lemma 19 and Theorem 17.

Lemma 20 *Let S be an inverse semigroup and let $a, b \in S$. Then the following conditions are equivalent:*

(i) $a\#aa^{-1}b$.
(ii) $a^{-1}a\#a^{-1}b$.

Theorem 18 *If S is an inverse semigroup, then $\not\leq_2$ is a right co-compatible co-order on it.*

Corollary 5 *Let S be an inverse semigroup and let $e \in E(S)$. Then $\langle e]_{\nleq_2}$ ($[e\rangle_{\nleq_2}$) is a qd-subset and $\langle e]_{\nleq_2} \subseteq \widetilde{e}$ ($[e\rangle_{\nleq_2} \subseteq \widetilde{e}$).*

5.3.1 Inverse Linked Relations

In what follows, the following notion will be needful. Let α and β be two relations defined on an inverse semigroup. A relation α and β are *inverse linked* of, shortly, *I-linked* if

$$\forall_{a,b\in S}\ ((a,b) \in \alpha \Leftrightarrow (a^{-1}, b^{-1}) \in \beta)$$

which will be denoted by $\alpha \Leftrightarrow \beta$.

Proposition 11 *Let α and β be two I-linked relations defined on an inverse semigroup, i.e., $\alpha \Leftrightarrow \beta$. Then $\alpha \cup \beta$ is an I-relation on S.*

Proof Straightforward. □

The main result of this section follows.

Theorem 19 *Let S be an inverse semigroup. Then:*

(i) $\nleq_1 \Leftrightarrow \nleq_2$.
(ii) $\nleq = \nleq_1 \cup \nleq_2$ *is an co-compatible I-co-order on S.*

Proof

(i) Let $a \nleq_2 b$. Then

$$\begin{aligned} a\#aa^{-1}b &\Leftrightarrow a^{-1}\#(aa^{-1}b)^{-1} = b^{-1}(a^{-1})^{-1}a^{-1} \\ &\Leftrightarrow a^{-1}\#b^{-1}(a^{-1})^{-1}a^{-1} \\ &\Leftrightarrow a^{-1} \nleq_1 b^{-1}. \end{aligned}$$

(ii) By Theorem 17 and Theorem 18, $\nleq = \nleq_1 \cup \nleq_2$ is a co-compatible co-order. By (i), $\nleq$ is an I-co-order.

□

The co-compatible I-co-order $\nleq$ is called *natural co-order* on an inverse semigroup with apartness. As a consequence of the previous theorem and the fact that idempotents commute, the definition of natural co-order on idempotents follows.

Corollary 6 *Let $E(S)$ be a semilattice of idempotents of an inverse semigroup with apartness. Then, the natural co-order on $E(S)$ is defined by*

$$\forall_{e,f\in E(S)}\ (e \nleq f \overset{def}{\Leftrightarrow} e\#ef).$$

A subset T of an inverse semigroup with apartness is *$[\rangle$-subset* if

$$\forall_{x,y\in S}\,(x \in T \Rightarrow x \nleq y \vee y \in T).$$

On the other hand, T is called *$\langle]$-subset* if

$$\forall_{x,y\in S}\,(x \in T \Rightarrow y \nleq x \vee y \in T).$$

Lemma 21 *Let S be an inverse semigroup. If inhabited, the set of a-idempotents $A(S)$ is an inverse $[\rangle$-qd-subset.*

Proof By Proposition 8(i), $A(S)$ is an inverse qd-subset. Let $a, b \in S$ such that $a \in A(S)$. Then $a^2\#a$, and, by co-transitivity of apartness, it follows

$$\begin{aligned} a^2\#(aa^{-1}b)(ba^{-1}a) &\vee (aa^{-1})b^2(a^{-1}a)\#(aa^{-1})b(a^{-1}a) \\ &\vee (aa^{-1})b(a^{-1}a)\#a = aa^{-1}a \\ &\Rightarrow a\#aa^{-1}b \vee a\#baa^{-1} \vee b^2\#b \vee a\#aa^{-1}b \\ &\Leftrightarrow a \nleq_1 b \vee a \nleq_2 b \vee b \in A(S) \\ &\Leftrightarrow a \nleq b \vee b \in A(S). \end{aligned}$$

Hence, $A(S)$ is an inverse $[\rangle$-qd-subset. □

6 Some Examples of Inverse Linked Co-congruences

A co-equivalence on a semigroup with apartness S is called a *left* (*right*) *co-congruence* if it is left (right) co-compatible. Clearly, a co-congruence is both, a left and a right co-congruence. A subset T of S is *co-full* if $T \subseteq_\sim E(S)$.

Theorem 20 *Let S be an inverse semigroup. Let M be an inverse co-full $\langle]$-co-subsemigroup of S. Then:*

(i) *Relation κ^l_M defined on S by*

$$(a, b) \in \kappa^l_M \overset{def}{\Leftrightarrow} a^{-1}b \in M$$

is left co-congruence on S.

(ii) *Relation κ^r_M defined on S by*

$$(a, b) \in \kappa^r_M \overset{def}{\Leftrightarrow} ab^{-1} \in M$$

is right co-congruence on S.

(iii) $\kappa^l_M \Leftrightarrow \kappa^r_M$.
(iv) $\kappa_M = \kappa^l_M \cup \kappa^r_M$ *is an I-co-congruence on S.*

Proof

(i) *Strong irreflexivity.* Let $(a, b) \in \kappa^l_M$, that is, $a^{-1}b \in M \subseteq_\sim E(S)$. Then $a^{-1}b\#a^{-1}a$, which implies $b\#a$.

Symmetry. Let $(a, b) \in \kappa^l_M$, i.e., $a^{-1}b \in M$. Then $((a^{-1}b)^{-1})^{-1} \in M$, which, as M is an inverse subset, implies $(a^{-1}b)^{-1} \in M$, i.e., $b^{-1}a \in M$. Hence, $(b, a) \in \kappa^l_M$.

Co-transitivity. Let $a, b, c \in S$ be such that $(a, c) \in \kappa^l_M$, that is, $a^{-1}c \in M$. Then, as M is a $\langle]$-subset, it follows that $a^{-1}bb^{-1}c \nleq a^{-1}c$ or $a^{-1}bb^{-1}c \in M$.

If $a^{-1}bb^{-1}c \nleq a^{-1}c$, then, by the definition of $\nleq$, $a^{-1}bb^{-1}c\#a^{-1}c(a^{-1}bb^{-1}c)^{-1}(a^{-1}bb^{-1}c)$ or $a^{-1}bb^{-1}c\#(a^{-1}bb^{-1}c)(a^{-1}bb^{-1}c)^{-1}a^{-1}c$. If

$$\begin{aligned} a^{-1}bb^{-1}c \;\#\; a^{-1}c(a^{-1}bb^{-1}c)^{-1}(a^{-1}bb^{-1}c) &= a^{-1}cc^{-1}bb^{-1}aa^{-1}bb^{-1}c \\ &= a^{-1}cc^{-1}bb^{-1}bb^{-1}aa^{-1}c \\ &= a^{-1}cc^{-1}bb^{-1}aa^{-1}c \\ &= a^{-1}bb^{-1}aa^{-1}cc^{-1}c \\ &= a^{-1}aa^{-1}bb^{-1}c \\ &= a^{-1}abb^{-1}c, \end{aligned}$$

which is a contradiction. From $a^{-1}bb^{-1}c\#(a^{-1}bb^{-1}c)(a^{-1}bb^{-1}c)^{-1}a^{-1}c$ in a similar manner can be also shown that $a^{-1}bb^{-1}c\#a^{-1}bb^{-1}c$. Therefore, $a^{-1}bb^{-1}c \in M$, which as M is a co-subsemigroup, implies $a^{-1}b \in M$ or $b^{-1}c \in M$, that is, $(a, b) \in \kappa^l_M$ or $(b, c) \in \kappa^l_M$.

Left co-compatibility. Let $a, b, u \in S$ be such that $(ua, ub) \in \kappa^l_M$, i.e., $(ua)^{-1}(ub) \in M$. So

$$\begin{aligned} (ua)^{-1}(ub) &= a^{-1}u^{-1}ub \\ &= a^{-1}aa^{-1}u^{-1}ub \\ &= (a^{-1}u^{-1}ua)a^{-1}b. \end{aligned}$$

M is a completely isolated subset so we have $a^{-1}u^{-1}ua \in M$ or $a^{-1}b \in M$. We have $a^{-1}u^{-1}ua = a^{-1}aa^{-1}u^{-1}uu^{-1}ua = (a^{-1}u^{-1}ua)(a^{-1}u^{-1}ua)$, so $a^{-1}u^{-1}ua \in E(S)$ which is, as M is a co-full subset, a contradiction. Hence, $a^{-1}b \in M$, or, equivalently, $(a, b) \in \kappa^l_M$.

Therefore, κ^l_M is a left co-congruence.

(ii) Similar to (i).

(iii) Let $a, b \in S$ be such that $(a^{-1}, b^{-1}) \in \kappa^l_M$. Then, by the definition of κ^l_M and as M is an inverse subset, it follows

$$(a^{-1})^{-1}b^{-1} \in M \Leftrightarrow (ba^{-1})^{-1} \in M \Leftrightarrow ab^{-1} \in M \Leftrightarrow (a, b) \in \kappa^r_M.$$

Therefore, $\kappa^l_M \Bumpeq \kappa^r_M$.

(iv) This is consequence of (iii) and Proposition 11.

□

6.1 Apartness Idempotent-separating Co-congruences

Recall in Sect. 4 we define *apartness idempotent-separating co-congruence* or shortly *a-idempotent-separating co-congruence* κ as a co-congruence satisfying:

- $\forall_{e,f \in E(S)} \, (e\#f \Rightarrow (e, f) \in \kappa)$.

Example 6 If $f : S \to T$ is an se-homomorphism between inverse semigroups with apartness S and T such that its restriction on the set of idempotent $f|_{E(S)}$ is a-injective, then I-co-congruence $coker f$ is a-idempotent separating on S. ◇

Let us consider special cases of Theorem 13

Corollary 7 *Let S be an inverse semigroup. Then:*

(i) *Relation μ_l defined on S by*

$$(a, b) \in \mu_l \overset{def}{\Leftrightarrow} \exists_{e \in E(S)} \, (a^{-1}ea\#b^{-1}eb)$$

is an a-idempotent-separating co-congruence.

(ii) *Relation μ_r defined on S by*

$$(a, b) \in \mu_r \overset{def}{\Leftrightarrow} \exists_{e \in E(S)} \, (aea^{-1}\#beb^{-1})$$

is an a-idempotent-separating co-congruence.

(iii) $\mu_l \Bumpeq \mu_r$.

(iv) *$\mu = \mu_l \cup \mu_r$ is an a-idempotent-separating I-co-congruence.*

Proof

(i) By Theorem 13 it follows that μ_l is a co-congruence on S. In a similar manner as in Theorem 14 we can prove that it is a-idempotent separating.

(ii) Similar to (i).

(iii) Let $(a^{-1}, b^{-1}) \in \mu_l$. Then, by the definition $(a^{-1})^{-1}ea^{-1}\#(b^{-1})^{-1}eb^{-1}$, for some $e \in E(S)$. Hence, $(aea^{-1})^{-1}\#(beb^{-1})^{-1}$ which implies $aea^{-1}\#beb^{-1}$, or, equivalently, $(a, b) \in \mu_r$. Thus, $\mu_l \Bumpeq\Bumpeq \mu_r$.

(iv) This is consequence of (iii) and Proposition 11.

□

There are relations defined on an inverse semigroup contained in μ which are a-idempotent-separating I-co-congruences too.

Theorem 21 *Let S be an inverse semigroup. Then:*

(i) *Relation κ_l defined on S by*

$$(a, b) \in \kappa_l \overset{def}{\Leftrightarrow} aa^{-1}\#bb^{-1}$$

is an a-idempotent-separating left co-congruence such that $\kappa_l \subseteq \mu_l \cap \mu_r$.

(ii) *Relation κ_r defined on S by*

$$(a, b) \in \kappa_r \overset{def}{\Leftrightarrow} a^{-1}a\#b^{-1}b$$

is an a-idempotent-separating right co-congruence such that $\kappa_l \subseteq \mu_l \cap \mu_r$.

(iii) $\kappa_l \Bumpeq \kappa_r$.

(iv) $\kappa = \kappa_l \cup \kappa_r$ *is an a-idempotent-separating I-co-congruence such that* $\kappa_l \subseteq \mu_l \cap \mu_r$.

Proof

(i) *Strong irreflexivity.* Let $(a, b) \in \kappa_l$. Then $aa^{-1}\#bb^{-1}$, and, by the co-transitivity of apartness, we have $aa^{-1}\#ab^{-1}$ or $ab^{-1}\#bb^{-1}$. It follows $a^{-1}\#b^{-1}$ or $a\#b$, and κ_l is strongly irreflexive.

Symmetry. Obvious.

Co-transitivity. Let $a, b, c \in S$ be such that $(a, b) \in \kappa_l$. Thus, $aa^{-1}\#bb^{-1}$, and we, further, have $aa^{-1}\#cc^{-1}$ or $cc^{-1}\#bb^{-1}$, i.e., we have $(a, c) \in \kappa_l$ or $(c, b) \in \kappa_l$. Thus, κ_l is co-transitive. We prove κ_l is a co-equivalence.

Left co-compatibility. $a, b, u \in S$ be such $(ua, ub) \in \kappa_l$, i.e., $(ua)(ua)^{-1}\#(ub)(ub)^{-1}$. Then we have $uaa^{-1}u^{-1}\#ubb^{-1}u^{-1}$, which implies $aa^{-1}\#bb^{-1}$, i.e., $(a, b) \in \kappa_l$. Therefore, κ_l is a left co-congruence on S.

A-idempotent-separating. Let $e, f \in E(S)$ be such that $e\#f$. Then $ee^{-1}\#ff^{-1}$ and we have $(e, f) \in \kappa_l$.

Let $(a, b) \in \kappa_l$. Then $aa^{-1}\#bb^{-1}$ and $aa^{-1}, bb^{-1} \in E(S)$. By Corollary 7, μ_l and μ_r are a-idempotent separating co-congruences, so $(aa^{-1}, bb^{-1}) \in \mu_l \cap \mu_r$ and $(a, b) \in \mu_l \cap \mu_r$ or $(a^{-1}, b^{-1}) \in \mu_l \cap \mu_r$. Thus, $(a, b) \in \mu_l \cap \mu_r$, and we have $\kappa_l \subseteq \mu_l \cap \mu_r$.

(ii) Similar to (i).

(iii) Almost obvious.

(iv) This is consequence of (iii) and Proposition 11.

□

Recall, in classical semigroup theory, a subset

$$Z_{E(S)} = \{z : \forall_{e \in E(S)}\ ez = ze\}$$

is called the *centralizer* of $E(S)$ in S.

Corollary 8 *Restriction of* κ_l (κ_r) *on the centralizer* $Z_{E(S)}$ *of* $E(S)$ *in* S *is an* I-*co-congruence.*

Proof Let $a \in Z_{E(S)}$. Then

$$aa^{-1} = (aa^{-1}a)a^{-1} = a^{-1}(aaa^{-1}) = a^{-1}aa^{-1}a = a^{-1}a,$$

which implies $(a, b) \in \kappa_l$ if and only if $(a, b) \in \kappa_r$. Thus, $\kappa_l = \kappa_r$ is a I-co-congruence on S. □

We are going to define two relations on an inverse semigroup with apartness which are contained in μ but are neither a-idempotent-separating nor co-congruences. For that purpose the following notion will be useful. A subset

$$K_E(S) = \{z \in S : \exists_{e\in E(S)} (ze\#ez)\}$$

is called the *co-centralizer* of $E(S)$ in S.

Lemma 22 *The co-centralizer* $K_E(S)$ *is an inverse qd-subset and* $K_E(S) \subseteq^{\sim} Z_E(S)$.

Proof Let $a, b \in S$ be such that $a \in K_E(S)$. Then, there exists an idempotent $e \in E(S)$ such that $ae\#ea$. By the co-transitivity of apartness, $ae\#be$ or $be\#eb$ or $eb\#ea$. Thus, $a\#b$ or $b \in K_E(S)$, and $K_E(S)$ is a qd-subset.

Let $a^{-1} \in K_E(S)$. Then, there exists an idempotent $e \in E(S)$ such that $a^{-1}e\#ea^{-1}$, i.e., $(ea)^{-1}\#(ae)^{-1}$. So, $ea\#ae$, that is, $a \in K_E(S)$, and $K_E(S)$ is an inverse subset.

Let $a, b \in S$ be such that $a \in K_E(S)$ and $b \in Z_E(S)$. Then, there exists an idempotent $e \in E(S)$ such that $ae\#ea$. Then $ae\#be$ or $be\#eb$ or $eb\#ea$. By the assumption $be\#eb$, i.e., $b \in K_E(S)$, is impossible. Thus, $a\#b$, and $a \in^{\sim} Z_E(S)$. □

A relation defined on S is called *co-tolerance* if it is strongly irreflexive and symmetric.

Theorem 22 *Let* S *be an inverse semigroup with apartness. Let* $K_E(S)$ *be the co-centralizer of* $E(S)$ *in* S*. Then:*

(i) *Relation* ν_l *defined on* S *by*

$$(a, b) \in \nu_l \overset{def}{\Leftrightarrow} ab^{-1} \in K_E(S)$$

is a left co-compatible co-tolerance and $\nu_l \subseteq \mu_l$.

(ii) *Relation* ν_r *defined on* S *by*

$$(a, b) \in \nu_r \overset{def}{\Leftrightarrow} a^{-1}b \in K_E(S)$$

is a right co-compatible co-tolerance and $\nu_r \subseteq \mu_r$.

Proof *Strong irreflexivity.* Let $(a, b) \in \nu_l$. Then there is $e \in E(S)$ such that $ab^{-1}e\#eab^{-1}$. Then, by the co-transitivity of apartness, $ab^{-1}e\#aa^{-1}e$ or $aa^{-1}e\#eab^{-1}$. Hence, $b^{-1}\#a^{-1}$ or $eaa^{-1}\#eab^{-1}$. In both cases we have $a\#b$.

Symmetry. Follows immediately due to the fact that $K_E(S)$ is an inverse subset.

Left co-compatibility. Let $a, b, u \in S$ be such that $(ua, ub) \in \nu_l$. Then $(ua)(ub)^{-1} \in K_E(S)$, that is, $(ua)(ub)^{-1}e\#e(ua)(ub)^{-1}$ for some $e \in E(S)$. So, $uab^{-1}u^{-1}e\#euab^{-1}u^{-1}$. By Lemma 16, there exists $f \in E(S)$ such that $eu = uf$ which implies $(eu)^{-1} = u^{-1}e = fu^{-1}$. We have

$$\begin{aligned} uab^{-1}u^{-1}e\#euab^{-1}u^{-1} &\Rightarrow uab^{-1}fu^{-1}\#ufab^{-1}u^{-1} \\ &\Rightarrow ab^{-1}f\#fab^{-1} \\ &\Rightarrow ab^{-1} \in K_E(S) \\ &\Leftrightarrow (a, b) \in \nu_l. \end{aligned}$$

Let $a, b \in S$ be such that $(a, b) \in \nu_l$, i.e., $ab^{-1} \in K_E(S)$. Then, there exists an idempotent $e \in E(S)$ such that $ab^{-1}e\#eab^{-1}$. Then

$$\begin{aligned} ab^{-1}e\#eab^{-1} &\Leftrightarrow ab^{-1}bb^{-1}e\#eaa^{-1}ab^{-1} \\ &\Rightarrow ab^{-1}ebb^{-1}\#aa^{-1}eab^{-1} \\ &\Rightarrow b^{-1}eb\#a^{-1}ea \\ &\Leftrightarrow (a, b) \in \mu_l. \end{aligned}$$

(ii). Similar to (i). □

Remark 11 There are a-idempotent-separating I-co-congruences on an inverse semigroup with apartness defined without using inverse linked relations and without being part of (any of) μ. A-idempotent-separating I-co-congruence η from Theorem 14 is an evidence for such types of co-congruences. ◇

7 Concluding Remarks

Mathematics is not a careful march down a well-cleared highway, but a journey into a strange wilderness, where the explorers often get lost. Rigour should be a signal to the historian that the maps have been made, and the real explorers have gone elsewhere. W. S. Anglin [2]

It is only recently that constructive mathematics has paid attention to semigroup theory. Inspired by results obtained from the use of interactive theorem-provers and the general theory of formal verification of constructive proofs, a new constructive algebraic theory, the so-called theory of semigroups with apartness, was developed by M. Mitrović and co-authors: M. N. Hounkonnou, M. A. Baroni, S. Crvenković, E. Darpö, D. A. Romano, and S. Silvestrov. See, for example, [12–14, 22, 23, 25, 26].

Based on the results presented in [12, 13], A. Cherubini and A. Frigeri in 2019 lay the foundations for the development of the theory of constructive inverse semigroups with apartness. Since then, M. Mitrović and D. A. Romano contributed to the area too [24, 28].

Much of our algebraic research of good quality on the class of constructive semigroups with apartness and their subclasses and related topics was and is concerned with problem-solving. But theory-building is also an important part of our scientific work. This chapter is a journey through abovementioned papers with constructive inverse semigroups with apartness as a main destination. We agree; sometimes it is the journey that teaches you a lot about your destination, for example, about further directions in development inverse semigroups with apartness (as well as semigroups with apartness in general) at the same time being aware about the problems posted (in the open, or somehow hidden way) along our way. We also agree that scientific work is a journey, not a destination. And, following Lao Tzu, an important figure in Chinese culture, any journey (longer or shorter) begins with one step. Our main intention was that step through this chapter to be made.

Appendix: Constructive Framework

> The belief in the universal validity of the principle of the excluded third in mathematics is considered by the intuitionists as a phenomenon of the history of civilization of the same kind as the former belief in the rationality of π, *or in the rotation of the firmament about the earth. L.E.J. Brouwer, source [15]*

Throughout this chapter constructive mathematics is understood as mathematics performed in the context of intuitionistic logic, that is, without the law of excluded middle, **LEM**. A presentation of constructive mathematics given within this section is broad rather than deep, and it is not intended to be comprehensive: we first sketch the history that motivated the early workers in the field of constructive mathematics, we describe informal intuitionistic logic, and we give the briefest outline on the constructively valid arguments. All material presented in this section is broad rather than deep, and it is not intended to be comprehensive. It is heavily based on the treatments in other standards on constructive mathematics books, such as, for example, [3, 5–8, 19, 31]. The main novelty is in the selection and arrangement of material.

Constructive Mathematics

In a words of A. Calder [10], constructive mathematics is founded on the belief that mathematics can only have real meaning if its concepts can be constructed by the human mind. This issue has divided mathematicians for more than a century.

The classical view of mathematics is essentially *descriptive*: we try to describe the facts about a static mathematical universe [...]. In opposition to this is the constructive view of mathematics, which focuses attention on *the dynamic interaction* of the individual with the mathematical universe. (See [19]) Following H. Wang [34], we may say that constructive mathematics is a "mathematics of doing" while classical mathematics is a "mathematics of being".

Constructive mathematics is often mis-characterized classical mathematics without the law of excluded middle, **LEM**, or as classical mathematics without the axiom of choice, **AC**. But seen from within the discipline, constructive mathematics is positively characterized by a strict provability requirement. Two fundamental principles underpin the constructivist approach:

- **Truth**: a proposition is considered true only when a proof for it is produced.
- **Existence:** the existence of an object is established through a proof that shows how to find it.

These principles lead to a rejection of proof by contradiction and skepticism about the meaning of many classical mathematics theorems. If we accept that existence should always be interpreted constructively, we must abandon the unrestricted use of the logical law of excluded middle, **LEM**.

In conclusion, classical and constructive mathematics should not be seen as rival domains where one must choose one for life. Instead, they should be viewed as complementary, helping to construct a more adequate overall picture. As G. Sommaruga [30] stated, they provide useful perspectives on the same grand structure. P. Bernays [4] also noted that it would be harmful to renounce either one, as they complement each other.

Constructive Mathematics: A Very Brief History

Understanding present-day constructive mathematics requires knowledge of its origins. In a sense, the work of those who preceded us is the foundation on which we built our own work. But, historical development of constructive mathematics is a subject in its own right. For the purpose of this chapter, we are going as briefly as possible through it.

Constructivity in Mathematics Before Brouwer It is widely accepted that nonconstructive proofs in mathematics are essentially a modern conception: with only a few exceptions, all mathematical proofs before 1880 were constructive. The very notion of "existence" in mathematics was, for all intents and purposes, taken to mean "constructive existence". There were, however, a few nonconstructive proofs, like Euler's proof in the eighteenth century of the infinitude of prime numbers, and Hilbert's nonconstructive proof of the finite basis theorem for polynomial ideals. Cantor's development of set theory is often marked as the beginning of the unrestricted use of nonconstructive arguments in mathematics. The German mathematician Kronecker, the most prominent of Cantor's intellectual opponents, observed in 1886 that "God made the natural numbers, everything else is the work of Man". That is, Leopold Kronecker had advocated a constructive approach to mathematics, i.e., a point of view that the work of the mathematician is to invent mathematics.

In his proof of the well-ordering theorem published in 1904 Zermelo had formulated and made essential use of the axiom of choice, **AC**. The "nonconstructive" character of this principle provoked the reactions of a number of prominent mathematicians of the time. Borel, for example, was particularly unhappy with Zermelo's use of the axiom of choice to make uncountably many arbitrary "choices", as was required when establishing the well-orderability of the continuum. The French mathematician Baire's objections went still further. For Baire, in the last analysis, everything in mathematics must be reduced to the finite. Lebesgue put the central question in a way that leaves no doubt in using constructive terms: Can the existence of a mathematical object be proved without at the same time defining it? Lebesgue's answer which was, in essence, negative made him to be considered as close to the constructive circles.

Among classical mathematicians, the term "constructive" is still sometimes used with the meaning "without making use of the axiom of choice".

Constructivity in Mathematics Since Brouwer The origin of modern constructive mathematics lies in the foundational debate at the turn of the twentieth century. At that time, the German mathematician D. Hilbert established some very deep existence theorems using nonconstructive methods. Almost at the same time, the Dutch mathematician L. E. J. Brouwer became philosophically convinced that the universal validity of contradiction proofs for existence proofs was unwarranted. Anyway, the story of modern constructivism really begins with the publication of the Brouwer's the doctoral thesis "On the Foundations of Mathematics" in 1907. "Briefly, Brouwer viewed mathematics as the activity of building constructions in the mind (of an ideal mathematician); mathematics is about such mental constructions, not about objects in some outside reality. For Brouwer, there is no platonistic universe of abstract ideas existing somewhere quite independently of human cognition" [32].

Despite continuing opposition, intuitionism survived and new constructive approaches to mathematics arose. In 1948–1949 in the former Soviet Union, A. A. Markov initiated a program of recursive constructive mathematics, **RUSS**—mathematics using intuitionistic logic and based on the Church-Markov-Turing thesis that all computable partial functions from the set $\mathbb{N}$ to itself are recursive. Kind of stagnancy can be noted by mid-1960s. Constructive mathematics was, when compared with its classical counterpart, virtually stagnant. Renaissance is caused in large part by the appearance of Errett Bishop's seminal book "*Foundations of Constructive Analysis*", **FCA**, in 1967. **FCA** is considered to be one of the most remarkable intellectual documents of the twentieth century. "Bishop demonstrated that pure mathematics can be developed from a constructive point of view while maintaining a continuity with classical terminology and spirit; much more of classical mathematics was preserved than had been though possible" ([8]).

In the intervening years there has been considerable activity in constructive analysis, algebra, and topology, in related foundational areas such as type theory, and in the relation between constructive mathematics and computer science.

(Informal) Intuitionistic Logic

Throughout this chapter, constructive mathematics, **CM**, is understood as mathematics practiced using intuitionistic logic. Logic in mathematics and computer science serves two primary roles: as a tool for applications and as a technique for establishing foundations.

Constructive mathematics is not based on a prior notion of logic. Instead, our understanding of logical connectives and quantifiers comes from mathematical intuition and experience. The point of departure is that a statement ϕ is considered true if there is a proof for it, meaning a mathematical construction rather than a deduction in some formal system. From the classical mathematics, **CLASS**, point of view, mathematics consists of a preexisting mathematical truth. From a constructive viewpoint, the judgement φ *is true* means that *there is a proof of* φ. In conclusion, "In actual building of constructive mathematics we do not need logic; nevertheless we find it convenient to use logical symbolism" [31].

Constructive reasoning differs from classical reasoning by assigning stronger meanings to some logical operators. Often, ideas in the history of mathematics need to be discovered multiple times by different people before becoming widely recognized. One such idea in the constructivist foundations of mathematics is the proof interpretation of intuitionistic logic, also known as the *Brouwer-Heyting-Kolmogorov* (**BHK**) interpretation. The **BHK** interpretation gives each of the logical symbols $\wedge$, $\vee$, $\Rightarrow$, $\neg$, $\forall$, $\exists$ a distinct meaning. Classically, the propositional connectivities can be defined from $\wedge$ and $\neg$, while $\exists$ can be defined from $\forall$ and $\neg$, so $\vee$, $\Rightarrow$, and $\exists$ are unnecessary. Intuitionistic logic, however, fully utilizes the expressive power of the formal language.

More on intuitionistic logics and **BHK** interpretation can be found, for example, in [5] or [31].

Constructivity and Too Many Principles

In constructive mathematics, there are a number of competing notions of constructivity, leading to various incompatible systems. The main exponents of constructivity include Brouwer's intuitionistic mathematics, **INT**; the constructive recursive mathematics of the Russian school of Markov, **RUSS**; and Bishop's constructive mathematics, **BISH**. Additionally, there are several other minor systems with specific constructive programs.

Every form has intuitionistic logic at its core. Different schools have different additional principles or axioms given by the particular approach to constructivism. For example, the notion of an *algorithm* or a *finite routine* is taken as primitive in **INT** and **BISH**, while **RUSS** operates with a fixed programming language and an algorithm is a sequence of symbols in that language. This diversity means that constructivism, in a broad sense, is not homogeneous. Even within a single school, or from a single mathematician at different times, views can vary significantly.

Errett Bishop's Style of Constructive Mathematics

Throughout this chapter *constructive mathematics* is understood as Bishop-style mathematics, **BISH**. There are at least two reasons for this.

First, **BISH** is consistent with the classical mathematics, **CLASS**, practiced by most mathematicians today: every proposition P in **BISH** has an immediate interpretation in **CLASS**, and a proof of P in **BISH** is also a proof of P in **CLASS**. This is not true in our other two varieties: Russian constructivism, **RUSS**, and Brouwer's intuitionism, **INT**. A second reason is that every proof of a proposition P in **BISH** is a proof of P in **RUSS** and in **INT**. Indeed, the last two varieties (as well as **CLASS**) may be regarded as extensions of **BISH**.

Furthermore, at least initially, classical mathematics is seen as a guide that helps the constructive mathematician develop new mathematics. For example, Errett Bishop [5] wrote: "We are not contending that idealistic mathematics is worthless from the constructive point of view. This would be as silly as contending that unrigorous mathematics is worthless from the classical point of view. Every theorem proved with idealistic methods presents a challenge: to find a constructive version, and to give it a constructive proof".

It is Heyting [16] who wrote in his famous book *Intuitionism, an Introduction*, 1956, that "In order to build up a definite branch of intuitionistic mathematics, it is necessary in the first place to have a thorough knowledge of the corresponding branch of classical mathematics, and in the second place to know by experience where the intuitionistic pitfalls lie".

Constructively Valid Arguments

According to Bishop [5], every classical theorem presents a challenge: to find a constructive version with a constructive proof. Bishop stated that the extent of the existence of good constructive substitutes for the theorems of classical mathematics demonstrates that classical mathematics has a substantial underpinning of constructive truth. In constructive mathematics, a "constructive theorem" is a theorem with a constructive proof. A classical theorem that is proven in a constructive manner is a constructive theorem. This constructive version can be obtained by strengthening the conditions and/or weakening the conclusion of the theorem. Although constructive theorems might look like the corresponding classical versions, they often have more complicated hypotheses and proofs. There are, often, several constructively different versions of the same classical theorem; see, for example, [22]. Some classical theorems are neither provable nor disprovable, that is, they are independent of **BISH**. For some classical theorems it is shown that they are not provable constructively.

Omniscience Principles and Brouwerian Counterexample

Towards the end of this section, we introduce the concepts of the omniscience principle and the Brouwerian counterexample. The law of excluded middle, **LEM**, can be regarded as the main source of nonconstructivity. It was Brouwer [9] who first observed that **LEM** was extended without justification to statements about infinite sets. Several consequences of **LEM** are not accepted in Bishop's constructivism. We will mention three such nonconstructive principles—the ones which were used through this chapter:

- **The limited principle of omniscience, LPO**: for each binary sequence $(a_n)_{n\geq 1}$, either $a_n = 0$ for all n, or else there exists n with $a_n = 1$.

- **The lesser limited principle of omniscience, LLPO**: if $(a_n)_{n\in\mathbb{N}}$ is a binary sequence containing at most one term equal to 1, then either $a_{2n} = 0$ for all $n \in \mathbb{N}$, or else $a_{2n+1} = 0$ for all $n \in \mathbb{N}$.

- **Markov's principle, MP**: for each binary sequence $(a_n)_{n\geq 1}$, if it is impossible that $a_n = 0$ for all n, then there exists n with $a_n = 1$.

Within constructive mathematics, a statement P, as in classical mathematics, can be disproved by giving a counterexample. However, it is also possible to give a Brouwerian counterexample to show that the statement is nonconstructive. A *Brouwerian counterexample to a statement* P is a constructive proof that P implies some nonconstructive principle, such as, for example, **LEM**, and its weaker versions **LPO**, **MP**. It is not a counterexample in the true sense of the word—it is just an indication that P does not admit a constructive proof.

Some Acronyms

CLASS	Classical mathematics, i.e., mathematics with classical logic
CM	Constructive mathematics, i.e., mathematics with intuitionistic logic
BISH	Errett Bishop's style of constructive mathematics
INT	Brouwer's intuitionistic mathematics
RUSS	Russian school of Markov
IL	Intuitionistic logic
BHK	Brouwer-Heyting-Kolmogorov
LEM	Law of excluded middle
AC	Axiom of choice
FCA	Foundations of Constructive Analysis [5]

References

1. Aczel, P., Rathjen, M.: Notes on Constructive Set Theory, REPORT No. 40, 2000/2001, ISSN 1103-467X, ISRN IML-R–40-00/01–SE, Institut Mittag-Leffler, The Royal Swedish Academy of Science (2000/2001)
2. Anglin, W.S.: Mathematics and History. Math. Intelligencer **14**, 6–12 (1992)
3. Bell, J.L.: Constructivity in mathematics. https://publish.uwo.ca/~jbell/constructivity.pdf
4. Bernays, P.: On Platonism in mathematics. In: Benacerraf, P., Putnam, H. (eds.) Philosophy of Mathematics: Selected Readings, pp. 274–286. Prentice-Hall, Englewood Cliffs (1964)
5. Bishop, E.: Foundations of Constructive Analysis. McGraw-Hill, New York (1967)
6. Bridges, D.S., Ishihara, H., Rathjen, M., Schwichtenberg H. (eds.): Handbook in Constructive Mathematics. Cambridge University Press, Cambridge (2023)
7. Bridges, D.S., Vîţă, L.S.: Apartness and Uniformity—A Constructive Development. CiE series on Theory and Applications of Computability. Springer, Berlin (2011)
8. Bridges, D.S., Richman F.: Varieties of Constructive Mathematics. Cambridge University Press, Cambridge (1987)
9. Brouwer, L.E.J.: De onbetrouwbaarheid der logische principes, Tijdschrijt voor wijsbegeerte **2**, 152–158, 1908. English translation: The Unreliability of the Logical Principles. In Heyting, A. (ed.) Collected Works, vol. 1, pp. 107–111 (North-Holland, Amsterdam, 1975)
10. Calder, A.: Constructive Mathematics. Sci. Am.**241**(4), 146–171 (1979)
11. Cherubini, A., Frigeri, A.: Inverse semigroups with apartness. Semigroup Forum **98**, 571–588 (2019). https://doi.org/10.1007/s00233-019-10019-y
12. Crvenković, S., Mitrović, M., Romano, D.A.: Semigroups with Apartness. Math. Logic Q. **59**(6), 407–414 (2013)
13. Crvenković, S., Mitrović, M., Romano, D.A.: Basic notions of (constructive) semigroups with apartness. Semigroup Forum **92**(3), 659–674 (2016)
14. Darpo, E., Mitrović, M.: Some results in constructive semigroup theory. arXiv:2103.07105
15. van, D.D (ed.):: Brouwer's Cambridge Lectures on Intuitionism (1951). Cambridge University Press, Cambridge (1981)
16. Heyting A.: Intuitionism, an Introduction. North-Holland, (1956)
17. Lawson M.V.: Inverse Semigroups: the Theory of Partial Symmetries. World Scientific, Singapore (1998)
18. Lombardi H.: Algebra in Bishop's style: a Course in Constructive Algebra. In: Bridges, D., Ishihara, H., Rathjen, M., Schwichtenberg, H. (eds.) Handbook of Constructive Mathematics, pp. 93–113. Cambridge University Press, Cambridge (2023)
19. Mines, R., Richman, F., Ruitenburg, W.: A Course of Constructive Algebra Springer, New York (1988)
20. Mitrović, M., Hounkonnou, M.N.: Constructive semigroups with apartness: a comprehensible survey for experts and non-experts. Innovative Mechanical Engineering, IME, University of Niš, Faculty of Mechanical Engineering **3**(3), 48–67 (2024). ISSN 2812–9229
21. Mitrović, M., Hounkonnou, M.N.: Constructive binary structures with apartness vs. classical ones. Book in progress
22. Mitrović, M., Hounkonnou, M.N., Baroni, M.A.: Theory of constructive semigroups with apartness—foundations, development and practice. Fundamenta Informaticae **184**(3), 233–271 (2021)
23. Mitrović, M., Hounkonnou M.N., Catarino P.: Constructive semigroups with apartness—a state of the art. In: Hounkonnou, M.N., Martinovic, D., Mitrović, M., Pattison, P. (eds.) Mathematics for Social Sciences and Arts – Algebraic Modeling (Semigroups, arts and social sciences: ubiquitous interactions). Mathematics in Mind. Springer, Berlin (2024)
24. Mitrović, M., Romano, D.A.: Co-congruence on an inverse semigroup with apartness that separates idempotents. https://www.researchgate.net/publication/353039175
25. Mitrović, M., Silvestrov, S.: (Apartness) Isomorphism theorems for basic constructive algebraic structures with special emphasize on constructive semigroups with apartness—an

overview. In Silvestrov, S., Malyarenko, M., Rančić, M. (eds.) Stochastic Processes and Algebraic Structures -From Theory Towards Applications, Volume II: Algebraic Structures and Applications, Västerås and Stockholm, Sweden, October 4–6. Springer, Berlin (2020)
26. Mitrović, M., Silvestrov, S., Crvenković, S., Romano, D. A.: Constructive semigroups with apartness: towards new algebraic theory. J. Phys. Conf. Ser. **1194**, 012076 (2019). https://doi.org/10.1088/1742-6596/1194/1/012076
27. Petrich, M.: Inverse Semigroups. John Wiley & Sons, New York (1984)
28. Romano, D.A.: Natural co-order and some co-congruences on inverse semigroup with apartness. Acta Universitatis Apulensis **73**, 101–138 (2023)
29. Ruitenburg, W.: Inequality in constructive mathematics. Notre Dame J. Formal Logic **32**, 533–553 (1991)
30. Sommaruga G.: Introduction. In Sommaruga G. (ed.) Foundational Theories of Classical and Constructive Mathematics. Springer, Berlin (2011)
31. Troelstra, A.S., van Dalen, D.: Constructivism in Mathematics, An Introduction, (two volumes). North-Holland, Amsterdam (1988)
32. Troelstra, A.S.: From constructivism to computer science. Theor. Comput. Sci. **211**, 233–252 (1999)
33. Uspenskii, V.A., Semenov, A.L. (eds.): Lecture Notes in Computer Science, vol. 122. Springer, Berlin (1981)
34. Wang, H.: Eighty Years of Foundational Studies. Dialectica **12**, 466–497 (1958)
35. Yengui, I.: Constructive Algebra: The Quillen-Suslin Theorem. In: Bridges, D., Ishihara, H., Rathjen, M., Schwichtenberg, H. (eds.) Handbook of Constructive Mathematics, pp. 114–149. Cambridge University Press, Cambridge (2023)

Part II
Semigroups in Forming Some Algebraic Structures

On Graph Inverse Semigroups

Yanhui Wang and Zhengpan Wang

Abstract A graph inverse semigroup $I(\Gamma)$ is a semigroup constructed from a directed graph Γ, where, roughly speaking, elements correspond to paths in the graph. The main purpose of this overview is to give an introduction to the basic research of graph inverse semigroups including Green's relations, idempotents, isomorphisms, congruences, and subsemigroups. In addition, this chapter also gives a short note to state much more research perspectives such as the connection with Cuntz-Krieger algebras, the connection with Leavitt path algebras, and the topological structures. This overview is not, of course, a comprehensive one.

Keywords Graph inverse semigroups · Isomorphisms · Congruences · Subsemigroups

1 Introduction

We recall that an *inverse semigroup* is a semigroup S such that for every $a \in S$ there exists a unique element $a^{-1} \in S$ such that $a = aa^{-1}a$ and $a^{-1} = a^{-1}aa^{-1}$. The book by Lawson [16] is a standard reference for the theory of inverse semigroups and their connections to other fields: any undefined notation and concepts about inverse semigroups that are used in this chapter may be found in [16]. In particular, we shall make use (often without comment) of the elementary fact that idempotents of an inverse semigroup commute. We shall also make use of the natural partial order on an inverse semigroup S (defined by $a \leq b$ if $a = eb$ for some idempotent e of S).

Y. Wang (✉)
College of Mathematics and Systems Science, Shandong University of Science and Technology, Qingdao, China
e-mail: yanhuiwang@sdust.edu.cn

Z. Wang
School of Mathematics and Statistics, Southwest University, Chongqing, China
e-mail: zpwang@swu.edu.cn

M. Mitrović, M. N. Hounkonnou (eds.), *Algebra Without Borders*, Mathematics in Mind, https://doi.org/10.1007/978-3-031-86477-3_4

All of the inverse semigroups that arise in this chapter (except nontrivial groups!) have a zero, which we denote by 0 (or 0_S if we need to specify the inverse semigroup S under consideration) and all homomorphisms under consideration will map 0 to 0. Thus, *(unless it is a group) an inverse semigroup S has a zero, and a homomorphism $f : S \to T$ between inverse semigroups S and T will be assumed to map 0_S onto 0_T.*

All graphs under consideration in this chapter will be *directed* graphs. A graph $\Gamma = (\Gamma^0, \Gamma^1, \mathbf{s}, \mathbf{r})$ consists of the *set of vertices* Γ^0 and the *set of edges* Γ^1, together with mappings $\mathbf{s}, \mathbf{r} : \Gamma^1 \to \Gamma^0$, called the *source mapping* and the *range mapping* for Γ, respectively. In what follows *we always assume that* $\Gamma = (\Gamma^0, \Gamma^1, \mathbf{s}, \mathbf{r})$ *is a graph*. The graph Γ is called *finite* if both sets Γ^0 and Γ^1 are finite and *infinite* in the other case.

Graph inverse semigroups were first introduced by Ash and Hall [5] (for a restricted class of directed graphs) in connection with their study of the partially ordered set of $\mathcal{J}$-classes of a semigroup. Graph inverse semigroups generalize the polycyclic monoids introduced by Nivat and Perrot [24] and arise very naturally in the extensive theories of graph C^*-algebras (see the paper by Kumjian et al. [15] or Paterson's paper [26] or his book [25] and Leavitt path algebras (see the book by Abrams et al. [3]). Graph inverse semigroups have been studied in their own right by several authors, for example, Costa and Steinberg [8]; Jones and Lawson [13]; Krieger [14], Lawson [17]; Luo and Wang [20]; Meakin et al. [21]; Mesyan and Mitchell [22]; and Wang [27].

2 Basic Properties of a Graph Inverse Semigroup

We start with some basic concepts. Let $\Gamma = (\Gamma^0, \Gamma^1, \mathbf{s}, \mathbf{r})$ be a graph. For each vertex $v \in \Gamma^0$, $\mathbf{s}^{-1}(v) = \{e \in \Gamma^1 : \mathbf{s}(e) = v\}$ and the *out-degree* of v is $|\mathbf{s}^{-1}(v)|$, the number of directed edges with source v; $\mathbf{r}^{-1}(v) = \{e \in \Gamma^1 : \mathbf{r}(e) = v\}$, and the *in-degree* of v is $|\mathbf{r}^{-1}(v)|$, the number of directed edges with range v. A vertex $v \in \Gamma^0$ with out-degree 0 is referred to as a *sink*. For an edge e, we allow for the possibility that $\mathbf{s}(e) = \mathbf{r}(e) = v \in \Gamma^0$, in which case e is a *loop* at v. A *path* in Γ is a finite sequence $p = e_1e_2 \ldots e_n$ of edges $e_i \in \Gamma^1$ with $\mathbf{r}(e_i) = \mathbf{s}(e_{i+1})$ for $i = 1, \ldots, n-1$. We define $\mathbf{s}(p) = \mathbf{s}(e_1)$ and $\mathbf{r}(p) = \mathbf{r}(e_n)$ and refer to p as a path from $\mathbf{s}(p)$ to $\mathbf{r}(p)$. We also consider a vertex v as being an *empty path* (i.e., a path with no edges) based at v and with $\mathbf{s}(v) = \mathbf{r}(v) = v$. We denote the set of paths of Γ by $\mathbf{Path}(\Gamma)$.

It is convenient to extend the notation so as to allow paths in which edges are read in either the direct or inverse direction. To do this, we associate with each edge e an "inverse edge" e^{-1} (sometimes called a "ghost edge" by some authors) with $\mathbf{s}(e^{-1}) = \mathbf{r}(e)$ and $\mathbf{r}(e^{-1}) = \mathbf{s}(e)$. We denote by Γ^{-1} the set $\{e^{-1} : e \in \Gamma^1\}$ and assume that $\Gamma^1 \cap \Gamma^{-1} = \emptyset$ and that the map $e \mapsto e^{-1}$ is a bijection from Γ^1 to Γ^{-1}. With this convention, for each path $p = e_1e_2 \ldots e_n$ in Γ we have an *inverse path* $p^{-1} = e_n^{-1} \dots e_2^{-1} e_1^{-1}$ and vice versa. As usual, $\mathbf{s}(p^{-1}) = \mathbf{s}(e_n^{-1}) = \mathbf{r}(e_n)$ and $\mathbf{r}(p) = \mathbf{r}(e_1^{-1}) = \mathbf{s}(e_1)$. The graph Γ is said to be *connected* if for all $v, w \in \Gamma^0$ there exist vertices $v_0, v_1, \dots, v_n \in \Gamma^0$ such that $v_0 = v$, $v_n = w$, and either $v_i = \mathbf{s}(e_i)$ and $v_{i+1} = \mathbf{r}(e_i)$ or $v_i = \mathbf{r}(e_i)$ and $v_{i+1} = \mathbf{s}(e_i)$ for some edge $e_i \in \Gamma^1$, $i = 0, 1, \dots, n-1$. While Γ is said to be *strongly connected* if for all $v, w \in \Gamma^0$ there is at least one path p with $\mathbf{s}(p) = v$ and $\mathbf{r}(p) = w$. A *strongly connected component* of Γ is a directed subgraph Σ maximal with respect to the property that for all $v, w \in \Sigma^0$ there is at least one path p in Σ with $\mathbf{s}(p) = v$ and $\mathbf{r}(p) = w$. A *cycle* is a path $c = e_1 e_2 \dots e_n$ such that $\mathbf{s}(c) = \mathbf{r}(c)$ and $\mathbf{s}(e_i) \neq \mathbf{s}(e_j)$ if $i, j \in \{1, 2, \dots, n\}$ and $i \neq j$. Two paths $p, q \in \mathbf{Path}(\Gamma)$ are *conjugate* if there are paths $r, s \in \mathbf{Path}(\Gamma)$ such that $p = rs$ and $q = sr$. Equivalently, there exists a path $t \in \mathbf{Path}(\Gamma)$ such that $tp = qt$. Conjugate paths must be circuits in Γ, and conjugacy amounts to the selection of an alternative initial edge. In particular, two cycles c_1 and c_2 are said to be *conjugate* if $c_1 = e_1 e_2 \dots e_n$ and $c_2 = e_i e_{i+1} \dots e_n e_1 \dots e_{i-1}$ for some i. Sometimes, we say that c_1 is a *cyclic permutation* of c_2. The graph Γ is *acyclic* if it has no nontrivial cycles. If $u = vw$ are paths, then v is called a *prefix* of u and a *proper prefix* if w is not the empty path, whereas w is called a *suffix* of u and a *proper suffix* if v is not the empty path. For any two paths u, v in Γ, we say that u, v are *prefix comparable* if one of u, v is a prefix of the other.

Define the *graph inverse semigroup* $I(\Gamma)$ of Γ as the semigroup generated by $\Gamma^0 \cup \Gamma^1 \cup \Gamma^{-1}$ together with a zero 0 subject to the relations

(V) $uv = \delta_{uv} u$ for all $u, v \in \Gamma^0$,

(E1) $\mathbf{s}(e)e = e = e\mathbf{r}(e)$ for each $e \in \Gamma^1$

(E2) $\mathbf{r}(e)e^{-1} = e^{-1} = e^{-1}\mathbf{s}(e)$ for each $e \in \Gamma^1$, and

(CK1) $e^{-1} f = \delta_{ef} \mathbf{r}(e)$ for all $e, f \in \Gamma^1$,

where δ is Kronecker delta.

We emphasize that condition (V) of the definition above implies that $v^2 = v$ for all $v \in \Gamma^0$, that is, the vertices of Γ are idempotents in $I(\Gamma)$.

To see that $I(\Gamma)$ is an inverse semigroup, we describe an inverse semigroup T over $\mathbf{Path}(\Gamma)$ together with 0. Set

$$T = \{pq^{-1} : p, q \in \mathbf{Path}(\Gamma), \mathbf{r}(p) = \mathbf{r}(q)\} \cup \{0\}.$$

We define an involution on T by the rule that

$$0^{-1} = 0 \text{ and } (pq^{-1})^{-1} = qp^{-1}$$

and a binary operation on T by the rule that

$$(pq^{-1})(rs^{-1}) = \begin{cases} pts^{-1} & \text{if } r = qt \text{ for some (possibly empty) path } t \\ p(st)^{-1} & \text{if } q = rt \text{ for some (possibly empty) path } t \\ 0 & \text{otherwise,} \end{cases}$$

for any $pq^{-1}, rs^{-1} \in T$. Note that for paths p, q, the expression pq is defined only when $\mathbf{r}(p) = \mathbf{s}(q)$ and, in that case, is just the path obtained in the obvious way by juxtaposition. For any path p, we specify that $\mathbf{s}(p)p = p = p\mathbf{r}(p)$. Note that the product of pq^{-1} and rs^{-1} is nonzero if and only if q and r are prefix comparable. It is easy to see that the set $E(T)$ of idempotents of T is $E(T) = \{pp^{-1} : p \in \mathbf{Path}(\Gamma)\} \cup \{0\}$. The product on $E(T)$ of two nonzero idempotents pp^{-1}, qq^{-1} is given by

$$(pp^{-1})(qq^{-1}) = \begin{cases} pp^{-1} & \text{if } q \text{ is a prefix of } p \\ qq^{-1} & \text{if } p \text{ is a prefix of } q \\ 0 & \text{otherwise.} \end{cases}$$

Hence, $E(T)$ is a semilattice.

Lemma 1 *The set T, with the above involution and binary operation, is an inverse semigroup with semilattice $E(T) = \{pp^{-1} : p \in \mathbf{Path}(\Gamma)\} \cup \{0\}$.*

Proof We first show that the binary operation on T is associative. If one of the terms is 0, then this is obvious. So let $p_iq_i^{-1} \in T$, $i = 1, 2, 3$, $\alpha = \big((p_1q_1^{-1})(p_2q_2^{-1})\big)(p_3q_3^{-1})$ and $\beta = (p_1q_1^{-1})\big((p_2q_2^{-1})(p_3q_3^{-1})\big)$. We need to show that $\alpha = \beta$.

The cases that give $\alpha \neq 0$ are as follows (for appropriate paths t_1, t_2):

(i) $q_2 = p_3t_1, q_1 = p_2t_2, \alpha = p_1(q_3t_1t_2)^{-1}$.
(ii) $q_2 = p_3t_1, p_2 = q_1t_2, \alpha = p_1t_2(q_3t_1)^{-1}$.
(iii) $p_3 = q_2t_1, q_1 = p_2t_1t_2, \alpha = p_1(q_3t_2)^{-1}$.
(iv) $p_3 = q_2t_1, p_2t_1 = q_1t_2, \alpha = p_1t_2q_3^{-1}$.

One checks directly that in each of these four cases, $\alpha = \beta$. (In case (iv), one needs to consider separately the cases $p_2 = q_1t$ and $q_1 = p_2t$). So if $\alpha \neq 0$, then $\alpha = \beta$. Similarly, one shows that if $\beta \neq 0$, then $\beta = \alpha$. The associative law then follows.

Next, we show that each element of T has a unique inverse element. It is trivial for zero element. For any $pq^{-1} \in T$, suppose that rs^{-1} is an inverse of pq^{-1}. Since $(pq^{-1})(rs^{-1})(pq^{-1}) = pq^{-1}$, it follows that q, r and s, p are prefix comparable, respectively. If p is a proper prefix of s, that is, $s = pt$ for some non-empty path t, then $(pq^{-1})(rs^{-1})(pq^{-1}) = (pq^{-1})(r(qt)^{-1})$, which is equal to either $pt_1(qt)^{-1}$ ($r = qt_1$ for some (possibly empty) path t_1) or else $p(qtt_2)^{-1}$ ($q = rt_2$ for some (possibly empty) path t_2). As t is non-empty path, it follows that $qt \neq q$ and $qtt_2 \neq q$. This contradicts $(pq^{-1})(rs^{-1})(pq^{-1}) = pq^{-1}$. Similarly, by considering $(rs^{-1})(pq^{-1})(rs^{-1}) = rs^{-1}$ we obtain that s is not a proper prefix of p. So $p = s$. Similarly, $r = q$ and $rs^{-1} = qp^{-1}$. Obviously, $(pq^{-1})(qp^{-1})(pq^{-1}) = pq^{-1}$, $(qp^{-1})(pq^{-1})(qp^{-1}) = qp^{-1}$. □

Proposition 1 *The map $\phi : I(\Gamma) \to T$ given by for any $v \in \Gamma^0$, $e \in \Gamma^1$ and the zero element 0,*

$$\phi(v) = vv^{-1},\ \phi(e) = e\mathbf{r}(e)^{-1},\ \phi(e^{-1}) = \mathbf{r}(e)e^{-1}\ \textit{and}\ \phi(0) = 0,$$

is a semigroup isomorphism.

Proof Let $W = \Gamma^0 \cup \Gamma^1 \cup \Gamma^{-1}$ and let F_W^0 be the free semigroup generated by W with a zero element 0. Then ϕ extends in the obvious way to a homomorphism from F_W^0 to T, denoted $\phi : F_W^0 \to T$. The map $\phi : F_W^0 \to T$ is onto since every $\alpha \in T$ is a product of $\phi(w)'s$: for example, if $p = p_1p_2\cdots p_k$ and $q = q_1q_2\cdots q_l$, and $pq^{-1} \in T$, then $pq^{-1} = \phi(p_1)\phi(p_2)\cdots\phi(p_k)\phi(q_l^{-1})\cdots\phi(q_2^{-1})\phi(q_1^{-1})$, where $p_1, p_2, \ldots, p_k, q_1, q_2, \ldots, q_l \in W$.

Let $\sim$ be the congruence on F_W^0 generated by relations (V), (E1), (E2), and (CK1) above. To prove that ϕ descends to a homomorphism $\tilde{\phi}$ from $I(\Gamma) = F_W^0/\sim$ onto T, we just have to check that the elements of $\phi(W)$ satisfy the relations in T corresponding to (V), (E1), (E2), and (CK1). This is straightforward.

For (V), $\phi(u)\phi(v) = (uu^{-1})(vv^{-1})$, which is equal to either u (if $u = v \in \Gamma^0$) or else 0 (if $u \neq v \in \Gamma$), that is, $\phi(u)\phi(v) = \delta_{\phi(u)\phi(v)}\phi(u)$.

For (E1), $\phi(\mathbf{s}(e))\phi(e) = (\mathbf{s}(e)(\mathbf{s}(e))^{-1})(e\mathbf{r}(e)^{-1}) = e\mathbf{r}(e)^{-1} = \phi(e)$ as $(\mathbf{s}(e))^{-1} = \mathbf{s}(e)$ for $e \in \Gamma^1$. Dually, we have $\phi(e)\phi(\mathbf{r}(e)) = \phi(e)$ for $e \in \Gamma^1$.

For (E2), $\phi(\mathbf{r}(e))\phi(e^{-1}) = (\mathbf{r}(e)(\mathbf{r}(e))^{-1})(\mathbf{r}(e)e^{-1}) = \mathbf{r}(e)e^{-1} = \phi(e^{-1})$ for each $e \in \Gamma^1$. Dually, we have $\phi(e^{-1})\phi(\mathbf{s}(e)) = \phi(e^{-1})$ for each $e \in \Gamma^1$.

(CK1) $\phi(e^{-1})\phi(f) = (\mathbf{r}(e)e^{-1})(f\mathbf{r}(f)^{-1})$ is equal to either 0 if $e \neq f$ or else $\mathbf{r}(f)^{-1}$ if $e = f$ for all $e, f \in \Gamma^1$.

To show that $\tilde{\phi} : I(\Gamma) \to T$ is an isomorphism, suppose that for some $w_1, w_2 \in F_W^0$ we have $\tilde{\phi}(w_1) = \tilde{\phi}(w_2)$. It is sufficient to show that $w_1 \sim w_2$. Using (V), (E1), (E2), and (CK1), we can find $w_3, w_4 \in F_W^0$, where $w_1 \sim w_3$, $w_2 \sim w_4$ and either $w_3 = w_4 = 0$, in which case $w_1 \sim w_2$, or, in an obvious notation, $w_3 = pq^{-1}$ and $w_4 = rs^{-1}$ for $p, q, r, s \in \mathbf{Path}(\Gamma)$ with $\mathbf{r}(p) = \mathbf{r}(q)$, $\mathbf{r}(r) = \mathbf{r}(s)$. Then

$$pq^{-1} = \tilde{\phi}(w_3) = \tilde{\phi}(w_1) = \tilde{\phi}(w_2) = \tilde{\phi}(w_4) = rs^{-1}$$

and $p = r$, $q = s$. So $w_3 = w_4$ and $w_1 \sim w_2$. □

According to Lemma 1 and Proposition 1, $I(\Gamma)$ is in fact an *inverse semigroup*. Moreover, every nonzero element of a graph inverse semigroup $I(\Gamma)$ may be uniquely written in the form pq^{-1} where p and q are (possibly empty) paths with $\mathbf{r}(p) = \mathbf{r}(q)$. We refer to this as the *canonical form* of a nonzero element of $I(\Gamma)$. The inverse of an element pq^{-1} is of course qp^{-1}. If pq^{-1} and rs^{-1} are nonzero elements of $I(\Gamma)$, then the product $(pq^{-1})(rs^{-1})$ is nonzero if and only if either q is a prefix of r (i.e., $r = qt$ for some (possibly empty) path t, in which case $(pq^{-1})(rs^{-1}) = pts^{-1}$), or else r is a prefix of q (i.e., $q = rt$ for some (possibly empty) path t, in which case $(pq^{-1})(rs^{-1}) = p(st)^{-1}$). The nonzero idempotents are of the form pp^{-1} for some (possibly empty) path p. The semilattice of idempotents is $I(\Gamma)$ is denoted by $E(I(\Gamma))$, that is, $E(I(\Gamma)) = \{pp^{-1} : p \in \mathbf{Path}(\Gamma)\} \cup \{0\}$.

If Γ is the graph with one vertex and one edge, then $I(\Gamma)$ is the bicyclic monoid with a (removable) zero. If $\Gamma = B_n$ is the bouquet of $n > 1$ circles (i.e., the graph with one vertex and n directed edges), then $I(\Gamma)$ is isomorphic to the polycyclic monoid P_n. This is the inverse monoid generated (as an inverse monoid) by a set A with $|A| = n$ subject to the defining relations $a^{-1}a = 1$ and $a^{-1}b = 0$ for all $a, b \in A$ with $b \neq a$. (Here we regard A as a set of labels for the edges of B_n.) The monoid P_n was introduced by Nivat and Perrot [24] as the syntactic monoid of the "correct parenthesis" language with n sets of parentheses. It was rediscovered in the operator algebra literature where it is referred to as the Cuntz monoid, used in the construction of the Cuntz algebra O_n (see Paterson's book [25] for details). The algebra constructed from the graph B_n in the original paper by Leavitt [19] is what is now referred to as the Leavitt path algebra of this graph (see (Abrams et al. [3]) for details).

Let S be a semigroup. The relation $\leq_{\mathcal{L}}$ on S is defined by the rule that for all $a, b \in S$,

$$a \leq_{\mathcal{L}} b \text{ if and only if } S^1a \subseteq S^1b.$$

It is clear that $\leq_{\mathcal{L}}$ is a pre-order, which is reflective and transitive. We denote the associated equivalence relation by $\mathcal{L}$, so that for all $a, b \in S$,

$$a\mathcal{L}b \text{ if and only if } S^1a = S^1b.$$

The relation $\leq_{\mathcal{R}}$ and $\mathcal{R}$ are the left-right duals of $\leq_{\mathcal{L}}$ and $\mathcal{L}$, that is, for all $a, b \in S$,

$$a \leq_{\mathcal{R}} b \text{ if and only if } aS^1 \subseteq bS^1$$

and

$$a\mathcal{R}b \text{ if and only if } aS^1 = bS^1.$$

The relation $\leq_{\mathcal{J}}$ on S is defined by the rule that for all $a, b \in S$,

$$a \leq_{\mathcal{J}} b \text{ if and only if } S^1aS^1 \subseteq S^1bS^1.$$

We denote the associated equivalence relation by $\mathcal{J}$, so that for all $a, b \in S$,

$$a\mathcal{J}b \text{ if and only if } S^1aS^1 = S^1bS^1.$$

We denote the intersection and the join of $\mathcal{L}$ and $\mathcal{R}$ by $\mathcal{H}$ and $\mathcal{D}$, respectively, that is,

$$\mathcal{H} = \mathcal{L} \cap \mathcal{R} \text{ and } \mathcal{D} = \mathcal{L} \vee \mathcal{R}.$$

As the composition of $\mathcal{L}$ and $\mathcal{R}$ commutes, we obtain that

$$\mathcal{D} = \mathcal{L} \circ \mathcal{R} = \mathcal{R} \circ \mathcal{L}.$$

The relations $\mathcal{L}$, $\mathcal{R}$, $\mathcal{J}$, $\mathcal{H}$, and $\mathcal{D}$ are known as Green's relations.

Lemma 2 *For any two nonzero elements $pq^{-1}, rs^{-1} \in I(\Gamma)$ we have that:*

(i) $pq^{-1} \leq_{\mathcal{L}} rs^{-1}$ *if and only if s is a prefix of q.*
(ii) $pq^{-1} \leq_{\mathcal{R}} rs^{-1}$ *if and only if r is a prefix of p.*
(iii) $pq^{-1} \leq_{\mathcal{J}} rs^{-1}$ *if and only if $\mathbf{s}(t) = \mathbf{r}(r)$ and $\mathbf{r}(t) = \mathbf{r}(p)$ for some path t in Γ.*

Proof

(i) Suppose that $pq^{-1}, rs^{-1} \in I(\Gamma)$. Then by the definition of $\leq_{\mathcal{L}}$, $pq^{-1} \leq_{\mathcal{L}} rs^{-1}$ in $I(\Gamma)$ if and only if $I(\Gamma)^1(pq^{-1}) \subseteq I(\Gamma)^1(rs^{-1})$. Since $r^{-1} \in I(\Gamma)$ and $\mathbf{r}(s) \in I(\Gamma)$, we deduce that $I(\Gamma)s^{-1} = I(\Gamma)\mathbf{r}(s)s^{-1} = I(\Gamma)r^{-1}rs^{-1} \subseteq I(\Gamma)^1(rs^{-1})$ and certainly we have $I(\Gamma)^1(rs^{-1}) \subseteq I(\Gamma)s^{-1}$. It follows that $I(\Gamma)s^{-1} = I(\Gamma)^{-1}(rs^{-1})$. Similarly, $I(\Gamma)^1(pq^{-1}) = I(\Gamma)q^{-1}$. So $I(\Gamma)^1 \subseteq I(\Gamma)^1(rs^{-1})$ if and only if $I(\Gamma)q^{-1} \subseteq I(\Gamma)s^{-1}$. It is equivalent to $q^{-1} \in I(\Gamma)s^{-1}$, which is in turn equivalent to $q^{-1} = t^{-1}s^{-1}$ for some path t in Γ, that is $q = st$, as required.
(ii) By the dual of the proof of (i), $pq^{-1} \leq_{\mathcal{R}} rs^{-1}$ if and only if $(pq^{-1})I(\Gamma)^1 \subseteq (rs^{-1})I(\Gamma)^1$, if and only if $p \in rI(\Gamma)$, if and only if $p = rt$ for some path t in Γ, as required.
(iii) Noticing that $rs^{-1} = r\mathbf{r}(r)s^{-1}$ and that $\mathbf{r}(r) = \mathbf{r}(r)\mathbf{s}(s) = r^{-1}(rs^{-1})s$, we get $I(\Gamma)(rs^{-1})I(\Gamma) = I(\Gamma)\mathbf{r}(r)I(\Gamma)$. Now suppose that there is a path t in Γ such that $\mathbf{s}(t) = \mathbf{r}(r)$ and $\mathbf{r}(t) = \mathbf{r}(p)$. Then $\mathbf{r}(p) = \mathbf{r}(t) = t^{-1}t = t^{-1}\mathbf{s}(t)t \in I(\Gamma)\mathbf{r}(r)I(\Gamma)$. Hence, $pq^{-1} = p\mathbf{r}(p)q^{-1} \in I(\Gamma)\mathbf{r}(r)I(\Gamma)$, and since $I(\Gamma)\mathbf{r}(r)I(\Gamma) = I(\Gamma)(rs^{-1})I(\Gamma)$, we have $pq^{-1} \in I(\Gamma)(rs^{-1})I(\Gamma)$. It follows that $pq^{-1} \leq_{\mathcal{J}} rs^{-1}$.

Conversely, if $pq^{-1} \leq_{\mathcal{J}} rs^{-1}$, then $pq^{-1} \in I(\Gamma)(rs^{-1})I(\Gamma)$, and therefore, $\mathbf{r}(p) = p^{-1}pq^{-1}q \in I(\Gamma(rs^{-1})I(\Gamma) = I(\Gamma)\mathbf{r}(r)I(\Gamma)$. Hence, $\mathbf{r}(p) = t_1t_2^{-1}t_3t_4^{-1}$ for some paths t_1, t_2, t_3, t_4 in Γ with $\mathbf{s}(t_2) = \mathbf{r}(r) = \mathbf{s}(t_3)$, $\mathbf{r}(t_2) = \mathbf{r}(t_1)$ and $\mathbf{r}(t_3) = \mathbf{r}(t_4)$. By the uniqueness of the representations of elements of $I(\Gamma)$ discussed above, for $t_1t_2^{-1}t_3t_4^{-1}$ to be a vertex we must have $t_1, t_4 \in \Gamma^0$. Hence, $t_1 = \mathbf{r}(p) = t_4$, and therefore, $\mathbf{r}(p) = t_2^{-1}t_3$. It follows that $t_2 = t_3$, and in particular, $\mathbf{s}(t_2) = \mathbf{r}(r)$ and $\mathbf{r}(t_2) = \mathbf{r}(t_1) = \mathbf{r}(p)$, as required. □

As a corollary of Lemma 2, we get the following description of Green's relations on a graph inverse semigroup (see [22]).

Corollary 1 *For any two nonzero elements $pq^{-1}, rs^{-1} \in I(\Gamma)$ we have that:*

(i) $pq^{-1}\mathcal{L}rs^{-1}$ *if and only if $q = s$.*
(ii) $pq^{-1}\mathcal{R}rs^{-1}$ *if and only if $p = r$.*
(iii) $pq^{-1}\mathcal{J}rs^{-1}$ *if and only if $\mathbf{r}(p)$ and $\mathbf{r}(r)$ are in the same strongly connected component of Γ.*
(iv) $pq^{-1}\mathcal{H}rs^{-1}$ *if and only if $pq^{-1} = rs^{-1}$.*
(v) $pq^{-1}\mathcal{D}rs^{-1}$ *if and only if $\mathbf{r}(p) = \mathbf{r}(r)$.*

Proof According to Lemma 2, (i)–(iii) are immediately obtained, while (iv) follows from (i) and (ii). For (v), we have $pq^{-1}\mathcal{D}rs^{-1}$ if and only if $pq^{-1}\mathcal{L}t_1t_2^{-1}$ and $t_1t_2^{-1}\mathcal{R}rs^{-1}$ for some paths t_1, t_2 in Γ such that $\mathbf{r}(t_1) = \mathbf{r}(t_2)$. By (i) and (ii), this is equivalent to $q = t_2$ and $t_1 = r$ for some path t_1, t_2 in Γ such that $\mathbf{r}(t_1) = \mathbf{r}(t_2)$ which is equivalent to $\mathbf{r}(r) = \mathbf{r}(q) = \mathbf{r}(p)$. □

Proposition 2 *Let* $\leq$ *be the natural partial order on* $I(\Gamma)$. *The following statements hold:*

(i) *For any two nonzero elements* $pq^{-1}, rs^{-1} \in I(\Gamma)$, $pq^{-1} \leq rs^{-1}$ *if and only if* $p = rt$ *and* $q = st$ *for some path* t *in* Γ.
(ii) *For any two nonzero idempotents* $qq^{-1} \leq pp^{-1}$ *on* $I(\Gamma)$ *if and only if* p *is a prefix of* q *in* Γ.
(iii) *The vertices of* Γ *are the maximal idempotents with respect to* $\leq$ *on* $I(\Gamma)$.
(iv) *An idempotent* g *is maximal in* $I(\Gamma)\backslash\Gamma^0$ *with respect to* $\leq$ *if and only if* $g = ee^{-1}$ *for some* $e \in \Gamma^1$.

Proof

(i) Suppose that $p = rt$ and $q = st$ for some path t in Γ. Then we have $pq^{-1} = rt(st)^{-1} = rtt^{-1}s^{-1} = rtt^{-1}\mathbf{r}(r)s^{-1} = (rtt^{-1}r^{-1})rs^{-1}$, which implies that $pq^{-1} \leq rs^{-1}$, since $rtt^{-1}t^{-1}$ is an idempotent.
Conversely, suppose that $pq^{-1} \leq rs^{-1}$. Then we have $pq^{-1} = (t_1t_1^{-1})rs^{-1}$ for some path t_1 in Γ. Since $tt^{-1}rs^{-1} \neq 0$, there is some path t in Γ such that either $r = t_1t$ or $t_1 = rt$. In the first case, $pq^{-1} = t_1t_1^{-1}rs^{-1} = rs^{-1}$, and hence, $p = rt$ and $q = st$, where $t = r = s$. In the second case, $uv^{-1} = t_1t_1^{-1}rs^{-1} = rtt^{-1}s^{-1}$, and hence, $p = rt$ and $q = st$, as required.
(ii) is immediately followed by (i).
(iii) Suppose that $u \in \Gamma^0$ and $g \in E(I(\Gamma))$ with $u \leq g$. Then $g \neq 0$, and hence, $g = tt^{-1}$ and $u = tss^{-1}t^{-1}$ for some paths t, s in Γ. Since u is a vertex, this can happen only if $g = t = s = u$, and hence, u is maximal.
Conversely, suppose that g is an idempotent maximal in $E(I(\Gamma))$. Then $u \neq 0$, and hence, $u = tt^{-1}$ for some path t in Γ. Thus, $u = tt^{-1} \leq \mathbf{s}(t)$ by (ii). Since u is maximal, it implies that $u = t = \mathbf{s}(t)$, and hence, $u \in \Gamma^0$.
(iv) Let $e \in \Gamma^1$, and suppose that $ee^{-1} \leq g$ for some $g \in E(I(\Gamma))\backslash\Gamma^0$. Then $g \neq 0$, and hence, $g = tt^{-1}$ and $ee^{-1} = tss^{-1}t^{-1}$ for some paths s, t in Γ. Since $e \in \Gamma^1$, this implies that either $e = t$ and $s = \mathbf{r}(e)$, or $e = s$ and $t = \mathbf{s}(e)$. In the second case, $g \in \Gamma^0$, contrary to the assumption. Thus, $e = t$, and therefore, $g = ee^{-1}$. Hence, ee^{-1} is maximal in $E(I(\Gamma))\backslash\Gamma^0$.

Conversely, suppose that g is an idempotent maximal in $E(I(\Gamma))\backslash\Gamma^0$. Then $g \neq 0$, and hence, $g = tt^{-1}$ for some path t in Γ. Since $tt^{-1} \notin \Gamma^0$, we can write $t = es$ for some $e \in \Gamma^1$ and some path s in Γ, and hence, $g = ess^{-1}e^{-1} \leq ee^{-1}$ by (ii). Since g is maximal in $E(I(\Gamma))\backslash\Gamma^0$, and $ee^{-1} \in E(I(\Gamma))\backslash\Gamma^0$, this implies that $s \in \Gamma^0$ and $g = ee^{-1}$. □

Recall that an inverse semigroup S with zero $0 \in S$ is said to be *E^*-unitary*, if whenever $e \in E(S)$, $s \neq 0$ and $s \in S$ with $e \leq s$, then $s \in E(S)$. It is clear that the graph inverse semigroup $I(\Gamma)$ is E^*-unitary.

Let $(E, \leq)$ be a partially ordered set, or poset for short. Given $g, h \in E$ we say that g *cover* h if $h \leq g$ and there is no $g_1 \in E$ such that $h \lneq g_1 \lneq g$. For each $g \in E$ define $\hat{g}$ to be the set of elements of E that are covered by g. An element $g \in E$ is said to be *0-minimal* if it covers the zero element.

Proposition 3 *For the graph inverse semigroup $I(\Gamma)$, $I(\Gamma)$ has no 0-minimal idempotents if and only if the out-degree of each vertex in Γ is at least* 1.

Proof Let u be a vertex with in-degree at least 1, and let e be an edge with $\mathbf{r}(e) = u$. Let t be a path with $\mathbf{s}(t) = u$, where we include the possibility that t is the empty path at u. Then $et(et)^{-1} \leq ee^{-1}$. It follows that if the in-degree of each vertex is at least 1, then there can be no 0-minimal idempotents. Now let u be a vertex with out-degree 0. Then u is a 0-minimal idempotent. □

3 Isomorphic Graph Inverse Semigroups

In this section, we investigate the coincidence between the isomorphism of graphs and isomorphism of the corresponding graph inverse semigroups (see [22]).

Let $\Gamma = (\Gamma^0, \Gamma^1, \mathbf{s}, \mathbf{r})$, $\Delta = (\Delta^0, \Delta^1, \mathbf{s}, \mathbf{r})$ be graphs. A mapping f from $\Gamma^0 \cup \Gamma^1$ to $\Delta^0 \cup \Delta^1$ is called a *graph homomorphism from Γ to Δ* if $f(\Gamma^0) \subseteq \Delta^0$, $f(\Gamma^1) \subseteq \Delta^1$, $f(\mathbf{s}(e)) = \mathbf{s}(f(e))$ and $f(\mathbf{r}(e)) = \mathbf{r}(f(e))$ for every $e \in \Gamma^1$. Let f be a graph homomorphism from Γ to Δ. Then f can be extended naturally to paths in Γ: for any nonempty path $p = e_1 e_2 \cdots e_n$ of Γ, define $f(p) = f(e_1) f(e_2) \cdots f(e_n)$. It is clear that a graph homomorphism preserves not only vertices and edges but also loops and paths together with their lengths. Furthermore, for any paths p_1, p_2 in Γ with $\mathbf{r}(p_1) = \mathbf{s}(p_2)$, we obviously have $f(p_1 p_2) = f(p_1) f(p_2)$. However, a graph homomorphism may map an edge which is not a loop to a loop. A bijective graph homomorphism f from Γ to Δ is a *graph isomorphism.* The graphs Γ and Δ are said to be *isomorphic* if there exists a graph isomorphism from Γ to Δ. It is straightforward to see that a graph isomorphism is extended to a bijection from the path set of Γ to the path set of Δ and that the inverse of a graph isomorphism is also a graph homomorphism and hence a graph isomorphism.

Theorem 1 *The graphs Γ and Δ are isomorphic if and only if the graph inverse semigroups $I(\Gamma)$ and $I(\Delta)$ are isomorphic.*

Proof First, let f be a graph isomorphism from Γ to Δ. We need to show that f can be extended to a semigroup isomorphism ϕ from $I(\Gamma)$ to $I(\Delta)$. As we mentioned in Sect. 2 every nonzero element in $I(\Gamma)$ can be uniquely written as pq^{-1}, where p, q are (possibly empty) paths with $\mathbf{r}(p) = \mathbf{r}(q)$. Now for any nonzero element $pq^{-1} \in I(\Gamma)$, define ϕ by $\phi(pq^{-1}) = f(p)(f(q))^{-1}$ (note that p or q may be empty); we also define $\phi(0) = 0$. It follows from the definition of a graph homomorphism and its extension to paths that ϕ is well defined. Next, we verify that ϕ is a semigroup homomorphism from $I(\Gamma)$ to $I(\Delta)$. We can see from the definition of ϕ that we only need to show that ϕ preserves the multiplication of nonzero elements. For any nonzero elements $pq^{-1}, rs^{-1} \in I(\Gamma)$, we observe that q is a prefix of r if and only if $f(q)$ is a prefix of $f(r)$ and that r is a prefix of q if and only if $f(r)$ is a prefix of $f(q)$ since both f and f^{-1} are graph homomorphisms. Hence, $(pq^{-1})(rs^{-1}) \neq 0$ if and only if $\phi(pq^{-1})\phi(rs^{-1}) = (f(p)(f(q))^{-1})(f(r)(f(s))^{-1}) \neq 0$. Moreover, if q is a prefix of r which means that there exists a path r_1 in Γ such that $r = qr_1$, then we obtain from condition (CK1) that

$$\begin{aligned}\phi((pq^{-1})(rs^{-1})) = \phi(pr_1s^{-1}) &= f(pr_1)(f(s))^{-1}\\ &= f(p)(f(q))^{-1}f(q)f(r_1)(f(s))^{-1}\\ &= f(p)(f(q))^{-1}f(qr_1)(f(s))^{-1} = \phi(pq^{-1})\phi(rs^{-1}).\end{aligned}$$

So ϕ is a semigroup homomorphism. The bijectivity of ϕ follows from the bijectivity of f.

Conversely, let ϕ be a semigroup isomorphism from $I(\Gamma)$ to $I(\Delta)$. Choose f to be the restriction of ϕ to $\Gamma^0 \cup \Gamma^1$. It follows from Proposition 2(iii) that $f(\Gamma^0) \subseteq \Delta^0$. For any $e \in \Gamma^1$, we suppose that $f(e) = rs^{-1}$ for some path r, s in Δ with $\mathbf{r}(r) = \mathbf{r}(s)$. Then we see from condition (CK1) that $f(\mathbf{r}(e)) = \phi(e^{-1}e) = (f(e))^{-1}f(e) = (sr^{-1})(rs^{-1}) = ss^{-1}$ which is a vertex by the previous discussion. Thus, we see that s is empty. Similarly, we have $\phi(ee^{-1}) = f(e)(f(e))^{-1} = (rs^{-1})(sr^{-1}) = rr^{-1}$. So we obtain from Proposition 2(iv) that rr^{-1} is maximal in $E(I(\Delta)) \setminus \Gamma^0$ which means that $r \in \Delta^1$. So we proved that $f(\Gamma^1) \subseteq \Delta^1$. Moreover, we see from conditions (E1) and (V) that $f(e) = \phi(\mathbf{s}(e)e) = f(\mathbf{s}(e))f(e)$ that $f(\mathbf{s}(e)) = \mathbf{s}(f(e))$. Similarly, we have $f(\mathbf{r}(e)) = \mathbf{r}(f(e))$. Therefore, f is a graph homomorphism. Furthermore, f is bijective since ϕ is. □

4 Congruences on Graph Inverse Semigroups

In this section, we proceed to describe congruences on $I(\Gamma)$ in terms of the underlying graph Γ (see [27]).

A subset H of Γ^0 is *hereditary* if $\mathbf{s}(e) \in H$ always implies $\mathbf{r}(e) \in H$ for $e \in \Gamma^1$. Let H be a hereditary subset of Γ^0, $\Gamma_1^0 = \Gamma^0 \setminus H$, $\Gamma_1^1 = \{e \in \Gamma^1 : \mathbf{r}(e) \notin H\}$, $\mathbf{s}_{\Gamma\setminus H} = \mathbf{s}|_{\Gamma_1^1}$ and $\mathbf{r}_{\Gamma\setminus H} = \mathbf{r}|_{\Gamma_1^1}$. Then we denote the graph $(\Gamma_1^0, \Gamma_1^1, \mathbf{s}_{\Gamma\setminus H}, \mathbf{r}_{\Gamma\setminus H})$ as $\Gamma \setminus H$.

Let ρ be a congruence on $I(\Gamma)$. Then clearly $H = 0\rho \cap \Gamma^0$ is a hereditary subset of Γ^0. Passing to the graph $\Gamma \setminus H$, we shall assume without loss of generality that $H = \emptyset$. So for graph inverse semigroups, we only need to describe these 0-*restricted congruences* where no vertex is equivalent to 0.

Let $p = e_1 \cdots e_n$ be a path in Γ and $v(p)$ denote the set $\{\mathbf{s}(e_i) : i = 1, \ldots, n\}$. If p is empty, we appoint that $v(p) = \emptyset$. An edge e is an *exit* to the path p if there exists i such that $\mathbf{s}(e) = \mathbf{s}(e_i)$ and $e \neq e_i$. The path p is said to be *no-exit* if it does not have an exit. Two cycles c_1 and c_2 are said to be *disjoint* if $v(c_1) \cap v(c_2) = \emptyset$. Let $\mathbb{N}$ be the set of all nonnegative integers. For any cycle c and any positive integer m, we appoint that $c^0 = \mathbf{s}(c)$ and $(c^{-1})^m = c^{-m}$. The next lemma directly follows from the related definitions.

Lemma 3 *For any no-exit cycle c and any path p in Γ with $\mathbf{s}(c) = \mathbf{s}(p)$, there exist paths p_1, p_2 in Γ and $k \in \mathbb{N}$ such that $p = c^k p_1$, $c = p_1 p_2$, and p_1 is a proper prefix of c.*

Lemma 4 *For any no-exit cycle c, any path p in Γ and $k \in \mathbb{Z}^+$, if $\mathbf{s}(c) = \mathbf{s}(p)$, then there exists a cyclic permutation c_1 of c such that*

(1) $p^{-1} c^k p = c_1^k$, *and*
(2) $c^k p p^{-1} = p c_1^k p^{-1}$.

Proof We see from Lemma 3 that $p = c^l p_1$ where $l \in \mathbb{N}$ and p_1 is a path in Γ such that $c = p_1 p_2$ for some path p_2 in Γ. Obviously, $c_1 = p_2 p_1$ is a cyclic permutation of c. It follows from (CK1) that

$$p^{-1} c^k p = p_1^{-1} (c^{-1})^l c^k c^l p_1 = p_1^{-1} c^k p_1 = p_2 c^{k-1} p_1 = c_1^k, \text{ and}$$
$$c^k p p^{-1} = c^k c^l p_1 p^{-1} = c^l p_1 c_1^k p^{-1} = p c_1^k p^{-1}.$$

That is, (1) and (2) hold. □

We write $\bar{\Gamma}^0 := \{v \in \Gamma^0 : |\mathbf{s}^{-1}(v)| = 1\}$, the set of vertices of out-degree one. For any $v \in \bar{\Gamma}^0$, we denote the unique edge in $\mathbf{s}^{-1}(v)$ as e_v. Let W be a subset of $\bar{\Gamma}^0$, $\mathbb{Z}^+$ be the set of all positive integers and $C(W)$ be the set of all cycles whose vertices lie in W. Since all vertices in W have index one, an arbitrary cycle $c \in C(W)$ is no-exit and any two cycles in $C(W)$ are either disjoint or cyclic permutations of each other. A *cycle function* $f : C(W) \to \mathbb{Z}^+ \cup \{\infty\}$ is a function that is invariant under cyclic permutations. A *congruence pair* (W, f) *of* Γ consists of a subset W of $\bar{\Gamma}^0$ and a cycle function f. Let H be a hereditary subset of Γ^0 and (W, f) be a congruence pair of $\Gamma \setminus H$. Then (H, W, f) is called a *congruence triple of* Γ.

Let us illustrate how a congruence gives rise to a congruence triple (respectively, a 0-restricted congruence gives rise to a congruence pair).

Let ρ be a congruence on $I(\Gamma)$, $H = 0\rho \cap \Gamma^0$ and $W = \{v \in \Gamma^0 \setminus H : ee^{-1} \,\rho\, v \text{ for } e \in \mathbf{s}_{\Gamma\setminus H}^{-1}(v)\}$. Later we shall see from Lemma 6 that in this case all vertices in W have index one in $\Gamma \setminus H$. Let $C(W)$ consist of all distinct cycles whose vertices lie in W. For $c \in C(W)$, let $f(c)$ be the smallest positive integer m such that $c^m \,\rho\, \mathbf{s}(c)$. If no power of c is equivalent to $\mathbf{s}(c)$, then we define $f(c) = \infty$. Then $T(\rho) = (H, W, f)$ is a congruence triple.

If ρ is a 0-restricted congruence, then $T(\rho) = (\emptyset, W, f)$. In this case, (W, f) is a congruence pair, which is still denoted by $T(\rho)$.

Conversely, let (W, f) be a congruence pair of Γ and let $\wp(W, f)$ denote the congruence generated by $\mathbf{R}$ consisting of all pairs $(e_v e_v^{-1}, v)$ for $v \in W$ and $(c^{f(c)}, \mathbf{s}(c))$ for $c \in C(W)$ with $f(c) \in \mathbb{Z}^+$.

Let (H, W, f) be a congruence triple of Γ and let $\wp(H, W, f)$ denote the congruence generated by all pairs $(v, 0)$ for $v \in H$, (ee^{-1}, w) for $w \in W$ with $\mathbf{s}(e) = w$ and $\mathbf{r}(e) \notin H$, and $(c^{f(c)}, \mathbf{s}(c))$ for $c \in C(W)$ with $f(c) \in \mathbb{Z}^+$.

Now let ρ be a 0-restricted congruence on $I(\Gamma)$ and $T(\rho) = (W, f)$. We first show that $\rho = \wp(W, f)$.

The next lemma directly follows from (V), (CK1) and the definition of a congruence on a semigroup.

Lemma 5 *Let $u, v \in \Gamma^0$ and p, q, t be paths in Γ.*

(1) *If $u \neq v$, then $(u, v) \notin \rho$.*
(2) *if $(pq^{-1}, v) \in \rho$, then $\mathbf{s}(p) = \mathbf{s}(q) = v$.*
(3) *$(pq, \mathbf{s}(p)) \in \rho$ if and only if $(qp, \mathbf{s}(q)) \in \rho$.*
(4) *if $(pqt^{-1}p^{-1}, \mathbf{s}(p)) \in \rho$, then $(qt^{-1}, \mathbf{s}(q) \in \rho$.*
(5) *if $(pp^{-1}, \mathbf{s}(p)) \in \rho$, $(qt^{-1}, \mathbf{s}(q)) \in \rho$ and $\mathbf{r}(p) = \mathbf{s}(q)$, then $(pqt^{-1}p^{-1}, \mathbf{s}(p)) \in \rho$.*

Lemma 6 *For any $pq^{-1} \in I(\Gamma)$, if $(pq^{-1}, \mathbf{s}(p)) \in \rho$, then both p and q are no-exit.*

Proof Suppose that $p = e_1 \cdots e_n$ and e is an exit to p with $\mathbf{s}(e) = \mathbf{s}(e_i)$. It follows from Lemma 5(2) and (CK1) that

$$(e_i \cdots e_n q^{-1} e_1 \cdots e_{i-1}, \mathbf{s}(e_i)) \in \rho.$$

We obtain from (E2) and (CK1) that

$$(0, \mathbf{r}(e)) = (0, e^{-1}e) = (e^{-1}e_i \cdots e_n q^{-1} e_1 \cdots e_{i-1}e, e^{-1}\mathbf{s}(e_i)e) \in \rho$$

which contradicts to the hypothesis $0\rho = \{0\}$. Hence, p is no-exit. We see that q is also no-exit since $(pq^{-1}, \mathbf{s}(p)) \in \rho$ implies that $(qp^{-1}, \mathbf{s}(q) \in \rho$. □

Lemma 7 *For any path p in Γ, if $(pp^{-1}, \mathbf{s}(p)) \in \rho$ and $p = rqs$ for some paths r, q, s in Γ, then $(qq^{-1}, \mathbf{s}(q)) \in \rho$.*

Proof We see from the hypothesis that $(rqss^{-1}q^{-1}r^{-1}, \mathbf{s}(r)) \in \rho$. Then it follows from Lemma 5(4) that $(qss^{-1}q^{-1}, \mathbf{s}(q)) \in \rho$ and $(ss^{-1}, \mathbf{s}(s)) \in \rho$. We see from $(ss^{-1}, \mathbf{s}(s)) \in \rho$ and the compatibility of ρ that $(qss^{-1}q^{-1}, qq^{-1}) \in \rho$. Hence, we obtain by the transitivity of ρ that $(qq^{-1}, \mathbf{s}(q)) \in \rho$. □

For any $c \in C(W)$, the semigroup $\langle c, c^{-1} \rangle$ generated by c and c^{-1} is the (infinite) bicyclic semigroup (see [12]), and $\langle c, c^{-1} \rangle / \rho$ is either the bicyclic semigroup or a cyclic group of order $f(c)$. This implies

Lemma 8 *For any $c \in C(W)$, if $f(c) \in \mathbb{Z}^+$, then for every $m \in \mathbb{Z}^+$, $(c^m, \mathbf{s}(c)) \in \rho$ implies that $f(c) \mid m$.*

A *closed path based at* v is a path $p = e_1 \cdots e_n$ such that $\mathbf{s}(p) = \mathbf{r}(p) = v$. A closed path $p = e_1 \cdots e_n$ based at v is called a *closed simple path based at* v if $\mathbf{s}(e_i) \neq v$ for every $i > 1$. A cycle is a closed simple path. As for closed paths and closed simple paths we have (see [2]).

Lemma 9 *For any $v \in \Gamma^0$ and any closed path p based at v, there exist unique closed simple paths $s_1, \ldots, s_m$ based at v such that $p = s_1 \cdots s_m$.*

Proof Let $p = e_1e_2 \cdots e_n$ with $e_1, e_2, \cdots, e_n \in \Gamma^1$, $K = \{i_k \in \{1, 2, \ldots, n\} : \mathbf{s}(e_{i_k}) = v\}$, $|K| = l$ and $1 = i_1 < i_2 < \ldots < i_l$. Take $s_j = e_{i_j} \cdots e_{i_{j+1}-1}$ for $j \in \{1, \ldots, l-1\}$ and $s_l = e_{i_l} \cdots e_n$. Then $s_1 \cdots s_l$ is a desired decomposition of p.

If $p = t_1 \cdots t_{l'}$ is a distinct decomposition into closed simple paths, then there exists j such that $s_1 = t_1, \ldots, s_{j-1} = t_{j-1}$ but $s_j \neq t_j$. It follows from (CK1) that

$$s_{j+1} \cdots s_l = s_j^{-1} s_{j-1}^{-1} \cdots s_1^{-1} s_1 \cdots s_l = s_j^{-1} s_{j-1}^{-1} \cdots s_1^{-1} t_1 \cdots t_{l'} = s_j^{-1} t_j \cdots t_{l'}.$$

Then we see from the multiplication of elements in $I(\Gamma)$ that s_j is a prefix of t_j or t_j is a prefix of s_j. However, if s_j is a proper prefix of t_j, then t_j cannot be a closed simple path, a contradiction. Similarly, t_j is not a proper prefix of s_j. So we get $s_j = t_j$, a contradiction. Thus, we prove the uniqueness. □

Lemma 10 *If p is a nonempty path in Γ and $(p, \mathbf{s}(p)) \in \rho$, then there exists a no-exit cycle c such that $p = c^m$ for some positive integer m.*

Proof We see from Lemma 5(2) that p is a closed path based at $\mathbf{s}(p)$. It follows from Lemma 9 that there exist unique closed simple paths $\mu_1, \ldots, \mu_m$ based at $\mathbf{s}(p)$ such that $p = \mu_1 \cdots \mu_m$. Then we obtain from Lemma 6 that $\mu_1 = \ldots = \mu_m = c$ for some no-exit cycle c. □

The following lemma can also be found in [22].

Lemma 11 *For any $v \in \Gamma^0$, every element in $v\rho$ is of the form psp^{-1} or $ps^{-1}p^{-1}$ for some paths p, s in Γ such that $\mathbf{s}(p) = v$ and $\mathbf{r}(p) = \mathbf{s}(s) = \mathbf{r}(s)$.*

Proof It follows from Lemma 5(2) that elements in $v\rho$ has form pq^{-1} with $\mathbf{s}(p) = \mathbf{s}(q)$. Moreover, we see from Lemma 6 that p, q are no-exit. Thus, either p is a prefix of q or q is a prefix of p. In the previous case, there exists a closed path s in Γ such that $pq^{-1} = ps^{-1}p^{-1}$; in the latter case, there exists a closed path s in Γ such that $pq^{-1} = qsq^{-1}$. □

Lemma 12 *For any $v \in \Gamma^0$, every element in $v\rho$ is of the form either pp^{-1} or $pc^{mf(c)}p^{-1}$, where p is a path in Γ, $\mathbf{s}(p) = v$, $v(p) \subseteq W$, and in the second case, $m \in \mathbb{Z} \setminus \{0\}$, $\mathbf{r}(p) = \mathbf{s}(c)$, $c \in C(W)$ with $f(c) \in \mathbb{Z}^+$.*

Proof Note Lemma 11 and take x in $v\rho$. Then either $x = psp^{-1}$ or $x = ps^{-1}p^{-1}$. We already have $\mathbf{r}(p) = \mathbf{s}(s)$, $\mathbf{s}(p) = v$ and $v(p) \subseteq W$ by Lemma 6. If s is empty, then it is clear $x = pp^{-1}$. If s is nonempty, then noticing $s = s(\mathbf{s}(s))^{-1}$, we see from Lemmas 5(4), (2), 10, and 8 that $x = pc^{mf(c)}p^{-1}$ where $m \in \mathbb{Z} \setminus \{0\}$ and $c \in C(W)$ with $f(c) \in \mathbb{Z}^+$. □

Proposition 4 $\wp(W, f) = \rho$.

Proof Recall that $\mathbf{R}$ consists of all pairs $(e_v e_v^{-1}, v)$ for $v \in W$ and $(c^{f(c)}, \mathbf{s}(c))$ for $c \in C(W)$ with $f(c) \in \mathbb{Z}^+$. Clearly, $\mathbf{R} \subseteq \rho$ so that $\wp(W, f) \subseteq \rho$. For any $pq^{-1}, rs^{-1} \in I(\Gamma) \setminus \{0\}$, if $(pq^{-1}, rs^{-1}) \in \rho$, we may suppose that p contains less edges than r without loss of generality. Then we see from Lemma 6 that $r = pr_1$ for some path r_1 in Γ. Concerning q and s, we have the following two cases. **Case 1.** $s = qs_1$ for some path s_1 in Γ. It follows from $(pq^{-1}, rs^{-1}) \in \rho$ and (CK1) that $(\mathbf{r}(p), r_1 s_1^{-1}) \in \rho$. Then we see from Lemmas 12 and 7 that $(\mathbf{r}(p), r_1 s_1^{-1}) \in \wp(W, f)$. Hence, we have $(pq^{-1}, rs^{-1}) = (pq^{-1}, pr_1 s_1^{-1} q^{-1}) \in \wp(W, f)$. **Case 2.** $q = sq_1$ for some nonempty path q_1 in Γ. It again follows from $(pq^{-1}, rs^{-1}) \in \rho$ and (CK1) that $(\mathbf{r}(p), r_1 q_1) \in \rho$. Then we observe from Lemma 10 that there exists $c \in C(W)$ with $f(c) \in \mathbb{Z}^+$ and such that $r_1 q_1 = c^m$ for some $m \in \mathbb{Z}^+$. Thus, it follows from Lemma 8 that there exists $k \in \mathbb{Z}^+$ such that $m = kf(c)$. Now we get from the definition of $\wp(W, f)$ that $(c^{kf(c)}, \mathbf{s}(c)) \in \wp(W, f)$ which means that $(\mathbf{r}(p), r_1 q_1) \in \wp(W, f)$. Noticing (CK1), we have

$$(r_1^{-1}, q_1) \in \wp(W, f) \Longrightarrow (q_1^{-1}, r_1) \in \wp(W, f) \Longrightarrow (pq_1^{-1}s^{-1}, pr_1 s^{-1}) \in \wp(W, f).$$

That is, $(pq^{-1}, rs^{-1}) \in \wp(W, f)$.

As a consequence, we obtain that $\rho \subseteq \wp(W, f)$ which leads to $\rho = \wp(W, f)$. □

Next, let (W, f) be a congruence pair of Γ and $\wp(W, f) = \varrho$. We shall prove that ϱ is a 0-restricted congruence and $T(\varrho) = (W, f)$.

Let S be a semigroup, let $\mathbf{T}$ be a relation on S, and denote the congruence generated by $\mathbf{T}$ as $\mathbf{T}^\sharp$. If c, d in S are such that $c = xay, d = xby$ for some $x, y \in S^1$, where either $(a, b) \in \mathbf{T}$ or $(b, a) \in \mathbf{T}$, we say that c is connected to d by an *elementary* $\mathbf{T}$-*transition* and write $c \to d$.

Remark 12 Recall that the relation $\mathbf{R}$ consists of all pairs $(e_v e_v^{-1}, v)$ for $v \in W$ and $(c^{f(c)}, \mathbf{s}(c))$ for $c \in C(W)$ with $f(c) \in \mathbb{Z}^+$ and that $\wp(W, f) = \varrho = \mathbf{R}^\sharp$. The definition of an elementary $\mathbf{R}$-transition $c \to d$ in $I(\Gamma)$ can be adjusted as $c = xay, d = xby$ for some $x, y \in I(\Gamma)$ (not adjoined with 1) where either $(a, b) \in \mathbf{R}$ or $(b, a) \in (R)$. In fact, if either x or y equals to 1, then either x or y can be replaced by some vertex since we have (V), (E1), and (E2).

Lemma 13 (Howie, Proposition 1.5.9) *Let* $\mathbf{T}$ *be a relation on a semigroup* S *and* $a, b \in S$. *Then* $(a, b) \in \mathbf{T}^\sharp$ *if and only if either* $a = b$ *or, for some* $n \in \mathbb{Z}^+$, *there exists a sequence*

$$a = z_1 \to z_2 \to \ldots \to z_n = b$$

of elementary $\mathbf{T}$-*transitions connecting* a *to* b.

Lemma 14 *For any $v \in \Gamma^0$, every element in $v\varrho$ is of the form either pp^{-1} or $pc^{kf(c)}p^{-1}$, where p is a path in Γ, $\mathbf{s}(p) = v$, $v(p) \subseteq W$, and in the second case, $k \in \mathbb{Z}$, $\mathbf{r}(p) = \mathbf{s}(c)$, $c \in C(W)$ with $f(c) \in \mathbb{Z}^+$.*

Proof If $w \in v\varrho$, then we know from Lemma 13 that there exists a sequence

$$v = z_1 \to z_2 \to \ldots \to z_n = w.$$

of elementary $\mathbf{R}$-transitions. We now prove by induction on n. Obviously, the result is true if $n = 1$. Suppose that z_{n-1} has one of the two forms in the lemma and that $z_{n-1} = pq^{-1}xrs^{-1}$, $z_n = pq^{-1}yrs^{-1}$ for some $pq^{-1}, rs^{-1} \in I(\Gamma) \setminus \{0\}$, where either $(x, y) \in \mathbf{R}$ or $(y, x) \in \mathbf{R}$. Then we have the following cases and subcases.
Case 1. $z_{n-1} = tt^{-1}$ for some path t in Γ. Note the multiplication of elements in $I(\Gamma)$ and that $v(t) \subseteq W$.
Subcase 1.1. $x \in W$ and $y = e_x e_x^{-1}$.
Subcase 1.1.1. $q = rq_1$ for some path q_1 in Γ. Then we get $z_{n-1} = pq_1^{-1}s^{-1} = tt^{-1}$ which means that $p = t = sq_1$. Thus, if q is empty, then we see that $z_n = te_xe_x^{-1}t^{-1}$. Clearly, $v(te_x) \subseteq W$. If q is nonempty, then we know that $z_n = pq^{-1}e_xe_x^{-1}rs^{-1} = pq^{-1}rs^{-1} = z_{n-1}$.
Subcase 1.1.2. $r = qr_1$ for some path r_1 in Γ. Then we get $z_{n-1} = pr_1s^{-1} = tt^{-1}$ which leads to $pr_1 = t = s$. One can similarly obtain that either $z_n = te_xe_x^{-1}t^{-1}$ or $z_n = z_{n-1}$ via discussing the two possibilities where r is empty or not.
Subcase 1.2. $x \in W$ and $y = c^{f(c)}$ for some $c \in C(W)$ with $\mathbf{s}(c) = x$ and $f(c) \in \mathbb{Z}^+$.
Subcase 1.2.1. $q = rq_1$ for some path q_1 in Γ. As in Subcase 1.1.1, we get $p = t = sq_1$. It follows from Lemma 4(1) that $z_n = pq_1^{-1}r^{-1}c^{f(c)}rs^{-1} = pq_1^{-1}c_1^{f(c_1)}s^{-1}$ for some cyclic permutation c_1 of c. By Lemma 3, there exist paths q_2, q_3 in Γ and $l \in \mathbb{N}$ such that $q_1 = c_1^l q_2$ and $c_1 = q_2q_3$. So we get $z_n = sc_1^lq_2q_2^{-1}(c_1^{-1})^lc_1^{f(c_1)}s^{-1}$. Take $c_2 = q_3q_2$. Clearly, c_2 is a cyclic permutation of c. If $l \geq f(c_1) = f(c_2)$, then we have

$$\begin{aligned} z_n &= sc_1^lq_2q_2^{-1}(c_1^{-1})^{l-f(c_1)}s^{-1} = sq_2c_2^l(c_2^{-1})^{l-f(c_2)}q_2^{-1}s^{-1} \\ &= (sq_2c_2^{l-f(c_2)})c_2^{f(c_2)}((c_2^{-1})^{l-f(c_2)}q_2^{-1}s^{-1}). \end{aligned}$$

Furthermore, $v(sq_2c_2^{l-f(c_2)}) \subseteq W$. If $l < f(c_1)$, then we have

$$z_n = sc_1^lq_2q_2^{-1}c_1^{f(c_1)-l}s^{-1} = sc_1^lq_2q_2^{-1}q_2q_3(c_1)^{f(c_1)-l-1}s^{-1} = sc_1^{f(c_1)}s^{-1}.$$

Obviously, $v(s) \subseteq W$.
Subcase 1.2.2. $r = qr_1$ for some path r_1 in Γ. As in Case 1.1.2, we also get $pr_1 = t = s$. It follows from Lemma 4(1) that $z_n = pq^{-1}c^{f(c)}qr_1s^{-1} = pc_1^{f(c_1)}r_1s^{-1}$ for some cyclic permutation c_1 of c. So we see from Lemma 4(2) that

$$z_n = pc_1^{f(c_1)} r_1 r_1^{-1} p^{-1} = pr_1 c_2^{f(c_2)} r_1^{-1} p^{-1}$$

for some cyclic permutation c_2 of c. We already have $v(pr_1) \subseteq W$.
Subcase 1.3. $x = e_y e_y^{-1}$ and $y \in W$. Then we get $z_{n-1} = pq^{-1} e_y e_y^{-1} rs^{-1}$. If q or r is nonempty, then we easily see that $z_{n-1} = pq^{-1} rs^{-1} = z_n$. So we only need to consider the case where both q and r are empty. In this case, we have $pe_y e_y^{-1} s^{-1} = tt^{-1}$. So we get $pe_y = t = se_y$ which means that $z_n = pp^{-1}$ with $v(p) \subseteq W$.
Subcase 1.4. $x = c^{f(c)}$ and $y \in W$ for some $c \in C(W)$ with $\mathbf{s}(c) = y$ and $f(c) \in \mathbb{Z}^+$.
Subcase 1.4.1. $q = rq_1$ for some path q_1 in Γ. Then we get $z_{n-1} = pq_1^{-1} r^{-1} c^{f(c)} rs^{-1}$ and $z_n = pq_1^{-1} s^{-1}$. It follows from Lemma 4(1) that there exists a cyclic permutation c_1 of c such that $z_{n-1} = pq_1^{-1} c_1^{f(c_1)} s^{-1}$. By Lemma 3, we see that $q_1 = c_1^l q_2$ where $l \in \mathbb{N}$ and $c_1 = q_2 q_3$ for some path q_3 in Γ. Clearly, $c_2 = q_3 q_2$ is a cyclic permutation of c. If $l \geq f(c)$, then we obtain that $z_{n-1} = pq_2^{-1} (c_1^{-1})^{l-f(c)} s^{-1} = tt^{-1}$ which leads to $p = t = sc_1^{l-f(c)} q_2$. Hence, we have

$$z_n = sc_1^{l-f(c_1)} q_2 q_2^{-1} (c_1^{-1})^l s^{-1} = sq_2 c_2^{l-f(c_2)} (c_2^{-1})^l q_2^{-1} s^{-1} = t(c_2^{-1})^{f(c_2)} t^{-1}$$

as required. If $l < f(c_1)$, then we see that

$$z_{n-1} = pq_2^{-1} c_1^{f(c_1)-l} s^{-1} = pq_3 c_1^{f(c_1)-l-1} s^{-1} = tt^{-1}$$

which leads to $pq_3 c_1^{f(c_1)-l-1} = t = s$. Hence, we have

$$z_n = pq_2^{-1} (c_1^{-1})^l (c_1^{-1})^{f(c_1)-l-1} q_3^{-1} p^{-1} = p(c_2^{-1})^{f(c_2)} p^{-1}.$$

Obviously, $v(p) \subseteq W$.
Subcase 1.4.2. $r = qr_1$ for some path r_1 in Γ. Then we get $z_{n-1} = pq^{-1} c^{f(c)} qr_1 s^{-1}$ and $z_n = pr_1 s^{-1}$. It follows from Lemma 4(1) that there exists a cyclic permutation c_1 of c such that $z_{n-1} = pc_1^{f(c_1)} r_1 s^{-1}$. So we see that $pc_1^{f(c)} r_1 = t = s$. Therefore, we obtain from Lemma 4(2) that

$$z_n = pr_1 r_1^{-1} (c_1^{-1})^{f(c_1)} p^{-1} = p(c_1^{f(c_1)} r_1 r_1^{-1})^{-1} p^{-1} = pr_1 (c_2^{-1})^{f(c_2)} r_1^{-1} p^{-1}$$

for some cyclic permutation c_2 of c. Moreover, $v(pr_1) \subseteq W$.
Case 2. $z_{n-1} = tc^{kf(c)} t^{-1}$, where t is a path in Γ, $\mathbf{s}(t) = v$, $v(t) \subseteq W$, $k \in \mathbb{Z}^+$, $\mathbf{r}(t) = \mathbf{s}(c)$ and $c \in C(W)$ with $f(c) \in \mathbb{Z}^+$.
Subcase 2.1. $x \in W$ and $y = e_x e_x^{-1}$.
Subcase 2.1.1. $q = rq_1$ for some path q_1 in Γ. Then we get $z_{n-1} = pq_1^{-1} s^{-1} = tc^{kf(c)} t^{-1}$ which means that $p = tc^{kf(c)}$ and $t = sq_1$. Thus, if q is empty, then we see from Lemma 4(2) that $z_n = tc^{kf(c)} e_x e_x^{-1} t^{-1} = te_x c_1^{kf(c_1)} e_x^{-1} t^{-1}$ for some

cyclic permutation c_1 of c. Obviously, $v(te_x) \subseteq W$. If q is nonempty, then we know that $z_n = pq^{-1}e_xe_x^{-1}rs^{-1} = pq^{-1}rs^{-1} = z_{n-1}$.
Subcase 2.1.2. $r = qr_1$ for some path r_1 in Γ. Then we get $z_{n-1} = pr_1s^{-1} = tc^{kf(c)}t^{-1}$ and similarly obtain that either $z_n = te_xc_1^{kf(c_1)}e_x^{-1}t^{-1}$ for some cyclic permutation c_1 of c or $z_n = z_{n-1}$ via discuss the two possibilities where r is empty or not.
Subcase 2.2. $x \in W$ and $y = d^{f(d)}$ for some $d \in C(W)$ with $\mathbf{s}(d) = x$ and $f(d) \in \mathbb{Z}^+$. It is not difficult to see that c and d are cyclic permutations of each other since they are no-exit.
Subcase 2.2.1. $q = rq_1$ for some path q_1 in Γ. As in Subcase 2.1.1, we get $p = tc^{kf(c)}$ and $t = sq_1$. It follows from Lemma 4(1) that $z_n = pq_1^{-1}r^{-1}d^{f(d)}rs^{-1} = pq_1^{-1}c_1^{f(c_1)}s^{-1}$ for some cyclic permutation c_1 of c. By Lemma 3, there exist paths q_2, q_3 in Γ and $l \in \mathbb{N}$ such that $q_1 = c_1^lq_2$ and $c_1 = q_2q_3$. Moreover, we get $z_n = tc^{kf(c)}q_2^{-1}(c_1^{-1})^lc_1^{f(c_1)}s^{-1}$. So we see that $c = q_3q_2$ since $\mathbf{r}(q_2) = \mathbf{s}(c)$. Hence, we see that

$$z_n = sc_1^lq_2c^{kf(c)}q_2^{-1}(c_1^{-1})^lc_1^{f(c_1)}s^{-1} = sq_2c^{kf(c)+l}q_2^{-1}(c_1^{-1})^lc_1^{f(c_1)}s^{-1}.$$

If $l \geq f(c)$, then we have

$$\begin{aligned} z_n &= sq_2c^{kf(c)+l}q_2^{-1}(c_1^{-1})^{l-f(c)}s^{-1} = sq_2c^{kf(c)+l}(c^{-1})^{l-f(c)}q_2^{-1}s^{-1} \\ &= (sq_2c^{l-f(c)})c^{(k+1)f(c)}((c^{-1})^{l-f(c)}q_2^{-1}s^{-1}). \end{aligned}$$

We observe that $v(sq_2c_1^{l-f(c)}) \subseteq W$. If $l < f(c)$, then we have

$$\begin{aligned} z_n &= sc_1^{kf(c_1)+l}q_2q_2^{-1}c_1^{f(c_1)-l}s^{-1} \\ &= sc_1^{kf(c_1)+l}q_2q_2^{-1}q_2q_3c_1^{f(c_1)-l-1}s^{-1} = sc_1^{(k+1)f(c_1)}s^{-1}. \end{aligned}$$

Obviously, $v(s) \subseteq W$.
Subcase 2.2.2. $r = qr_1$ for some path r_1 in Γ. As in Subcase 2.1.2, we also get $z_{n-1} = pr_1s^{-1} = tc^{kf(c)}t^{-1}$ so that $pr_1 = tc^{kf(c)}$ and $s = t$. It follows from Lemma 4(1) that $z_n = pq^{-1}d^{f(d)}qr_1s^{-1} = pc_1^{f(c_1)}r_1s^{-1}$ for some cyclic permutation c_1 of c. By Lemma 3, there exist paths r_2, r_3 in Γ and $l \in \mathbb{N}$ such that $r_1 = c_1^lr_2$ and $c_1 = r_2r_3$. Let $c_2 = r_3r_2$. Clearly, c_2 is a cyclic permutation of c_1 and $r_1 = r_2c_2^l$. It follows from $pr_1 = tc^{kf(c)}$ that $c = c_2$. Thus, we observe that

$$z_n = pc_1^{f(c_1)+l}r_2s^{-1} = pr_2c^{f(c)+l}s^{-1} = pr_1c^{f(c)}s^{-1} = tc^{(k+1)f(c)}t^{-1}.$$

We already have $v(t) \subseteq W$.
Subcase 2.3. $x = e_ye_y^{-1}$ and $y \in W$. Then we get $z_{n-1} = pq^{-1}e_ye_y^{-1}rs^{-1}$. If q or r is nonempty, then we easily see that $z_{n-1} = pq^{-1}rs^{-1} = z_n$. So we only need

to consider the case where both q and r are empty which means that $pe_ye_y^{-1}s^{-1} = tc^{kf(c)}t^{-1}$. So we get $pe_y = tc^{kf(c)}$ and $t = se_y$ which lead to $pe_y = se_yc^{kf(c)} = sc_1^{kf(c_1)}e_y$. Hence, we get $p = sc_1^{kf(c)}$ so that $z_n = ps^{-1} = sc_1^{kf(c_1)}s^{-1}$ with $v(s) \subseteq W$.

Subcase 2.4. $x = d^{f(d)}$ and $y \in W$ for some $d \in C(W)$ with $\mathbf{s}(d) = y$ and $f(d) \in \mathbb{Z}^+$. Similar to Subcase 2.2, we observe that c and d are cyclic permutations of each other.

Subcase 2.4.1. $q = rq_1$ for some path q_1 in Γ. Then we get $z_{n-1} = pq_1^{-1}r^{-1}d^{f(d)}rs^{-1}$ and $z_n = pq_1^{-1}s^{-1}$. It follows from Lemma 4(1) that there exists a cyclic permutation c_1 of c such that $z_{n-1} = pq_1^{-1}c_1^{f(c_1)}s^{-1}$. By Lemma 3, we see that $q_1 = c_1^lq_2$ where $l \in \mathbb{N}$ and $c_1 = q_2q_3$ for some path q_3 in Γ. If $l \geq f(c_1)$, then we obtain that $z_{n-1} = pq_2^{-1}(c_1^{-1})^{l-f(c_1)}s^{-1} = tc^{kf(c)}t^{-1}$ which leads to $p = tc^{kf(c)}$ and $t = sc_1^{l-f(c_1)}q_2$. These also imply that $c = q_3q_2$. Hence, we have

$$\begin{aligned}z_n &= sc_1^{l-f(c_1)}q_2c^{kf(c)}q_2^{-1}(c_1^{-1})^ls^{-1} = sq_2c^{l-f(c)}c^{kf(c)}(c^{-1})^lq_2^{-1}s^{-1}\\&= (sq_2c^l)c^{(k-1)f(c)}((c^{-1})^lq_2^{-1}s^{-1})\end{aligned}$$

and $v(sq_2c_1^l) \subseteq W$. If $l < f(c_1)$, then we see that

$$z_{n-1} = pq_2^{-1}c_1^{f(c_1)-l}s^{-1} = pq_3c_1^{f(c_1)-l-1}s^{-1} = tc^{kf(c)}t^{-1}$$

so that $pq_3c_1^{f(c_1)-l-1} = tc^{kf(c)}$, $t = s$ and $c_1 = c$. So we get $pq_3 = tc^{(k-1)f(c)+l+1}$ which means that $p = tc^{(k-1)f(c)+l}q_2$. Clearly, $c_2 = q_3q_2$ is a cyclic permutation of c. Hence, we have

$$z_n = tc^{(k-1)f(c)+l}q_2q_2^{-1}(c^{-1})^lt^{-1} = (tq_2c_2^l)c_2^{(k-1)f(c_2)}((c_2^{-1})^lq_2^{-1}t^{-1})$$

and $v(tq_2c_2^l) \subseteq W$.

Subcase 2.4.2. $r = qr_1$ for some path r_1 in Γ. Then we get $z_{n-1} = pq^{-1}d^{f(d)}qr_1s^{-1}$ and $z_n = pr_1s^{-1}$. It follows from Lemma 4(1) that there exists a cyclic permutation c_1 of c such that $z_{n-1} = pc_1^{f(c_1)}r_1s^{-1} = tc^{kf(c)}t^{-1}$. So we see that $pc_1^{f(c_1)}r_1 = tc^{kf(c)}$ and $t = s$. Noticing Lemma 3, one has $pc_1^{f(c_1)}r_1 = pr_1c^{f(c)} = tc^{kf(c)}$ which leads to $pr_1 = tc^{(k-1)f(c)}$. Therefore,. we obtain that $z_n = tc^{(k-1)f(c)}t^{-1}$.

Case 3. $z_{n-1} = t(c^{-1})^{kf(c)}t^{-1}$, where t is a path in Γ, $\mathbf{s}(t) = v$, $v(t) \subseteq W$, $k \in \mathbb{Z}^+$, $\mathbf{r}(t) = \mathbf{s}(c)$ and $c \in C(W)$ with $f(c) \in \mathbb{Z}^+$. With a similar discussion as in Case 2, we can see that z_n has the required form. □

Corollary 2 *The congruence* ϱ *is* 0*-restricted.*

Remark 13 The relation $\mathbf{R}$ used in the proof of Proposition 14 can be chosen smaller. $\mathbf{R}$ only needs to consist of all pairs $(e_v e_v^{-1}, v)$ for $v \in W$ and $(c^{f(c)}, \mathbf{s}(c))$ for $c \in C_0$ such that $f(c) \in \mathbb{Z}^+$ and C_0 contains exactly one cyclic permutation of d for every $d \in C(W)$. This can be seen from (CK1).

Proposition 5 $T(\varrho) = (W, f)$.

Proof Let $T(\varrho) = (W_1, f_1)$. It is not difficult to see that $W \subseteq W_1$. Conversely, if $v \in W_1$, then we certainly have $(v, e_v e_v^{-1}) \in \varrho$. It follows from Lemma 14 and the uniqueness of the expression of every nonzero element in $I(\Gamma)$ that $v \in W$. So we get $W_1 = W$.

For f and f_1, arbitrarily taking $c \in C(W)$ with $f(c) \in \mathbb{Z}^+$, we observe from the definition of ϱ and Lemma 8 that $f_1(c) \mid f(c)$ since $(c^{f(c)}, \mathbf{s}(c)) \in \varrho$. On the other hand, let $f_1(c) \in \mathbb{Z}^+$. Again noticing Lemma 14 and uniqueness of the expression of every nonzero element in $I(\Gamma)$, we obtain that there must exist $k \in \mathbb{Z}^+$ such that $c^{f_1(c)} = c^{kf(c)}$ since $(\mathbf{s}(c), c^{f_1(c)}) \in \varrho$. Thus, we get $f(c) \mid f_1(c)$. In conclusion, we have $f_1 = f$. □

Up to now, we have completed the proof of the following main result.

Theorem 2 *The mapping T from the set of all 0-restricted congruences on $I(\Gamma)$ to the set of all congruence pairs of Γ and the mapping $\wp$ from the set of all congruence pairs of Γ to the set of all 0-restricted congruences on $I(\Gamma)$ are inverses. In particular, there exists a one-to-one correspondence between 0-restricted congruences on $I(\Gamma)$ and congruence pairs of Γ.*

Theorem 2 immediately implies

Theorem 3 *The mapping T from the set of all congruences on $I(\Gamma)$ to the set of all congruence triples of Γ and the mapping $\wp$ from the set of all congruence triples of Γ to the set of all congruences on $I(\Gamma)$ are inverses. In particular, there exists a one-to-one correspondence between congruences on $I(\Gamma)$ and congruence triples of Γ.*

To end this part, we give some applications of the description of congruences on graph inverse semigroups. Let S be a semigroup. S is called *congruence-free* if the identity relation and the universal relation are the only congruences on S. If S has a zero element, S is called *0-simple* if S has only $\{0\}$ and S as its ideals. Let $\Gamma = (\Gamma^0, \Gamma^1, r, s)$ be a graph. Notice the correspondence of the set of ideals in $I(\Gamma)$ and the set of hereditary subsets of Γ^0. The next corollary is obvious.

Corollary 3 *The following are equivalent on $I(\Gamma)$:*

(1) *$I(\Gamma)$ is 0-simple.*
(2) *The only hereditary subsets of Γ^0 are $\emptyset$ and Γ^0.*
(3) *Γ is a strongly connected graph.*

The equivalence of (1) and (3) in the following corollary is proved in [22] [**Theorem 10**].

Corollary 4 *The following are equivalent:*

(1) *Every congruence on $I(\Gamma)$ is a Rees congruence.*
(2) *For every hereditary subset H of Γ^0, $\bar{\Gamma}^0{}_H = \emptyset$.*
(3) *For every $e \in \Gamma^1$, there exists a path p in Γ such that e is an exit to p, $\mathbf{s}(e) = \mathbf{s}(p)$, and $\mathbf{r}(e) = \mathbf{r}(p)$.*

Proof Again noticing the correspondence of the set of ideals in $I(\Gamma)$ and the set of hereditary subsets of Γ^0, we see from the definition of congruence triples, Theorem 3 and its proof that (1) and (2) are equivalent.

Assume that there exists some $e_0 \in \Gamma^1$ such that either e_0 is not an exit to any path, or any path p to which e_0 is an exit and satisfies $\mathbf{s}(e_0) = \mathbf{s}(p)$ does not end at $\mathbf{r}(e_0)$. Taking $H_0 = \{\mathbf{r}(p) : p$ is a nonempty path in Γ which is distinct to e_0, $\mathbf{s}(e_0) = \mathbf{s}(p)\}$, we see that H_0 is a hereditary subset of Γ^0 (probably empty). However, $\bar{\Gamma}^0{}_{H_0} \neq \emptyset$ since $\mathbf{s}(e_0) \in \bar{\Gamma}^0{}_{H_0}$. Hence, we obtain that (2) implies (3). By the definition of hereditary subsets and $\bar{\Gamma}^0{}_H$, one easily sees that (3) implies (2). □

The equivalence of (1) and (3) in the next corollary is proved in [5][**Theorem 3**] for graphs where there exists at most one edge from a vertex to another and in [22] [**Corollary 11**] for general cases.

Corollary 5 *The following are equivalent:*

(1) *The semigroup $I(\Gamma)$ is congruence-free.*
(2) *$\emptyset$ and Γ^0 are the only hereditary subsets and $\bar{\Gamma}^0 = \emptyset$.*
(3) *The graph Γ is strongly connected and no vertex of Γ has out-degree one.*

Proof It follows from Theorem 3 and Corollary 4 that (1) and (2) are equivalent. Moreover, noticing that no vertex of Γ has index one if and only if $\bar{\Gamma}^0 = \emptyset$, we see that (2) is equivalent to (3) from Corollary 3. □

Theorem 4 *Let ρ be a 0-restricted congruence on $I(\Gamma)$ determined by the congruence pair (W, f). Then $I(\Gamma)/\rho$ is isomorphic to a graph inverse semigroup if and only if*

(1) *$W \subseteq \{v \in \Gamma^0 : v$ has out-degree 1, e_v is a loop at $v\}$.*
(2) *For any $v \in W$, $f(e_v) = 1$.*

Proof *Sufficiency.* Suppose that conditions (1) and (2) are satisfied. We proceed to prove that $I(\Gamma)/\rho$ is isomorphic to the graph inverse semigroup $I(\Delta)$, where Δ is the graph with $\Delta^0 = \Gamma^0$, $\Delta^1 = \Gamma^1 \setminus \{e_v : v \in W\}$, and the source mapping for Δ is the restriction of the source mapping for Γ and the range mapping for Δ is the restriction of the range mapping for Γ. By conditions (1) and (2) and Theorem 2, ρ is generated by all pairs $(e_v e_v^{-1}, v)$ and (e_v, v) where $v \in W$. However, in the inverse semigroup $I(\Gamma)$, the relation $(e_v, v) \in \rho$ implies the relation $(e_v e_v^{-1}, v) \in \rho$. Hence, ρ is generated by all pairs (e_v, v) where $v \in W$. Let ϕ be the function that maps a loop e_v of Γ at a vertex v in W to the vertex v and fixes all other vertices and edges of Γ. Then ϕ is a function that maps the generators of $I(\Gamma)$ to the generators of $I(\Delta)$. This function ϕ extends to a homomorphism which we again denote by ϕ from $I(\Gamma)$ onto $I(\Delta)$. To see this, note that if pq^{-1} is a nonzero element of $I(\Gamma)$, then $\phi(pq^{-1}) = pq^{-1}$ if neither p nor q contains an edge e_v that is a loop at some vertex $v \in W$. If pq^{-1} does contain such a loop e_v, then we must have $p = e_1e_2 \cdots e_ne_v^k$ and $q = f_1f_2 \cdots f_me_v^l$ for some $k, l \geq 0$ (with at least one of k or t greater than 0). Then we see that $\phi(pq^{-1})$ is obtained from pq^{-1} by removing the path $e_v^k(e_v^{-1})^l$ at the vertex $r(p) = r(q) = v$. Then from the definition of the multiplication of elements in $I(\Gamma)$ it is easy to see that ϕ is a homomorphism from $I(\Gamma)$ onto $I(\Delta)$. But then since ρ is generated by the pairs (e_v, v) where e_v is a loop at some vertex $v \in W$, it follows that the kernel of ϕ is ρ and so the inverse semigroups $I(\Gamma)/\rho$ and $I(\Delta)$ are isomorphic.

Necessity. We may assume that W is nonempty or else the congruence ρ determined by the pair (W, f) is the identity congruence. Suppose that $I(\Gamma)/\rho \cong I(\Delta)$ for some graph Δ.

Recall first that the idempotents of a graph inverse semigroup $I(\Gamma)$ are of the form pp^{-1} for some path p and that the maximal idempotents in the partial order correspond to the vertices of Γ by Proposition 2. Recall also that in any homomorphism between inverse semigroups, idempotents lift, and so the idempotents of the inverse semigroup $I(\Gamma)/\rho$ are of the form $(pp^{-1})\rho$ for some path p in Γ. Suppose that $(pp^{-1})\rho \geq v\rho$ for some vertex v and path p in Γ. Then $(pp^{-1}v)\rho = (vpp^{-1})\rho = v\rho$. Since ρ is 0-restricted, $v\rho \neq 0\rho$ and so we must have $s(p) = v$, in which case it follows that $(pp^{-1})\rho = v\rho$. Hence, $v\rho$ is maximal in the partial order in the graph inverse semigroup $I(\Delta) \cong I(\Gamma)/\rho$, and so we may view $v\rho$ as a vertex of Δ for each vertex v of Γ.

Now suppose that condition (1) fails. Then there exists a vertex v in W such that $s(e_v) = v$ and $r(e_v)v = 0$. We have $e_v^{-1}e_v = r(e_v)$ and $(e_ve_v^{-1}, v) \in \rho$, and so $v\rho \, \mathcal{D} \, r(e_v)\rho$ in $I(\Gamma)/\rho$. But since ρ is 0-restricted, we cannot have $v\rho = r(e_v)\rho$ since $r(e_v)v = 0$ in $I(\Gamma)$. Hence, $v\rho$ and $r(e_v)\rho$ are distinct vertices of Δ that are $\mathcal{D}$-related in $I(\Delta) \cong I(\Gamma)/\rho$. This, together with condition (1) of the theorem and Theorem 2, contradicts Corollary 1, and so condition (1) must hold. Then for any vertex $v \in W$ the edge e_v is a loop at v. This implies that $(e_ve_v^{-1})\rho = (e_v^{-1}e_v)\rho = v\rho$, and so $e_v\rho$ is in the $\mathcal{H}$-class of the idempotent $v\rho$ in the graph inverse semigroup $I(\Delta)$. But it is routine to see that if $pq^{-1}qp^{-1} = qp^{-1}pq^{-1}$ in a graph inverse semigroup, then $p = q$ and so $pq^{-1} = pp^{-1}$, and so graph inverse semigroups are combinatorial. From this it follows that $e_v\rho = v\rho$, that is, $f(e_v) = 1$. Hence, condition (2) must also hold. □

5 Subsemigroups of Graph Inverse Semigroups

In this section, we discuss some typical subsemigroups of a graph inverse semigroup (see [4] and [6]).

We first remark that two nonzero elements pq^{-1} and rs^{-1} of $I(\Gamma)$ are $\mathcal{D}$-equivalent if and only if $\mathbf{r}(p) = \mathbf{r}(q) = \mathbf{r}(r) = \mathbf{r}(s)$ by Corollary 1. It follows that each nonzero $\mathcal{D}$-class contains exactly one vertex of Γ. We use D_v to denote the $\mathcal{D}$-class which contains vertex $v \in \Gamma^0$ and set $D_v^0 = D_v \cup \{0\}$.

Proposition 6 *For each vertex v of Γ, D_v^0 is an inverse subsemigroup of $I(\Gamma)$.*

Proof Let $v \in \Gamma^0$ and $pq^{-1}, rs^{-1} \in D_v^0$. Then $\mathbf{r}(p) = \mathbf{r}(q) = \mathbf{r}(r) = \mathbf{r}(s) = v$. If the product $(pq^{-1})(rs^{-1})$ is zero, we have $(pq^{-1})(rs^{-1}) \in D_v^0$. Assume that the product $(pq^{-1})(rs^{-1})$ is nonzero. Then either q is a prefix of r (i.e., $r = qt$ for some (possibly empty) path t, in which case $(pq^{-1})(rs^{-1}) = pts^{-1}$), or else r is a prefix of q (i.e., $q = rt$ for some (possibly empty) path t, in which case $(pq^{-1})(rs^{-1}) = p(st)^{-1}$). In the first case we have $\mathbf{r}(pt) = \mathbf{r}(s) = v$ and in the second case we have $\mathbf{r}(p) = \mathbf{r}(st) = v$. Both of them lead to $(pq^{-1})(rs^{-1}) \in D_v^0$. Hence, D_v^0 is a subsemigroup of $I(\Gamma)$.

Notice that if $pq^{-1} \in D_v^0$, then $qp^{-1} \in D_v^0$, that is, $(pq^{-1})^{-1} \in D_v^0$. So D_v^0 is an inverse subsemigroup of $I(\Gamma)$. □

According to Corollary 1, two nonzero elements pq^{-1} and rs^{-1} of $I(\Gamma)$ are $\mathcal{J}$-equivalent if and only if there exist paths t_1, t_2 in Γ such that $\mathbf{s}(t_1) = \mathbf{r}(p) = \mathbf{r}(t_2)$ and $\mathbf{r}(t_1) = \mathbf{r}(r) = \mathbf{s}(t_2)$. There exists a one to one correspondence between the set of strongly connected components of a directed graph Γ and nonzero $\mathcal{J}$-classes of $I(\Gamma)$. For each nonzero $\mathcal{J}$-class J of $I(\Gamma)$, $J \cap \Gamma^0$ is a strongly connected component of Γ. Hence, we denote the $\mathcal{J}$-class which contains a strongly connected component Δ of Γ by J_Δ. Further, we have

Proposition 7 *For each strongly connected component Δ of Γ, $\mathcal{J}_\Delta^0 = J_\Delta \cup \{0\}$ is an inverse subsemigroup of $I(\Gamma)$.*

Proof Notice that for each strongly connected component Δ of Γ, $\mathcal{J}_\Delta = \cup_{v \in \Delta} D_v$. According to Proposition 6 it is sufficient to show that $\mathcal{J}_\Delta$ is closed with respect to the multiplication on $I(\Gamma)$. Suppose that $v_1, v_2 \in \Delta$, and let $pq^{-1} \in D_{v_1}^0$ and $rs^{-1} \in D_{v_2}^0$. Then $\mathbf{r}(p) = \mathbf{r}(q) = v_1$ and $\mathbf{r}(r) = \mathbf{r}(s) = v_2$. If the product $(pq^{-1})(rs^{-1})$ is zero, we have $(pq^{-1})(rs^{-1}) \in J_\Delta^0$. Assume that the product $(pq^{-1})(rs^{-1})$ is nonzero. Then either q is a prefix of r (i.e., $r = qt$ for some (possibly empty) path t, in which case $(pq^{-1})(rs^{-1}) = pts^{-1}$), or else r is a prefix of q (i.e., $q = rt$ for some (possibly empty) path t, in which case $(pq^{-1})(rs^{-1}) = p(st)^{-1}$). In the first case we have $\mathbf{r}(pt) = \mathbf{r}(s) = v_2$, which follows that $(pq^{-1})(rs^{-1}) \in D_{v_2}^0 \subseteq J_\Delta^0$. In the second case we have $\mathbf{r}(p) = \mathbf{r}(st) = v_1$, which implies that $(pq^{-1})(rs^{-1}) \in D_{v_1}^0 \subseteq J_\Delta^0$. □

Let S be an inverse semigroup with semilattice of idempotents $E(S)$. A subset $A \subset S$ is *closed* if we have $s \in A$ when $a \in A$ and $a \leq s$. The *closure* $B^\uparrow$ of a subset $B \subseteq S$ is defined as

$$B^\uparrow = \{s \in S : s \geq b \text{ for some } b \in B\}.$$

In particular if B is a singleton, that is, $B = \{b\}$, then $B^\uparrow$ is simply denoted by $b^\uparrow$.

A closed inverse subsemigroup H of an inverse semigroup S is said to be *proper* if it satisfies the following two conditions.

(P1) $0 \notin H$.
(P2) For each nonzero $s \in S$ there exists $a \in S$ such that $s^{-1}s$, $(sa)^{-1}sa \in H$.

Lemma 15 (Lawson [16], Proposition 5.13) *Let S be an inverse semigroup whose idempotents form a finite chain. Then S is a Clifford semigroup.*

Theorem 5 *Each proper closed inverse subsemigroup of $I(\Gamma)$ belongs to exactly one of the following classes:*

(i) *Finite chain type: it consists of a finite chain of idempotents.*
(ii) *Infinite chain type: it consists of an infinite chain of idempotents.*
(iii) *Cycle type: it is of the form*

$$\begin{aligned} L_{p,d} =& \{rs^{-1} : r = dp^i t, s = dp^j t, i, j \geq 0, t \text{ is a prefix of } p\} \\ & \cup \{qq^{-1} : q \text{ is a prefix of } d\}, \end{aligned}$$

where p is a circuit in Γ, d is a path in Γ with $\mathbf{r}(d) = \mathbf{r}(p)$, and where p, d have no nontrivial suffix in common. $L_{p,d}$ is the smallest closed inverse subsemigroup of $I(\Gamma)$ containing dpd^{-1}.

Proof Let H be a proper closed inverse subsemigroup of $I(\Gamma)$. If p_1 and p_2 are paths being segments of elements of H and are not prefix comparable, then the product of the idempotents $p_1p_1^{-1}$ and $p_2p_2^{-1}$ in H is equal to 0, and so $0 \in H$ and by closure $H = I(\Gamma)$. Hence, if H is proper, any two paths being segments of elements of H are prefix comparable and hence have the same source. By definition, if $rs^{-1} \in I(\Gamma)$, then $\mathbf{r}(r) = \mathbf{r}(s)$; hence, if $rs^{-1} \in L$, then $\mathbf{s}(r) = \mathbf{s}(s)$ and $\mathbf{r}(r) = \mathbf{r}(s)$. It follows that $E(H)$ is a linearly ordered set. There are now two possibilities: this linearly ordered set is either finite or infinite. If it is finite, then it is a finite chain of groups by Lemma 15. However, the only groups available are the trivial ones. It follows that if $E(H)$ is finite, then $H = E(H)$ is just a finite chain of idempotents and so is the closed inverse subsemigroup determined by its smallest idempotent. This accounts for both finite and infinite chain types.

In the light of the above, we need now only describe those proper closed inverse subsemigroups that contain non-idempotent elements. We first show that $L_{p,d}$ is closed with respect to the multiplication on $I(\Gamma)$ and so $L_{p,d}$ is a subsemigroup of $I(\Gamma)$. For any $q_1q_1^{-1}$, $q_2q_2^{-1}$ with q_1 and q_2 are prefixes of d, respectively, we have

$$(q_1q_1^{-1})(q_2q_2^{-1}) = \begin{cases} q_1q_1^{-1} & \text{if } q_2 \text{ is a prefix of } q_1 \\ q_2q_2^{-1} & \text{if } q_1 \text{ is a prefix of } q_2 \\ 0 & \text{otherwise,} \end{cases}$$

that is, $(q_1q_1^{-1})(q_2q_2^{-1}) \in L_{p,d}$. Suppose that $rs^{-1} \in L_{p,d}$ where $r = dp^it, s = dp^jt, i, j \geq 0$, and t is a prefix of p. Suppose that $qq^{-1} \in L_{p,d}$ where q is prefix of d. We have that $(qq^{-1})(rs^{-1}) = (qq^{-1})(dp^it(dp^jt)^{-1}) = dp^it(dp^jt)^{-1} = (dp^it(dp^jt)^{-1})(qq^{-1}) = (rs^{-1})(qq^{-1})$ as q is a prefix of d, which follows that $(qq^{-1})(rs^{-1}) = (rs^{-1})(qq^{-1}) \in L_{p,d}$. Now suppose that $r_1s_1^{-1}, r_2s_2^{-1} \in L_{p,d}$ where $r_1 = dp^{i_1}t_1, s_1 = dp^{j_1}t_1, r_2 = dp^{i_2}t_2, s_2 = dp^{j_2}t_2, i_1, i_2, j_1, j_2 \geq 0$, and t_1, t_2 are prefixes of p. As t_1 is a prefix of p, we write $p = t_1t_3$ for some path t_3 in Γ. Suppose that $j_1 \lneq i_2$. We deduce that

$$\begin{aligned}
(r_1s_1^{-1})(r_2s_2^{-1}) &= (dp^{i_1}t_1(dp^{j_1}t_1)^{-1})(dp^{i_2}t_2(dp^{j_2}t_2)^{-1}) \\
&= dp^{i_1}t_1t_1^{-1}(p^{j_1})^{-1}d^{-1}dp^{i_2}t_2(dp^{j_2}t_2)^{-1} \\
&= dp^{i_1}t_1t_1^{-1}(p^{j_1})^{-1}p^{i_2}t_2(dp^{j_2}t_2)^{-1} \\
&= dp^{i_1}t_1t_1^{-1}p^{i_2-j_1}t_2(dp^{j_2}t_2)^{-1} \\
&= dp^{i_1}t_1t_1^{-1}pp^{i_2-j_1-1}t_2(dp^{j_2}t_2)^{-1} \\
&= dp^{i_1}t_1t_1^{-1}t_1t_3p^{i_2-j_1-1}t_2(dp^{j_2}t_2)^{-1} \qquad (p = t_1t_3) \\
&= dp^{i_1}t_1t_3p^{i_2-j_1-1}t_2(dp^{j_2}t_2)^{-1} \\
&= dp^{i_1}pp^{i_2-j_1-1}t_2(dp^{j_2}t_2)^{-1} \\
&= dp^{i_2-j_1+i_1}t_2(dp^{j_2}t_2)^{-1} \in L_{p,d},
\end{aligned}$$

and a similar calculation applies if $j_1 \gneq i_2$. If $j_1 = i_2$ and t_1 is a prefix of t_2, say $t_2 = t_1t_4$ for some path t_4 in Γ, we get that

$$\begin{aligned}
(r_1s_1^{-1})(r_2s_2^{-1}) &= (dp^{i_1}t_1(dp^{j_1}t_1)^{-1})(dp^{i_2}t_2(dp^{j_2}t_2)^{-1}) \\
&= dp^{i_1}t_1t_1^{-1}(p^{j_1})^{-1}d^{-1}dp^{i_2}t_2(dp^{j_2}t_2)^{-1} \\
&= dp^{i_1}t_1t_1^{-1}(p^{j_1})^{-1}p^{i_2}t_2(dp^{j_2}t_2)^{-1} \\
&= dp^{i_1}t_1t_1^{-1}p^{i_2-j_1}t_2(dp^{j_2}t_2)^{-1} \\
&= dp^{i_1}t_1t_1^{-1}t_2(dp^{j_2}t_2)^{-1} \qquad (j_1 = i_2) \\
&= dp^{i_1}t_1t_1^{-1}t_1t_4(dp^{j_2}t_2)^{-1} \qquad (t_2 = t_1t_4) \\
&= dp^{i_1}t_1t_4(dp^{j_2}t_2)^{-1} \\
&= dp^{i_1}t_2(dp^{j_2}t_2)^{-1} \in L_{p,d} \qquad (t_2 = t_1t_4)
\end{aligned}$$

and a similar calculation applies if $j_1 = i_2$ and t_2 is a prefix of t_1. Hence, $L_{p,d}$ is a subsemigroup of $I(\Gamma)$. Since the inverse of an element of $L_{p,d}$ is clearly also in $L_{p,d}$, we deduce that $L_{p,d}$ is an inverse subsemigroup of $I(\Gamma)$.

Since we ascend in the natural partial order in $L_{p,d}$ by deleting identical suffixes from the path dp^it and dp^jt, or from a given prefix of d, it is also clear that $L_{p,d}$ is a closed inverse subsemigroup of $I(\Gamma)$.

Conversely, we show that any proper closed inverse subsemigroup of $I(\Gamma)$ which contains non-idempotent elements has the form $L_{p,d}$. If F is a closed inverse subsemigroup of $I(\Gamma)$ and contains dpd^{-1}, then for any $m, n \geq 0$ we have

$$(dpd^{-1})^n(d(dp)^{-1})^m = (dp^nd^{-1})(d(dp^m)^{-1}) = dp^n(dp^m)^{-1} \in F.$$

Ascending in the natural partial order, we may obtain any element of $L_{p,d}$, and so $L_{p,d} \subseteq F$.

Suppose that $H \neq E(H)$ is a closed inverse subsemigroup of $I(\Gamma)$. Then there exists $rs^{-1} \in H$ with $r \neq s$, and we may assume that the path s is shorter than the path r. Hence, s is a prefix of r and so $r = sp$ for some path t in Γ. Since $\mathbf{s}(r) = \mathbf{s}(r)$ and $\mathbf{r}(r) = \mathbf{r}(s)$, p must be a circuit in Γ. If p and s share a common suffix, with $p = p_1 t$ and $s = s_1 t$, then $sp_1 s_1^{-1} \geq sp_1 t(s_1 t)^{-1} = sps^{-1}$, and so by closure $sp_1 s_1^{-1} \in H$.

Amongst the non-idempotent elements $sp_1 s_1^{-1} \in H$, choose $s = d$ to have smallest possible length, and then having chosen d, choose p to be a non-empty circuit of smallest possible length. Then d, p do not share a nontrivial suffix. Now for any $m \geq 0$ we have $dp^m(dp^m)^{-1} \in E(H)$ and so, if $r_1 r_2^{-1} \in H$, each r_i is a prefix of some path dp^{m_i}. Since H is closed, every prefix of d is in H and by minimality of $|d|$, every element of H that contains prefix of d is an idempotent qq^{-1}. Hence, if $|r_i| < |d|$ we have $r_1 = r_2$. So we may now assume that for $i = 1, 2$ we have $|r_i| \geq |d|$ and so $r_1 = dp^i t$, $r_2 = dp^j t$ for some $i, j \geq 0$ and prefix t of p. □

The following result may immediately conclude from Theorem 5.

Corollary 6 *If the graph Γ contains no circuit, then every proper closed inverse subsemigroup of $I(\Gamma)$ is a chain of idempotents.*

At the end of this section, we classify the closed inverse subsemigroups of a graph inverse semigroup up to conjugacy.

The following lemma is due to Lawson [18] and Alali and Gilbert [4].

Lemma 16 *Let S be an E^*-unitary inverse semigroup. If H and K are conjugate closed inverse subsemigroups of S with $H \neq S \neq K$ and $H \subseteq E(S)$, then $K \subseteq E(S)$. Moreover, if H has a minimum idempotent, then so does K.*

Theorem 6

(i) *Let L be a closed inverse subsemigroup of $I(\Gamma)$ of finite chain type, Then all closed inverse subsemigroups conjugate to L are of finite chain type. Two closed inverse subsemigroups $L = qq^{-1\uparrow}$ and $K = pp^{-1\uparrow}$ are conjugate in $I(\Gamma)$ if and only if $\mathbf{r}(p) = \mathbf{r}(q)$.*

(ii) *Let L be a closed inverse subsemigroup of $I(\Gamma)$ of infinite chain type. Then all closed inverse subsemigroups conjugate to L are also of infinite chain type. Two closed inverse subsemigroups of infinite chain type are conjugate if and only if there are idempotents $rr^{-1} \in L$ and $ss^{-1} \in K$ such that for all paths t in Γ, we have that $rt(rt)^{-1} \in L$ if and only if $st(st)^{-1} \in K$.*

(iii) *Let L be a closed inverse subsemigroup of $I(\Gamma)$ of cycle type. The only closed inverse subsemigroups conjugate to L are of cycle type. Moreover, $L_{p,d}$ is conjugate to $L_{q,c}$ if and only if p and q are conjugate circuits in Γ.*

Proof

(i) If follows from Lemma 16 that if L has finite chain type, then so does every closed inverse subsemigroup conjugate to L.

Suppose that $L = qq^{-1\uparrow}$ and $K = pp^{-1\uparrow}$ are such that $\mathbf{r}(p) = \mathbf{r}(q)$. Then $pq^{-1} \in I(\Gamma)$, and for any prefix t_1 of q we have $(pq^{-1})(t_1t_1^{-1})(qp^{-1}) = pp^{-1} \in K$. Similarly, for any prefix t_2 of p, $(qp^{-1})(t_2t_2^{-1})(pq^{-1}) = qq^{-1} \in L$. Hence, L and K are conjugate.

Conversely, suppose that L and K are conjugate, with conjugating element $rs^{-1} \in I(\Gamma)$, so that for any prefixes t_3 of q and t_4 of p we have

$$(rs^{-1})(t_3t_3^{-1})(sr^{-1}) \in K \text{ and } (sr^{-1})(t_4t_4^{-1})(rs^{-1}) \in L.$$

Then $(rs^{-1})(\mathbf{s}(q)\mathbf{s}(q)^{-1})(sr^{-1}) \in K$, so that s and $\mathbf{s}(q)$ are prefix-comparable, hence $\mathbf{s}(s) = \mathbf{s}(q)$, and $(rs^{-1})(\mathbf{s}(q)\mathbf{s}(q)^{-1})(sr^{-1}) = rr^{-1} \in K$. Therefore, r is a prefix of p. Similarly, s is a prefix of q.

Let $p = rt_5$ for some path $t_5 \in \mathbf{Path}(\Gamma)$. We have

$$(sr^{-1})(pp^{-1})(rs^{-1}) = (sr^{-1})(rt_5(rt_5)^{-1})(rs^{-1}) = st_5(st_5)^{-1} \in L$$

and so st_5 is a prefix of q. Let $q = st_5t_6$ for some path $t_6 \in \mathbf{Path}(\Gamma)$. We obtain that

$$(rs^{-1})(qq^{-1})(sr^{-1}) = (rs^{-1})(st_5t_6(st_5t_6)^{-1})(sr^{-1}) = rt_5t_6(rt_5t_6)^{-1} \in K$$

and so rt_5t_6 is a prefix of p. But $p = rt_5$ and so $t_6 \in \Gamma^0$, $t_6 = \mathbf{r}(p)$ and $q = st_5$. Hence, $\mathbf{r}(p) = \mathbf{r}(q)$.

(ii) By Lemma 16 any closed inverse subsemigroup K conjugate to L must be chain type, and by part (i) K must be infinite. Suppose that $(sr^{-1})L(rs^{-1}) \subseteq K$ and $(rs^{-1})K(sr^{-1}) \subseteq L$. Since $0 \notin K$ we have, for all $pp^{-1} \in L$, that is, r is prefix comparable with p and similarly for all $qq^{-1} \in K$, that is, s is prefix comparable with q. If we consider p with $|p| \geq |r|$, then r must be a prefix of p and by closure of L we have $rr^{-1} \in L$. Similarly, $ss^{-1} \in K$. Suppose that $rt(rt)^{-1} \in L$. Then $(sr^{-1})(rt(rt)^{-1})(rs^{-1}) = st(st)^{-1} \in K$, and similarly if $st(st)^{-1} \in K$, then $rt(rt)^{-1} \in L$.

Conversely, if r and s exist as in the Theorem and $t_1t_1^{-1} \in L$, then r is a prefix comparable with t_1. If t_1 is a prefix of r, with $r = t_1t_2$ for some path $t_2 \in \mathbf{Path}(\Gamma)$, then

$$(sr^{-1})(t_1t_1)(rs^{-1}) = (s(t_1t_2)^{-1})(t_1t_1^{-1})(t_1t_2s^{-1}) = ss^{-1} \in K$$

and if r is a prefix of t_1 with $t_1 = rt_3$ for some path $t_3 \in \mathbf{Path}(\Gamma)$ then $rt_3(rt_3)^{-1} \in L$ and so

$$(sr^{-1})(t_1t_1^{-1})(rs^{-1}) = (sr^{-1})(rt_3(rt_3)^{-1})(rs^{-1}) = (st_3(st_3)^{-1}) \in K.$$

Similarly, $(rs^{-1})K(sr^{-1}) \subseteq L$, and L and K are conjugate.

(iii) By part (i) and (ii), any closed inverse subsemigroup of $I(\Gamma)$ that is conjugate to $L_{p,d}$ must be of cycle type. Suppose that the closed inverse subsemigroups $L_{p,d}$ and $L_{q,c}$ are conjugate in $I(\Gamma)$, and so there exists $sr^{-1} \in I(\Gamma)$ such that

$$(sr^{-1})L_{p,d}(rs^{-1}) \subseteq L_{q,c} \tag{5.1}$$

and

$$(rs^{-1})L_{q,c}(sr^{-1}) \subseteq L_{p,d}. \tag{5.2}$$

Since $L_{q,c}$ is closed and $L_{p,d}$ is the smallest closed inverse subsemigroup of $I(\Gamma)$ containing dpd^{-1}, then (5.1) is equivalent to $(sr^{-1})(dpd^{-1})(rs^{-1}) \in L_{q,c}$. Also, since $0 \notin L_{q,c}$ we must have r prefix-comparable with t_1 and t_2 whenever $t_1t_2^{-1}$ is an element of $L_{p,d}$. Hence, $rr^{-1} \in L_{p,d}$, and similarly $ss^{-1} \in L_{q,c}$.

First, suppose that $r = dp^it_2$ and $s = dq^jt_1$ for some $i, j \geq 0$, where t_2 is a prefix of p and t_1 is a prefix of q. Write $p = t_2p_1$. We obtain that

$$\begin{aligned}(sr^{-1})(dpd^{-1})(rs^{-1}) &= (dq^jt_1(dp^it_2)^{-1})(dpd^{-1})(dp^it_2(dq^jt_1)^{-1}) \\ &= dq^jt_1p_1t_2(dq^jt_1)^{-1} \in L_{q,c}.\end{aligned}$$

It follows that $dq^jt_1p_1t_2 = dq^kt_1$ for some $k \geq 0$. Comparing lengths of these paths, we see that $k > j$, and then after cancellation we obtain $t_1p_1t_2 = q^{k-j}t_1$. Hence, p_1t_2 is conjugate to some power of q, and since p_1t_2 is a conjugate of p, we conclude that p is conjugate to some power of q.

Now suppose that r is a prefix of d and write $d = rt_3$ and that s is a prefix of c and write $c = st_4$. We now obtain

$$(sr^{-1})(dpd^{-1})(rs^{-1}) = (sr^{-1})(rt_3p(rt_3)^{-1})(rs^{-1}) = st_3p(st_3)^{-1} \in L_{q,c}.$$

Since by assumption p is not the empty path, we have $st_3 = cq^it_4$ and $st_3p = cq^jt_4$ for some prefix t_4 of q and some $i, j \geq 0$. Again comparing lengths, we see that $j > i$, and then $st_3p = cq^it_4p = cq^jt_4$. After cancellation we obtain that $t_4p = q^{j-i}t_4$ and again p is a conjugate to a power of q.

Hence, for each possibility of r, we deduce from (5.1) that p is conjugate to some power of q. Using equation (5.2) we deduce similarly that q is conjugate to a power of p. Again comparing lengths, we conclude that p and q are conjugate.

Conversely, if p, q are conjugate, suppose that $p = q_1q_2$ and $q = q_2q_1$. Then it is easy to check that setting $r = c$ and $s = dq_1$ furnishes a pair (r, s) satisfying (5.1) and (5.2). □

6 Notes

The polycyclic inverse monoid P_n is constructed from the free monoid on an n-letter alphabet. Such a monoid can be viewed as the free category on the directed graph consisting of one vertex and n loops. Replacing free monoids by free categories, we get graph inverse semigroups first introduced by Ash and Hall [5]. Cuntz and Krieger introduced a class of C^*-algebras, constructed from suitable directed graphs, known as *Cuntz-Krieger algebras* [9]. As an analogues of the Cuntz-Krieger algebras, Abrams and Pino introduced what they called *Leavitt path algebras* [2]. In general, there are three angles to deeply investigate graph inverse semigroups: the connection with Cuntz-Krieger algebras [13], the connection with Leavitt path algebras [11, 21], and the topological structures [6] and [21].

An inverse semigroup is said to be *distributive* if it has joins of all finite non-empty compatible subsets and multiplication distributes over those finite joins that exist. An inverse semigroup is said to be *Boolean* if it is distributive and its semilattice of idempotents is a generalized Boolean algebra.

Let Γ be a directed graph satisfying the condition that the in-degree of each vertex is at least 2 and finite. Define the *Cuntz-Krieger inverse semigroup* CK_Γ of Γ in the following way:

(CKI1) It is Boolean.

(CKI2) It contains a copy of $I(\Gamma)$ and every element of CK_Γ is a join of a finite, non-empty compatible subset of $I(\Gamma)$.

(CKI3) $g = \bigvee_{g' \in \hat{g}} g'$ for each maximal idempotent g of $I(\Gamma)$, where $\hat{g}$ is the set of idempotents immediately covered by g.

(CKI4) It is the freest distributive inverse semigroup satisfying the above conditions.

The Cuntz-Krieger inverse semigroup CK_Γ of Γ is the ample semigroup of a topological groupoid associated with Γ, and the semigroup analogue of the Leavitt path algebra of Γ. There exists a homomorphism from $I(\Gamma)$ to CK_Γ and $I(\Gamma)$ is embedded into CK_Γ if and only if Γ has the additional property that the in-degree of each vertex is either 0 or at least 2 [13].

Another class of semigroups closely related to graph inverse semigroups and Leavitt path algebras is referred as *Leavitt inverse semigroups* [21].

Let F be a field and let Γ be a row finite graph; that is, $|\mathbf{s}^{-1}(v)|$ is finite for very vertex v in Γ. Recall (see (Abrams et al. [3]) that the Leavitt path algebra $L_F(\Gamma)$ is the F-algebra generated by the set $\Gamma^0 \cup \Gamma^1 \cup \Gamma^{-1}$ subject to the relations (V), (E1), (E2), and (CK1) defining the graph inverse semigroup $I(\Gamma)$ and the additional "Cuntz-Krieger" relations

(CKL) $v = \Sigma_{e \in \mathbf{s}^{-1}(v)} ee^{-1}$ for all $v \in \Gamma^0$ such that v is not a sink.

We refer the reader to the survey article by Abrams [1] or the book by Abrams et al. [3] for much information about Leavitt path algebras.

We denote by $LI(\Gamma)$ the multiplicative subsemigroup of L_F generated by $\Gamma^0 \cup \Gamma^1 \cup \Gamma^{-1}$. Of course $LI(\Gamma)$ is a proper subset of $L_F(\Gamma)$ since the addition and scalar multiplication operations are not used in constructing elements of the subsemigroup. In fact $LI(\Gamma)$ is an inverse semigroup, named the *Leavitt inverse semigroup* of the graph Γ [21]. Meakin et al. [21] gave a presentation for this semigroup (as an inverse semigroup) by generators and relations up to isomorphism.

Let $L(\Gamma)$ be the semigroup generated by the set $\Gamma^0 \cup \Gamma^1 \cup \Gamma^{-1}$ subject to the relations (V), (E1), (E2), and (CK1) used to define the graph inverse semigroup $I(\Gamma)$ and the additional relations:

(LI) $e_v e_v^{-1} = v$ for each vertex $v \in \Gamma^0$ of out-degree 1.

Clearly, $L(\Gamma)$ is an inverse semigroup since it is a homomorphic image of $I(\Gamma)$. Moreover, for each graph Γ, $LI(\Gamma) \cong L(\Gamma)$.

There exists an insight to investigate topologies that turn $I(\Gamma)$ into a topological semigroup. A *topological (inverse) semigroup* is a topological space with a continuous semigroup operation (and a continuous inversion). If S is a semigroup (an inverse semigroup) and τ is a topology on S such that (S, τ) is a topological (inverse) semigroup, then we shall call τ an *(inverse) semigroup topology* on S. Mesyan et al. [23] investigated topologies that turn $I(\Gamma)$ into a topological semigroup. In any such topology that is Hausdorff, $I(\Gamma)\backslash\{0\}$ must be discrete for any graph Γ. It generalizes the result that the bicyclic monoid (without zero) is discrete in any semigroup topology [10]. If Γ is a finite graph, then the only locally compact Hausdorff semigroup topology on $I(\Gamma)$ is the discrete topology [23]. Bardyla [6] generalized this result to all graphs and describes all graph inverse semigroups which allow only discrete locally compact semigroup topology.

Theorem 7 (Bardyla [6], Theorem 4 or 6) *The discrete topology is the only locally compact semigroup topology on $I(\Gamma)$ if and only if Γ contains a finite number of vertices and there does not exist a pair of vertices $u, v \in \Gamma^0$ such that the set $\{p : p \in \mathbf{Path}(\Gamma) with\ \mathbf{r}(p) = u\}$ is finite and the set $\{e \in \Gamma^1 : \mathbf{s}(e) = u \text{ and } \mathbf{r}(e) = v\}$ is infinite.*

Corollary 7 (Bardyla [6], Corollary 5) *There exists a non-discrete locally compact semigroup topology on $I(\Gamma)$ if and only if there exists a non-discrete topology τ such that $(I(\Gamma), \tau)$ is a locally compact metrizable topological inverse semigroup.*

At the end of this section, we change our angle a little to give a short discussion about the semigroup generated by a high-order logic network, which is an analogue of graph inverse semigroups. To see this, we recall some notation related to networks.

A *network* $\Gamma = (\Gamma^0, \Gamma^1)$ consists of a set Γ^0 of vertices and a set Γ^1 of edges (termed relations in the following). It is a correlation (resp. logic) network if each relation in Γ^1 is a subset of distinct vertices (resp. an ordered pair of disjoint subsets) of Γ^0. A correlation (resp. logic) network is called a traditional undirected (resp. directed) graph if each relation in Γ^1 is a subset of two distinct vertices (resp. an ordered pair of disjoint singleton subsets) of Γ^0. Clearly, any relation in a traditional undirected or directed graph is related only to two vertices. A network $\Gamma = (\Gamma^0, \Gamma^1)$ is a *higher-order network* if there is $t \in \Gamma^1$ related more than two vertices in Γ^0 [7]. A relation in Γ^1 is said to be *logic* if it is an ordered pair of disjoint subsets of Γ^0, denoted by T_A^B where A and B are disjoint subsets of Γ^0. Here A and B are called the *source* and the *range* of T_A^B, denoted as $\mathbf{s}(T_A^B)$ and $\mathbf{r}(T_A^B)$, respectively. A logic relation $e \in \Gamma^1$ is called an *m-n order logic relation* if $|\mathbf{s}(e)| = m$ and $|\mathbf{r}(e)| = n$, $m, n \in \mathbb{N}$. In particular, an m-n-order logic relation is called an *mth-order logic relation* if $n = 1$. Notice that any relation in a traditional directed graph is a first-order logic relation. The *first-order logic network* means a traditional directed graph and a *higher-order logic network* is a logic and higher-order network [28].

It is worth noting that the graph Γ involved in the graph inverse semigroup $I(\Gamma)$, the Cuntz-Krieger inverse semigroup CK_Γ, and the Leavitt inverse semigroup $LI(\Gamma)$ is a first-order logic network where the path $p = e_1e_2\cdots e_n$ of relations $e_i \in \Gamma^1$ is such that $\mathbf{r}(e_i) = \mathbf{s}(e_{i+1})$ for $i = 1, \ldots, n-1$. While in a higher-order logic network Γ, $\mathbf{r}(e)$ and $\mathbf{s}(e)$ may not be singleton for $e \in \Gamma^1$. Hence, there exists a sequence $p = e_1e_2\cdots e_n$ of relations $e_i \in \Gamma^1$ with $\mathbf{r}(e_i) \neq \mathbf{s}(e_{i+1})$ but $\mathbf{r}(e_i) \cap \mathbf{s}(e_{i+1}) \neq \emptyset$ for $i = 1, \ldots, n-1$ in a higher-order logic network. Naturally, what algebraic system does such sequences generate? Wang et al. [29] use this kind of sequences to generate a right ample semigroup S_Γ with a zero element, which belongs to the class of right abundant semigroups, named *network right ample semigroups*. In particular, if a higher-order logic network is replaced by a first-order logic network, S_Γ is a graph inverse semigroup.

Acknowledgments The research is supported by Natural Science Foundation of China (Grant No:12271442, 11501331). The first author was supported by the Cultivation Project of Young and Innovative Talents in Universities of Shandong Province.

References

1. Abrams, G.: Leavitt path algebras: the first decade. Bull. Math. Sci. **5**, 59–120 (2015)
2. Abrams, G., Pino, G.A.: The Leavitt path algebra of a graph. J. Algebra **293**, 319–334 (2005)
3. Abrams, G., Ara, P., Molina, M.S.: Leavitt Path Algebras. Springer Lecture Notes in Mathematics. Springer, Berlin (2017)
4. Alali, A., Gilbert, N.D.: Closed inverse subsemigroups of graph inverse semigroups. Commun. Algebra **45**(11), 4667–4678 (2017)
5. Ash, C.J., Hall, T.E.: Inverse semigroups on graphs. Semigroup Forum **11**, 140–145 (1975)
6. Bardyla, S.: On locally compact topological graph inverse semigroups. Topol. Appl. **267**, 106873 (2019)
7. Bianconi, G.: Higher-Order Networks: An Introduction to Simplicial Complex. Cambridge University Press, Cambridge (2022)
8. Costa, A., Steinberg, B.: A categorial invariant of flow equivalence of shifts. Ergodic Theory Dyn. Syst. **36**, 470–513 (2016)
9. Cuntz, J., Krieger, W.: A class of C^*-algebras and topological Markov chains. Invent. Math. **56**, 251–268 (1980)
10. Eberhart, C., Selden, J.: On the closure of the bicyclic semigroup. Trans. Am. Math. Soc. **144**, 115–126 (1969)
11. Fan, H.B., Wang, Z.P.: On Leavitt inverse semigroups. J. Algebra Appl. **20**, 2150158 (2021)
12. Howie, J.M.: Fundamentals of Semigroup Theory. Clarendon Press, Oxford (1995)

13. Jones, D.G., Lawson, M.V.: Graph inverse semigroups: their characterization and completion. J. Algebra **409**, 444–473 (2014)
14. Krieger, W.: Subshifts and semigroups. Bull. London Math. Soc. **38**, 617–624 (2006)
15. Kumjian, A., Pask, D., Raeburn, I.: Cuntz-Krieger algebras of directed graphs. Pacific J. Math. **184**(1), 161–174 (1998)
16. Lawson, M.V.: Inverse Semigroups: The Theory of Partial Symmetries. World Scientific, Singapore (1998)
17. Lawson, M.V.: E^*-unitary inverse semigroups. In: Semigroups, Algorithms, Automata, and Languages (Coimbra, 2001), pp. 195–214. World Scientific Publishing, River Edge (2002)
18. Lawson, M.V.: Primitive partial permutation representations of the polycyclic monoids and branching function systems. Periodica Math. Hungarica **58**(2), 189–207 (2009)
19. Leavitt, W.G.: The module type of a ring. Trans. Am. Math. Soc. **103**, 113–130 (1962)
20. Luo, Y.L., Wang, Z.P.: Semimodularity in congruence lattices of graph inverse semigroups. Commun. Algebra **49**(6), 2623–2632 (2021)
21. Meakin, J.C., Milan, D., Wang, Z.P.: A class of inverse semigroups related to Leavitt path algebras. Adv. Math. **384**, 107729 (2021)
22. Mesyan, Z., Mitchell, J.D.: The structure of a graph inverse semigroup. Semigroup Forum **93**, 111–130 (2016)
23. Mesyan, Z., Mitchell, J.D., Morayne, M., Péresse, Y.H.: Topological graph inverse semigroups. Topol. Appl. **208**, 106–126 (2016)
24. Nivat, M., Perrot, J-F.: Une généralisation du monoïde bicyclique. Comptes Rendus de l'Academie des Sciences de Paris **271**, 824–827 (1970)
25. Paterson, A.L.T.: Groupoids, Inverse Semigroups, and Their Operator Algebras. Birkhäuser, Boston (1998)
26. Paterson, A.L.T.: Graph inverse semigroups, groupoids and their C^*-algebras. J. Oper. Theory **48**, 645–662 (2002)
27. Wang, Z.P.: Congruences on graph inverse semigroups. J. Algebra **534**, 51–64 (2019)
28. Wang, Y. H., Meng, D.Z.: Quasi-semilattices on networks. Axioms **12**, 943 (2023)
29. Wang, Y.H., Gao, P., Ren, X.M.: Network right $\star$-abundant semigroups. Semigroup Forum, submitted

Unification via Projectivity in Varieties of Hoops

Paolo Aglianò and Sara Ugolini

Abstract This chapter is a survey on projective algebras in varieties of hoops and their connection to unification problems for the logics of which they are the equivalent algebraic semantics.

Keywords Unification · Projectivity · Mathematical fuzzy logic · Generalized rotations · Product logic

1 Introduction

Hoops are a particular variety of residuated monoids related to logic which were defined in an unpublished manuscript by Büchi and Owens, inspired by the work of Bosbach on partially ordered monoids [24]. Hoops indeed correspond to commutative monoids endowed with a partial order $\leq$ definable by means of the residuum $\rightarrow$ of the monoid operation as $a \leq b$ if and only if $a \rightarrow b = 1$, and such an order is *natural*, in the sense that $a \leq b$ if and only if there exists c such that $a = b \cdot c$. The first systematic study of hoops has been carried out by Ferreirim in her PhD thesis [36] and in her works with Blok [19, 20], while in her work with Aglianò and Montagna [6] they investigate the connection between hoops and the (fuzzy) logics in Hájek framework of logics arising from a continuous t-norm [42].

In order to give more details to the interested readers, we remark that hoops are algebras of logic in a very precise sense. By a *logic* $\mathcal{L}$ we mean a substitution invariant consequence relation $\vdash_{\mathcal{L}}$ on the term algebra $\mathbf{T}_\rho(\omega)$ (also called *algebra of formulas*) of some algebraic language ρ. In loose terms, whenever the logical equivalence between formulas (i.e., provable equivalence according to $\vdash_{\mathcal{L}}$) is

P. Aglianò
DIISM, University of Siena, Siena, Italy
e-mail: agliano@unisi.it

S. Ugolini (✉)
Artificial Intelligence Research Institute, IIIA - CSIC, Barcelona, Spain
e-mail: sara@iiia.csic.es

M. Mitrović, M. N. Hounkonnou (eds.), *Algebra Without Borders*, Mathematics in Mind, https://doi.org/10.1007/978-3-031-86477-3_5

a congruence on the algebra of formulas, then we can form the *Lindenbaum-Tarski algebra* as the quotient of provably equivalent formulas of the algebra of formulas. If the quasivariety $\mathsf{Q}_{\mathcal{L}}$ generated by the Lindenbaum-Tarski algebra satisfies some further conditions, then it is the class of algebraic models of $\mathcal{L}$ and $\mathcal{L}$ is *algebraizable* with *equivalent algebraic semantics* $\mathsf{Q}_{\mathcal{L}}$; Blok and Pigozzi [21] formalized this connection and gave necessary and sufficient conditions for a logic to be algebraizable. Essentially, one needs a finite set of one-variable equations $\tau = \{\delta_i \approx \varepsilon_i : i = 1, \dots, n\}$ in the language of $\mathsf{Q}_{\mathcal{L}}$ and a finite set of formulas of $\mathcal{L}$ in two variables $\Delta(x, y) = \{\varphi_1(x, y), \dots, \varphi_m(x, y)\}$ that allow to transform equations, quasiequations, and universal sentences in $\mathsf{Q}_{\mathcal{L}}$ into formulas, rules, and clauses of $\mathcal{L}$ and vice versa; moreover, this transformation must respect both the consequence relation and the semantical consequence. That is to say, for all sets of formulas Γ of $\mathcal{L}$ and formulas $\varphi \in \mathbf{T}_\rho(\omega)$,

$$\Gamma \vdash \varphi \text{ if and only if } \{\delta_i(\Gamma)\approx\varepsilon_i(\Gamma) : i=1,\dots,n\} \vDash_{\mathsf{Q}_{\mathcal{L}}} \{\delta_i(\varphi)\approx\varepsilon_i(\varphi) : i=1,\dots,n\}$$

where of course $\delta_i(\Gamma) \approx \varepsilon_i(\Gamma)$ is a shorthand for $\delta_i(\psi) \approx \varepsilon_i(\psi)$ for all $\psi \in \Gamma$ and also

$$\mathsf{Q}_{\mathcal{L}} \vDash (x \approx y) \Leftrightarrow (\bigwedge_{i=1}^{n} \bigwedge_{j=1}^{m} (\delta_i(\varphi_j(x, y)) \approx \varepsilon_i(\varphi_j(x, y))).$$

A quasivariety Q is a *quasivariety of logic* if it is the equivalent quasivariety semantics for some algebraizable logic $\mathcal{L}_{\mathsf{Q}}$; the Galois connection between algebraizable logics and quasivarieties of logic is given by

$$\mathcal{L}_{\mathsf{Q}_{\mathcal{L}}} = \mathcal{L} \qquad \mathsf{Q}_{\mathcal{L}_{\mathsf{Q}}} = \mathsf{Q}.$$

We stress that not every logic is algebraizable and not every quasivariety is a quasivariety of logic; however, recently Moraschini and Raftery showed that every quasivariety is categorically equivalent to a quasivariety of logic [49].

Nonetheless, the variety H of hoops is a (quasi)variety of logic (and so are all its subquasivarieties). In particular, let us consider the variety of *semilinear* (i.e., generated by totally ordered algebras) hoops, called *basic hoops*. The latter are the equivalent algebraic semantics of the 0-free fragment of *Hajek's Basic Logic* $\mathcal{BL}$ [42], $\mathcal{BL}^+$. This among other things implies that the axiomatic extensions of $\mathcal{BL}^+$ are in one-to-one correspondence with the subvarieties of H and the finitary extensions of $\mathcal{BL}^+$ are in one-to-one correspondence with the subquasivarieties of H.

This chapter studies varieties of hoops and their class of *projective algebras*. From a purely algebraic standpoint, projective algebras are interestingly connected with *splittings*. The notion of splitting comes from lattice theory; a *splitting pair* [60] in a lattice $\mathbf{L}$ is a pair $(a, b) \in \mathbf{L}^2$ such that $a \not\leq b$ and for all $c \in L$, either $c \leq a$ or $b \leq c$. It is clear that in this case $\mathbf{L}$ is the disjoint union of the principal ideal generated by a and the principal filter generated by b. The concept of splitting can be applied to lattices of subvarieties $\Lambda(\mathsf{V})$ of a variety V; an algebra $\mathbf{A} \in \mathsf{V}$ is *splitting* in V if there is a subvariety $\mathsf{W}_{\mathbf{A}} \subseteq \mathsf{V}$ such that $(\mathsf{W}_{\mathbf{A}}, \mathbf{V}(\mathbf{A}))$ is a splitting pair in $\Lambda(\mathsf{V})$. It was first observed by McKenzie [48] that there is a close connection between projective algebras in a variety and splitting algebras in its lattice of subvarieties. McKenzie proved the following for varieties of lattices but it holds in general (for a proof the reader can consult [12], Lemma 3.2).

Theorem 1 *Let* V *be any variety and let* $\mathbf{A}$ *be an algebra that is subdirectly irreducible and projective in* V*. Then:*

(i) $\mathsf{U} = \{\mathbf{B} \in \mathsf{V} : \mathbf{A} \notin \mathbf{S}(\mathbf{B})\}$ *is a subvariety of* V.
(ii) *For any subvariety* W *of* V *either* $\mathsf{W} \subseteq \mathsf{U}$ *or* $\mathsf{V}(\mathbf{A}) \subseteq \mathsf{W}$.
(iii) *There is an equation* σ *in the language of* V *such that* $\mathbf{B} \in \mathsf{U}$ *if and only if* $\mathbf{B} \models \sigma$.

Hence, $(\mathsf{U}, \mathsf{V}(\mathbf{A}))$ *constitutes a splitting pair in the lattice of subvarieties of* V.

Even though we will not deal specifically with splitting hoops in this chapter (for reasons of space), we point out that splitting algebras in the variety of hoops (and in many of its subvarieties) have been studied and classified in [2].

Projective algebras are interesting also from the logical perspective; indeed, they are connected to unification problems for the corresponding logics via Ghilardi's algebraic approach to unification [40].

The start of the study of unification problems can be traced back to Herbrand's doctoral thesis and in the pioneering work of Robinson on resolution [55]. The usual syntactic unification problem starts with a pair of terms (s, t) over some language, and a solution or *unifier* is a uniform replacement of the variables occurring in s and t by other terms in the same language that makes s and t identical. If syntactical identity is replaced by equality modulo an equational theory, one speaks of E-unification; since we will only deal with the latter, we will just refer to E-unification as unification in what follows. Ghilardi shows that, in an algebraizable logic, the study of a unification problem modulo logical equivalence can be carried out purely in the algebraic framework. Moreover, the *unification type*, which essentially measures the cardinality of the set of "best solutions" to a unification problem, can be studied algebraically as well. In more detail, if a logic has a variety V as equivalent algebraic semantics, a unification problem corresponds to a *finitely presented* algebra in V (a finitely generated quotient of a finitely generated free algebra), and a unifier consists in a homomorphism to a *projective* algebra in V (that is, a retract of a free algebra in V). Thus, the study of projective algebras in varieties corresponding to a logic gives relevant information on the logical unification problems.

After a section of preliminaries, we recall basic facts about projective algebras in (quasi)varieties, and we survey results about projective algebras in relevant varieties of hoops and bounded hoops (i.e., hoops with an extra constant 0 that is the bottom element in the order), contained, for instance, in [11, 12, 28, 29, 47, 58]. Then we recall Ghilardi's approach to algebraic unification, detailing the connections between the symbolic and algebraic approach. Finally, we apply the known results about projectivity in varieties of hoops to the study of unification problems.

2 Preliminaries

For the unexplained basic notions from universal algebra, we refer the reader to [25]. We start by recalling the needed facts about hoops; for more details and further references the reader can consult [6, 20, 36].

A *hoop* is an algebra $\mathbf{A} = (A, \cdot, \rightarrow, 1)$ such that $(A, \cdot, 1)$ is a commutative monoid and:

(H1) $x \rightarrow x = 1$.
(H2) $x \cdot (x \rightarrow y) = y \cdot (y \rightarrow x)$,
(H3) $(x \cdot y) \rightarrow z = x \rightarrow (y \rightarrow z)$.

We call a hoop **A** *bounded* if it has an extra constant 0 in the signature that is the least element: $a \wedge 0 = 0$ for each $a \in A$.

Every hoop carries a natural order structure definable as

$$x \leq y \text{ iff } x \rightarrow y = 1$$

that is, in general a meet-semilattice ordering, where $\wedge$ is definable as

$$x \wedge y = x \cdot (x \rightarrow y).$$

Moreover, $(\cdot, \rightarrow)$ form a *residuated pair* with respect to this lattice order, i.e.,

$$x \cdot y \leq z \text{ iff } y \leq x \rightarrow z.$$

In fact, hoops can be seen as particular residuated structures. A *commutative residuated* $\wedge$*-semilattice* is an algebra $\mathbf{A} = (A, \cdot, \rightarrow, \wedge, 1)$ where:

(i) $(A, \cdot, 1)$ is a commutative monoid.
(ii) $(A, \wedge)$ is a semilattice.
(iii) $(\cdot, \rightarrow)$ form a residuated pair with respect to the semilattice order.

We call a commutative residuated (semi)lattice *integral* if 1 is the top element in the (semi)lattice ordering and *bounded* if it is integral and there is also an extra element 0 that is the bottom element of the (semi)lattice. Hoops fit in the residuated semilattice framework since they are term equivalent to the variety of commutative integral residuated $\wedge$-semilattices satisfying the divisibility condition:

$$x \wedge y = x \cdot (x \rightarrow y). \tag{div}$$

A commutative residuated lattice is an algebra $\mathbf{A} = (A, \cdot, \rightarrow, \wedge, \vee, 1)$ where $(A, \cdot, \rightarrow, \wedge, 1)$ is a commutative residuated $\wedge$-semilattice, and $(A, \wedge, \vee)$ is a lattice. Thus, hoops are the $\wedge$-free subreducts of divisible commutative integral residuated lattices. We call a totally ordered hoop a *chain*. *Basic hoops* constitute the subvariety of hoops generated by chains, and bounded basic hoops are called *BL-algebras* since they are the equivalent algebraic semantics of Hájek Basic Logic [42].

With respect to their structure theory, all varieties of residuated $\wedge$-semilattices are *ideal-determined* in the sense of [13], and the role of ideals is played by the *congruence filters*. A congruence filter is a non-empty semilattice filter closed under products. In particular, there is a lattice isomorphism between the lattice of congruence filters and the congruence lattice. Given a hoop $\mathbf{A}$, the isomorphism is given by the following two maps, where θ is any congruence of $\mathbf{A}$ and F is any congruence filter of $\mathbf{A}$:

$$\theta \mapsto F_\theta = \{a \in A : (a, 1) \in \theta\}, \qquad F \mapsto \theta_F = \{(a, b) : a \to b, b \to a \in F\}.$$

Finitely generated (or, equivalently, compact) congruences are then in bijection with finitely generated congruence filters. Now, given $\mathbf{A}$ a hoop, and $X \subseteq A$, the congruence filter generated by X in $\mathbf{A}$ is the set

$$\mathrm{Fil}_{\mathbf{A}}(X) = \{b \in A : a_1 \cdot \ldots \cdot a_n \leq b,\ a_1, \ldots, a_n \in X,\ n \in \mathbb{N}\}.$$

It then easily follows that finitely generated congruence filters correspond to principal congruence filters (it suffices to take the meet of the generators), and analogously finitely generated congruences are principally generated.

2.1 The Ordinal Sum

The ordinal sum construction has proven to be a key tool to understand hoops. Most of the content of this section is taken from [7]. Let $\langle I, \leq\rangle$ be a totally ordered set. For all $i \in I$ let $\mathbf{A}_i$ be a hoop such that for $i \neq j$, $A_i \cap A_j = \{1\}$. Then $\bigoplus_{i\in I} \mathbf{A}_i$ (the *ordinal sum* of the family $(\mathbf{A}_i)_{i\in I}$) is the structure whose base set is $\bigcup_{i\in I}(A_i \setminus \{1\}) \cup \{1\}$ and the operations are

$$x \to y = \begin{cases} x \to^{\mathbf{A}_i} y & \text{if } x, y \in A_i \\ y & \text{if } x \in A_i \text{ and } y \in A_j \text{ with } i > j \\ 1 & \text{if } x \in A_i \setminus \{1\} \text{ and } y \in A_j \text{ with } i < j \end{cases}$$

$$x \cdot y = \begin{cases} x \cdot^{\mathbf{A}_i} y & \text{if } x, y \in A_i \\ y & \text{if } x \in A_i \text{ and } y \in A_j - \{1\} \text{ with } i > j \\ x & \text{if } x \in A_i \setminus \{1\} \text{ and } y \in A_j \text{ with } i < j \end{cases}$$

Note that the ordinal sum of hoops is always a hoop; if in addition I has a minimum, say 0, and $\mathbf{A}_0$ is a bounded hoop, then the ordinal sum is a bounded hoop. The subalgebras of the ordinal sum of a family of hoops are easily characterized.

Proposition 1 *Let* $(\mathbf{A}_i)_{i\in I}$ *be a family of hoops. Then the subhoops of* $\bigoplus_{i\in I}\mathbf{A}_i$ *are exactly the hoops of the form* $\bigoplus_{i\in I}\mathbf{B}_i$, *where* $\mathbf{B}_i$ *is a (possibly trivial) subhoop of* $\mathbf{A}_i$. *Moreover, if* $\bigoplus_{i\in I}\mathbf{A}_i$ *is a BL-algebra (and so* I *has a minimum and* $\mathbf{A}_0$ *is a BL-algebra), then the basic subalgebras of* $\bigoplus_{i\in I}\mathbf{A}_i$ *are exactly the subalgebras of the form* $\bigoplus_{i\in I}\mathbf{B}_i$, *where* $\mathbf{B}_0$ *is a basic subalgebra of* $\mathbf{A}_0$ *and* $\mathbf{B}_i$ *is a subhoop of* $\mathbf{A}_i$.

It is easy to see that the ordinal sum of finitely many hoops is associative; hence, we feel free to write $\mathbf{A}_0 \oplus \cdots \oplus \mathbf{A}_n$ without parentheses. In this case, something more can be said.

Proposition 2 *Let* $\mathbf{A}_0, \ldots, \mathbf{A}_n$ *be hoops. Then the set of homomorphic images of* $\mathbf{A}_0 \oplus \cdots \oplus \mathbf{A}_n$ *is*

$$\mathbf{H}(\mathbf{A}_0) \cup \{\mathbf{A}_0 \oplus \mathbf{B} : \mathbf{B} \in \mathbf{H}(\mathbf{A}_1)\} \cup \cdots \cup \{\mathbf{A}_0 \oplus \cdots \oplus \mathbf{A}_{n-1} \oplus \mathbf{B} : \mathbf{B} \in \mathbf{H}(\mathbf{A}_n)\}.$$

Proof We shall sketch the proof for the ordinal sum of two hoops, since the general form follows easily by induction. Let h a homomorphism of $\mathbf{A}\oplus\mathbf{B}$ and consider the set $F_h = \{x : h(x) = 1\}$. Then F_h is a filter of $\mathbf{A} \oplus \mathbf{B}$. Suppose first that $F_h \subseteq B$; then for $a_1, a_2 \in A$, $h(a_1) = h(a_2)$ if and only if $h(a_1 \to a_2) = h(a_2 \to a_1) = 1$. Thus, $a_1 \to a_2, a_2 \to a_1 \in F_h \subseteq B$ and, by the definition of ordinal sum, this is possible if and only if $a_1 \to a_2 = a_2 \to a_1 = 1$. Thus, $a_1 = a_2$ and h restricted to $\mathbf{A}$ is an isomorphism. Now it is easily seen that $h(\mathbf{A}\oplus\mathbf{B}) \cong \mathbf{A}\oplus h(\mathbf{B})$. Suppose now that $F_h \cap A \setminus \{1\} \neq \emptyset$ and let $a \in A$, $a \neq 1$, such that $h(a) = 1$. Then for all $b \in B$, $a \leq b$ and thus $1 = h(a) \leq h(b)$. Hence, $B \subseteq F_h$ and clearly $h(\mathbf{A} \oplus \mathbf{B}) \cong h_{|\mathbf{A}}(\mathbf{A})$. □

Proposition 3 (cf. [23], Lemma 5.2) *Let* $\mathbf{A}^i_0, \ldots, \mathbf{A}^i_n$, $i \in I$ *be hoops. Then* $\mathbf{P}_u(\{\mathbf{A}^i_0 \oplus \cdots \oplus \mathbf{A}^i_n : i \in I\})$ *consists of algebras of the form* $\mathbf{B}_0 \oplus \cdots \oplus \mathbf{B}_n$, *where* $\mathbf{B}_j \in \mathbf{IP}_u(\mathbf{A}^j_i : i \in I)$.

Proof Let $\mathbf{A} = \prod_{i\in I}(\mathbf{A}^1_i \oplus \cdots \oplus \mathbf{A}^n_i)$, let U be any ultrafilter on I, and let $\mathbf{A}_U = \mathbf{A}/U$. Clearly, $\mathbf{A}_U \in \mathbf{P}_u(\mathbf{A}^1_i \oplus \cdots \oplus \mathbf{A}^n_i : i \in I)$ and any member of $\mathbf{P}_u(\mathbf{A}^1_i \oplus \cdots \oplus \mathbf{A}^n_i : i \in I)$ is of the form $\mathbf{A}_U$ for some ultrafilter U on I. For $b \in A$, let b_U be the class of b modulo U. For $j = 1, \ldots, n-1$, let $X(b, j) = \{i \in I : b_i \in A^j_i - \{1\}\}$ and let $X(b, n) = \{i \in I : b_i \in A^n_i\}$. Since the $X(b, j)$ form a partition of I, there is exactly one j_b such that $X(b, j_b) \in U$. Of course j_b really depends on b_U, in that if $b_U = c_U$, then $j_b = j_c$.

Now for $j = 1, \ldots, n$, let $Y^j = \{b_U : b \in A \text{ and } j_b = j\}$; it is easy to check that $Y^j \cup \{1\}$ is the universe of a subhoop $\mathbf{B}^j$ of $\mathbf{A}_U$. If we take $b_U \in Y^j$ and $c_U \in Y^h$ with $j < h$, then $X(b, j) \cap X(c, h) \in U$ and for $i \in X(b, j) \cap X(c, h)$ one has $b^j_i c^j_i = c^j_i$ and $c^j_i \to b^j_i = b^j_i$. It follows that $b_U c_U = b_U$ and $c_U \to b_U = b_U$, and hence, $\mathbf{A}_U = \mathbf{B}^1 \oplus \cdots \oplus \mathbf{B}^n$.

It remains to show that $\mathbf{B}^j$ is isomorphic with $(\prod_{i\in I}\mathbf{A}^j_i)/U$. If $b_U \in Y^j \cup \{1\}$, then the set $\{i \in I : b_i \in A^j_i\} \in U$. If c is defined by

$$c_i = \begin{cases} b_i & \text{if } b_i \in A_i^j \\ 1 & \text{otherwise} \end{cases}$$

then $c \in \prod_{i \in I} A_i^j$ and $c_U = b_U$. Let now c_U^j be the equivalence class of c modulo U in $\prod_{i \in I} A_i^j$; one can easily verify that the map $c_U^i \longmapsto b_U = c_U$ is an isomorphism from $\prod_{i \in I} \mathbf{A}_i^j$ to $\mathbf{B}^j$, as desired. □

Corollary 1 *If* $\mathbf{A}_1, \ldots, \mathbf{A}_n$ *are basic hoops, then*

$$\mathbf{P}_u(\mathbf{A}_1 \oplus \cdots \oplus \mathbf{A}_n) \subseteq \{\mathbf{B}_1 \oplus \cdots \oplus \mathbf{B}_n : \mathbf{B}_i \in \mathbf{P}_u(\mathbf{A}_i)\}.$$

It is obvious that the conclusions of Proposition 3 and Corollary 1 hold in case $\mathbf{A}_0$ is a bounded hoop.

If $\mathbf{A} = \bigoplus_{i \in I} \mathbf{A}_i$, then each of the $\mathbf{A}_i$'s is *a component* of $\mathbf{A}$ and the set $\{\mathbf{A}_i : i \in I\}$ is a *decomposition* of $\mathbf{A}$. One can easily show that (cf. [7], p.112):

(i) The collection Δ of all decompositions of $\mathbf{A}$ is a set.
(ii) Δ can be partially ordered; if $C, D \in \Delta$, then $C \leq D$ is D is a refinement of C in the usual sense.

A hoop is *sum irreducible* if it cannot be expressed nontrivially as an ordinal sum; in other words $\mathbf{A}$ is sum irreducible if and only if it is nontrivial and whenever $\mathbf{A} \simeq \bigoplus_{i \in I} \mathbf{A}_i$ then $\mathbf{A} \cong \mathbf{A}_i$ for some $i \in I$ (and all the other components are trivial).

Theorem 2 *Any hoop is the ordinal sum of sum irreducible hoops.*

Proof This is really a set theoretic proof, not an algebraic proof. Given a hoop $\mathbf{A}$, let $\langle \Delta, \leq \rangle$ the poset of its decompositions and note that Δ is nonempty since $\{\mathbf{A}\} \in \Delta$. Let Γ be a totally ordered subset of Δ and for any $a \in A$ let

$$A_a = \bigcap_{D \in \Gamma} \{S : \mathbf{A} \text{ is the unique component in } D \text{ containing } a\}.$$

It is easy to check that $A_a \cup \{1\}$ is the universe of a subalgebra $\mathbf{A}_a$; we claim that $C = \{\mathbf{A}_a : a \in A \setminus \{1\}\}$ is the least upper bound of Γ. This is immediate, since C is clearly a decomposition and moreover $A_a = A_b$ if and only if a and b lie in the same component of all the decompositions in Γ. Hence, Γ is a refinement of them all and hence their least upper bound. So by Zorn lemma $\langle \Delta, \leq \rangle$ has maximal elements and this elements are all decompositions of $\mathbf{A}$ into an ordinal sum of sum irreducible hoops. □

If $\mathbf{A}$ is a subdirectly irreducible hoop, then we have the following important result.

Theorem 3 ([20]) *A hoop* $\mathbf{A}$ *is subdirectly irreducible if and only if it is isomorphic with* $\mathbf{B} \oplus \mathbf{C}$*, where* $\mathbf{B}$ *is a hoop and* $\mathbf{C}$ *is a totally ordered hoop.*

2.2 Relevant Subvarieties of Hoops

We denote by H the variety of hoops and by H_b the variety of bounded hoops; clearly, H is the class of zero-less subreducts of the variety H_b of bounded hoops, so any set of equations not involving 0 defines a subvariety of either variety. A (bounded) hoop is n-potent for some n if it satisfies the equation $x^n \approx x^{n+1}$; we recall also that a variety V is *locally finite* if any finitely generated algebra in V is finite. From now on **2** will denote the two-element hoop; it is immediate to see that it is term equivalent to the two-element generalized Boolean algebra (and if it is bounded to the two-element Boolean algebra).

Theorem 4 ([20]) *For any variety* V *of (bounded) hoops the following are equivalent:*

(i) V *is locally finite.*
(ii) V *consists entirely on* n*-potent (bounded) hoops for some* $n \in \mathbb{N}$.

Let us now show how basic hoops (BL-algebras) (i.e., [bounded] hoops that are subdirect products of chains) can be characterized equationally. We denote the variety of basic hoops by BH and the variety of BL-algebras by BL. The following is the *prelinearity* equation:

$$(x \to y) \to z \leq ((y \to x) \to z) \to z. \tag{prel}$$

Theorem 5 ([6]) *For a variety* V *of hoops the following are equivalent:*

(i) V *satisfies (prel).*
(ii) *Every hoop in* V *is a subdirect product of totally ordered hoops.*

Hence, V *satisfies (prel) if and only if every subdirectly irreducible member of* V *is totally ordered.*

Proof To show that (1) implies (2) it is clearly enough to show that, if (1) holds, then every subdirectly irreducible hoop in V is totally ordered. If $\mathbf{A}$ is subdirectly irreducible, then by Theorem 3, it is isomorphic with $\mathbf{B} \oplus \mathbf{C}$ where $\mathbf{C}$ is totally ordered. Hence, if $a, b \in A$ and $a \not\leq b$ and $b \not\leq a$, then $a, b \in B$. Let now $c \in C \setminus \{1\}$; then by definition of ordinal sum $a \to b, b \to a \leq c$. But then prel fails in $\mathbf{A}$ with the substitution $x = a, y = b, z = c$. Therefore, either $a \leq b$ or $b \leq a$ and $\mathbf{A}$ is totally ordered.

Conversely, it is enough to see that every totally ordered hoop satisfies (prel), but this is obvious because of the equations holding in hoops (for instance, $x \leq (y \to x) \to x$). □

It was observed first in [36] that basic hoops (and BL-algebras) have a term definable lattice structure.

Proposition 4 *In any basic hoop (*BL*-algebra)* $\mathbf{A}$ *the join of two elements is definable; for all* $a, b \in A$

$$a \vee b := ((a \to b) \to b) \wedge ((b \to a) \to a).$$

The proof is not hard; the reader can consult [4] for the proof of an even more general statement; however, from now on when dealing with basic hoops and BL-algebras, we will use the $\vee$ freely.

Subvarieties of basic hoops (and BL-algebras) have been studied extensively in the past twenty years ([1, 3, 7, 8, 10, 14]). A *Wajsberg hoop* is a hoop satisfying the identity

$$(x \to y) \to y \approx (y \to x) \to x; \tag{T}$$

(T) is often called *Tanaka's equation* thanks to Tanaka's work in this framework [57], although this particular identity seems to appear later in [61]. A bounded Wajsberg hoop is called a *Wajsberg algebra*; the varieties of Wajsberg hoops and Wajsberg algebras will be denoted by WH and WA, respectively. The latter are term-equivalent to the more well-known *MV-algebras*, equivalent algebraic semantic of infinite-valued Łukasiewicz logic [31].

A hoop is *cancellative* if its underlying monoid is cancellative in the usual sense; it is easily seen that a hoop is cancellative if it satisfies the equation

$$x \to yx \approx y. \tag{canc}$$

Hence, cancellative hoops form a variety which will be denoted by C; note that no bounded hoop can be cancellative.

Lemma 1 ([20])

(i) *Every Wajsberg hoop is a basic hoop.*
(ii) *Every cancellative hoops is a Wajsberg hoop.*
(iii) *A totally ordered Wajsberg hoop is either cancellative or bounded.*

If $(\mathbf{A}_i)_{i\in I}$ is a family of totally ordered hoops, then $\bigoplus_{i\in I} \mathbf{A}_i$ is a totally ordered hoop; this simple observation paved the way for the following fundamental result.

Theorem 6 ([7]) *For a totally ordered hoop* $\mathbf{A}$ *the following are equivalent:*

(i) $\mathbf{A}$ *is sum irreducible.*
(ii) $\mathbf{A}$ *is a Wajsberg hoop.*

Hence, any totally ordered hoop is the ordinal sum of Wajsberg hoops and any bounded Wajsberg hoop is the ordinal sum of a family $(\mathbf{A}_i)_{i\in I}$ *of Wajsberg hoops, such that* I_0 *is a Wajsberg algebra.*

A basic hoop is a *Gödel hoop* if it satisfies

$$x \wedge y \approx xy;$$

if it is also bounded, it is called a *Gödel algebra*. The varieties of Gödel hoops and Gödel algebras are denoted by GH and GA, respectively. Gödel algebras are subdirect products of totally ordered Heyting algebras, the equivalent algebraic semantics of intuitionistic logic.

A basic hoop is a *product hoop* [6] if it satisfies

$$(y \to z) \vee ((y \to xy) \to x) \approx 1;$$

a bounded product hoop is called a *product algebra* and the varieties of product hoops and product algebras are denoted by PH and PA, respectively.

Theorem 7

(i) *The varieties* GH *and* GA *are locally finite.*
(ii) *a totally ordered Gödel hoop (algebras) is an ordinal sum of copies of* $\mathbf{2}$ *(the first of which is bounded);*
(iii) *A totally ordered product hoop is either cancellative or it is isomorphic to* $\mathbf{2} \oplus \mathbf{C}$ *where* $\mathbf{C}$ *is a (possibly trivial) totally ordered cancellative hoop.*
(iv) *A totally product algebra is isomorphic to* $\mathbf{2} \oplus \mathbf{C}$ *where* $\mathbf{C}$ *is a (possibly trivial) totally ordered cancellative hoop and* $\mathbf{2}$ *is bounded.*

The varieties of Wajsberg, Gödel, and product algebras are arguably considered the most relevant subvarieties of BL-algebras.

3 Projectivity

Given a class K of algebras, an algebra $\mathbf{A} \in \mathsf{K}$ is *projective* in K if for all $\mathbf{B}, \mathbf{C} \in \mathsf{K}$, any homomorphism $h\colon \mathbf{A} \longmapsto \mathbf{C}$, and any surjective homomorphism $g\colon \mathbf{B} \longmapsto \mathbf{C}$, there is a homomorphism $f\colon \mathbf{A} \longmapsto \mathbf{B}$ such that $h = gf$.

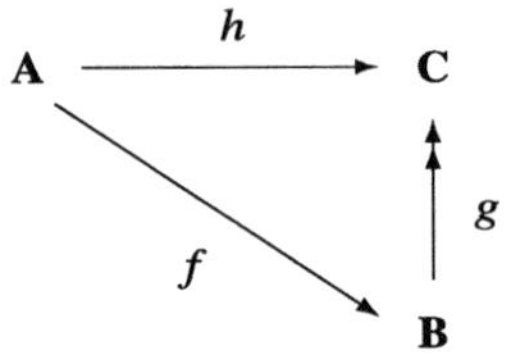

Determining the projective algebras in a class is usually a challenging problem, especially in a general setting. If however K contains all the free algebras on K (in particular, if K is a variety of algebras), projectivity admits a simpler formulation. We call an algebra $\mathbf{B}$ a *retract* of an algebra $\mathbf{A}$ if there is a homomorphism $g\colon \mathbf{A} \longmapsto \mathbf{B}$ and a homomorphism $f\colon \mathbf{B} \longmapsto \mathbf{A}$ with $gf = \mathrm{id}_{\mathbf{B}}$ (and thus, necessarily, f is injective and g is surjective). The following theorem was proved first by Whitman for lattices [59] but it is well known to hold for any class of algebras.

Theorem 8 *Let* K *be a class of algebras containing all the free algebras in* K *and let* **A** $\in$ K. *Then the following are equivalent:*

(i) **A** *is projective in* K.
(ii) **A** *is a retract of a free algebra in* K.
(iii) **A** *is a retract of a projective algebra in* K.

In particular every free algebra in K *is projective in* K.

Proof It is clear that (2) and (3) are equivalent; we prove the equivalence between (1) and (2).

Suppose **A** is projective in K, and let **F** be the free algebra of which **A** is (isomorphic to) a quotient. Then by the definition of projectivity we can close the following diagram:

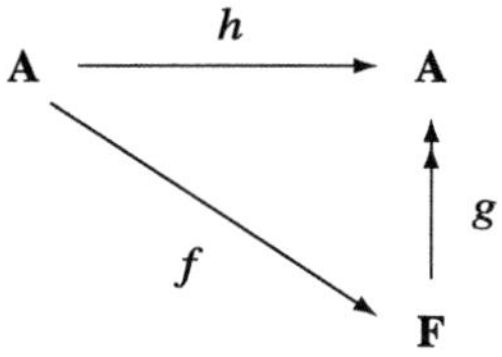

The homomorphisms f, g show that **A** is a retract of **F**.

Vice versa, assume that **A** is a retract of a free algebra **F** $\in$ K generated by a set X; in particular, let the retraction be testified by homomorphisms $i : \mathbf{A} \to \mathbf{F}$ and $j : \mathbf{F} \to \mathbf{A}$. Consider now **B**, **C** $\in$ K, a homomorphism $h\colon \mathbf{A} \longmapsto \mathbf{C}$, and a surjective homomorphism $g\colon \mathbf{B} \longmapsto \mathbf{C}$. We want to show that we can define $f : \mathbf{A} \to B$ such that $gf = h$. For each $x \in X$, let b_x be an element of **B** such that $g(b_x) = hj(x)$. Since **F** is free in K over X, this assignment extends to a homomorphism $k : \mathbf{F} \to \mathbf{B}$ such that $gk = hj$. Let now $f = ki$. Then $gf = gki = hji = h$, and thus, **A** is projective. □

It follows immediately that if V is a variety and **A** is projective in V, then it is projective for any subvariety W of V to which it belongs; equivalently, if **A** $\in$ W is not projective in W, then it cannot be projective in any supervariety of W.

Let us also observe that the algebraic definition of projectivity we have given corresponds to a categorical notion for the corresponding algebraic category (where the objects are the algebras in K and whose morphisms are the homomorphisms between algebras in K), where surjective homomorphisms are regular epimorphisms. It therefore follows that projectivity is preserved by categorical equivalence: the projective objects in one category are exactly the images, through the functor yielding the equivalence, of the projective objects in the other category.

In general, determining all projective algebras in a variety is a complicated task. As examples of success in characterizing projective algebras in interesting classes we can quote:

- A finite lattice is projective in the variety of all lattices if and only if it semidistributive and satisfies Whitman's condition (W) (see [51] for details).

- A finite distributive lattice is projective in the variety of distributive lattices if and only if the meet of any two meet irreducible elements is again meet irreducible [16].
- A Boolean algebra is projective in the variety of distributive lattices if and only if it is finite [16]; hence, every finite Boolean algebra is projective in the variety of Boolean algebras.
- An abelian group is projective in the variety of abelian groups if and only if it is free.
- If $\mathsf{M}_{\mathbf{D}}$ is the variety of left modules on a principal ideal domain $\mathbf{D}$, then a module is projective in $\mathsf{M}_{\mathbf{D}}$ if and only if it is free (this is a consequence of the so-called Quillen-Suslin theorem; see [54]).
- Even more generally [53], if $\mathbf{A}$ is a finite quasi-primal algebra, then if $\mathbf{A}$ has no minimal nontrivial subalgebra, each finite algebra in $\mathbf{V}(\mathbf{A})$ is projective; alternatively a finite algebra in $\mathbf{V}(\mathbf{A})$ is projective if and only if it admits each minimal subalgebra of $\mathbf{A}$ as a direct decomposition factor.

In what follows, we will be mainly interested in projective algebras that are finitely generated. We will see that in many interesting cases, these will correspond to *finitely presented* algebras in the variety. An algebra is *finitely presented* if it has a finite presentation (X, Σ) given by a finite set of generators X and a finite set of identities Σ, written in the appropriate language and involving as variables the elements in X. Given a quasivariety Q and a set X, we write $\mathbf{F}_{\mathsf{Q}}(X)$ for the free algebra in Q over X. Given a finite presentation in the language of Q, the corresponding finitely presented algebra in Q is a quotient of the kind $\mathbf{F}_{\mathsf{Q}}(X)/\theta_\Sigma$, where θ_Σ is the *relative* compact congruence generated by the pairs $\{(t_i, s_i) : t_i \approx s_i \in \Sigma\}$ (that is, the smallest congruence for which the quotient belongs to Q, see [41]).

Note that if a quasivariety Q has finite type, every finite algebra in Q is finitely presented, and if Q is locally finite, every finitely presented algebra in Q is finite. We remark that the notion of finitely presented algebra is a categorical notion [38], and thus, it is preserved under categorical equivalence. The proof of the following theorem is standard (the reader can see [40]).

Theorem 9 *For a finitely presented algebra* $\mathbf{A} \in \mathsf{Q}$ *the following are equivalent:*

(i) $\mathbf{A}$ *is projective in* Q.
(ii) $\mathbf{A}$ *is projective in the class of all finitely presented algebras in* Q.
(iii) $\mathbf{A}$ *is a retract of a finitely generated free algebra in* Q.

There is another general observation that is relevant:

Proposition 5 *Let* K *be a class of algebras. If every algebra* $\mathbf{B} \in \mathsf{K}$ *is a retract of every algebra* $\mathbf{A} \in \mathsf{K}$ *of which it is an homomorphic image, then every algebra in* K *is projective in* K.

Proof Let K be a class of algebras satisfying the hypothesis of the proposition. Consider any algebra $\mathbf{C} \in \mathsf{K}$, and suppose there is an homomorphism $h\colon \mathbf{C} \to \mathbf{B}$ and a surjective homomorphism $g : \mathbf{A} \to \mathbf{B}$. Since $\mathbf{B}$ is a homomorphic image of $\mathbf{A}$, it is also its retract. Thus, there exists an injective homomorphism $f' \colon \mathbf{B} \to \mathbf{A}$ such that $gf' = id_{\mathbf{B}}$. Then we can consider the homomorphism $f \colon \mathbf{C} \to \mathbf{A}$, $f = f'h$. We get $gf = gf'h = h$, and thus, $\mathbf{C}$ is projective in K. □

Observe that if V is a locally finite variety of finite type, then the concepts of finitely generated, finitely presented, and finite coincide. Hence:

Corollary 2 *Let* V *be any variety; if every finite algebra in* V *is projective in the class of finite algebras in* V*, then for any locally finite variety* $\mathsf{W} \subseteq \mathsf{V}$ *every finitely presented algebra in* W *is projective in* W*.*

3.1 Projectivity in Varieties of Hoops

Most of the results in this section are contained in [5, 11, 12, 58].

3.1.1 Locally Finite Varieties

Lemma 2 *Let* $\mathbf{A}$ *be a hoop and let* $u \in A$ *be idempotent; then for all* $a, b \in A$

$$u \to (a \to b) = (u \to a) \to (u \to b) \tag{1}$$

$$u \to ab = (u \to a)(u \to b). \tag{2}$$

The two properties above, albeit similar, are quite different in nature; the first one can be proved through a standard computation [22], while the second involves very complex calculations: it was first proved in [56] via a computer-assisted proof. We can now prove the following.

Theorem 10 *Every finite hoop is finitely projective in the variety of hoops; hence, in any locally finite variety of hoops, every finitely presented (equivalently, finite) algebra is projective.*

Proof Let $\mathbf{A}, \mathbf{B}$ be finite hoops and let $g \colon \mathbf{B} \longrightarrow \mathbf{A}$ be a surjective homomorphism. Let $\theta = \ker(g)$ and $F = 1/\theta$; then F is a filter of $\mathbf{B}$ and, since $\mathbf{B}$ is finite, it has a minimum u which is idempotent. Let $f(x) \colon u \to x$; then by Lemma 2 f is an endomorphism of $\mathbf{B}$. It is also idempotent since for $b \in B$

$$f(f(b)) = u \to (u \to b) = u^2 \to b = u \to b = f(b).$$

Now if $b \in F$, then $f(b) = u \to b = 1$; conversely, if $f(b) = 1$ then $u \to b = 1$, i.e., $u \leq b$ and so $b \in F$. In conclusion $F = 1/\ker(f)$ and thus $\ker(f) = \theta$; if we set $\mathbf{C} = f(\mathbf{B})$, which is a subalgebra of $\mathbf{B}$, then

$$\mathbf{A} \cong \mathbf{B}/\theta = \mathbf{B}/\ker(f) \cong \mathbf{C}.$$

Let now $h: \mathbf{A} \longmapsto \mathbf{C}$ be the resulting isomorphism, then h is an injective homomorphism from $\mathbf{A}$ to $\mathbf{B}$. Moreover, by definition $h(a) = f(b)$ where b is such that $g(b) = a$; observe also that the fact that f is idempotent implies that $(b, f(b)) \in \ker(f) = \theta$, thus $g(b) = g(f(b))$. Hence, if $a \in A$ and $b \in B$ with $g(b) = a$, we have

$$g(h(a)) = g(f(b)) = g(b) = a.$$

We have just proved that **A** is a retract of **B**; by Proposition 5 the conclusion follows. □

In particular any finite Brouwerian semilattice (i.e., an idempotent hoop) is finitely projective in the variety of Brouwerian semilattices. Since the latter is locally finite [46] we get at once that all the finitely presented Brouwerian semilattices are projective (a fact already noted in [40]).

The class of $\rightarrow$-subreducts of the variety of commutative and integral residuated lattice is the (proper) quasivariety BCK of BCK-algebras; those algebras were introduced by K. Iseki [43] and they have been widely studied since; the fact that they are indeed subreducts of commutative and integral residuated lattices is again a consequence of the quoted result in [52]. In [37] the class of BCK-algebras that are implicative subreducts of hoops has been described: it is a variety called HBCK and the algebras therein are called HBCK-algebras. An analogous of Theorem 10 follows with the same proof in the reduced language.

Theorem 11 *Every finite HBCK-algebra is finitely projective in the variety of HBCK-algebras; hence, in any locally finite variety of HBCK-algebras every finitely presented (equivalently, finite) algebra is projective.*

In order to transfer our argument to full hoops or to HL-algebras we have to take care of the join. The easiest case is the one in which the join is definable, which is equivalent to saying that the hoop satisfies the equation (PJ) (see [4]). Thus, prelinear hoops are full hoops (i.e., they are residuated lattices) and the conclusion applies.

Corollary 3 *Every finite basic hoop is finitely projective in the variety of basic hoops; hence, in any locally finite variety of basic hoops every finitely presented (equivalently, finite) algebra is projective.*

If we remove the hypothesis of being locally finite, the previous result does not hold. Indeed, for instance, not all finitely presented Wajsberg hoops are projective, as shown in [58].

3.1.2 Wajsberg Hoops

Let us now consider some nonlocally finite varieties of hoops. As a particularly relevant example, let us start with cancellative hoops; the latter do not have nontrivial finite models, since a finite semilattice is necessarily lower bounded, hence not cancellative. Therefore, they are not locally finite. Cancellative hoops can also be seen as negative cones of lattice ordered abelian groups (or *ℓ-groups* for short), and actually, they are categorically equivalent to ℓ-groups [35]. Now, projective and finitely generated ℓ-groups coincide with finitely presented ℓ-groups, as shown in [18]. Just as the properties of being projective and finitely presented, in every variety also the concept of an algebra being finitely generated is categorical, i.e., it can be described in the abstract categorical setting by properties of morphisms, and thus, all these notions are preserved by categorical equivalences. Therefore, we have the following result.

Proposition 6 *Finitely presented cancellative hoops are exactly the finitely generated and projective in their variety.*

Let us move to a larger variety of hoops, still nonlocally finite and containing cancellative hoops: Wajsberg hoops. Projectivity in Wajsberg hoops has been studied in [58], using geometrical methods; in particular, it is shown that finitely presented Wajsberg hoops (in analogy with finitely presented MV-algebras) can be studied by means of rational polyhedra in real unit cubes $[0, 1]^n$. While the entire theory is too complex to be reproduced here, we summarize some of the main results, that are relevant for our purposes. The following is in analogy with the case of MV-algebras.

Theorem 12 ([58]) *Finitely generated and projective Wajsberg hoops are finitely presented, but not all finitely presented Wajsberg hoops are projective.*

Moreover, it is shown that finitely presented subdirectly irreducible Wajsberg hoops are exactly the algebras $\mathbf{W}_n$, for $n \geq 1$,[58, Proposition 4.10] which directly implies that $\mathbf{C}_\omega$ cannot be projective in WH. Moreover, it is observed that no bounded Wajsberg hoop can be projective in the variety of Wajsberg hoops since the description of finitely generated free Wajsberg hoops (first shown in [9]) shows that they do not have idempotents except for 1. Therefore:

Corollary 4 ([58]) *The subdirectly irreducible Wajsberg hoops* $\mathbf{C}_\omega$, $\mathbf{C}_n^\omega$, $\mathbf{W}_n$ *for* $n \geq 2$ *are not projective in* WH, *nor in any variety of residuated lattices that contains* WH.

In particular, the (0-free reduct of the) 2-element Boolean algebra is not projective in the variety of Wajsberg hoops, while it is the only finite projective algebra in MV-algebras [34]. As examples of projective Wajsberg hoops, one can consider the free algebra on one generator $\mathbf{F}_{\mathsf{WH}}(x)$; then all the subalgebras generated by $x \to x^n$ for a fixed $n \in \mathbb{N}$ are projective in WH (see Example 4.13 in [58]).

Finally, let us consider (nonlocally finite) proper varieties of Wajsberg hoops. First, we recall that Wajsberg hoops have a *canonical presentation*: let $\mathbf{G}$ be a lattice ordered abelian group; by [50], if u is a strong unit of $\mathbf{G}$ we can construct a bounded Wajsberg hoop $\Gamma(\mathbf{G}, u) = \langle [0, u], \to, \cdot, 0, u \rangle$ where $ab = \max\{a + b - u, 0\}$ and $a \to b = \max\{u - a + b, u\}$. The main result of [50] is that any bounded Wajsberg hoop can be presented in this way (really there is a categorical equivalence between the category of abelian ℓ-groups with strong unit and the category of bounded Wajsberg hoops). Let now $\mathbb{Z} \times_l \mathbb{Z}$ denote the lexicographic product of two copies of $\mathbb{Z}$. In other words, the universe is the cartesian product, the group operations are defined componentwise, and the ordering is the lexicographic ordering (w.r.t. the natural ordering of $\mathbb{Z}$); then $\mathbb{Z} \times_l \mathbb{Z}$ is a totally ordered abelian group and we can apply Γ to it. A *Wajsberg chain* is a totally ordered Wajsberg hoop. Let us define some useful Wajsberg chains:

- The finite Wajsberg chain with $n+1$ elements $\mathbf{Ł}_n = \Gamma(\mathbb{Z}, n)$.
- The infinite finitely generated Wajsberg chain $\mathbf{Ł}_n^\infty = \Gamma(\mathbb{Z} \times_l \mathbb{Z}, (n, 0))$.
- The infinite finitely generated Wajsberg chain $\mathbf{Ł}_{n,k} = \Gamma(\mathbb{Z} \times_l \mathbb{Z}, (n, k))$.

A *v-presentation* is a triple $\{I, J, K\}$ where I, J are finite subsets of $\mathbb{N}$ and $K \subseteq \{0\}$; a v-presentation is *reduced* if:

- $I \cup K \cup J \neq \emptyset$.
- If $K = \{0\}$, then $J \neq \emptyset$.
- No $m \in I$ divides any $m' \in (I \setminus \{m\}) \cup J$.
- No $t \in J$ divides any $t' \in J \setminus \{t\}$.

Theorem 13 ([9]) *The proper subvarieties of* WH *are in one-to-one correspondence with the reduced triples via the mapping*

$$\{I, J, \emptyset\} \longmapsto \mathbf{V}(\{\mathbf{Ł}_m : m \in I\} \cup \{\mathbf{Ł}_t^\infty : t \in J\})$$

$$\{I, \emptyset, \{0\}\} \longmapsto \mathbf{V}(\{\mathbf{Ł}_m : m \in I\} \cup \{\mathbf{C}_\omega\}).$$

Now if we denote by $\mathbf{V}(I, J, K)$ the proper subvariety of WH associated to the reduced triple, from the results in [5] we can deduce that:

Theorem 14

(i) *If* $\mathsf{V} = \mathbf{V}(I, \emptyset, \{0\})$, *then every totally ordered algebra in* V *is projective in* V.
(ii) *If* $\mathsf{V} = \mathbf{V}(I, \{1\}, \emptyset)$, *then every totally ordered algebra in* V *is projective in* V.
(iii) *For any* $\mathsf{V} = \mathbf{V}(I, J, K)$, $\mathbf{C}_\omega$ *is projective in* V.
(iv) *For any* $\mathsf{V} = \mathbf{V}(I, J, K)$, $\mathbf{L}_n$ *is projective in* V *if and only if* $n \mid i$ *for some* $i \in I \cup J$.
(v) *For any* $\mathsf{V} = \mathbf{V}(I, J, K)$, $\mathbf{L}_n^\infty$ *is projective in* V *if and only if* $n = 1$.

3.2 Projectivity in Varieties of Bounded Hoops

For bounded hoops we have the following lemma that limits greatly the number of projective bounded hoops.

Lemma 3 ([12]) *Let* V *be any variety of bounded hoops; if* $\mathbf{A}$ *is projective in* V, *then* $\mathbf{2}$ *is a homomorphic image of* $\mathbf{A}$.

As mentioned above, in some cases **2** is the only projective algebra, e.g., in the variety of MV-algebras. In the latter, projective algebras have been studied with geometrical methods (which in fact inspired and are utilized for the previously mentioned results about Wajsberg hoops in [58]) by Cabrer and Mundici [28–30], grounding on the connection between finitely presented MV-algebras and rational polyhedra (which lifts to the often-called Marra-Spada duality [47]). A full characterization of finitely generated and projective MV-algebras in terms of rational polyhedra is shown in an unpublished manuscript by Cabrer [27]. From the algebraic perspective, we have that all finitely generated and projective MV-algebras are finitely presented, but not all finitely presented MV-algebras are projective. The only finite projective MV-algebra, as mentioned above, is the 2-element Boolean algebra (this is also observed in [34]).

In parallel with the previous subsection, let us now consider locally finite varieties of bounded hoops.

3.2.1 Locally Finite Varieties

If in a bounded hoop **A** we define $\neg a := a \to 0$, then any bounded hoop having an element a such that $\neg a = a$ cannot have **2** as a homomorphic image, so it cannot be projective in any variety of bounded hoops. Luckily for locally finite varieties the condition in Lemma 3 is sufficient as well. First, we observe

Lemma 4 *The variety of bounded hoops satisfies the quasiequation:*

$$x^2 \approx x \quad \Longrightarrow \quad (x \to y) \to \neg\neg x \approx \neg\neg x. \tag{Q}$$

The proof can be found in [12] and it is basically an induction on the size of the subdirectly irreducible algebras in BL.

If **A** is any bounded hoop, then the double negation is a $\cdot$-homomorphism, that is, for all $a, b \in A$

$$\neg\neg(a \cdot b) = \neg\neg a \cdot \neg\neg b.$$

This has been shown by Cignoli and Torrens in [32, Theorem 4.8]; it follows that if $u \in A$ is idempotent, so is $\neg\neg u$.

Lemma 5 *If* **A** *is a bounded hoop,* $a, b, u \in \mathbf{A}$ *and* u *is idempotent, then*

$$ua = u \wedge a \tag{U1}$$

$$u \to ua = u \to a \tag{U2}$$

$$\neg\neg u((u \to a) \to (u \to b)) = (u \to a) \to \neg\neg u(u \to b) \tag{U3}$$

$$\begin{aligned}(u \to a) \to (u \to b) &= \neg\neg u(u \to a) \to \neg\neg u(u \to b)\\ &= \neg\neg u(u \to a) \to (u \to b).\end{aligned} \tag{U4}$$

The proof is fairly standard; we observe the quasiequation (Q) is paramount in showing that if u is idempotent and $a \in A$, then

$$(u \to a) \to \neg\neg u = \neg\neg u.$$

Lemma 6 *Let* $\mathbf{A}$ *be a bounded hoop, and let* $\mathbf{2}$ *denote its subalgebra with domain* $\{0, 1\}$. *If* $\varphi : \mathbf{A} \longrightarrow \mathbf{2}$ *is a homomorphism and* $u \in A$ *is idempotent, then the map*

$$f(x) = (u \to x)(\neg u \to \varphi(x))$$

is an endomorphism of $\mathbf{A}$.

Now we can prove the characterization result.

Theorem 15 ([12]) *A finite bounded hoop* $\mathbf{A}$ *is finitely projective in the variety of bounded hoops if and only if there exists a homomorphism* $\varphi : \mathbf{A} \longrightarrow \mathbf{2}$.

Corollary 5 *Let* V *be a locally finite variety of bounded hoops; then a finite* $\mathbf{A} \in \mathsf{V}$ *is projective in* V *if and only if there is a homomorphism* $\varphi : \mathbf{A} \longrightarrow \mathbf{2}$.

The case of BL-algebras is managed in the same way; since the join is definable by prelinearity, all the arguments go through without change.

Corollary 6 *Let* V *be a locally finite variety of BL-algebras; then a finite* $\mathbf{A} \in \mathsf{V}$ *is projective in* V *if and only if there is a homomorphism* $\varphi : \mathbf{A} \longrightarrow \mathbf{2}$.

Since every finite algebra $\mathbf{A}$ in a variety is a subdirect product of finite subdirectly irreducible algebras, and all the subdirect factors are homomorphic images of $\mathbf{A}$, we can sharpen our results some more.

Theorem 16 *Let* V *be a locally finite variety of bounded hoops or* BL*-algebras such that every finite subdirectly irreducible in* V *has* $\mathbf{2}$ *as homomorphic image. Then every finitely presented nontrivial algebra in* V *is projective.*

The theorem above is (yet another) reason why every finitely presented (i.e., finite) nontrivial Boolean algebra is projective: the variety of Boolean algebras is locally finite and the only subdirectly irreducible is $\mathbf{2}$. As a final observation, we stress that the local finiteness observation in Corollary 5 is necessary; the variety of Wajsberg algebras is not locally finite and the only finite projective algebra in WA is $\mathbf{2}$, as already observed above.

3.2.2 Rotations

In this subsection we refer to [11]. Let us start with an example: a variety of BL-algebras is *Stonean* if it satisfies the equation

$$\neg x \vee \neg\neg x \approx 1.$$

It is the straightforward consequence of the characterization of the subdirectly irreducible BL-algebras in [7] that a finite subdirectly irreducible algebra in a Stonean variety is of the form $\mathbf{A} \cong \mathbf{2} \oplus \mathbf{B}$, where $\mathbf{B}$ is a totally ordered hoop. Since $\mathbf{B}$ is a filter of $\mathbf{A}$, we can collapse it and get $\mathbf{2}$ as a homomorphic image of $\mathbf{A}$. Hence, locally finite Stonean varieties of BL-algebras fall under the scope of Theorem 16. Stonean BL-algebras are a particular instance of varieties of bounded hoops that can be obtained from varieties of hoops using a particular construction.

This construction can be (and has been) applied to integral residuated lattices in general (see [11]) but we will introduce it for basic hoops in the spirit of the earlier [10]. However, we have to be careful here, since the construction may take us out of the realm of BL-algebras; namely, while the prelinearity equation is always preserved by this construction, divisibility might not. So we will end up in the variety of prelinear, bounded, and commutative residuated lattices, a.k.a. MTL-*algebras* (if we start from basic hoops) or in the variety of integral, bounded, and commutative residuated semilattices.

The *disconnected rotation* [44, 45] of a hoop $\mathbf{A}$ is the MTL-algebra $\mathbf{A}^*$ whose lattice reduct is given by the union of A and its disjoint copy $A' = \{a' : a \in A\}$ with dualized order, placed below A: for all $a, b \in A$,

$$a' < b, \text{ and } a' \leq b' \text{ iff } b \leq a.$$

In particular, the top element of $\mathbf{A}^*$ is the top 1 of $\mathbf{A}$ and the bottom element of $\mathbf{A}^*$ is the copy $0 := 1'$ of the top 1. $\mathbf{A}$ is a subalgebra, the products in A' are all defined to be the bottom element $0 = 1'$, and furthermore, for all $a, b \in A$,

$$a \cdot b' = (a \to b)', \quad b' \to a = 1;$$

$$a \to b' = (b \cdot a)', \quad a' \to b' = b \to a.$$

A *nucleus* on a residuated lattice $\mathbf{A} = (A, \wedge, \vee, \cdot, \to, 1)$ is a closure operator δ on $\mathbf{A}$ that satisfies $\delta(x)\delta(y) \leq \delta(xy)$, for all $x, y \in A$. It is known that then $\mathbf{A}_\delta = (\delta[A], \wedge, \vee_\delta, \cdot_\delta, \to, \delta(1))$ is a residuated lattice, where $x \vee_\delta y = \delta(x \vee y)$ and $x \cdot_\delta y = \delta(xy)$.

A *rotation operator* for an algebra $\mathbf{A}$ is a map $\delta : A \to A$ that is a nucleus and a lattice homomorphism. Given such a rotation operator, the *generalized δ-rotation* $\mathbf{A}^\delta$ (defined in [15]) differs from the disconnected rotation above in that it replaces A' with $\delta[A]' = \{\delta(a)' : a \in A\}$, where $\delta(a)'$ is short for $(\delta(a))'$. Moreover, for all $a \in A, b \in \delta[A]$, the only changed operation is:

$$a \to b' = (\delta(ba))'.$$

The *generalized n-rotation via δ* (see [15, 26]), $\mathrm{Rot}_n^\delta(\mathbf{A})$ or $\mathbf{A}_n^\delta$, of $\mathbf{A}$ with respect to a rotation operator δ and a natural number $n \geq 2$ is defined on the disjoint union of A^δ and $\{\ell_i : 0 < i < n-1\}$. We also set $\ell_0 = 0$ and $\ell_{n-1} = 1$, the bounds of $\mathbf{A}^\delta$. The order extends the order of $\mathbf{A}^\delta$ by

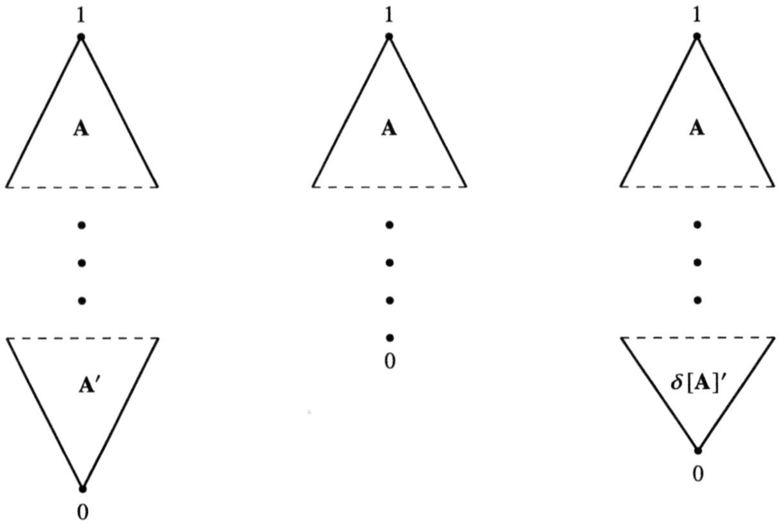

Fig. 1 Respectively: a disconnected 5-rotation, a 5-lifting, and a generalized 5-rotation of a hoop **A**

$$b < \ell_1 < \ldots < \ell_{n-2} < a,$$

for all $a \in A$ and $b \in \delta[A]'$. The operations extend those of $\mathbf{A}^{\delta}$, of the n-element Łukasiewicz chain $Ł_n$, where $0 = \ell_0 < \ell_1 < \ldots < \ell_{n-2} < \ell_{n-1} = 1$, and for $0 < i < n-1$, $a \in A$ and $b \in \delta[A]'$:

$$a\ell_i = \ell_i, \quad b\ell_i = 0.$$

The resulting structure is an MTL-algebra [10, 26] if **A** is a basic hoop. In particular, if $\delta = id$, we call the construction a *disconnected n-rotation*, and if $\delta(x) = 1$ we call the construction an *n-lifting* (see Fig. 1).

We mention that the generalized n-rotation construction has been generalized to noncommutative structures in [39]. We have the following interesting fact:

Proposition 7 *Let* **A** *be a basic hoop; then:*

(i) *Its disconnected n-rotation is a* BL*-algebra if and only if* $n = 2$ *and* **A** *is cancellative.*
(ii) *For any n its n-lifting is a* BL*-algebra.*

Proof For (1) we observe that, by definition, the universe of the disconnected n-rotation of **A** is given by $A \cup A' \cup \{\ell_i : 0 < i < n-1\}$; if $n > 2$, then divisibility fails. In fact if $\ell_i \neq 0, 1$ and $a \in A \setminus 1$, then $a' < \ell_i$, but

$$(\ell_i \to a')\ell_i = 1' \neq a' = a' \wedge \ell_i.$$

Conversely, if $\mathbf{A}$ is cancellative and $n = 2$, easy calculations show that $\mathbf{A}_2^\delta$ is a **BL** algebra.

The proof of (2) is the simple verification that divisibility is preserved. □

If $\mathbf{A}$ is a basic hoop, let us denote by $\mathbf{A}^\delta$ its disconnected rotation and by $\mathbf{A}^{\lambda_n}$ its n-lifting. Let now, for any variety V of basic hoops,

$$\mathsf{V}^{\delta_n} = \mathbf{V}(\{\mathbf{A}_n^\delta : \mathbf{A} \in \mathsf{V}\}) \qquad \mathsf{V}^{\lambda_n} = \mathbf{V}(\{\mathbf{A}^{\lambda_n} : \mathbf{A} \in \mathsf{V}\});$$

by Proposition 7 they are all varieties of MTL-algebras.

If C is the variety of cancellative hoops, then C^δ is the variety generated by *perfect* MV-algebras, sometimes called DLMV (since it was axiomatized by Di Nola and Lettieri); it is easily seen that they are all Wajsberg algebras (since cancellative hoops are Wajsberg hoops). On the other hand, it is easy to show that C^{λ_2} is the variety PA of product algebra and the variety BL^{λ_2} is the variety of Stonean BL-algebras encountered at the beginning of the section.

Through a detailed study of free algebras in varieties that are generalized rotations of (basic) hoops, it is possible to discover a link between a variety of hoops and its disconnected rotation or n-lifting.

Theorem 17 ([11]) *Let* V *be a variety of basic hoops; then every finitely generated projective algebra in* V_n^δ *(* V^{λ_n}*) is isomorphic to some finite direct product* $\mathbf{A}_1^{\delta_{k_1}} \times \cdots \times \mathbf{A}_m^{\delta_{k_m}}$ *(*$\mathbf{A}_1^{\lambda_{k_1}} \times \cdots \times \mathbf{A}_m^{\lambda_{k_m}}$*), for* $k_j - 1$ *dividing* $n - 1$*) where* $\mathbf{A}_1, \ldots, \mathbf{A}_m$ *are finitely generated and projective in* V.

The same also holds for finitely presented algebras.

Lemma 7 ([11]) *Let* V *be a variety of basic hoops; then every nontrivial finitely presented algebra in* V^{δ_n} *(* V^{λ_n}*) is isomorphic to some finite direct product* $\mathbf{A}_1^{\delta_{k_1}} \times \cdots \times \mathbf{A}_m^{\delta_{k_m}}$ *(*$\mathbf{A}_1^{\lambda_{k_1}} \times \cdots \times \mathbf{A}_m^{\lambda_{k_m}}$*), for* $k_j - 1$ *dividing* $n - 1$*) where* $\mathbf{A}_1, \ldots, \mathbf{A}_m$ *are finitely presented in* V.

To connect finitely generated and finitely presented algebras we need two further results. First:

Theorem 18 ([11]) *Let* V *be a variety of basic hoops such that every finitely generated projective algebra is finitely presented. Then every finitely generated projective algebra in* V_n^δ *(* V^{λ_n}*) is finitely presented.*

Conversely,

Theorem 19 ([11]) *Let* V *be a variety of basic hoops such that every is finitely presented algebra is projective. Then a finitely presented algebra in* V_n^δ *(* V^{λ_n}*) is projective if and only if it has* **2** *as homomorphic image.*

It follows that every nontrivial finitely presented product algebra is projective in PA, every nontrivial finitely presented perfect Wajsberg algebra is projective in the variety of perfect Wajsberg algebra, and if V is a locally finite variety of Stonean BL-algebras, then every nontrivial finitely presented algebra in V is projective in V.

4 Unification

In this section we recall Ghilardi's approach to unification problems via the study of projective algebras [40]. In particular, we observe that Ghilardi's work focuses on the study of the unification type from the symbolic and algebraic point of view. For the sake of the reader, we here rewrite the proofs so to first show the correspondence between the symbolic and algebraic approach with respect to the existence of solutions, and in a second moment we show the equivalence of the unification types.

The classical syntactic unification problem, given two terms s, t (built from function symbols and variables), finds a *unifier* for them, that is, a uniform replacement of the variables occurring in s and t by other terms that makes s and t identical. When the latter syntactical identity is replaced by equality modulo a given equational theory E, one speaks of E-unification, which is what we study here. Thus, the unification problems we consider are up to logical equivalence. Such a unification problem for a quasivariety that is the equivalent algebraic semantics of a logic interprets unification problems for the logic up to logical equivalence.

Definition 1 By a *symbolic unification problem* over a language $\mathcal{L}$ we mean a finite set of identities:

$$\Sigma = \{s_i \approx t_i \mid s_i, t_i \in \mathbf{T}_{\mathcal{L}}(X),\ X \text{ finite set of variables}, i = 1 \dots n\}.$$

A solution or *symbolic unifier* to Σ in a class of algebras K is a substitution $\sigma : \mathbf{T}_{\mathcal{L}}(X) \to \mathbf{T}_{\mathcal{L}}(\omega)$ such that

$$\models_{\mathsf{K}} \sigma(s_i) \approx \sigma(t_i)$$

for all identities $s_i \approx t_i \in \Sigma$.

We have the following observation.

Proposition 8 *A unification problem in the language of a quasivariety* Q *has a solution in* Q *if and only if it has a solution in* $\mathsf{V}(\mathsf{Q})$.

We will now show the core idea of Ghilardi's approach to algebraic unification, that is: a symbolic unification problem corresponds to a finitely presented algebra, and its unifier corresponds to a homomorphism of the finitely presented algebra to some projective algebra. But first, let us clarify what we mean by a finitely presented algebra in a quasivariety.

Definition 2 Given a quasivariety Q we say that an algebra is *finitely presented in* Q if there exists a finite set X and a finite set H of pairs of terms over X such that $\mathbf{A} \cong \mathbf{F}_{\mathsf{Q}}(X)/\theta_{\mathsf{Q}}(H)$, where $\theta_{\mathsf{Q}}(H)$ is the least relative congruence $\theta_{\mathsf{Q}}(H)$ on $\mathbf{A}$ containing H.

We observe that the least relative congruence $\theta_{\mathsf{Q}}(H)$ in the previous definition exists given the results in [41].

Definition 3 Let Q be a quasivariety. An *algebraic unification problem* for Q is a finitely presented algebra $\mathbf{F}_{\mathsf{Q}}(X)/\theta_{\mathsf{Q}}(H) \in \mathsf{Q}$. A solution or *algebraic unifier* is a homomorphism $u : \mathbf{F}_{\mathsf{Q}}(X)/\theta_{\mathsf{Q}}(H) \to \mathbf{P}$, for some projective algebra $\mathbf{P} \in \mathsf{Q}$.

Given a set of identities Σ over variables in X, by $\mathbf{F}_{\mathsf{Q}}(X)/\theta_{\vec{Q}}(\Sigma)$ we mean the finitely presented algebra defined by the congruence $\theta_{\mathsf{Q}}(\Sigma)$ generated by the set of identities $\{(s_i, t_i) : s_i \approx t_i \in \Sigma\}$. The following proof appears also in [11].

Theorem 20 *A symbolic unification problem Σ has a solution in a quasivariety Q if and only if the algebraic unification problem $\mathbf{F}_{\mathsf{Q}}(X)/\theta_{\mathsf{Q}}(\Sigma)$ has a solution in Q.*

Proof Suppose $\sigma : \mathbf{F}_{\mathsf{Q}}(X) \to \mathbf{F}_{\mathsf{Q}}(Y)$ is a solution to the symbolic unification problem. Consider the natural epimorphism $\pi_\Sigma : \mathbf{F}_{\mathsf{Q}}(X) \to \mathbf{F}_{\mathsf{Q}}(X)/\theta_{\mathsf{Q}}(\Sigma)$. Since $\theta_{\mathsf{Q}}(\Sigma)$ is the least congruence of $\mathbf{F}_{\mathsf{Q}}(X)$ containing the set of pairs $S = \{(p, q) : p \approx q \in \Sigma\}$ and given that $S \subseteq \ker(\sigma)$, by the second homomorphism theorem we can close the following diagram with the homomorphism u_σ defined by $u_\sigma(t/\theta_{\mathsf{Q}}(\Sigma)) = \sigma(t)$.

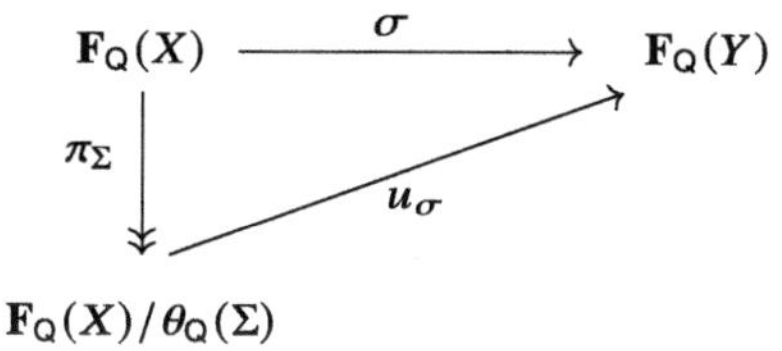

$u_\sigma : \mathbf{F}_{\mathsf{Q}}(X)/\theta_{\mathsf{Q}}(\Sigma) \to \mathbf{F}_{\mathsf{Q}}(Y)$ is then an algebraic unifier for $\mathbf{F}_{\mathsf{Q}}(X)/\theta_{\mathsf{Q}}(\Sigma)$.

Conversely, let $\mathbf{A} = \mathbf{F}_{\mathsf{Q}}(X)/\theta_{\mathsf{Q}}(\Sigma)$, and consider a unifier $u : \mathbf{A} \to \mathbf{P}$, where $\mathbf{P}$ is projective and a retract of $\mathbf{F}_{\mathsf{Q}}(Y)$ witnessed by an embedding $i : \mathbf{P} \to \mathbf{F}_{\mathsf{Q}}(Y)$; we show that the substitution

$$\sigma_u : x \longmapsto i(u(x/\theta_{\mathsf{Q}}(\Sigma)))$$

is a unifier for Σ in Q.

Indeed, since σ_u is defined by a composition of homomorphism, and as above the set of pairs $S = \{(p, q) : p \approx q \in \Sigma\}$ is contained in its kernel, we get that σ_u is a unifier for Σ in Q. □

If a (symbolic or algebraic) unification problem has a solution, we say that it is *unifiable*. Let $\mathbf{F}_{\mathsf{Q}}$ be the smallest free algebra in Q (either 0- or 1-generated). The following is an interesting observation.

Proposition 9 *The unification problem* $\mathbf{F}_{\mathsf{Q}}(X)/\theta_{\mathsf{Q}}(\Sigma)$ *has a solution if and only if there is a homomorphism from* $\mathbf{F}_{\mathsf{Q}}(X)/\theta_{\mathsf{Q}}(\Sigma)$ *to* $\mathbf{F}_{\mathsf{Q}}$.

Proof $\mathbf{F}_{\mathsf{Q}}$ is projective (since it is free). Moreover, it is a homomorphic image of every free algebra in Q, and since projective algebras embed into some free algebra, there is a homomorphism to $\mathbf{F}_{\mathsf{Q}}$ from any projective algebra in Q. □

We are now going to introduce two notions of *unification type* from the symbolic and algebraic point of view, which is meant to represent the cardinality of the set of best solutions to a problem and show that they are equivalent. The idea is that a unifier is better than another one if it is more *general*, in a natural sense that we now make precise.

Definition 4 Let Σ be a symbolic unification problem and σ_1, σ_2 two unifiers for it. Then σ_1 is more general than σ_2 if the latter can be obtained from σ_1 by further substitution. That is, there is a substitution τ such that $\sigma_2 = \tau \circ \sigma_1$. In this case we write $\sigma_2 \preceq \sigma_1$.

Lemma 8 ([40]) *Let* Σ *be a symbolic unification problem; then* σ *is a preorder on its set of unifiers* $\mathcal{U}_\Sigma$.

We now consider the associated partial order on the set of unifiers U_Σ that are "equally general" ($\sigma_1 \sim \sigma_2$ iff $\sigma_1 \preceq \sigma_2, \sigma_2 \preceq \sigma_1$), which we denote by the same symbol without danger of confusion. Let us now define the corresponding algebraic notion.

Definition 5 Consider the algebraic unification problem $\mathbf{F}_{\mathsf{Q}}(X)/\theta_\Sigma$. If u_1, u_2 are two different unifiers, with projective targets $\mathbf{P}_1$ and $\mathbf{P}_2$, respectively, we say that u_1 is *more general* than u_2 if there exists a homomorphism $m : \mathbf{P}_1 \longmapsto \mathbf{P}_2$ such that $m \circ u_2 = u_1$. We write $u_2 \sqsubseteq u_1$.

Lemma 9 ([40]) $\sqsubseteq$ *is a preorder on the set of unifiers of* $\mathbf{F}_{\mathsf{Q}}(X)/\theta_\Sigma$, $\mathsf{U}_{\mathbf{F}_{\mathsf{Q}}(X)/\theta_\Sigma}$.

We again consider the corresponding partial order on the set of algebraic unifiers that are equally general.

Given a unification problem (symbolic or algebraic), its *unification type* is given by the cardinality of complete maximal elements in the partial order and is said:

- *Unitary*, if there is one maximal element (and not more than one); *strong unitary* if the maximal unifier on the algebraic side is the identity map
- *Finitary*, if there are finitely many maximal elements (but not infinitely many)
- *Infinitary*, if there are infinitely many
- *Nullary*, otherwise

The type of a quasivariety Q is the worst unification type occurring in the finitely presented algebras of V(Q). Using the translation between symbolic and algebraic unification problems detailed above, it is then easy to show the following.

Theorem 21 ([40]) *Consider a finite set of identities* Σ *written in the language of a quasivariety* Q*, over variables in a finite set* X*. The partially ordered sets* $(\mathcal{U}_\Sigma, \preceq)$ *and* $(\mathsf{U}_{\mathbf{F}_\mathsf{Q}(X)/\theta_\Sigma}, \sqsubseteq)$ *are isomorphic.*

Corollary 7 ([40]) *The symbolic and algebraic unification type of a quasivariety* Q *coincide.*

4.1 Unification in Hoops

Given the results shown in the previous sections, we have seen that several interesting varieties of hoops have the property that all finitely presented algebras are projective; this immediately implies that they have strong unitary unification type.

Theorem 22 *Every locally finite variety of basic hoops and HBCK-algebras has strong unitary unification type.*

For bounded hoops, the result we obtained was less strong; indeed we had that a finitely presented bounded hoop in a locally finite variety of bounded hoop is projective if and only if it has **2** homomorphic image. However, this is enough to conclude that the unification type is strong unitary. Indeed, a finitely presented algebra represents a *solvable* unification problem if and only if it has a homomorphism to the 2-element Boolean algebra (Proposition 9), and the unification type only considers the cardinality of the set of best solutions of a unifiable problem.

Theorem 23 *Every locally finite variety of bounded basic hoops has strong unitary unification type.*

With respect to varieties of bounded hoops obtained by generalized rotation constructions, an immediate consequence of Theorem 19 is the following.

Theorem 24 ([11]) *Let* V *be a variety of basic hoops with strong unitary unification type. Then* V_n^δ *and* V^{λ_n}*, for any* $n \geq 2$*, have strong unitary unification type.*

Not only locally finite varieties have this property, indeed clearly:

Theorem 25 *Cancellative hoops have strong unitary unification type.*

Corollary 8 *The following varieties have strong unitary unification type: product algebras, the variety generated by perfect MV-algebras, Gödel algebras, and nilpotent minimum algebras (with negation fixpoint).*

Finally, we have also mentioned varieties of hoops where projective algebras had a more complex description, actually given in terms of geometrical objects; we refer to MV-algebras and Wajsberg hoops. The former have been shown in [47] to have nullary unification type, and the same holds for Wajsberg hoops as shown in [58].

5 Conclusions and Open Problems

In this chapter we have summarized the main existing results about projective algebras in varieties of hoops, with some detour in specific varieties of residuated lattices. There are many problems that are still open; our techniques apply mainly to locally finite varieties and in particular we have no description of the finitely generated projective algebras in the whole variety of basic hoops or BL-algebras. However, some recent results, still in the making, seem to show that there are other nonlocally finite varieties of hoops for which a description of the finitely generated projective algebras is possible.

Another possible (and much harder) path is to abandon the realm of hoops and consider varieties of divisible (integral and commutative) residuated lattices or divisible FL_{ew}-algebras. It is well known that divisible residuated lattices (divisible FL_{ew}-algebras) coincide with hoops (bounded hoops) only in the prelinear case. However, some results can be obtained also in this case: for instance, in [12] it has been shown that every finite projective algebra in the variety of divisible FL_{ew}-algebras is of the form $\mathbf{A} \oplus \mathbf{2}$ for some finite $\mathbf{A}$. In case of Heyting algebras, which can be seen as idempotent divisible FL_{ew} algebras), a more precise description is possible: a finite projective Heyting algebra is $\mathbf{A} \oplus \mathbf{2}$ where $\mathbf{A}$ is an ordinal sum of finitely many copies of $\mathbf{2}$ and $\mathbf{4}$ (the four-element diamond). This was originally observed in [17], but see [12] for a more modern approach. In case of divisible residuated lattices there is even less information; for instance, we do not have any description of the projective algebras is the variety of Brouwerian lattices (zero-less subreducts of Heyting algebras). However, some recent results in [33] seem to hint that a description is possible.

References

1. Aglianò, P.: Varieties of BL-algebras I, revisited. Soft Comput. **21**, 153–163 (2017)
2. Aglianò, P.: Splittings in GBL-algebras II: the representable case. Fuzzy Sets Syst. **373**, 19–36 (2019)
3. Aglianò, P.: Varieties of BL-algebras III: splitting algebras. Studia Logica **107**, 1235–1259 (2019)
4. Aglianò, P.: A short note on divisible residuated semilattices. Soft Comput. **24**, 259–266 (2020)
5. Aglianò, P.: Quasivarieties of Wajsberg hoops. Fuzzy Sets Syst. **465**, 108514 (2023)
6. Aglianò, P., Ferreirim, I.M.A., Montagna, F.: Basic hoops: an algebraic study of continuous t-norms. Studia Logica **87**(1), 73–98 (2007)
7. Aglianò, P., Montagna, F.: Varieties of BL-algebras I: general properties. J. Pure Appl. Algebra **181**, 105–129 (2003)
8. Aglianò, P., Montagna, F.: Varieties of BL-algebras II. Studia Logica **106**(4), 721–737 (2018)
9. Aglianò, P., Panti, G.: Geometrical methods in Wajsberg hoops. J. Algebra **256**, 352–374 (2002)
10. Aglianò, P., Ugolini, S.: MTL-algebras as rotations of basic hoops. J. Logic Comput. **29**(5), 763–784 (2021)

11. Aglianò, P., Ugolini, S.: Projectivity and unification in substructural logics of generalized rotations. Int. J. Approx. Reason. **153**, 172–192 (2023)
12. Aglianò, P., Ugolini, S.: Projectivity in (bounded) commutative integral residuated lattices. Algebra Universalis **84**, 2 (2023)
13. Aglianò, P., Ursini, A.: Ideals and other generalizations of congruence classes. J. Aust. Math. Soc. **53**, 103–115 (1992)
14. Aguzzoli, S., Bianchi, M.: Strictly join irreducible varieties of BL-algebras: The missing pieces. Fuzzy Sets Syst. **418**, 84–100 (2021)
15. Aguzzoli, S., Flaminio, T., Ugolini, S.: Equivalences between subcategories of MTL-algebras via Boolean algebras and prelinear semihoops. J. Log. Comput. **27**, 2525–2549 (2017)
16. Balbes, R.: Projective and injective distributive lattices. Pacific J. Math. **21**, 405–420 (1967)
17. Balbes, R., Horn, A.: Injective and projective Heyting algebras. Trans. Am. Math. Soc. **148**, 549–559 (1970)
18. Beynon, W.M.: Applications of duality in the theory of finitely generated lattice-ordered abelian groups. Can. J. Math. **29**, 243–254 (1977)
19. Blok, W.J., Ferreirim, I.M.A.: Hoops and their implicational reducts (abstract). Algebraic Methods in Logic and Computer Science, vol. 28, pp. 219–230. Banach Center Publications (1993)
20. Blok, W.J., Ferreirim, I.M.A.: On the structure of hoops. Algebra Universalis **43**, 233–257 (2000)
21. Blok, W.J., Pigozzi, D.: Algebraizable Logics. Memoirs of the American Mathematical Society, vol. 396. American Mathematical Society, Providence, Rhode Island (1989)
22. Blok, W.J., Pigozzi, D.: On the structure of varieties with equationally definable principal congruences III. Algebra Universalis **32**, 545–608 (1994)
23. Blok, W.J., Raftery, J.G.: Varieties of commutative residuated integral pomonoids and their residuation subreducts. J. Algebra **190**, 280–328 (1997)
24. Bosbach, B.: Komplementäre halbgruppen. Axiomatik und arithmetik. Fund. Math. **64**, 257–287 (1969)
25. Burris, S., Sankappanavar, H.P.: A course in universal algebra. Graduate Texts in Mathematics. Springer, Berlin (1981)
26. Busaniche, M., Marcos, M., Ugolini, S.: Representation by triples of algebras with an MV-retract. Fuzzy Sets Syst. **369**, 82–102 (2019). https://doi.org/10.1016/j.fss.2018.10.024
27. Cabrer, L.: Rational simplicial geometry and projective unital lattice-ordered abelian groups (2014). https://arxiv.org/pdf/1405.7118.pdf
28. Cabrer, L.: Simplicial geometry of unital lattice-ordered abelian groups. Forum Mathematicum **27**, 1309–1344 (2015)
29. Cabrer, L., Mundici, D.: Projective MV-algebras and rational polyhedra. Algebra Universalis **62**, 63–74 (2009)
30. Cabrer, L., Mundici, D.: Rational polyhedra and projective lattice-ordered abelian groups with order unit. Commun. Contemp. Math. **14**, 1250017 (2012)
31. Cignoli, R., D'Ottaviano, I.M.L., Mundici, D.: Algebraic Foundations of Many-Valued Reasoning. Kluwer, Dordecht (2000)
32. Cignoli, R., Torrens, A.: Glivenko like theorems in natural expansions of BCK-logic. Math. Log. Q. **50**, 111–125 (2004)
33. Citkin, A.: Hereditarily structurally complete positive logics. Rev. Symbolic Logic **13**, 483–502 (2020)
34. Di Nola, A., Grigolia, R., Lettieri, A.: Projectiv MV-algebras. Int. J. Approx. Reason. **47**, 323–332 (2008)
35. Di Nola, A., Lettieri, A.: Perfect MV-algebras are categorically equivalent to abelian ℓ-groups. Studia Logica **53**, 417–432 (1994)
36. Ferreirim, I.M.A.: On varieties and quasivarieties of hoops and their reducts. Ph.D. Thesis, University of Illinois at Chicago, 1992
37. Ferreirim, I.M.A.: On a conjecture by Andrzej Wroński for BCK-algebras and subreducts of hoops. Scient. Math. Japonica **53**, 119–132 (2001)

38. Gabriel, P., Ullmer, F.: Lokar präsentiertbare Kategorien. Lecture Notes in Mathematics, vol. 221. Springer, Berlin (1971)
39. Galatos, N., Ugolini, S.: Gluing residuated lattices. Order (2023). https://doi.org/10.1007/s11083-023-09626-w.
40. Ghilardi, S.: Unification through projectivity. J. Logic Comput. **7**, 733–752 (1997)
41. Gorbunov, V.: Algebraic Theory of quasivarieties. Plenom, New York (1998)
42. Hájek, P.: Metamathematics of fuzzy logic. Trends in Logic—Studia Logica Library, vol. 4. Kluwer Academic Publisher, Dordrecht/Boston/London (1998)
43. Iséki, K.: An algebra related with a propositional calculus. Proc. Japan Acad. **41**, 26–29 (1962)
44. Jenei, S.: Structure of left-continuous triangular norms with strong induced negations (i) Rotation construction. J. Appl. Non-Class. Log. **10**, 83–92 (2000)
45. Jenei, S.: On the structure of rotation-invariant semigroups. Arch. Math. Logic **42**, 489–514 (2003)
46. Köhler, P.: Brouwerian semilattices. Trans. Am. Math. Soc. **268**(1), 103–126 (1981)
47. Marra, V., Spada, L.: Duality, projectivity, and unification in łukasiewicz logic and MV-algebras. Ann. of Pure and Appl. Logic **164**, 192–210 (2013)
48. McKenzie, R.: Equational bases and nonmodular lattice varieties. Trans. Am. Math. Soc. **174**, 1–43 (1972)
49. Moraschini, T., Raftery, J.G.: On prevarieties of logic. Algebra Universalis, vol. 80 (2019)
50. Mundici, D.: Interpretation of AFC^*-algebras in Łukasiewicz sentential calculus. J.Funct. Anal. **65**, 15–63 (1986)
51. Nation, J.B.: Finite sublattices of a free lattice. Trans. Am. Math. Soc. **269**, 311–337 (1982)
52. Ono, H., Komori, Y.: Logics without the contraction rule. J. Symb. Log. **50**, 169–201 (1985)
53. Quackenbush, R.W.: Demi-semi-primal algebras and Mal'cev type conditions. Math. Z. **122**, 166–176 (1971)
54. Quillen, D.: Projective modules over polynomial rings. Inventiones Mathematicae **36**, 167–171 (1976)
55. Robinson, J.A.: A machine-oriented logic based on the resolution principle. J. ACM **12**, 23–41 (1965)
56. Spinks, M., Veroff, R.: On a homomorphism property of hoops. Bull. Sect. Logic **33**, 135–142 (2004)
57. Tanaka, S.: On $\wedge$-commutative algebras. Math. Seminar Notes Kobe Univ. **3**, 59–64 (1975)
58. Ugolini, S.: The polyhedral geometry of Wajsberg hoops. J. Logic Comput. **34**(3), 557–589.
59. Whitman, P.M.: Free lattices. Ann. Math. **42**, 325–330 (1941)
60. Whitman, P.M.: Splittings of a lattice. Am. J. Math. **65**, 179–196 (1943)
61. Yutani, H.: On a system of axioms of a commutative BCK-algebra. Math. Seminar Notes Kobe Univ. **5**, 255–256 (1977)

From Krasner's Graded to Krasner–Vuković's Paragraded Groups and Rings

Mirjana Vuković

Abstract Although graded rings, the grades of which are embedded in an abelian group, have been studied by Chevalley (The Construction and Study of Certain Important Algebras, 1955), the first relatively general definition of graded groups and graded rings was given by Bourbaki (1958). However, that definition was not general enough because it was based on an abelian graded group. M. Krasner started from that definition (Krasner 1980), giving up the original unnecessary restriction. Krasner's definition, leaving aside the hypothesis of commutativity, shows that the structure of a graded group is characterized by both the underlying abstract group and its homogeneous subset (or even only by the homogeneous subset). It is well known that the classical Bourbaki–Krasner's graded structures belong to the category that is not closed with respect to the direct product and the direct sum. The main subject of Krasner's and my joint papers and monograph (Krasner and Vuković 1986a,b, 1987a,b) was to introduce the structures (groups, rings, modules) that generalize the corresponding graded structures and have, in each of these three cases, the property of closure with respect to the direct sum and the direct product. We called the obtained structures para- and extra-graded. The goal of this chapter is to review the basic definitions and assertions concerning graded, para-, and extra-graded structures mentioned in the title. Within the framework of Bourbaki–Krasner graduations and para- and extra-graduations, these structures were studied by M. Krasner, M. Chadeyras, E. Halberstadt, myself, and my students. All results mentioned here have been published or will be published in the papers cited in the bibliography.

Keywords Structures: group (ring): graded · para- (extra-) graded; (graded) fields; (quasi)corpoid; quasi- · para- · extra-graduation; Homogeneous part: homogroupoid · aneid; Homomorphism: (quasi) homogeneous

---

M. Vuković (✉)
Academy of Sciences and Arts of Bosnia and Herzegovina, Sarajevo, Bosnia and Herzegovina
e-mail: mvukovic@anubih.ba

M. Mitrović, M. N. Hounkonnou (eds.), *Algebra Without Borders*, Mathematics in Mind, https://doi.org/10.1007/978-3-031-86477-3_6

Introduction: Review of the Historical Development

The chapter contains an introduction with a short history of the notion of graduation, a review of the Krasner's fundamental notions on a corpoid which originates from [10] (see also [11–16]), and general graded groups and graded rings [4, 8, 17] (see also [22, 24]) (Ch. I, Sections 1.1. and 1.2.) as well as a finish with para- and extra-graded groups and paragraded rings which first appeared in [19–22] (see also [28–33, 33–40]) as a solution of an important problem in the category of general graded structures, known as the problem of closure of graded structures (groups, rings, modules) with respect to the direct sum and the direct product of homogeneous parts of factors (Ch. II).

The notion of graduation is, at least in the light of certain examples, quite an old one [1, 5, 6]. It was introduced explicitly into mathematics during the eighteenth century when *Euler* defined the notion of homogeneous polynomials and, more generally, homogeneous real functions [7] (see also [18]).

However, when applied to the case of one variable, it was already known to mathematicians well before *Euler* (*Diophantus*, some *Arab mathematicians*, *Viète*, *Descartes* [5]), while the germ of the notion of homogeneity can be seen in early Greek mathematics ("multiplication of segments" that preceded *Grassmann's Ausdehnungslehre theory*) and in some documents from *Babylonian times* [18] (see also [1, 6, 9]).

Other notions of homogeneity and its corresponding grades (which is analogous to the notion of degree) were introduced after Euler in various algebraic structures. Some of these structures are well known as weights of polynomials and functions, dimension of geometric and topological objects, and order of differential operators, and later, the more general notion of Z-graded ring as well as more generally Z^2-graded, Z^3-graded,..., was introduced (see, for example, *Samuel* and *Zariski* [25]).

Although *Chevalley* also examined graded rings, the grades of which are embedded in an abelian group [3], the first relatively general definition of graded groups and graded rings was given in *Bourbaki* [2]. But this definition was unnecessarily based on the notion of the abelian graded group and was not general enough. Despite the fact that there was no justification for restricting it to the commutative case, it was a very good definition.

However, a little is known that all started much earlier – before Bourbaki, with the abstract notion of corpoid, introduced during 1940s by *Mark Krasner*, while he investigated valued fields and observed a connection to their valuation rings through the equivalence of valuations. More precisely, a corpoid first appeared in a series of M. Krasner's *Comptes Rendus* notes (of French Academy of Sciences), in 1944 and 1945 [10–12]; his paper "*Annéaux gradués généraux*" [17], starting from Bourbaki's definition of a graduation and giving up the original unnecessary restrictions, arrived at the most general graded groups and rings.

From Krasner's definition, which does not assume commutativity, it can be seen that the structures of a graded group and graded ring are characterized by both the underlying abstract group and the homogeneous part (or even only by the homogeneous part), with the partial operation[1] induced from the group.

The characterization axioms give way to three study methods of graded groups (in principle equivalent) and of other more complex graded structures (rings and modules):

(i) The nonhomogeneous methods that consider the structures as such.
(ii) The semihomogeneous methods that consider the underlying nongraded structures together with homogeneous subsets satisfying some additional axioms.
(iii) The homogeneous methods that consider only the homogeneous subset endowed with a partial composition operation satisfying some additional axioms, or in the case of "richer" structures, those endowed with other composition operations generally defined everywhere.

Under certain conditions, filtered structures of the same type can be obtained from these graded structures, with a totally ordered filtrations (but not necessarily a Z-filtration), which are called grades of the corresponding filtered structures.

Often, however, the homogeneous graded subset is the only interesting subset for the study of the filtered structure from which it originated.

Furthermore, Mark Krasner, removing the unnecessary restrictions from Bourbaki's definitions, introduced adequate notions of *graded groups*, *graded rings*, and *graded modules*, together with their *homogeneous subsets*, which he called "*groupoids*," "*aneids*," and "*moduloids*,"[2] respectively [17].

1. M. Krasner also introduced the graded homogeneous analogue of a field, which he named a *corpoid*, in 1944 [10] (see also [8, 14]). He defined this notion starting from valued fields, of which the homogeneous subsets are logically graded through the filtrations of their valuations (which can also be *Krull's* valuations).

   These "*corpoids*," called *skeletons* (of the corresponding valued fields), have the property that the group of their grades[3] is without torsion.

   Moreover, the commutativity of the valued field implies the commutativity of its skeleton.
2. If q is a corpoid, then a unitary q-moduloid V is called a q-vector space. Regular q-vector spaces have a theory of linear independence and dimension, which is, formally, a similar (and even richer than) theory of vector spaces over fields.

   If Q is a supercorpoid of q, it can be considered, in the usual way, as a q-vector space, which is always regular, allowing us in this way to define the grade (to right or left) $[Q : q]$ of the corpoidal extension Q / q as the dimension of this space (on the same side).

[1] The homogeneous part equipped with a partial operation was named "homogroupoid" by M. Krasner, which represents a special case of structures in the theory of groups known as amalgams (see, for example, [26]).

[2] The corresponding nonhomogeneous notion is called a graded field.

[3] The grades of a corpoid form a group.

3. General facts about corpoids (including definition of a skeleton) and their extensions as well as the theory of regular vector spaces on over these structures formulated by Krasner in [14] were presented in 1954, also by M. Krasner (see [15]).
4. Commutative without torsion corpoids[4] have a theory of extensions (based on the theory adequately constructed from polynomials over these corpoids) that is similar to *Steinitz's theory* for field extension, but is much richer, and a good *Galois' theory* (see [16]), in which the theorem on primitive element is no longer true for the extension of separable corpoids with finite grade, but despite that, the fundamental theorem of Galois theory can be proved. M. Krasner presented this theory in a brief note [11] of 1944, but the work was never published in its entirety. Its statement can only be found in the preprint [16].

These theories have important applications in the theory of valued fields, in particular in the theory of ramification, as presented in M. Krasner's notes [11] and [12] from 1944 to 1945, under the hypothesis of a finite grade, presented in part in the preprint [16], but without this additional hypothesis.

The homogeneous theory of commutative graded rings, from the noetherian point of view, was studied by *Marcel Chadeyras* [4] in his doctoral thesis, while *Emanuel Halbershtadt*, in his doctoral thesis [8], studied the homogeneous theory of generally noncommutative regular rings, mainly from the artinian point of view – both Krasner's pupils.

It is known that the various graded structures (groups, rings, modules) compose the categories that are not closed with respect to the direct sum and the direct product.

In our joint papers [19–21] and the monograph "*Structures paragraduées* (*groups, rings, modules*)" [22], M. Krasner and I introduced algebraic structures called *paragraded* (groups, rings, modules), which are already called Krasner–Vuković paragraded structures after us. These structures present a generalization of the corresponding classical *Bourbaki–Krasner's* graded structures and have the property that, in each of these three cases, their categories are closed with respect to the direct sum and the direct product, so that the support of the homogeneous part of that product is a direct product restriction, respectively the direct product of the homogeneous parts of the components.

In this way, M. Krasner and I have developed a general theory of graded structures that generalizes not only the theory of graded structures as it is defined by Bourbaki [2], but also the previous papers of Marc Krasner (see [17]).

[4] Therefore, their group of grades is without torsion [14].

1 Graded Groups and Rings

1.1 Graded Groups in Krasner's Sense

The abstract notion of a corpoid [10] (see also [35]) led Krasner to the development of a general graded theory. Krasner, starting from his corpoid and Bourbaki's definition of graduation, giving up his hypothesis of commutativity – original unnecessary condition, discovered general theory of graded structures (groups, rings, modules) and together with them the theory of *homogroupoids* (*homogeneous parts of groups*), *anneids* (*homogeneous parts of rings*), and *moduloids (homogeneous parts of modules*), where a *corpoid*, as a special case of an anneid, is viewed as a homogeneous part of a graded field [17].

However, graded fields are not to be understood as fields with graduation, but fields are to be understood as graded fields with trivial graduation.

Note that Krasner's approach to graduation differs from Bourbaki's approach: Krasner did not restrict himself to the case when the set of grades is a group, but he assumed it to be only a nonempty set in which the associativity, the commutativity, and the existence of the neutral element are also not assumed.

As is well known, the notion of rings relies on the notion of groups. That is why we are going to start by presenting definitions of general graded groups in Krasner's sense, including axiomatization from his work [17] and some essential properties of these notions, before introducing Krasner–Vuković's general graded – the so-called paragraded groups and rings which represent the content of the aforementioned monographic work [22] and papers [19–21], written in collaboration with M. Krasner.

Now we will give definitions of a graded group by Krasner called *nonhomogeneous aspect* and of a homogroupoid which are essential notions in this part of the chapter.

Definition 1.1.1 Let G be a multiplicative group with the neutral element e and let be a nonempty set. A graduation of the group G is a mapping

$$\gamma : \Delta \to Sg(G),\ \gamma(\delta) = G_\delta\ (\delta \in \Delta), \tag{1.1.1}$$

such that

$$G = \bigoplus_{\delta \in \Delta} G_\delta,$$

where $Sg(G)$ is the set of all subgroups of G.

A group G, with graduation γ, is called a graded group.

A graded group is called abelian if it is abelian as an abstract group.

Remark 1.1.1 In case when G is commutative group, the operation will be denoted additively and neutral element by 0.

The essential features of the graded groups from which some will be expressed through the following remarks and definitions:

Remark 1.1.2 From (1.1.1) follows, in particular, that G_δ, $\delta \in \Delta$ are normal subgroups of the group G, and if $\delta, \delta' \in \Delta$ and $\delta \neq \delta'$, then

elements $x \in G_\delta$ and $x' \in G_{\delta'}$ commute, and $G_\delta \cap G_{\delta'} = \{e\}$.

Definition 1.1.2 The graduation γ of the group G is called:

- Strict if $\gamma(\delta) = G_\delta \neq \{e\}$, for all $\delta \in \Delta$.
- Trivial if $\gamma(\delta) = G_\delta = \{e\}$, i.e., Δ is singleton.

Remark 1.1.3 Every abstract group may be considered as a trivial group.

Definition 1.1.3 Let G be a graded group with graduation (1.1.1). If

$$\Delta^* = \{\delta \in \Delta \mid G_\delta \neq \{e\}\},$$

then the mapping $\gamma^* = \gamma|_{\Delta^*}$ is a strict graduation of G, and it is called the strict kernel of graduation (1.1.1).

Definition 1.1.4 Let G be a graded group with graduation γ defined by (1.1.1). The subset $H = \cup_{\delta \in \Delta} G_\delta$ $(= \cup_{\delta \in \Delta^*} G_\delta)$ of the group $G(\neq \{e\})$ is called the homogeneous part of G (for graduation γ) and elements $x \in H$ are called homogeneous elements of the group G.

In other words, H is the set of all products of finitely many elements $y_{\delta^1}, \ldots, y_{\delta^{(n)}}$, $n \geq 1$, $y_{\delta^{(i)}} \in H \cap G_{\delta^{(i)}}$ where all grades $\delta^{(i)}$ are supposed to be distinct.

Remark 1.1.4 For a homogeneous element $x \neq e$, there is the unique $\delta \in \Delta^*$ for which $x \in G_\delta$ is called the grade (degree) of x and denoted by $\delta(x)$ while the corresponding G_δ are called homogeneous components.

Remark 1.1.5 Element e in general does not have a grade (degree), but if *the empty grades* exist, i.e., if $\Delta \neq \Delta^*$, we can associate one of *the empty grades* to the neutral element e which we denote by 0 ("*zero*") and call it a *zero grade* [17] (see also [28, 35]). Sometimes, when there are no empty grades, it is useful to adjoin to a special "ideal" grade denoted 0 and to put $G_0 = \{e\}$.

Definition 1.1.5 If $\Delta = \Delta^* \cup \{0\}$ and if we put $\delta(e) = 0$, then the graduation γ is called proper [17].

Definition 1.1.6 The graduation $\gamma : \Delta \to Sg(G)$ is said to be strict if $\Delta = \Delta^*$.

Notice, when two subgroups G_δ and $G_{\delta'}$ are different, they have only e as common element, i.e., $\delta \neq \delta' \Rightarrow G_\delta \cap G_{\delta'} = \{e\}$. So, every $x \in H$, $x \neq e$, belongs to one and only one G_δ, $\delta \in \Delta$.

It is easy to see that the following lemmas hold:

Lemma 1.1.1 *The graduation $\gamma : \Delta \rightarrow Sg(G)$, with $\gamma(\delta) = G_\delta$ ($\delta \in \Delta$), is injective mapping if and only if γ is strict or proper.*

Lemma 1.1.2 *Let G be a graded group with graduation (1.1.1). The mapping $\delta : H \rightarrow \Delta$, with $\delta(x) = \delta$ ($x \in H$) is surjective if and only if graduation γ is proper.*

As we have already mentioned, another important property of Krasner's graded structures is that these structures are characterized by both the underlying abstract group and the homogeneous part (even only by the homogeneous part) equipped with the operation(s) induced by operation(s) of the structure as well as that characterization axioms give a way to three study methods of graded structures in principle equivalent: (*i*) the *nonhomogeneous*, (*ii*) the *semihomogeneous*, and (*iii*) the *homogeneous* method, which is not defined for Bourbaki's graded structures.

Definition 1.1.7 ([17], see also [28, 35]) Let G be a graded group with two graduations γ defined by (1.1.1) and

$$\gamma' : \Delta' \rightarrow Sg(G), \quad \text{with } \gamma'(\delta) = G'_\delta \ (\delta \in \Delta') \tag{1.1.2}$$

and

$$\Delta^* = \{\delta \in \Delta \mid \gamma(\delta) \neq \{e\}\}, \quad \Delta'^* = \{\delta' \in \Delta' \mid \gamma'(\delta) \neq \{e\}\}.$$

We say that the graduations γ and γ' are:

(i) Equivalent if there exists a bijective mapping
$\varphi : \Delta \rightarrow \Delta'$ such that $\gamma(\delta) = \gamma'(\varphi(\delta))$, for every $\delta \in \Delta$,
i.e., if its proper parts are equal up to the names of grades.

(ii) Weakly equivalent if there is a bijective mapping
$\varphi : \Delta \rightarrow \Delta'$ such that $\gamma^*(\delta) = \gamma'^*(\varphi(\delta))$, for every $\delta \in \Delta$,
i.e., if their strict kernels are equivalent, where
$$\gamma^* = \gamma|_{\Delta^*}, \ \gamma'^* = \gamma'|_{\Delta'^*}.$$

Obviously, strict graduation is determined up to the equivalence by the set $\{G_\delta \mid \delta \in \Delta\}$, while a graduation is determined up to the weak equivalence by the set $\{G_\delta \mid \delta \in \Delta^*\}$.

The homogeneous part H of a group G, together with the group structure of G, determines the corresponding graduation up to the weak equivalence. Indeed, if
$a \in H^* = H \backslash \{e\}, b \in H$ and if $a \in G_\delta$, then $b \in G_\delta$, iff $ab \in H$, i.e.,
iff ab is a homogeneous element.

Hence, if for $a \in H^*$, we define $G(a) = \{x \in H \mid ax \in H\}$, and then sets
$\{G_\delta \mid \delta \in \Delta^*\}$ and $\{G(a) \mid a \in H^*\}$ coincide.

This observation was for M. Krasner a motivation for defining a graded group from semihomogeneous aspect as an ordered pair (G, H), where $H \subseteq G$ is the homogeneous part of G with respect to some graduation of G in the previous sense.

This, however, required the characterization of the homogeneous part of the graded group with the following system of six axioms:

Theorem 1.1.1 ([17] see also [22, 28]) *A nonempty subset H of a group G is the homogeneous part of G, with respect to some graduation γ of G if and only if the following conditions are satisfied:*

1° $e \in H$.
2° *If* $x \in H$, *then* $x^{-1} \in H$.
3° *If* $x, y, z, xy, yz \in H$ *and* $y \neq e$, *then* $xz \in H$.
4° *If* $x, y \in H$ *and* $xy \notin H$, *then* $xy = yx$.
5° H *generates* G.
6° *If* $n \geq 2$ *and if* $x_1, \ldots, x_n \in H$ *are such that for all* i, j, $1 \leq i, j \leq n$, $i \neq j$, *we have* $x_i x_j \notin H$, *then* $x_1 x_2 \ldots x_n \neq e$.

The theorem just stated could, in the case of commutative groups, be formulated as follows:

Theorem 1.1.2 ([17] see also [22, 28]) *A nonempty subset H of a group G is the homogeneous part of the commutative graded group G, with respect to some graduation γ of G if and only if the following conditions are satisfied:*

1° $0 \in H$.
2° *If* $x \in H$, *then* $-x \in H$.
3° *If* $x, y, z, x + y, y + z \in H$ *and* $y \neq 0$, *then* $x + z \in H$.
4° H *generates* G.
5° *If* $n \geq 2$ *and* $x_1, \ldots, x_n \in H$ *are such that for all* i, j, $1 \leq i, j \leq n$, $i \neq j$, $x_i + x_j \notin H$, *then* $x_1 + x_2 + \ldots + x_n \neq 0$.

It means that for a given pair $(G, \ H \subseteq G)$, *where G is a group*, and if H satisfies the axioms $1° - 6°$, we can reconstruct a corresponding graduation γ (up to an equivalence).

Taking into account this theorem, the graded group may be defined in the following way:

Definition 1.1.8 The ordered pair $(G, \ H \subseteq G)$, where G is a group and $H \subseteq G$ satisfies the axioms $1° - 6°$, is called the graded group.

1.2 Homogroupoids

If H is a homogeneous part of the graded group G, for some graduation γ, then the multiplicative operation on G induces a partial operation on H, as follows:

$$\text{if } x, y \in H, \text{ then } xy \text{ is defined in } H \text{ iff } xy \in G \text{ is the element from } H \qquad (*)$$

and in that case the result is the same and we write it in the same way.

If this situation occurs, we say that the elements x, y are *composable* (*addible* in the case of an additive operation) and we denote it by $x\#y$ ([17], see also [22, 28]).

It is easy to see that

$x\#y$ iff x and y are from the same subgroup $G(a)$, where
$G(a) = \{x \in H \mid ax \in H\}$, $a \in H^* = H \backslash \{e\}$.

Such partial structures $(H; \cdot)$ are called *homogroupoids* which is specified by the following definition:

Definition 1.2.1 The structure $(H; \cdot)$, where H is a homogeneous part of graded group G for some graduation γ with a partial operation induced from G according to $(*)$, is called a homogroupoid, and its corresponding graded group $G \supseteq H$ is called (graded) group closure, or its enveloping group.

Remark 1.2.1 Elements $x (\neq e)$ from different subgroups G_δ are not composable.

Now we can prove the following statement:

Theorem 1.2.1 *If the structure $(H; \cdot)$, obtained for some graduation γ of the group G, with the partial operation "$\cdot$," can also be obtained in the same way from some another graduation γ' of the group $G' \supseteq H$, then the group G' is H-isomorph to G.*

In other words, there exists an isomorphism from the group G' to the group G that induces the identity mapping e_H on H.

Definition 1.2.2 The pair (G, H) (from Theorem 1.2.1), determined up to an H-isomorphism, is called the linearization of the structure $(H; \cdot)$ and is denoted by $\overline{(H; \cdot)}$ or simply $\overline{H}$ if there is no danger of confusion.

Thus, when H, with the induced operation from G, is given, we may reconstruct G up to H-isomorphism. Indeed,
if $a \in H^*$, then $G(a)$ can be defined as $G(a) = \{x \in H \mid a\#x\}$.

In this case G is the direct sum of different subgroups $G(a)$ and H is obviously the homogeneous part of G. The group G, obtained in this way, is called the linearization of the structure $(H; \cdot)$ which we denote by $\overline{(H; \cdot)}$), or simply by $\overline{H}$, where

$$\overline{H} = \oplus_{a \in H^*} G(a) \quad ([17], \text{see also } [22]).$$

There is a natural idea now to define a graded group using the corresponding homogeneous part, at least up to H-isomorphism. In order to do that, we need to characterize the structure $(H; \cdot)$ which is the homogeneous part of some graded group, with the partial operation induced from the operation of that group.

This characterization by M. Krasner called *homogeneous aspect* can be pronounced by the following theorem:

Theorem 1.2.2 ([17]) *The structure $(H; \cdot)$ is the homogeneous part of some graded group G with partial operation induced from that group, if and only if the following conditions hold:*

1°° There exists an element $e \in H$ such that
for every $x \in H$, $x\#e$ and $x \cdot e = x$ (axiom of neutral element).

$2^{\circ\circ}$ *For every* $x \in H$, $x\#x$ *(axiom of autocomposability).*
$3^{\circ\circ}$ *If for all* $x, y, z \in H$, $x\#y$, $y\#z$, *and* $y \neq e$, *then* $x\#z$ *(axiom of almost transitivity of composability).*
$4^{\circ\circ}$ *If for all* $a \in H^*$, $H(a) = \{x \in H \mid a\#x\}$ *is a group with respect to multiplicative operation, where* $x\#y$ *means that* $x \cdot y$ *exists.*

Now we will introduce the notion of homogroupoid from homogeneous aspect.

Definition 1.2.3 ([17]) The structure $(H; \cdot)$ that satisfies the conditions $1^{\circ\circ} - 4^{\circ\circ}$ of the previous theorem is called a homogroupoid.

Therefore, the structures occurring are coincident (up to an H-isomorphism) with the linearizations of the homogroupoids $(H; \cdot)$.

A homogroupoid $(H; \cdot)$ is commutative if all its composable elements mutually commute, that is, the following proposition holds:

Proposition 1.2.1 *A necessary and sufficient condition for a graded group* G *to be abelian is that each set* $H(a)$ *is an abelian group, meaning*

$$a\#b \ \text{ implies } \ ab = ba.$$

In fact, the homogroupoid $(H; \cdot)$ *is an amalgam of the groups* G_δ, $\delta \in \Delta^*$, *such that*

$$\delta \neq \delta' \ \text{ implies } \ G_\delta \cap G_{\delta'} = \{e\}.$$

Definition 1.2.4 We say that a subgroup g of the group G is a homogeneous subgroup of the group G if it is generated by $h = g \cap H$.

Definition 1.2.5 The subset h of the homogroupoid H is called subhomogroupoid of H, if it is the homogeneous part of the homogeneous subgroup g of the linearization $\overline{H}$ $(G = H)$ of the homogroupoid $(H; \cdot)$.

Proposition 1.2.2 *The set* $h \subseteq H$ *is subhomogroupoid of homogroupoid* H *if and only if:*

1° *For* $x \in h$ *it follows that* $x^{-1} \in h$.
2° *For* $x, y \in h$ *and* $x\#y$ (in $(H; \cdot)$) *it follows that* $xy \in h$.

Let h $(\subseteq H)$ be a subhomogroupoid of the homogroupoid H. Obviously, h is normal in H if and only if:

for all $a \in h \backslash \{e\}$ and $x \in H$, $a\#x$ (which implies $x\#a$) implies $ax = xa$.

Proposition 1.2.3 *A homogeneous subgroup* g *of the group* G *is a normal subgroup of* G *if and only if* $h = g \cap H$ *is a normal subgroup of* H.

1.3 Graduations γ_g *and* $\gamma_{G/g}$ *Induced by the Graduation* γ. *Factor Homogroupoid*

1. If g is a homogeneous subgroup of the group G, then the graduation γ of the group G determines a graduation $\gamma_g : \Delta \to Sg(G)$ of g, with the same set of grades Δ, which we call the graduation induced by γ (also denoted as γ) and which is given by

$$\gamma_g(\delta) = G_g \cap g \quad (\delta \in \Delta).$$

The homogeneous part of g, with respect to γ_g, is equal to $h = g \cap H$.
2. If g is a normal subgroup of the group G, then the graduation γ induces the graduation $\gamma_{G/g}$ of the group G/g (also denoted as γ), defined by

$$\gamma_{G/g}(\delta) = G_\delta g/g \quad (\delta \in \Delta).$$

Remark 1.3.1 Since $G_\delta g/g$ is canonically isomorph to

$$(G_\delta/g) \cap g_\delta = G_\delta/g_\delta, \quad \text{where} \quad g_\delta = \gamma_\delta(\delta),$$

$\gamma_{G/g}(\delta)$ can be identified with G_δ/g_δ. In this case, G/g is identified with $\oplus_{\delta\in\Delta}$ (G_δ/g_δ), and the homogeneous part Hg/g of G/g, with respect to $\gamma_{G/g}$, is identified with $\cup_{\delta\in\Delta}$ (G_δ/g_δ), in the direct sum of groups G_δ/g_δ, $\delta \in \Delta$.

Considering that this union and the induced partial operation defined on G/g (using the previous identification) can also be defined directly through the elements of homogroupoid $(H; \cdot)$ and its normal subgroup h, defined in this way, we call it the factor homogroupoid of $(H; \cdot)$ by h and denote it by H/h.

In the case when the graduation is strict or proper, the graduations γ_g and $\gamma_{G/g}$ are not necessarily such. This is one of the reasons why our considerations are not limited to strict and proper graduations only.

1.4 Direct Sum and Direct Product of Graded Groups

Assume that G is the graded group with graduation (1.1.1) and that $(g^{(i)} \mid i \in I)$ is the family of its homogeneous subgroups. Let $G = \oplus_{i\in I} g^{(i)}$ be the direct sum or $G = \Pi_{i\in I} g^{(i)}$ is the direct ("*internal*") product of the family $(g^{(i)} \mid i \in I)$ – its homogeneous subgroups. Let, for $x \in G$,

$\mu : x \to (x^{(i)} \mid i \in I)$ be the canonical mapping

of the group G to the direct sum respectively direct ("*external*") product of subgroups $g^{(i)}$ $(i \in I)$, and let $h^{(i)}$ be homogeneous parts of $g^{(i)}$ and $e^{(i)}$ the corresponding neutral elements.

Definition 1.4.1 If $e^{(i)} \in A^{(i)} \subseteq g^{(i)}$ $(i \in I)$, then the sets $\oplus^s_{i\in I} A^{(i)}$ and $\Pi^s_{i\in I} A^{(i)}$ are called the set products if they consist of elements

$$(a^{(i)} \in A^{(i)} \mid i \in I)$$

where $a^{(i)} = e^{(i)}$ for almost all $i \in I$, i.e., only finitely many elements $a^{(i)} \neq e^{(i)}$, $i \in I$.

Remark 1.4.1 The condition that $a^{(i)} = e^{(i)}$, for almost all $i \in I$, need not be satisfied for the latter set product.

It can be seen that the homogroupoid $(H; \cdot)$ is the direct product of homogroupoids $(h^{(i)}; \cdot)$, $i \in I$, in G. However, homogroupoid $(H; \cdot)$ is not determined by homogroupoids $(h^{(i)}; \cdot)$ up to an isomorphism, but depends on the order of subgroups $g^{(i)}$ in G.

Therefore, there is no (up to an isomorphism) unique product of the family of homogroupoids $\{(h^{(i)}; \cdot) \mid i \in I\}$. Moreover, the supports of product and set product of homogroupoids $(h^{(i)}; \cdot)$, $i \in I$, need not coincide.

Starting from the operations in homogroupoids $(h^{(i)}; \cdot)$, $i \in I$, can be defined the partial multiplication (which is itself interesting) as follows:

$(x^{(i)} \mid i \in I) = x\#y = (y^{(i)} \mid i \in I)(x^{(i)}, y^{(i)} \in h^{(i)})$ iff $x^{(i)}\#y^{(i)}$ for all $i \in I$. In that case we define $x \cdot y = (x^{(i)} \cdot y^{(i)} \mid i \in I)$.

However, such a structure, which is called *multigroupoid*, does not have to be homogroupoid, because its composability is not almost transitive, as can be seen from the following example:

Example 1.4.1 Let $I = \{1, 2\}, a, a' \in h^{(1)}$ the elements which are not composable, $b \in h^{(2)}$. $b \neq e^{(2)}$, and

$$x = (a, e^{(2)}),\ \ y = (e^{(1)}, b)\ z = (a', e^{(2)}).$$

Then

$$y \neq e = (e^{(1)}, e^{(2)}),\ \ x\#y,\ \ y\#z,\ \text{ but }\ x\#z\ \text{ does not hold.}$$

1.5 *Aspects of the Theory of Graded Groups*

According to M. Krasner [17] (see also [28]), there are three aspects for studying both graded groups and richer graded structures such as rings and modules, which are essentially equivalent. These are:

1. *The nonhomogeneous aspect* that considers graded groups as such.
2. *The semihomogeneous aspect* that considers the underlying nongraded groups G together with a homogeneous subset $H \subseteq G$ satisfying some additional axioms.
3. *The homogeneous aspect* that considers only the homogeneous subset H endowed with a partial composition operation satisfying some additional axioms or, in the case of more complex structures, these endowed with other composition operation generally defined everywhere.

   It is difficult to say which of these three methods is "better" in each of these cases. When we are interested in all elements of a graded structure, which is primarily the case when the set of grades is totally ordered (because in this case the grades can be defined for nonhomogeneous elements, too), the aspect of nonhomogeneity is more adequate than the other two. On the other hand, if we are interested only in homogeneous elements and homogeneous objects, then the semihomogeneous aspect and especially the homogeneous aspect is more adequate and they are the best in that case. They are also in general very closely related.

1.6 Homogeneous and Quasihomogeneous Homomorphisms. Homomorphisms and Quasihomomorphisms

Let (G, H) and (G', H') be two graded groups, with *corresponding homogroupoids H and H' and neutral elements e and e', respectively.*

Definition 1.6.1 For a homomorphism $\varphi : G \to G'$ we say that it is quasihomogeneous if $\varphi(H) \subseteq H'$.

Let $\varphi : G \to G'$ *be a quasihomogeneous homomorphism. If $x, y \in H$ are such that $\varphi(x) \neq e'$, and $\varphi(y) \neq e'$, then, by definition,*

$$\delta(x) = \delta(y) \text{ implies } \delta(\varphi(x)) = \delta(\varphi(y)).$$

Definition 1.6.2 Let $\varphi : G \to G'$ be a quasihomogeneous homomorphism and let $x, y \in H$ be such that $\varphi(x) \neq e'$, and $\varphi(y) \neq e'$. The homomorphism φ is called homogeneous if

$$\delta(\varphi(x)) = \delta(\varphi(y)) \text{ implies } \delta(x) = \delta(y).$$

Every homomorphism φ of group G to group G' is determined, in fact, by its restriction $\psi = \varphi|_H$, since H generates *the group* G, and $\psi : H \to G'$ can be extended to a homomorphism $\varphi : G \to G'$ iff

$$\text{for all } x, y \in H, x\#y \text{ implies } \psi(xy) = \psi(x)\psi(y),$$

since every element $x \in G$ can be uniquely represented (to the permutation of the factors) as the product of elements that are not pairwise composable.

Definition 1.6.3 If φ is a quasihomogeneous (homogeneous) homomorphism, its restriction $\psi = \varphi|_H : H \to H'$ is called quasihomomorphism (homomorphism).

It is clear that the following lemma holds:

Lemma 1.6.1 *The mapping $\psi : H \to H'$ is:*

(1) Quasihomomorphism iff

$$\textit{for all } x, y \in H, x\#y \textit{ implies } \psi(x)\#\psi(y) \textit{ and } \psi(xy) = \psi(x)\psi(y)$$

(2) Homomorphism if

$$\textit{for all } x, y \in H, \psi(x) \neq e', \psi(y) \neq e', \textit{ and } \psi(x)\#\psi(y) \textit{ implies } x\#y$$

1.7 Partial Operation $x^{(\varphi)}y$. Agglutination and its Kernel

Let $\varphi : G \to G'$ *be quasihomogeneous homomorphism.* Then, using φ, we can define an operation $x^{(\varphi)}y$ of composability for (φ) on $H^{(\varphi)} = \varphi^{-1}(H')$, or simply φ-composability:

$$\text{for } x, y \in H^{(\varphi)};\ x\#^{(\varphi)}y, \text{ iff } \varphi(x)\#\varphi(y) \text{ in } (H'; \cdot).$$

Definition 1.7.1 The set $(H^{(\varphi)}; \cdot)$, where we take $x^{(\varphi)}y = xy$, using xy as shorthand for $x^{(\varphi)}y$, is called the agglutination of H for φ.

The agglutination $(H^{(\varphi)};\cdot)$ is determined by $\psi = \varphi\,|_H$, up to an H-isomorphism itself a homogroupoid $(H;\cdot)$, since G can be reconstructed and from that homogroupoid up to such an isomorphism (similarly, G' from $(H';\cdot)$).

It can be easily seen that the H-isomorphism transforms the extension φ from ψ to G', and $H^{(\varphi)} = \varphi^{-1}(H')$, and $x^{(\varphi)}y$ is consistent with G. Therefore, if

$\psi : H \to H'$ *is a quasihomomorphism* and

$\varphi : \overline{(H;\cdot)} \to \overline{(H';\cdot)}$ is a *quasihomogeneous homomorphism* which extends ψ, then $(H^{(\varphi)};\cdot)$ depends only on φ, and not on the choice φ, up to an H-isomorphism.

Definition 1.7.2 If $\psi : H \to H'$ is a quasihomomorphism and $\varphi\overline{(H;\cdot)} \to \overline{(H';\cdot)}$ is an extension of ψ, then $(H^{(\varphi)};\cdot)$ is denoted as $(H^{(\psi)};\cdot)$ and is called the agglutinate of H with ψ, while the phenomenon itself is called agglutination.

In general, agglutinates are not homogroupoids. More precisely, they are homogroupoids only in two cases as stated in the following proposition:

Proposition 1.7.1 *The agglutinate* $(H^{(\varphi)};\cdot)$ *of* H *is a homogroupoid only in the case when:*

(1) The kernel of φ, $N^{(\varphi)} = \{e\}$, *i.e.,* φ, *as an homomorphism of abstract groups, is an isomorphism.*
(2) The operation $x^{(\varphi)}y$ *is defined everywhere on* $H^{(\varphi)}$, *i.e., the product* xy *in* H' *is defined on* $\varphi(H)$, *if and only if exists* $\delta' \in \Delta'$ *such that* $\varphi(H) \subseteq G'_{\delta}$.

Definition 1.7.3 The set $N^{(\varphi)} = C$, which generally consists of elements $c \in H^{(\varphi)}$ that are φ- composable with every element $x \in H^{(\varphi)}$, is called the core of $(H^{(\varphi)};\cdot)$.

The core is a normal subgroup of the group G. Considering that

$$C \subseteq \varphi^{-1}(G'_{\delta'}) = G^{(\varphi)}_{\delta'} \quad (\delta' \in \Delta')$$

is a subgroup of G, contained in $H^{(\varphi)}$, we can define the factor structure:

$$G/C = \oplus_{\delta' \in \Delta'}(G^{(\varphi)}_{\delta'}/C) \text{ and } H^{(\varphi)} = \cup_{\delta' \in \Delta'} G^{(\varphi)}_{\delta'},$$

which is, to a large extent, analogous to the case of the graded group, but more general.

In his work [17] M. Krasner referred to such a factor structure as a quasigraded group, and M. Chadeyras defined analogous structures in the case of rings [4].

1.8 Quasigraduation as an Analog of Graduation

Let us assume that the graduation of the group G' is proper given by $\Delta" = \Delta' \cup \{\omega\}$, where Δ' is the set of grades of that graduation. Let $\kappa : \Delta" \to Sg(G)$ be a mapping such that

$$\kappa(\omega) = C, \quad \text{and} \quad \kappa(\delta') = G_{\delta'}^{(\varphi)}, \quad \text{or} \ \{e\}, \ (\delta' \in \Delta')$$

depending on whether $\delta' \neq 0$ or $\delta' = 0$.

In the case when $x \in H^{(\varphi)}$, but it is not in the core C, x belongs to only one of the subgroups $G_{\delta'}^{(\varphi)}$ and then we define $\delta^{(\varphi)}(x) = \delta'$. In the case when $x \in C$, we define $\delta^{(\varphi)}(x) = \omega$, or 0, depending on whether $x \neq e$ or $x = e$.

Now we will provide a characterization of quasigraded groups in terms of semihomogeneity:

Definition 1.8.1 A pair $(G, H \subseteq G)$, where G is a group and $H \subseteq G$ (referred to as the homogeneous part of the group G), is called a quasigraded group if it satisfies the following conditions:

1° There exists a normal subgroup $C \subseteq H$ of the group G such that $CH = H$.
2° If $x \in H$, then $x^{-1} \in H$.
3° If $x, y, z, xy, yz \in H$ and $y \neq C$ then $xz \in H$.
4° If $x, y \in H$ and $xy \notin H$, then $[x, y] = xyx^{-1}y^{-1} \in C$.
5° H generates group G.
6° If $n \geq 2$ and $x_1, \ldots, x_n \in H$, such that $x_i x_j \notin H$, for $i \neq j$ and $i, j = 1, 2, \ldots, n$, then $x_1 x_2 \cdots x_n \notin C$.

The following proposition provides a characterization of quasigraded groups in terms of homogeneity.

Proposition 1.8.1 *The structure $(H; \cdot)$, where H is the homogeneous part of a quasigraded group G and xy is the induced partial operation on H corresponding to the operation in G, is characterized by the following axioms from the aspect of homogeneity:*

1°° *$x\#x$.*
2°° *If $x\#y$, $y\#z$, and $y \notin C$, then $x\#z$.*
3°° *If $a \notin C$, then xy is defined everywhere on $H(a) = \{x \in H \mid a\#x\}$, and $(H(a); \cdot)$ is a group with $C \subseteq H(a)$ as a normal subgroup, where $C = \{c \in H \mid \text{for all } x \in H, x\#c\}$ is the core of $(H; \cdot)$.*

Definition 1.8.2 The homogeneous part $(H; \cdot)$ of the quasigraded group G is called a groupoid (analogous to a homogroupoid in the case of a group).

If $(H; \cdot)$ and $(H'; \cdot)$ are two groupoids, a quasihomomorphism $\varphi : H \to H'$ can be identified with an element of the multigroupoid:

$$(\Pi_{a \in H} H'_a = H; \cdot) \text{ denoted as } (H'; \cdot)^H \text{ such that}$$
$$\{a \to x_a = \varphi(a) \mid a \in H\}.$$

The elements in $(H'; \cdot)^H$, identified in this way with quasihomomorphisms, are characterized by the condition:

$$a\#b \text{ implies } (xa\#xb \text{ and } x_{ab} = x_a x_b).$$

These defined groupoids, when they are commutative, are denoted *additively* as $(H; +)$ and $(H'; +)$, and the *composability* is called *addibility*. In that case, the set of quasihomomorphisms from H to H' is denoted as $Hom(H, H')$. The set $Hom(H, H')$ is closed under the partial operation (called addition here) in $(H; +)^H$.

Often, a stricter additivity is defined on $Hom(H, H')$ compared to that in $(H'; +)^H$, so that when elements $\varphi, \psi \in Hom(H, H')$, are additively defined, their sum $\varphi + \psi$ is the same as in $(H'; +)^H$.

1.9 Graded Rings in Krasner's Sense

Let $R = (R; +, \cdot)$ be a ring (not necessarily associative) and let

$$\gamma : \Delta \to Sg(R; +), \quad \text{with } \gamma(\delta) = R_\delta \ \ (\delta \in \Delta) \qquad (*)$$

be the graduation of its additive group $(R; +)$.

Definition 1.9.1 The graduation γ is called the graduation of ring R, if

$$\text{for every } \ \xi, \eta \in \Delta, \ \ \text{there exists } \ \zeta \in \Delta \ \text{ such that } \ R_\xi R_\eta \subseteq R_\zeta . \qquad (**)$$

In this case, we say that the ring R is graded.

Remark 1.9.1 Let R be a graded ring with graduation $(*)$. If:

- $R_\xi R_\eta \neq \{0\}$, then ζ from $(**)$ is unique and is denoted $\xi\eta$
- $R_\xi R_\eta = \{0\}$, then $R_\xi R_\eta \subseteq R_\zeta$ for every $\zeta \in \Delta$.

Definition 1.9.2 We say that the grades $\xi, \eta \in \Delta$ are composable and write: $\xi \# \eta$ if $R_\xi R_\eta \neq \{0\}$ and in this case the product of elements ξ and η is denoted $\xi\eta$ and $R_\xi R_\eta \subseteq R_{\xi\eta}$.

In other words, the product $\xi \cdot \eta$ is defined, in the same way as in the previous definition for ξ, η for which $R_\xi R_\eta \neq \{0\}$, but it may not be defined, or it can be arbitrary for ξ, η for which $R_\xi R_\eta = \{0\}$.

Definition 1.9.3 The graduation γ of the ring R is called strict if it is strict as a graduation of its additive graded group $(R; +)$, i.e.:

(i) $\gamma = \gamma^*$ where $\gamma^* = \gamma \mid_{\Delta^*}$, $\Delta^* = \{\delta \in \Delta \mid R_\delta \neq \{0\}\}$
and if
(ii) for every $\xi, \eta \in \Delta$ such that $\xi \# \eta$ implies $R_\xi R_\eta \neq \{0\}$.

Definition 1.9.4 The graduation γ of the ring R is called normal if:

(i) For every $\xi, \eta \in \Delta$, $\xi \# \eta$.
(ii) For every $\xi, \eta \in \Delta$, $R_\xi R_\eta = \{0\}$ implies $\xi \cdot \eta = 0$.

Remark 1.9.2 If R is an associative ring and if:

$$R_\xi R_\eta R_\zeta \neq \{0\} \ \ (\xi, \eta, \zeta \in \Delta), \ \text{ then } \ (\xi\eta)\zeta = \xi(\eta\zeta).$$

However, in the case when $R_\xi R_\eta R_\zeta = \{0\}$, all possible cases can occur, even when graduation γ is strict. It may happen that one of the products $(\xi\eta)\zeta$, $\xi(\eta\zeta)$ is defined while the other is not, or both are defined but unequal, i.e., $(\xi\eta)\zeta \neq \xi(\eta\zeta)$ (M. Chadeyras, [4]).

Definition 1.9.5 The homogeneous part of the graded ring R with the graduation γ is the set $A = h(R) = \cup_{\delta\in\Delta} R_\delta$.

According to this definition, the grades $\delta(x)$ of elements $x \in A$, different from zero, are defined in the same way as in the case of a group.

Let us recall, if the elements $x, y \in A$, are different from 0, then

$$\delta(x) = \delta(y) \text{ if and only if } x + y \in A.$$

For graded rings, the following statement holds:

Proposition 1.9.1 *In a graded ring R, the condition:*

$$(\forall\xi, \eta \in \Delta)\ (\exists\zeta \in \Delta) \textit{ such that } R_\xi R_\eta \subseteq R_\zeta$$

is equivalent to the following two conditions:

I *The product of two homogeneous elements is homogeneous, i.e.,* $A^2 \subseteq A$.
II *If* $x, y \in A$ *and* $xy \neq 0$, *then* $\delta(xy)$ *depends only of* $\delta(x)$ *and* $\delta(y)$.

It can be shown that condition I implies II.

Since the graduations of the graded ring $R = (R; +, \cdot)$ are some of the graduations of its additive group $(R; +)$, they are defined (assuming the multiplication of grades is strict or proper, regardless of the name of the essential grades), using the pair (R, A).

Like in case of graded groups, graded rings may be observed from the semihomogeneous aspect, i.e., the graded ring R may be observed as the ordered pair $(R, A \subseteq R)$, where A satisfies certain axioms.

Now let us switch to the semihomogeneous aspect. The conditions for the subset A of *ring* R under which the pair (R, A) is a graded ring from that aspect are stated in the following proposition:

Proposition 1.9.2 *For the structure* $(A, +, \cdot)$ *to be the homogeneous part of the graded ring R, it is necessary and sufficient for A to satisfy the following conditions:*

1° $0 \in A$.
2° $x \in A$ *implies* $-x \in A$.
3° *If* $x, y, z, x + y, y + z \in A$ *and* $y \neq 0$, *then* $x + z \in A$.
4° *A generates the additive group* $(R, +)$ *of the ring R.*
5° *If* $n \geq 2$, $x_1, x_2, \ldots, x_n \in A$, *and, if for all* $i, j : 1 \leq i, j \leq n, i \neq j$ *for which* $x_i + x_j \notin A$, *then* $x_1 + x_2 + \ldots + x_n \neq 0$.
6° $A^2 \subseteq A$.

Definition 1.9.6 The ordered pair $(R, A \subseteq R)$, where R is the ring, is called the graded ring if A satisfies conditions $1° - 6°$ of the previous proposition.

Definition 1.9.7 For the graded ring $(R; +, \cdot)$, we say it is unitary if it has a homogeneous identity.

According to 6°, multiplicative operation on the ring induces multiplicative operation on A, which is defined everywhere on A. This motivates us to introduce the following notion:

Definition 1.9.8 If $A^* = A \setminus \{0\}$ is a multiplicative group, the graded ring (R, A) is called the graded division ring.

Now the natural questions arise:

1. *What happens in the case when we consider it from the aspect of homogeneity?* *A homogeneous part* A of a ring R should be regarded as a structure with two operations inherited from the ring:
   (a) Partial addition "+," whose composability here is called addibility and denoted by #
   (b) The multiplication operation "·" which is defined everywhere (because $A^2 \subseteq A$)
2. *When* $(A; +, \cdot)$ *is such a structure, is it the homogeneous part of some (nonassociative) graded*[5] *ring?*

   Before answering these questions, we will introduce the notion of aneid.

1.10 Aneids and Corpoids

In homogeneous aspect, our interest lies in observation of structures $(A, +, \cdot)$ which represent homogeneous parts of graded rings $(R, +, \cdot)$ and whose operations are induced by ones of rings R. These structures are called *aneids*, while the corresponding graded rings, which are unique up to A-isomorphism, are called *linearizations of aneids* $(A, +, \cdot)$ and are denoted $\overline{(A, +, \cdot)}$.

The next proposition gives the characterization of aneids, but at the same time it also gives the answer to the questions raised:

Proposition 1.10.1 *For* $(A; +, \cdot)$ *to be a nonassociative aneid – the homogeneous part of a graded ring* R *with partial "addition" and multiplication inherited from that ring – it is necessary and sufficient for the following conditions to hold:*

I *The multiplication on* A *is everywhere defined and there exists an element denoted by* $0 \in A$*, which is called zero, such that*
$0x = x0 = 0$ (for $x \in A$).

II *For all* $x \in A$*:* *1.* $x\#0$*;* *2.* $x\#x$ *and*
3. for all $x, y, z \in A$*,* $x\#y$*,* $y\#z$ *and* $y \neq 0$ *implies* $x\#z$*.*

[5] Here it is implied that the ring is associative.

III *If* $a \neq 0$*, addition is everywhere defined on*

$$A(a) = \{x \in A \mid a\#x\}$$

and $A(a)$ *is an abelian group.*

IV *If* $x\#y$*, then* $zx\#zy$*,* $xz\#yz$*, and* $z(x + y) = zx + zy$*,* $(x + y)z = xz + yz$*.*

Definition 1.10.1 A structure $(A; +, \cdot)$ that satisfies conditions I–IV of the previous proposition is called an aneid. If multiplication is associative, or a nonassociative, aneid is associative or not associative.

Definition 1.10.2 The ring $R = (R; +, \cdot)$ that corresponds to the aneid $(A; +, \cdot)$ is determined up to the A-isomorphism, and we call it the linearization of $(A; +, \cdot)$, denoted by $\overline{(A; +, \cdot)}$ (or simply $\overline{A}$).

Definition 1.10.3 We call the graded ring R a graded field (denoted as Q), if the set A^* of its homogeneous nonzero elements is a multiplicative group. The homogeneous part $(A^*; \cdot)$ is then called a corpoid.

Remark 1.10.1 Axioms II and III (Proposition 1.10.1) can be replaced by weaker axioms II′ and III′, where III is replaced by an appropriate condition for $H(e)$. Corpoides are characterized by the same axiom systems as aneids (axioms I–IV), i.e., hold:

Proposition 1.10.2 *The structure* $(A; +, \cdot)$ *is a corpoid if and only if the following conditions are satisfied:*

I' A^* *is a multiplicative group*[6] *(with a neutral element zero, denoted as* 0*, and a two-sided neutral element* e*).*

II' 1′) $e\#0$; 2′) $e\#e$; 3′) ($x\#e$ *and* $e\#y$) *implies* $x\#y$.

III' *The addition is everywhere defined on*

$$H(e) = \{x \in H \mid e\#x\}$$

and $H(e)$ *is an abelian group with respect to addition.*

IV' *If* $x\#y$*, then* $zx\#zy$*,* $xz\#yz$*,* $z(x + y) = zx + zy$*, and* $(x + y)z = xz + yz$*.*

Definition 1.10.4 The group $A(e)^*$ is called the corpoid group of A. The corpoid is called the corpoid without torsion if its corpoid group is without torsion.

Definition 1.10.5 The aneid A is regular from right (from left) if

$$xz \neq 0,\ yz \neq 0, \text{ and } xz\#yz \quad (zx \neq 0,\ zy \neq 0 \text{ and } zx\#zy) \text{ imply } x\#y,$$

and it is regular if it is both right-regular and left-regular. The aneid is called regular if it is regular from both left and right.

Definition 1.10.6 An associative commutative aneid, which is regular and has no nontrivial divisors, is called the integroid.

Definition 1.10.7 If (R, A) is a graded ring, and $B \subseteq A$ its subobject, then that subobject is called homogeneous if it is generated by $B \cap A$.

---

[6] Graded corps are not viewed as corps by which graduations are defined, but as a special case of a graded corps with a trivial graduation.

The corresponding substructure $B \cap A$ of the aneid $(A; +, \cdot)$ is then called a subaneid, right, left, or two-sided idealoid (or simply an ideal).

Remark 1.10.2 If h is a two-sided idealoid of the aneid $(A; +, \cdot)$, then naturally, we can define the factor structure $(A/h; +, \cdot)$ of A by h, which is also an aneid and represents the homogeneous part of the factor ring $A/\overline{h}$, where A and $\overline{h}$ denote the linearization of the aneid A and the idealoid h, respectively.

For aneids, the concepts of idealoids (right, or two-sided) and properties such as being prime, primary, maximal, etc. are defined similarly as in the case of graded rings, and various results are obtained, such as:

1. A/h has no zero divisors if and only if the idealoid h is prime.
2. A/h is corpoid if and only if $A^2 \not\subseteq h$ and h is a maximal ideal, and so on.

1.11 Direct Sum and Direct Product of Graded Rings

Let (R, A) be a graded ring. We say that it is a direct product or a direct sum of the family $(R_i \mid i \in I)$ of its idealoids if it satisfies this as an abstract ring, and its idealoids are also homogeneous.

However, the support of A, except in trivial cases, is not a direct sum or a direct product limited to the homogeneous parts A_i of R_i.

Just like in the case of graded groups, for a family $\{(R_i, A_i) \mid i \in I\}$ of graded rings and a family $(A_i \mid i \in I)$ of ideals,[7] we can define sets of the direct sums and the direct products, but there is no unique sum that depends only on that family. Therefore, the category of rings is not closed with respect to the direct product, just as it is not closed under the direct product bounded to support.

However, it is possible to define structures called multianeids, whose support is equal to the product of aneids $(A_i; +)$, $i \in I$, where multiplication (everywhere) is defined as follows:

$$(x_i \in A_i \mid i \in I)\,(y_i \in A_i \mid i \in I) = (x_i \cdot y_i \in A_i \mid i \in I).$$

1.12 Quasihomogeneous Homomorphisms of Quasigraded Rings. Agglutinations

Let $(R; A)$ and $(R'; A')$ be two graded rings with the corresponding aneids $A = (A; +, \cdot)$ and $A' = (A'; +, \cdot)$.

[7] So we will call the union of a group and a two-sided zero element.

Definition 1.12.1 A mapping $\varphi : R \to R'$, or $\varphi : A \to A'$, is called a quasihomogeneous homomorphism from (R, A) to (R', A'), or a quasihomomorphism from $(A; +, \cdot)$ to $(A'; +, \cdot)$, if:

(1) φ is a homomorphism of the additive graded groups $(R; +)$ to $(R'; +)$, or a quasihomomorphism from $(A; +)$ to $(A'; +)$.
(2) Multiplicative homomorphism, i.e.,

$$\varphi(xy) = \varphi(x)\varphi(y), \text{ for all } x, y \in A.$$

Definition 1.12.2 The structure $(A^{(\varphi)}; +, \cdot)$ with multiplication induced by the linearization of $R = A$ is defined everywhere on $A^{(\varphi)}$ (and is associative if multiplication on A is associative). We call it an agglutination of the aneid $A = (A; +, \cdot)$ with respect to φ if $(A^{(\varphi)}; +)$ is an "additive" agglutination of the homogenous groupoid $(A; +)$.

These agglutinations are aneids only in some special cases, for example, when $C = N^{(\varphi)} = \{0\}$. In that case, they represent structures called anels.

As motivation for introducing the concept of quasigraduation on the ring R, we used agglutinations of groupoids and agglutinations of aneids.

Definition 1.12.3 For a ring $R = (R; +, \cdot)$ we say that it is quasigraded if there is a given ideal C (called the core) and a family of subgroups $(R_\delta \mid \delta \in \Delta)$ of the additive group of the ring R, such that

$$R/C = \bigoplus_{\delta \in \Delta} (R_\delta / C).$$

Now we will give the definition of a quasigraded ring from the aspect of semihomogeneity and state a proposition that provides necessary and sufficient conditions under which a set $(A; +, \cdot)$ is an anel, i.e., the homogeneous part of a quasigraded ring R.

Definition 1.12.4 An ordered pair $(R; A \subseteq R)$, where R is a ring, is called a quasigraded ring if the set A satisfies the following conditions:

1° There exists an ideal $C \subseteq A$, of the ring R, such that $C + A = A$.
2° $x \in A$ implies $-x \in A$.
3° For all $x, y, z \in A$, $x + y \in A$, $y + z \in A$, and $y \notin C$ imply $x + z \in A$.
4° A generates the group $(R; +)$.
5° If $n \geq 2$ and $x_1, \ldots, x_n \in A$ such that $x_i + x_j \notin A$, for $i \neq j$ $(i, j = 1, \ldots, n)$, then $x_1 + \ldots + x_n \notin C$.
6° $A^2 \subseteq A$.

The concept of anel A for a ring R is analogous to the concept of groupoid H for a group G and can be characterized axiomatically as follows:

Definition 1.12.5 The structure $(A; +, \cdot)$ is an anel if the following conditions are satisfied:

1°° $x \in A$ implies $x\#x$.

$2^{\circ\circ}$ For $x, y, z \in A$: $x\#y$, $y\#z$, $x \neq z$, and $y \notin C$ imply $x\#z$.
$3^{\circ\circ}$ If $a \notin C$, the addition on $A(a) = \{x \in A \mid a\#x\}$ is defined everywhere, and $A(a)$ is an abelian additive group.
$4^{\circ\circ}$ Multiplication is defined everywhere.
$5^{\circ\circ}$ $AC \subseteq C$ and $CA \subseteq C$.
$6^{\circ\circ}$ $x\#y$ implies $zx\#zy$, $xz\#yz$, and $z(x+y) = zx + zy$, and $(x+y)z = xz + yz$, where $C = \{c \in A \mid (\forall x \in A)\ x\#c\}$.

1.13 Applications

It can be proven that proper analogues of commutative fields, from the aspect of extension theory and *Galois theory*, are not commutative corpoids but rather commutative corpoids without torsion. Specifically, thanks to the theory of polynomials, for commutative corpoids without torsion, the theory of extensions closely resembles Steiniz's theory of field extensions, but it is richer because it involves simultaneously extending both the field and its group of units when extending a corpoid. This theory of extensions contains Galois theory, which essential results coincide (with some additional precision) with the corresponding results of Galois field theory, although the proofs are not entirely the same. Thus:

1. The theory of primitive element does not hold for separable extensions of such corpoids.
2. The proof of the second part of Galois' theorem for finite, algebraic, normal extensions of such corpoids is based not only on the analogous theorem for fields but also on the fact that the first Galois cohomology group for the multiplicative group of such field extensions is trivial, as well as on the duality with abelian groups.

If k is a valuation field, then a commutative (if the field k is commutative) corpoid without torsion can be defined, denoted as $s(k)$, which is the synthesis of its residue field $\overline{k}$ (which is a field of $s(k)$) and its valuation group $\Gamma(k)$ (which is the group of essential grades of $s(k)$).

The theory of commutative corpoids without torsion (and, in particular, its theory of extensions and Galois theory) has significant applications in commutative valuation fields, especially in their theory of ramification [12] and [13].

Interestingly, in the case of commutative and even noncommutative (when the group of grades is commutative), Krasner's graded fields (mostly corpoids) are encompassed (as a special case), by the theory of graded rings in the sense of Bourbaki. However, neither Bourbaki nor his works are cited, even though Krasner introduced the concept of corpoids and a large part of the theory related to them as early as 1944 [10] and [11], and he announced and published them together with proofs in works [13] and [14] in 1949 and 1953–1954, respectively, considerably earlier than Bourbaki.

2 Extra- and Para-graded Groups and Rings

There were many significantly open problems that arose naturally in the theory of graded structures, but there is no doubt that for more than twenty years one of the very interesting problems in the theory of graduations has been the problem of closure of the direct product and the direct sum.

Therefore, the aim of the research in my joint work with M. Krasner was to find the graded structures: groups, rings, and modules, which generalize the classical graded structures and, at the same time, have in each of the three cases: the property of closure with respect to the direct sum and the direct product in the sense that the support of the homogeneous subset of the composition (direct sum or direct product) is the restricted direct sum or direct product of the components.

Dreaming of new paths in the theory of graded structures helped M. Krasner and me to overcome seemingly difficult obstacles, guiding us to solve this problem and thus discover and laid the foundations of a fascinating new theory – the most general graded structures called paragraded which generalizes not only the theory of graded structures as is exposed by *Bourbaki* [2], but also the previous results of *M. Krasner* [7], *M. Chadeyras* [4], etc.

This section is dedicated to the nonhomogeneous aspect of the paragraded groups which appeared for the first time in my papers [19–21] and a monograph [22] written together with the famous French mathematician Marc Krasner.

2.1 Extra- and Para-graded Groups

First, we will provide a review of the basic definitions and results that constitute the main motivation for paragraded groups and then we will pass to the introduction of three examples of paragraduations that are not graduations.

This section highlights a nonhomogeneous aspect of the paragraded groups.

Definition 2.1.1 ([22], see also [19, 28]) Let G be a group (multiplicative), with e as neutral element. The mapping

$$\pi\colon \Delta \to Sg(G),\ \pi(\delta) = G_\delta\ (\delta \in \Delta),$$

of a partially ordered set $(\Delta, <)$, which is from bellow a complete semi-lattice and from above inductively ordered, to the set $Sg(G)$ of subgroups of the group G is called a paragraduation if it satisfies the following six-axiom system:

(i) $\pi(0) = G_0 = \{e\}$, where $0 = \inf\{\Delta\}$; $\delta < \delta' \Rightarrow G_\delta \subseteq G_{\delta'}$.

Remark 2.1.1

r1. $H = \cup_{\delta\in\Delta} G_\delta$ is called the homogeneous part of the group G with respect to π, and elements $x \in H$ are called homogeneous elements of G.

r2. If $x \in H$ and $\Delta(x) = \{\delta \in \Delta \mid x \in G_\delta\}$, we say that $\delta(x) = \inf\{\Delta(x)\}$ is a grade of x. We have $\delta(x) = 0$ iff $x = e$. Elements $\delta(x), x \in H$, are called the principal grades and they form a set denoted by Δ_p which is called the principal part of Δ. Restriction $\pi\mid_{\Delta_p}$ is called the proper kernel of π.

(ii) If $\theta \subseteq \Delta$, then $\bigcap_{\delta\in\Delta} G_\delta = G_{\inf\theta}$.
(iii) If $x, y \in H$ and $yx = zxy$, then $z \in H$ and $\delta(z) \leq \inf(\delta(x), \delta(y))$.

Remark 2.1.2 $z = yxy^{-1}x^{-1}$ is a commutator $[y, x]$ of the elements y and x, more precisely denoted by $z(x, y)$.

(iv) The homogeneous part H is a generating set of G.
(v) Let $A \subseteq H$ be a subset such that for all $x, y \in A$ there exists an upper bound for $\delta(x), \delta(y)$. Then there exists an upper bound for all $\delta(x), x \in A$;
vi) G is generated by H with the set **R** of H-inner and left commutation relation:

- $xy = z$ (H-inner relations).
- $yx = z(x, y)xy$ (left commutation relations).

The group with paragraduation is called paragraded group.

Definition 2.1.2 ([12] see also [19]) The mapping π from Definition 2.1.1 which satisfies axioms (i)-(v) and instead vi)

(vi′) If $\delta_1, \dots, \delta_s \in \Delta_p$ are pairwise incomparable and if $x_i, x_i' \in H$, $i = 1, \dots, s$ are such that
$x_1x_2\dots x_s = x_1'x_2'\dots x_s'$ *and* $x_i, x_i' \in G_\delta$, $\forall i = 1, \dots, s$, then $\delta(x_i^{-1}x_i') < \delta_i$

is called extragraduation.

The group with extragraduation is called an extragraded group.

Remark 2.1.3 Axiom (v) is equivalent to the axiom:

(v ′) *Let $A \subseteq H$ be a subset such that for all $x, y \in A$, we have $xy \in H$. Then there exists $\delta \in \Delta$ such that $A \subseteq G_\delta$.*

Later we will see that every extragraduation is a paragraduation.

Definition 2.1.3 ([22]) Let $g \subseteq H$ be a subgroup of G. Such a subgroup g is called the subgroup of the homogeneous part H and the set of all subgroups of the homogeneous part H is denoted by $Sg(H)$.

Definition 2.1.4 The subgroup $g \in Sg(H)$ is called saturated with respect to π if

$$x \in g \Rightarrow G_{\delta(x)} \subseteq g.$$

Remark 2.1.4 ([22]) Every subgroup G_δ, $\delta \in \Delta$, is saturated.

Now we will introduce some notions terminology that we will use in the proof of Theorem (2.1.1).

Definition 2.1.5 Every finite sequence $x_1, \dots, x_s$ of the elements from H or elements which inverses are from H is called a word of H.

Remark 2.1.5 Let m be the word of H. We can associate to word m the product of its elements (in a given order) in G. That product is denoted by (m).

Definition 2.1.6 If $m = x_1 \dots x_s$ is a word, then the array

$$d(m) = (\delta(x_1), \dots, \delta(x_s))$$

is called the array of grades of that word.

Definition 2.1.7 The word $x_1 \dots x_s$ is called reduced if for every pair x_i, x_j there exists no common upper bound of elements $\delta(x_i), \delta(x_j), i \neq j$ (in other words, if for all $i, j = 1, \dots, s, i \neq j$, we have $x_i x_j \notin H$).

The following theorem is very important.

Theorem 2.1.1 (Krasner–Vuković [22], see also [35, 38]) *An extragraded group is a paragraded group, i.e., an extragraded group G is generated by H with set relations* **R**.

Proof Let P be the set of elements of the form $x_1 \dots x_s, x_i \in H$, observed as words of H. We will assume that two words from P are equal if they are equal as elements of G. Let $(\text{mod } R)$ be the congruence in P and C some class of that congruence. It is obvious that all words from C are equal and that two classes are equal if their words are equal.

Suppose that $m = x_1 \dots x_s$ is a word and that $d(m) = (\delta(x_1), \dots, \delta(x_s))$ is the array of grades of that word.

Let $m = x_1 \dots x_s \in C$ be the shortest word. We claim that it is reduced. Assume otherwise, i.e., assume that there exist elements $x_i, x_j, i < j$, such that $x_i x_j \in H$.

Let

$$x_i x_q = x'_q x_i, \text{ for } i < q < j .$$

Hence, $x'_q \in H$ and

$$x_1 \dots x_{i-1} x_i x_{i+1} \dots x_j \dots x_s \equiv x_1 \dots x_{i-1} x'_{i+1} x_i x_{i+2} \dots x_j \dots x_s \ (\text{mod } R),$$

$$x_1 \dots x_{i-1} x_i x_{i+1} \dots x_j \dots x_s \equiv x_1 \dots x_{i-1} x'_{i+1} \dots x'_q x_i x_{q+1} \dots x_j \dots x_s \ (\text{mod } R),$$

and, for $q = j - 1$

$$x_1 \dots x_{i-1} x_i x_{i+1} \dots x_j \dots x_s \equiv x_1 \dots x_{i-1} x'_{i+1} \dots x'_{j-1} x_i x_j \dots x_s \ (\text{mod } R) .$$

Since $x_i x_j \in H$, the last word is congruent modulo R to the word $y_1 \dots y_{s-1}$, which length is less than s, where

$$y_q = x_q, \ \textit{if} \ 1 \leq q < i, \ y_q = x'_{q+1}, \ \textit{if} \ i \leq q < j - 1, \ \text{and}$$
$$y_{j-1} = x_i x_j, \ y_q = x_{q-1}, \ \textit{if} \ j \leq q < s .$$

Thus, in case when m is not a reduced word, it does not belong to C and is not the shortest one.

On the other hand, let us notice that for word $m \in C$ there exists a word $m' \in C$ which array of grades $d(m')$ is obtained from the array $d(m) = (\delta_1, \dots, \delta_s)$ of the word $m = x_1 \dots x_s$, where $\delta_i = \delta(x_i)$, by applying an arbitrary permutation on it.

Indeed, if

$1 \le i < s$ and $x_i x_{i+1} = x'_{i+1} x_i$, then the word $m = x_1 \dots x_{i-1} x'_{i+1} x_i x_{i+2} \dots x_s$,

has an array of grades $(\delta_1, \dots, \delta_{i-1}, \delta'_{i+1}, \delta_i, \delta_{i+2}, \dots, \delta_s)$ which is obtained from $d(m)$ by transposing arbitrary neighboring elements δ_i, δ_{i+1}. The set

$$\Delta(m) = \{\delta(x_i) \mid i = 1, \dots, s\} \textit{ is called the set of grades of } m.$$

Let us prove that equally reduced words $m = x_1 \dots x_s$ and $n = y_1 \dots y_t$ have the same length, i.e., that $s = t$ and that the corresponding sets of grades $\Delta_1 = \Delta(m)$ and $\Delta_2 = \Delta(n)$ are totally ordered in sense that the corresponding elements are comparable.

Let Δ^* be the set of maximal elements of $\Delta_1 \cup \Delta_2$. Observe some total ordering on Δ^*, with $(\delta_1^*, \dots, \delta_u^*)$ as the array of elements of Δ^* is written with respect to the ordering.

If $\delta \in \Delta_1 \cup \Delta_2$ is not from Δ^*, then there exist elements $\delta_i^* \in \Delta^*$ which are an upper bound of δ. Let $\delta^*(\delta)$ be the first of those upper bounds.

If $\delta, \delta' \in \Delta_1 \cup \Delta_2$, but $\delta, \delta' \notin \Delta^*$, then we cannot have $\delta^*(\delta) = \delta^*(\delta')$. Namely, for otherwise, δ and δ' would be from one of the sets Δ_1, Δ_2 and would have a common upper bound, which is impossible, since m and n are reduced words.

Let $\delta^*(\delta) = \delta$ if $\delta \in \Delta^*$. Aline the elements $\delta \in \Delta_1$ in the same way as the corresponding $\delta^*(\delta)$ are and do the same for the elements of the set Δ_2. We notice that there exist words

$$m' = x'_1 \dots x'_s \equiv m \pmod R \quad \text{and} \quad n' = y'_1 \dots y'_s \equiv n \pmod R$$

which arrays of grades $d(m')$ and $d(n')$ are the arrays of grades of elements of the set Δ_1, respectively Δ_2 and $m' = n'$.

If $1 \le j \le u$,

put $\overline{x}_j = x'_i$, *if* $\delta^*(\delta(x'_i)) = \delta_j^*$ *and put* $\overline{x}_j = 1$, *if such i does not exist*,

and then, in the same way, define elements $\overline{y}_j$.

Since the grades of Δ^* are elements of Δ_p and pairwise incomparable, grades $\delta_1^*, \dots, \delta_u^*$ and elements $\overline{x}_j, \overline{y}_j, j = 1, \dots, u$ satisfy the assumptions vi') of Definition 2.1.2. On the other hand, since $\delta_j^* \in \Delta_1 \cup \Delta_2$, one of the common upper bounds of elements $\delta(\overline{x}_j)$, $\delta(\overline{y}_j)$ is δ_j^*, and hence, $\delta(\overline{x}_j^{-1}\overline{y}_j) < \delta_j^*$. However, if one of the elements $\overline{x}_j, \overline{y}_j$ is equals e, then

$$\delta(\overline{x}_j^{-1}\overline{y}_j) = \max\{\delta(\overline{x}_j),\ \delta(\overline{y}_j)\} = \delta_j^*,$$

contrary to the previous inequality.

Hence, none of the elements $\overline{x}_j, \overline{y}_j$ equals e, and every δ_j^* has the form $\delta^*(\delta)$, for some $\delta \in \Delta_1$. Thus, the mapping $\delta \to \delta^*(\delta)$ is surjection; hence, bijection $\Delta_1 \to \Delta^*$ and $s = u$. From the reasons, $t = u$. We conclude that $s = t$. It is clear that the corresponding elements of sets Δ_1, Δ_2, alined in this way are comparable.

Assume there exist equal words which are not congruent modulo R. According to already proven, there exist reduced words of the same length for which the corresponding elements of their arrays of grades are comparable.

Let $m = x_1 \dots x_s$ and $m' = x'_1 \dots x'_s$ be two such words with shortest length. Then s ≥ 2, since two words of length 1 are equal only if they coincide. Put $\delta_i =$

$\max\{\delta(x_i), \delta(x_i')\}$. Then the words m and m' satisfy $vi\,'$) of Definition 2.1.2 for array $\delta_1, \ldots, \delta_s$. Hence,

$$\text{if } y_1 = x_1^{-1}x_1' \text{ and if we put } n = x_1x_2\ldots x_s \text{ and } n' = y_1x_2'\ldots x_s'$$

then, since

$$m = m',\ \textit{we have}\ n = n',\ \textit{but}\ n \not\equiv n' \ (\text{mod}\, R),$$

for otherwise, we would have $m \equiv m'$ (mod R), because

$$m \equiv x_1 n \ (\text{mod}\, R)\ \ \textit{and}\ \ m' \equiv x_1 n' \ (\text{mod}\, R),\ \ \textit{hence},\ \ n = x_1^{-1}m\ \ \textit{and}\ \ n' = x_1^{-1}m\text{'}.$$

The word n is reduced with length $s - 1$.

The word $n' = n$ cannot be reduced, since it has length $s > s - 1$, but it can be proven that there exists a word $n'' \equiv n'$(mod R), which is reduced, of length $s - 1$, for which one can assume that the corresponding elements of arrays $d(n)$ and $d(n'')$ are comparable. Contradiction, since the least length of a word is s. □

This theorem leads us to the conclusion that every extragraduation is at the same time a paragraduation. The converse statement is obviously not true (see [22]).

It is known that there is a large class of extragraduations (paragraduations) which are graduations [22]. However, there are examples of paragraduations which are not graduations.

First, we provide two examples of classes of paragraduations that are graduations [22].

2.2 *Examples of Extra- (Para-) graduations that are and that are Not Graduations*

Example 2.2.1 Assume that $\Delta = \Delta^* \cup \{0\}$ is the set in which all elements $\delta \in \Delta^*$ are greater than 0 and pairwise incomparable. We thus have

$$\inf(\delta, \delta') = 0, \quad \text{for all}\ \ \delta \neq \delta'\ (\delta, \delta' \in \Delta^*).$$

Hence,

$$G_\delta \cap G_{\delta'} = \{e\}\ \textit{and, for}\ x \in G_\delta,\ \ y \in G_{\delta'},\ \ \delta(z(x, y)) = 0 \Rightarrow z(x, y) = e,\ \ \textit{i.e.},\ \ yx = xy.$$

Also, if $\delta_1, \delta_2, \ldots \delta_n$ are distinct elements of Δ^*, they are pairwise incomparable and if

$$x_1x_2\ldots x_n = x_1'x_2'\ldots x_n'\ \ (x_i, x_i' \in H)\ \ \textit{and}\ \ \delta(x_i), \delta(x_i') < \delta_i\ \ (i = 1, \ldots, n),$$

then, according to Definition 2.1.2 $vi\,'$), we have

$$\delta(x_i^{-1}x_i') < \delta_i \Rightarrow \delta(x_i^{-1}x_i') = 0,\ \ \textit{and}\ \ x_i^{-1}x_i' = e\ \Rightarrow x_i = x_i'.$$

This means that G is a direct sum

$$G = \bigoplus_{\delta\in\Delta^*} G_\delta = \bigoplus_{\delta\in\Delta} G_\delta$$

thus meaning that such extragraduation is a graduation which empty grades are assumed to be the grades of e.

Conversely, a graduation which has all homogeneous elements with grades is an extragraduation with appropriate set of grades.

Example 2.2.2 Assume that $\Delta = \Delta^* \cup \{\omega, 0\}$, where all elements $\delta \in \Delta^*$ are greater than ω, pairwise incomparable where $\omega > 0$. Hence, if elements $\delta, \delta' \in \Delta^*$, are distinct, then

$$\inf(\delta, \delta') = \omega, \quad \text{so,} \quad G_\delta \cap G_{\delta'} = G_\omega,$$

and hence

$$\text{if } x \in G_\delta,\, y \in G_{\delta'},\ \delta(z(x, y)) \leq \omega \Rightarrow z(x, y) \in G_\omega.$$

So, we have that G_ω commutes with all elements of G_δ, for every $\delta \in \Delta^*$, and so does with all elements of $H = \cup_{\delta \in \Delta^*} G_\delta$, and since H is a generating set of G, it commutes with all elements of G. This implies that G_ω is normal subgroup of G.

On the other hand, all elements $x, y \in G$ commute modulo G_ω. Hence,

$$G/G_\omega = \bigoplus_{\delta \in \Delta^*} (G_\delta/G_\omega).$$

This means that G is a quasigraded group which center is $C = G_\omega$ and

$$\pi : \Delta_p \to Sg(G) \text{ its quasigraduation.}$$

On the other hand, it is obvious that an extragraduation of quasigraded group has considered form.

Example 2.2.3 Let A be a ring and observe the ring of upper triangular matrices:

$$R = \begin{pmatrix} A & A \\ A & 0 \end{pmatrix},$$

and let

$$R_{\delta_1} = \begin{pmatrix} A & 0 \\ 0 & 0 \end{pmatrix},\ R_{\delta_2} = \begin{pmatrix} 0 & A \\ 0 & 0 \end{pmatrix},\ R_{\delta_3} = \begin{pmatrix} A & A \\ 0 & 0 \end{pmatrix},\ R_{\delta_4} = \begin{pmatrix} 0 & 0 \\ 0 & A \end{pmatrix}.$$

If

$$R_0 = \begin{pmatrix} 0 & 0 \\ 0 & 0 \end{pmatrix},$$

denote by Δ the set $\{0, \delta_1, \delta_2, \delta_3, \delta_4\}$ where for convenience, $\delta_0 = 0$. The set Δ is a partially ordered set, which is from below complete semi-lattice and from above inductively ordered, with respect to

$$\delta_i < \delta_j \Leftrightarrow R_{\delta_i} \subseteq R_{\delta_j} \quad i, j = 0, 1, 2, 3.$$

It is easy to see that the mapping

$$\pi : \delta_i \to R_{\delta_i},\, \delta_i \in \Delta\,, \text{ is a paragraduation,}$$

but it is not a graduation, since, for instance,

$$R_{\delta_1} \cap R_{\delta_3} = R_{\delta_1} \neq R_0.$$

From these examples we saw that there are paragraduations which are and which are not graduations.

However, each graduation is a special paragraduation. Namely, it is sufficient that the set of empty grades is well ordered, taking the empty grade 0 as its first element and after all of them to put mutually incomparable essential grades of the given graduation.

2.3 Direct Sum and Direct Product of Paragraded Groups

For the family of the paragraded groups, the direct product and the direct sum, in the natural way, are defined. Those are again paragraded groups, and the direct sum is a homogeneous subgroup of the direct product.

The direct product and the direct sum of the paragraded groups are extragraded groups, if and only if all of the factors are extragraded groups. However, the direct product of the graded groups is the only extragraded group, except if it is at most one of the factors with nontrivial graduation, that is $\Delta_\alpha \neq \{0\}$.

Now we will introduce the direct product and the direct sum of paragraded groups.

Let $\{(G_\alpha, \cdot, \pi_\alpha) \mid \alpha \in A\}$ be a family of paragraded groups with paragraduations $\pi_\alpha : \Delta_\alpha \to Sg(G_\alpha)$, $\alpha \in A$, where $(\Delta_\alpha, <)$ is the set of grades of $(G_\alpha, \cdot, \pi_\alpha)$ and e_α its neutral element.

Let us observe

$$\text{the direct product } G = \prod_{\alpha\in A} G_\alpha \text{ and the direct sum } \bigoplus_{\alpha\in A} G_\alpha$$

of paragraded groups $(G_\alpha | \ \alpha \in A)$, respectively, with $e = (e_\alpha | \ \alpha \in A)$ neutral element.

Obviously, both, the direct product and the direct sum, are groups with respect to the operation "$\cdot$," defined for

$$x = (x_\alpha \mid \alpha \in A),\ y = (y_\alpha \mid \alpha \in A) \in G \ \ with \ \ x \cdot y = (x_\alpha \cdot y_\alpha \mid \alpha \in A).$$

Next, we focus on question whether these groups are paragraded and according to the following theorem (formulated in [22] without proof, see also [35]), they are.

Theorem 2.3.1 *Groups* $\prod_{\alpha\in A} G_\alpha$ *and* $\bigoplus_{\alpha\in A} G_\alpha$ *are paragraded with*

$$\pi : \Delta = \prod_{\alpha\in A} \Delta_\alpha \to Sg\Big(\prod_{\alpha\in A}(G_\alpha, +)\Big),$$

where

$$\pi(\delta) = (\pi_\alpha(\delta_\alpha) \mid \alpha \in A) = \prod_{\alpha\in A} \pi_\alpha(\delta_\alpha),$$

as their paragraduation mappings.

Before passing on to the proof of the aforementioned theorem, we will introduce some more notions and prove the following lemma:

Lemma 2.3.1 *Let* $\Delta = \prod_{\alpha\in A} \Delta_\alpha$ *be the direct product of sets* $\Delta_\alpha, \alpha \in A$, *ordered as follows: if* $\delta = (\delta_\alpha \in \Delta_\alpha \mid \alpha \in A)$, $\delta' = (\delta'_\alpha \in \Delta_\alpha \mid \alpha \in A) \in \Delta$, *then*

$$\delta \leq \delta' \Leftrightarrow \delta_\alpha \leq \delta'_\alpha \; (\forall \alpha \in A) \, .$$

The set $(\Delta, <)$ *is then as ordered as* Δ_α, $\alpha \in A$.

Proof Indeed, if $\overline{\Delta} \subseteq \Delta$, then

$$\{\inf \{\delta_\alpha \mid \delta = (\delta_\beta \mid \beta \in A) \in \overline{\Delta}\} \mid \alpha \in A\} = \inf_{\delta \in \overline{\Delta}} \delta,$$

and if m_α, $\alpha \in A$, is the upper bound of

$$\{\delta_\alpha \mid \delta = (\delta_\beta \mid \beta \in A) \in T\} \, ,$$

where T is the chain in Δ, then $m = (m_\alpha \mid \alpha \in A)$ is the upper bound of T. □

Let us go back to proof of Theorem 2.3.1.

Proof It remains to prove that mapping $\pi : \Delta \to Sg(G)$ satisfies the axioms (i)–(vi) of Definition 2.1.1.

$$\pi(0_\alpha \in \Delta_\alpha \mid \alpha \in A) = \Pi_{\alpha \in A} \, \pi_\alpha(0_\alpha) = \Pi_{\alpha \in A} \, e_\alpha = (e_\alpha \mid \alpha \in A) = e \, .$$

Let $\delta = (\delta_\alpha \in \Delta_\alpha \mid \alpha \in A)$, $\delta' = (\delta'_\alpha \in \Delta_\alpha \mid \alpha \in A) \in \Delta$ and $\delta < \delta'$. Then

$$\delta_\alpha < \delta'_\alpha \; (\alpha \in A), \quad \text{and thus } \; \pi_\alpha(\delta_\alpha) \subseteq \pi_\alpha(\delta'_\alpha), \; \alpha \in A.$$

Hence,

$$\pi(\delta) \subseteq \pi(\delta')$$

and thus (i) holds.

Δ is a partially ordered set of grades, which is from below a complete semi-lattice and from above inductively ordered and

$$\bigcap_{\delta_\alpha \in \theta_\alpha} G_{\delta_\alpha} = G_{\inf \theta_\alpha} \; \text{ because } \; \theta = \prod_{\alpha \in A} \theta_\alpha \subseteq \Delta,$$

and thus (ii) holds as well.

Let $x, y \in H = \Pi_{\alpha \in A} H_\alpha$ and

$$yx = zxy, \; x = (x_\alpha \mid \alpha \in A), \; y = (y_\alpha \mid \alpha \in A), \; z = (z_\alpha \mid \alpha \in A).$$

Then we have

$$y_\alpha x_\alpha = z_\alpha x_\alpha y_\alpha \;\; (\alpha \in A),$$

and thus

$$z_\alpha \in H_\alpha \; (\alpha \in A) \; \textit{and} \; \delta(z_\alpha) \leq \inf(\delta(x_\alpha), \delta(y_\alpha)) \; (\alpha \in A).$$

It follows that

$$\delta(z) \leq \inf(\delta(x), \delta(y)),$$

which means that (iii) is satisfied.

Since the sets H_α are generating sets of G_α, we have that H is the generating set of G. Thus, (iv) holds.

Let $A \subseteq H$ be a subset such that

$$xy \in H, \; x = (x_\alpha \mid \alpha \in A), \; y = (y_\alpha \mid \alpha \in A) \in H.$$

Hence, $x_\alpha y_\alpha \in H_\alpha$ $(\alpha \in A)$, and thus, there exists $\delta_\alpha \in \Delta_\alpha$ such that $A_\alpha \subseteq G_{\delta_\alpha}$. Thus, we have

$$A \subseteq G_\delta, \; \delta = \Pi_{\alpha \in A} \, \delta_\alpha,$$

and $v') \Leftrightarrow v)$ is satisfied.

Every group G_α is generated by H_α with a set of relations $\boldsymbol{R}$, which means that the same set of relations generates G. □

We will now state and prove one of the main properties of paragraded groups.

Theorem 2.3.2 *The homogeneous part H of the direct product and the direct sum of paragraded groups is the direct product and the direct sum of corresponding homogeneous parts, respectively.*

Proof Really, let H_α be a homogeneous part of G_α, $\alpha \in A$. Then we have

$$H = \cup_{\delta\in\Delta} \pi(\delta) = \cup_{\delta\in\Delta} (\pi_\alpha(\delta_\alpha) \mid \alpha \in A) = \cup_{\delta\in\prod_{\alpha\in A}\Delta_\alpha} (\pi_\alpha(\delta_\alpha) \mid \alpha \in A)$$
$$= \prod_{\alpha\in A} \cup_{\delta_\alpha\in\Delta_\alpha} (\pi_\alpha(\delta_\alpha)) = \prod_{\alpha\in A} H_\alpha ,$$

which is contrary to the category of graded groups [14]. □

2.4 Paragraded Rings

In this section we will first gather up a few less known necessary notions from the theory of paragraded rings. We will start with the definition of the paragraded ring:

Definition 2.4.1 ([22, 28]) A ring R is called paragraded ring if there exists a mapping

$$\pi : \Delta \to \mathrm{Sg}(R, +), \ \pi(\delta) = R_\delta \ (\delta \in \Delta),$$

of a partially ordered set $(\Delta. <)$ which is from below complete semi-lattice and from above inductively ordered, to the set $\mathrm{Sg}(R, +)$ of subgroups R_δ of additive group $(R, +)$, called paragraduation if the following axioms are satisfied:

(r1) $\pi(0) = R_0 = \{0\}$, where $0 = \inf \Delta$; $\delta < \delta' \Rightarrow R_\delta \subseteq R_{\delta'}$.

Remark 2.4.1 $A = \bigcup_{\delta\in\Delta} R_\delta$ is called *the homogeneous part* of R with respect to π, and elements of A are called *homogeneous elements*.

Remark 2.4.2 If $x \in A$, we say that $\delta(x) = \inf\{\delta \in \Delta \mid x \in R_\delta\}$ is the *degree of* x. We have $\delta(x) = 0$ if and only if $x = 0$. Elements $\delta(x)$, $x \in A$, are called *principal degrees* and they form a set which is denoted by Δ_p.

(r2) $\theta \subseteq \Delta \Rightarrow \bigcap_{\delta\in\theta} R_\delta = R_{\inf\theta}$.
(r3) Homogeneous part A is a generating set of $(R, +)$ with the set of A-inner relations: $x + y = z$.
(r4) Let $B \subseteq A$ be a subset such that for all $x, y \in B$ there exists an upper bound for $\delta(x), \delta(y)$. Then there exists an upper bound for all $\delta(x)$, $x \in B$.
(r5) For all $\xi, \eta \in \Delta$ there exists $\zeta \in \Delta$ such that $R_\xi R_\eta \subseteq R_\zeta$.

As usual, we denote $\Delta^* = \Delta \backslash \{0\}$.

It can easily be seen that the paragraded ring can be introduced via the paragraded group as follows:

Definition 2.4.2 ([22]) The ring $(R, +, \cdot)$ is called paragraded if its additive group $(R, +)$ is a paragraded group, with paragraduation π and set of grades Δ, and if

$$(\forall \xi, \eta \in \Delta)\ (\exists \zeta \in \Delta) \text{ such that } R_\xi R_\eta \subseteq R_\zeta, \tag{$**$}$$

where $R_\delta = \pi(\delta)$.

Remark 2.4.3 From this definition follows the existence of a binary operation on Δ:

$$\xi\eta = \sup\{\delta(x) \mid x \in R_\xi R_\eta\},$$

called the minimal multiplication [19, 22, 28], and so, $R_\xi R_\eta \subseteq R_\zeta$.

Definition 2.4.3 ([22]) If R is a paragraded ring with paragraduation π, then the mapping $(\xi, \eta) \to \xi\eta$ from $\Delta \times \Delta$ to Δ is called Δ-multiplication of grades if the following holds:

1. $R_\xi R_\eta \subseteq R_{\xi\eta}$.
2. $(\forall \xi, \xi', \eta, \eta' \in \Delta)\ \xi \leq \xi'$ and $\eta \leq \eta' \Rightarrow \xi\eta \leq \xi'\eta'$.

Remark 2.4.4 If R is a paragraded ring with paragraduation π, then there exists

$$\zeta = \sup\{\delta(x) \mid x \in R_\xi R_\eta\}.$$

If we put $\zeta = \xi\eta$, we will get Δ-multiplication and we call it the minimal multiplication [22].

Lemma 2.4.1 ([22, 28]) *If A is a homogeneous part of the additive group $(R, +)$ of paragraded ring $(R, +, \cdot)$ and $\delta(x)$ the grade of an element $x \in A$, then $(**)$ holds if and only if the following conditions are satisfied:*

(i) $x \in A,\ y \in A \Rightarrow xy \in A$ $(A^2 \subseteq A)$.
(ii) *If $\xi, \eta \in \Delta$, then for $x, x', y, y' \in A$ for which $\delta(x) \leq \xi$, $\delta(x') \leq \xi$, $\delta(y) \leq \eta$, $\delta(y') \leq \eta$, we have $xy + x'y' \in A$.*

Proof It is clear that conditions (i) and (ii) follow from $(**)$. Conditions (i) and (ii) imply that $R_\xi R_\eta \subseteq A$ and that for all $z, z' \in R_\xi R_\eta$ we have $z + z' \in A$. Hence, according to (r5), there exists an element $\zeta \in \Delta$ such that $R_\xi R_\eta \subseteq R_\zeta$. □

Remark 2.4.5 In the case of graded rings, analogous of conditions (i) and (ii) are not equivalent. It may only be stated that (i) implies (ii) [4].

Lemma 2.4.2 ([22]) *If π is a paragraduation of the additive group $(R, +)$ of paragraded ring $(R, +, \cdot)$, then the following conditions are equivalent:*

(b) $(\forall \xi, \xi', \eta, \eta' \in \Delta)\ \xi \leq \xi'$ *and* $\eta \leq \eta' \Rightarrow \xi\eta \leq \xi'\eta'$.
(b′) $(\forall x, x', y, y' \in A)\ x + x' \in A$ *and* $y + y' \in A \Rightarrow xx' + yy' \in A$.

Proof It is clear that (b') follows from (b). Let us assume that (b) holds and that elements $x + x' \in A$, $y + y' \in A$, for $x, x', y, y' \in A$, satisfy assumptions of (b'). According to (r4), there exist $\xi, \eta \in \Delta$ such that $\{x, x'\} \subseteq R_\xi$, and $\{y, y'\} \in R_\eta$, and by (b), $xy + x'y' \in A$. □

Remark 2.4.6 Let $(R_j)_{j\in J}$ be a family of paragraded rings with paragraduations

$$\pi_j : \Delta_j \to \mathrm{Sg}(R_j, +),$$

and let us observe the direct product $\prod_{j\in J} R_j$ and the direct sum $\bigoplus_{j\in J} R_j$ of that family. If

$$x = \{x_j \mid j \in J\} \ \textit{and} \ \ y = \{y_j \mid j \in J\},$$

then we put

$$x + y = \{x_j + y_j \mid j \in J\} \ \textit{and} \ \ x \cdot y = \{x_j \cdot y_j \mid j \in J\}.$$

Thus, we obtain that both the direct product and the direct sum of this family of rings are rings.

The following theorem shows that both of these rings are paragraded.

Theorem 2.4.1 ([18, 19]) *Rings* $\prod_{j\in J} R_j$ *and* $\bigoplus_{j\in J} R_j$ *are paragraded with respect to the mapping:*

$$\pi : \Delta = \prod_{j\in J} \Delta_j \to \mathrm{Sg}(\prod_{j\in J} R_j, +),\ \pi(\delta) = (\pi_j(\delta_j) \mid j \in J),\ (\delta = (\delta_j \mid j \in J))$$

as their paragraduation.

Proof It is easy to prove that axioms of Definition 2.4.1. are satisfied for the mapping π. □

Remark 2.4.7 Contrary to the category of graded rings, homogeneous part A of the direct product and the direct sum of paragraded rings is the direct product and the direct sum of corresponding homogeneous parts, respectively [14, 18, 19].

Really, if A_j is a homogeneous part of a paragraded ring R_j, $j \in J$, then we have

$$\begin{aligned} A &= \bigcup_{\delta\in\Delta} \pi(\delta) = \bigcup_{\delta\in\Delta} (\pi_j(\delta_j) \mid j \in J) \\ &= \bigcup_{\delta\in\prod_{j\in J}\Delta_j} (\pi_j(\delta_j) \mid j \in J) = \prod_{j\in J} \bigcup_{\delta_j\in\Delta_j} \pi_j(\delta_j) \\ &= \prod_{j\in J} A_j\,. \end{aligned}$$

As we saw in Chapter I the characterization axioms of graded structures (groups, rings, modules) in Krasner's sense give way to three study methods in principle equivalent. Analogous aspects exist for the most general graded structures introduced by M. Krasner and myself, called paragraded (or Krasner–Vuković's) structures (groups, rings, modules) which, for rings, can be expressed as follows:

1. *Nonhomogeneous* where R is considered as a ring with a determined paragraduation.
2. *Semihomogeneous*, where R is considered as a pair (R, A) of a ring R and its homogeneous part $A \subseteq R$ where a paragraduation π of R can be reconstructed up to an equivalence.
3. *Homogeneous* where a paraanneid $(A; \cdot)$ is considered and where its linearization $\overline{(A; \cdot)}$ can be reconstructed, i.e., the paragraded ring for which $(A; \cdot)$ is the homogeneous part.

First, we will now explain what a linearization is and introduce a paraanneid.

Remark 2.4.8 Let A be a homogeneous part of some paragraded ring R. If A is also a homogeneous part of some other paragraded ring R', the rings R and R' are A - *isomorphic*, i.e., a ring with the homogeneous part A is determined (up to an A-isomorphism), and when it occurs, it will be denoted by $\overline{A}$ and called *a linearization of* A (see [22, 28]).

A paragraded ring R from a *semihomogeneous* aspect will be the object of the next part of section, i.e., when R is considered as the pair (R, A) *of the paragraded ring* R, *and its homogeneous part* $A(\subseteq R)$ and perhaps the paragraduation π of R is reconstructed up to the equivalence.

Definition 2.4.4 Let R be a paragraded ring and A its homogeneous part with operations (addition and multiplication) restricted from R. Then, for induced addition, which is partial,

$$x\#y \Leftrightarrow x + y \in A \quad (\forall x, y \in A). \tag{1}$$

The obtained structure we call a paraanneid [22, 28] and for elements $x, y \in A$, for which is $x\#y$, we say that they are addible.

Proposition 2.4.1 *If* $x \in A$, *let* $A(x) = \{y \in A \mid x\#y\}$. *Paraanneid* A *satisfies the following axioms:*

(a1) *There exists an element* $0 \in A$ *such that* $A = A(0)$ *and such that for all* $x \in A$ *we have* $0 + x = x$.
(a2) *If* $a \in A$, $x + y$ *is always defined on* $g(a) = \{x \in A \mid A(x) \supseteq A(a)\}$, *and* $(g(a), +)$ *is an abelian subgroup.*
(a3) *If* $B \subseteq A$ *is such that for all* $x, y \in B$, *we have* $x\#y$, *then there exists a subset* $g \subseteq A$ *such that* $x + y \in g$, *for all* $x, y \in g$, $x \in g \Rightarrow g(x) \subseteq g$ *and* $B \subseteq g$.
(a4) $A^2 \subseteq A$.
(a5) $x\#y$ *and* $x'\#y' \Rightarrow xx'\#yy'$ $(\forall x, x'y, y' \in A)$.

Definition 2.4.5 Structure $(A, +, \cdot)$ which satisfies axioms (a1)–(a5) is called a **quasianneid** [21, 22].

Remark 2.4.9 Quasianneid does not have to be a paraanneid, but it is under a few more assumptions (see [21, 22]) in which case A can be *linearized* [22] to a paragraded ring, denoted by $\overline{A}$, which homogeneous part it is.

Remark 2.4.10 $(R, +, \cdot)$ is the paragraded ring with respect to some paragraduation π and the homogeneous part A; as we have seen, the addition and multiplication defined on R induce on A the same operations which we denote in the same way.

Since $(R, +, \cdot)$ is a ring and, since according to (a4)

$$x\#y \Rightarrow (\forall z \in A)\ zx\#zy \ \ and \ \ xz\#yz,$$

we also have

(a6) For every $x, y, z \in A$,

$$x\#y \Rightarrow z(x + y) = zx + zy \ \ and \ \ (x + y)z = xz + yz.$$

We can now extend the previous Definition 1.9.1 of a paragraded ring and so we will get:

Definition 2.4.6 The paragraded ring $(R, +, \cdot)$ is a commutative, or associative, or Lie ring if $(A, +, \cdot)$ satisfies the following conditions:

Commutativity: $xy = yx$ $(\forall x, y \in A$.

Associativity: $(xy)z = x(yz)$ $(\forall x, y, z \in A)$.

Lie conditions:

L1 $x^2 = 0$ $(\forall x \in A)$.

L2 $xy = -yx$ $(\forall x, y \in A)$.

L3 elements $(xy)z$, $(yz)x$, $(zx)y$ $(\forall x, y, z \in A)$ are mutually addible and we have

$$(xy)z + (yz)x + (zx)y = 0,$$

respectively.

Definition 2.4.7 A quasianneid is called commutative, associative, or Lie if it satisfies conditions of commutativity, associativity, or L1–L3, respectively.

Remark 2.4.11 In order for $A \subseteq R$ to be a homogeneous part of a ring R with respect to some paragraduation π of a ring $(R, +, \cdot)$ it is necessary and sufficient that A is a homogeneous part of a commutative group $(R, +)$ and that conditions (a) and (b′) (Definition 2.4.3, Lemma 2.4.2) are satisfied for A.

In other words, the following theorem holds:

Theorem 2.4.2 *Let $(R; A \subseteq R)$ be a given pair of a paragraded ring R and its subset A. A is a homogeneous part of a paragraded ring R with respect to some paraduation π if and only if the system of the following five axioms is satisfied:*

1° *For all $a \in A$, $R(x) = \{y \in A \mid A(y) \supseteq A(x)\}$ is a subgroup of the additive group $(R, +)$.*

2° *If $B \subseteq A$ is such that $x + y \in A$, for all $x, y \in B$, then there exists a subgroup g of the additive group $(R, +)$, such that $B \subseteq g \subseteq A$, hence saturated.*

3° *$(R, +)$ is a commutative group which is generated by A with the set of relations: $R_j = \{x + y - z = 0 \mid x, y, z \in A,\ x + y = z \in (A, +)\}$.*

4° $A^2 \subseteq A$.
5° *For all* $x, y, x', y' \in A$, *such that* $x + x', y + y' \in A$, *we have* $xy + x'y' \in A$.

2.5 Open Questions

As we have already pointed out, for a long time one of the most interesting questions in the theory of graded structures (groups, rings, modules) was the question of when these structures are closed with respect to the direct product and the direct sum. Marc Krasner and I answered that question in the late 1980s, in three Krasner–Vuković's Proceedings of Japan Academy notes [19–21] and a monograph [22].

In this way we developed our theory of paragraded structures which generalizes not only the theory of graded structures as is exposed in Bourbaki [2], but also the previous results of M. Krasner [17] and M. Chadeyras [4].

But there are still many important questions in the theory of (para) graded structures, such as:

(i) Examples of graded structures in algebra and modular theory (perhaps quantum groups?)
(ii) Examples of paragraded structures and their applications (see [33–35] and [23])
(iii) Relations with algebraic geometry (direct image, inverse image)
(iv) Sheaves of paragraded modules, *End* and *Hom*
(v) Graded algebra of p-adic modular form
(vi) Radicals of paragraded rings (see [33–35], and [37])
(vii) Paragraded groups and profinite groups
(viii) The connection between Krasner–Vuković' s paragraded structures (groups, rings, modules) and Lie super-algebras

But the most challenging question is perhaps a question pointed out by Alexei Panchishkin in his comment [27] related to the project of Noèmi Combe:

"Noémie Combe's project for the CIRM Conference 2024, related to her earlier joint work with Yu. I. Mannin on Frobenius varieties, used Eisenstein's series and perhaps Eisenstein's measure. A further project after the CIRM Conference 2024 extending Yuri Manin's work on δ-rings and their p-adic applications comes from the paragraded rings by Krasner-Vuković, introduced in [22], and extended by Mirjana Vuković in Brown-McCoy and large Brown-McCoy radicals of paragraded rings [40]."

References

1. Archimedes, Works (translated by Heath T.L.). Cambridge University Press, Cambridge (1897)
2. Bourbaki, N.: Algèbre, Chap. I, Hermann et Cie, Paris, 2e edit. et 3e edit. (1958)

3. Chevalley, C.: The Construction and Study of Certain Important Algebras. Publ. of the Math. Soc. of Japan, Tokyo (1955)
4. Chadeyras, M.: Essai d'une théorie noetherienne homogène pour les annaux commutatifs dont la graduation est aussi générale que possible. Bull. Soc. Math. France Suppl. Mém. **22**, 1–143 (1970)
5. Descartes, R.: La géometrie. (imp. Leyde I. M.), 1637
6. Euclid: Elements (transl. Heath J. L.), 2e edit. Dover Publications, Nev York (1956)
7. Euler, L.: Arithmetique universelle. Saint Pétersbourg (1768–1769)
8. Halberstadt, E.: Théorie artinienne homogène des anneaux gradués non commutatifs a grades réguliers (Thèse de doctorat). Université Pierre et Marie Curie, Paris (1971)
9. Heath, T.L.: A History of Greek Mathematics, vol. 1. Oxford (1931)
10. Krasner, M.: Une généralisation de la notion de corps-corpoïde, Un corpoïde, remarquable pour la théorie des corps valués. C. R., Acad. Sci., Paris, Sér. I, Math. **219**, 345–347 (1944)
11. Krasner, M.: Hypergroupes moduliformes et extramoduliformes. C.R. Acad. Sci., Paris, Sér. I, Math. **219**, 473–476 (1944)
12. Krasner, M.: Théorie de la ramification dans les exensions finies des corps valués: hypergroupe d'inertie et de ramification; Théorie extrinsèque de la ramification. C.R. Acad. Sci. Paris, Sér. I, Math. **220**, 28–30 (1945)
13. Krasner, M.: Quelques méthodes nouvelles dans la théorie des corps valués complets, Algèbre et théorie des nombres, Colloque de l'Institut du CNRS. No. 24, Paris (1949), Édit. CNRS, Paris (1950)
14. Krasner, M.: Congruences multiplicatives. Squelettes et corpoïdes, Sém. Krasner 1953–54, Vol.1, exp. No.4, Secrétariat Math. de la Faculté des Sci., Paris
15. Krasner, M.: Théorie élémentaire des corpoïdes commutatifs sans torsion, Séminaire Krasner 1953–54, Vol. 2, exp. No 5, Secrétariat Math. de la Faculté des Sci., Paris,1956
16. Krasner, M.: Théorie de Galois des corpoïdes commutatifs sans torsion et ses applications à la théorie de la ramification des extensions algébriques des corps valués. Preprint de l'Université de Paris VI (1978)
17. Krasner, M.: Anneaux gradués généraux, Colloque d'algèbre, pp. 209–308. Université Renns 1 (1980)
18. Krasner, M.: Le vieux qui est neuf, pp. 443–472. Revue Roumain de Mathématiques pures et appliquées, XXVII (1982)
19. Krasner, M., Vuković, M.: Structures paragraduées (groupes, anneaux, modules) I. Proc. Jpn. Acad., Ser. A **62**(9), 350–352 (1986). https://projecteuclid.org/euclid.pja/1195514019
20. Krasner, M., Vuković, M.: ibd. II: 62 (1986), Ser. A, No.10, pp. 389–391. https://projecteuclid.org/euclid.pja/1195514064
21. Krasner, M., Vuković, M.: ibd. III: 63 (1987), Ser. A, No. 1, pp. 10–12. https://projecteuclid.org/euclid.pja/1195514122
22. Krasner, M., Vuković, M.: Structures paragraduées (groupes, anneaux, modules) (monograph), Queen's Papers in Pure and Applied Mathemat. No.77, pp. 1–163. Queen's University, Kingston, Ontario (1987)
23. Курош, А.Г.: Теория групп, издат. третье, дополненное, Наука, Москва (1967)
24. Ribenboim, P.: Il mondo Krasneriano, Queen's Preprint No. 1983, 12, pp. 1–158. Queen's University Kingston, Ontario (1987)
25. Samuel, P., Zarisky, O.: Commutative Algebra, vol.2. D.Van Nostrand, Prinston (1960)
26. Serre, J.P.: Arbres, amalgames, SL_2, Astérisque No. 46. Soc. Math. de France, Paris (1977)
27. Panchihkin, A.: Further Development on Krasner-Vuković Paragraded Structures and p-adic Interpolation of Yubo Jin L-values. Sarajevo J. Math. **20**(33), No.1, 13–23 (2024). https://sjm.anubih.ba/index.php/sjm/article/view/474/472
28. Vuković, M.: Structures graduées et paragraduées, pp. 1–40. Prepublication de l'Inst. Fourier, Univ. Grenoble I, No. 536, Grenoble (2001). https://www-fourier.univ-grenoble-alpes.fr/sites/default/files/ref_536.pdf
29. Vuković, M.: Krasner's and Vuković's Paragraduations. In: International Conference, pp. 432–436. MIT. 2009, Kopaonik, 27–31.08.2009, Zbornik radova. http://www.yt3w.info/zbornik.pdf

30. Vuković, M.: History of graded structures (Lectures at postgraduate studies), University of East Sarajevo 2008 and University of Sarajevo (2010)
31. Vuković, M.: (& E.Ilić Georgijević). Primary Decomposition of General Graded Structures. Buletunul Acad. de Ştiinţe a Republicii Moldova. Matematica, 1, 77, 87–96 (2015). https://www.math.md/files/basm/y2015-n1/y2015-n1-(pp87-96).pdf
32. Vuković, M.: (& E.Ilić Georgijević). A Note on Radicals of Paragraded Rings. Sarajevo J. Math. **12**(25), No. 2, Suppl. 307–316 (2016). https://sjm.anubih.ba/index.php/sjm/article/view/172/171
33. Vuković, M.: (&E.Ilić Georgijević). A Note on General Radicals of Paragraded Rings. Sarajevo J. Math. **12**(25), No. 2, Suppl. 317–324 (2016)
34. Vuković, M.: (& E.Ilić Georgijević). The Wedderburn–Artin Theorem for Paragraded Rings. J. Math. Sci., **221**(3), 391–400 (2017) (Translation from Fundam. Prikl. Mat., Moscow, T 19, No. 6, 125–139 (2014)). https://link.springer.com/article/10.1007/s10958-017-3233-7
35. Vuković, M.: From Krasner's Corpoid and Bourbaki's Graduations to Krasner's Graduations and Krasner-Vuković's Paragraduations. Sarajevo J. Math. **14**(27), No. 2, 175–190 (2018). https://sjm.anubih.ba/index.php/sjm/article/view/72/71
36. Vuković, M.: On noncommutative paragraded rings. Sarajevo J. Math. **16**(1), 5–11 (2020). https://sjm.anubih.ba/index.php/sjm/article/view/33/32
37. Vuković, M.: Radicals of Paragraded Rings. dedicated to A.V. Mikhalev Радикалы, Springer's J. Math. Sci., Vol. **275**, No. 4, 379-392 (2023) (Translation from Fund. Prikl. Mat., Moscow, T. 24, No. 2, pp. 3–22 (2022)). https://link.springer.com/article/10.1007/s10958-023-06689-3?fromPaywallRec=true
38. Vuković, M.: Panoramic view of graded structures from Euler and Bourbaki-Krasner to Krasner – Vuković. Springer's J. Math. Sciences, vol. 283, No. 6 (2024), pp. 838–848 (Translation from Fund. Prikl. Mat., T. 24 (2023), No. 3, pp. 23–37. http://mech.math.msu.su/~fpm/ps/k23/k233/k23303.pdf
39. Vuković, M., Ilić-Georgijević, E.: Paragraded Rings and their ideals. Springer's J. Math. Sci. **191**(5), 654–660 (2013) (Translation from Fund. Prikl. Mat., T. 17 (2011/12), No. 4, pp. 83–93). http://mech.math.msu.su/~fpm/ps/k1112/k114/k11405.pdf
40. Vuković, M.: Brown-McCoy and large Brown-McCoy Radicals of Paragraded Rings accepted for publication)

Part III
Applications in (Natural and Life) Sciences

Transformation Semigroups and Their Applications

Katarzyna Pichór and Ryszard Rudnicki

Abstract In this chapter we present transformation semigroups and their applications. We begin with Klein's approach to geometry based on invariants of transformation groups. Then, we present symmetry groups in chemistry and in classical mechanics. Next we introduce one-parameter semigroups of transformations and their applications in ergodic theory. Our main subject are one-parameter semigroups of operators, in particular stochastic semigroups. We present general results on their existence and long-time behavior. We also give examples of one-parameter semigroups related to Markov chains, diffusion, and processes with jumps. We focus on the applications of semigroups of operators in biology. Among other things, we study models of DNA evolution; growth of erythrocyte population; gene expression; cell cycle; and the movement of bacteria and insects. We also consider models with stochastic noise and different population models.

Keywords Transformation semigroup · Stochastic semigroup · Markov process · Asymptotic stability · Biological model

1 Introduction

Transformation semigroups are used today in many scientific fields because they allow a unified description of various natural processes and convenient formulation of theorems concerning their properties. The intensive study of semigroups began in the early twentieth century as a generalization of the notion of a group. The concept of a group was introduced by Évariste Galois in the 1830s to study polynomial equations. The abstract definition of a group was formulated by Arthur Cayley [19]

K. Pichór
Institute of Mathematics, University of Silesia in Katowice, Katowice, Poland
e-mail: katarzyna.pichor@us.edu.pl

R. Rudnicki (✉)
Institute of Mathematics, Polish Academy of Sciences, Katowice, Poland
e-mail: ryszard.rudnicki@us.edu.pl

M. Mitrović, M. N. Hounkonnou (eds.), *Algebra Without Borders*, Mathematics in Mind, https://doi.org/10.1007/978-3-031-86477-3_7

in 1854, but, as in the case of Galois, it referred only to finite groups. In 1872 Felix Klein in his famous the *Erlangen program* presented a novel approach to geometry based on the invariants of transformation groups (Sect. 2.2).

Transformation groups are used to study symmetries of some geometrical or chemical objects. In Sect. 2.3 we present a short introduction to molecular and crystal symmetries. The idea of using transformation groups has been successfully applied to describe Newtonian mechanics [3]. In Sect. 2.4 we present the group of Galilean transformations which, in combination with Newton's principle of determinacy, allow the derivation of equations of motion in inertial systems and the fundamental laws of classical mechanics.

One-parameter semigroups of transformations of some space X are called dynamical systems. Different types of dynamical systems are studied depending on the structure of the space X. For example, if X is a topological space, then we study continuous dynamical systems on X and if X is a differentiable manifold—smooth dynamical systems. In ergodic theory measure-preserving dynamical systems are considered. Such systems consist of measurable transformations on a probability space (X, Σ, μ) invariant with respect to the measure μ [21, 47, 49, 64]. Properties of dynamical systems can be studied using their representations on the space $L^1(X, \Sigma, \mu)$. In Sect. 2.5 we introduce a semigroup of Frobenius–Perron operators on L^1, and we show the relations between ergodic properties of a dynamical system and asymptotic behavior of the semigroup of Frobenius–Perron operators.

Among the transformation semigroups, the one-parameter semigroups of operators play an important role, as they are used to describe and study processes that change over time. They are usually associated with evolution equations of the form $x'(t) = Ax(t)$, where A is some linear (often unbounded) operator in a Banach space. We introduce in Sect. 3.1 the notion of C_0-semigroups of operators, and we show their relation with evolution equations.

Our main subject are stochastic semigroups. Stochastic semigroups are semigroups of positive linear operators defined on the space L^1 preserving the set of densities. These semigroups are generated by partial differential equations of different types and describe the behavior of the distributions of Markov processes like diffusion processes, piecewise deterministic Markov processes (PDMPs), and hybrid stochastic processes [80]. They are also used to study ergodic properties of dynamical systems [49]. In Sect. 3.2 we introduce the notions of stochastic and substochastic semigroups, and we state general theorems on their existence. In the following sections, we present basic examples of stochastic semigroups related to Markov chains (Sect. 3.3), diffusion (Sect. 3.4), processes with jumps (Sect. 3.5), processes with random switching (Sect. 3.6), and stochastic billiards (Sect. 3.7).

Long-time behaviour of semigroups of operators is studied in Part 4. We begin with a continuous version of the classical Perron–Frobenius theorem (Sect. 4.1). This theorem concerns semigroups of linear operators on the space $\mathbb{R}^n$. The study of such semigroups is a good starting point for introducing asymptotic properties of semigroups in Banach spaces and provides simple examples. One of the tools in studying the behavior of a quasi-compact semigroup of operators when $t \to \infty$ is its spectral decomposition (see Sect. 4.2). In particular, if the semigroup $\{T(t)\}_{t\geq 0}$ on a Banach lattice E is also positive and irreducible, then there exist $r \in \mathbb{R}$, nonzero $x^* \in E$, and $\alpha \in E^*$ such that

$$\lim_{t\to\infty} e^{-rt}T(t)x = x^*\alpha(x) \quad \text{for } x \in E. \tag{1}$$

Asymptotic properties of stochastic semigroups are studied in the next four sections. We begin with a theorem concerning uniformly continuous stochastic semigroups and with the lower function theorem of Lasota and Yorke [50] (see Sect. 4.3). The main results concern stochastic semigroups bounded from below by integral operators called partially integral semigroups (see Sect. 4.4). If such a semigroup $\{P(t)\}_{t\geq 0}$ has a unique and positive density f^*, then it is asymptotically stable, i.e., $P(t)f \to f^*$ for each density f. Strengthening slightly the condition for partial integrability, we obtain a theorem about the decomposition of a semigroup into asymptotically stable parts and a sweeping part (see Sect. 4.5). This theorem and its conclusions are convenient for the study of many advanced stochastic semigroups. We show in Sect. 4.6 that if a stochastic semigroup has some compact properties, then it has also an asymptotic decomposition.

Part 5 is devoted to applications in biological models. Our aim is to demonstrate both the wealth of applications of semigroups in biological models and the usefulness of the methods presented for their study. Among other things, we study models of DNA evolution; growth of erythrocyte population; population growth with stochastic noise; gene expression; and the movement of bacteria and insects. We also study structured population models and more advanced models based on semi-Markov processes—kangaroo movement and a cell cycle model [76].

2 Transformation Semigroups Everywhere

2.1 Basic Definitions and Examples

A *semigroup* $(\mathbf{G}, \cdot)$ is an arbitrary set $\mathbf{G}$ together with a binary operation $\cdot$ on $\mathbf{G}$ with the property of associativity, i.e., $(x \cdot y) \cdot z = x \cdot (y \cdot z)$ for all $x, y, z \in \mathbf{G}$. An element $e \in \mathbf{G}$ is called *identity* if $e \cdot x = x \cdot e = x$ for all $x \in \mathbf{G}$. From the property of associativity, it follows that a semigroup has at most one identity element. A *monoid* is a semigroup with an identity. An element x of the monoid $(\mathbf{G}, \cdot, e)$ has two-sided inverse $x^{-1} \in \mathbf{G}$, shortly called *inverse*, if $x \cdot x^{-1} = x^{-1} \cdot x = e$. A monoid is called a *group* if each element of $\mathbf{G}$ has an inverse. A semigroup $(\mathbf{G}, \cdot)$ is called *commutative* if $x \cdot y = y \cdot x$ for all $x, y \in \mathbf{G}$.

A semigroup of elements acting as transformations of some space X is called a *transformation semigroup* or a *semigroup action on a set*. In this case the binary operation on $\mathbf{G}$ is the composition $\circ$ of transformations and the identity element is the identity transformation I.

A *topological semigroup* $(\mathbf{G}, \cdot)$ is a topological space that is also a semigroup such that the semigroup operation $(x, y) \mapsto x \cdot y$ is continuous. If $(\mathbf{G}, \cdot)$ is a group, then it is called a *topological group* if we additionally assume that the inversion map $x \mapsto x^{-1}$ is continuous.

One-parameter semigroup of transformations is a family $\mathbf{G} = \{g_t\}_{t\in T}$ of transformations of a set X such that T is one of the sets $\mathbb{N} = \{0, 1, \dots\}$, $\mathbb{Z}$, $\mathbb{R}$, $\mathbb{R}_+ = [0, \infty)$; $g_s \circ g_t = g_{s+t}$ for $s, t \in T$ and $g_0 = I$. If X is a topological space and the map $\varphi\colon T \times X \to X$ given by $\varphi(t, x) = g_t(x)$ is continuous, then family $\mathbf{G}$ is called a C_0*-semigroup*. If $T = \mathbb{N}$, then there is a function g such that g_t is the t-th iterate of g, and if $T = \mathbb{Z}$, then additionally $g_{-t} = g_t^{-1}$ for $t > 0$. If $T = \mathbb{N}$ and $g\colon X \to X$ is a continuous function, then $\mathbf{G}$ is a C_0-semigroup; if $T = \mathbb{Z}$ and g is a homeomorphism of X, then $\mathbf{G}$ is a C_0-group. One-parameter semigroup of transformations is also called a *dynamical system*. In the theory of dynamical systems, a one-parameter semigroup is called *endomorphism* if $T = \mathbb{N}$; *automorphism* if $T = \mathbb{Z}$; *flow* if $T = \mathbb{R}$; and *semiflow* if $T = \mathbb{R}_+$. If X has some additional structure, then a dynamical system should be compatible with this structure, e.g., if X is equipped with a topology, then $\mathbf{G}$ should be a C_0-semigroup.

Transformation semigroups provide a rich source of many examples and counterexamples of the concepts introduced earlier. The basic examples are linear maps of $\mathbb{R}^n$ or $\mathbb{C}^n$. Such a map is of the form $x \mapsto Ax$ for some $A \in \mathrm{M}^n(K)$, where $\mathrm{M}^n(K)$ denotes the set of all $n \times n$ matrices over the field $K = \mathbb{R}$ or $K = \mathbb{C}$, respectively. Since the composition of linear maps corresponds to the matrix multiplication, the set $\mathrm{M}^n(K)$ with operation of multiplication and with identity matrix is a monoid. Since $\mathrm{M}^n(K)$ is isomorphic to K^{n^2} and multiplication is a continuous function, $\mathrm{M}^n(K)$ is a topological semigroup. The subset of $\mathrm{M}^n(K)$ that consists of all invertible matrices denoted by $\mathrm{GL}(n, K)$ is a topological group because the operation $A \mapsto A^{-1}$, $A \in \mathrm{GL}(n, K)$, is continuous. By $\mathrm{O}(n)$ we denote the subset of $\mathrm{GL}(n, \mathbb{R})$ that consists of all *orthogonal matrices*, i.e., A is orthogonal if $A^{\mathrm{T}} = A^{-1}$, where A^{T} is the transpose of A. The set $\mathrm{O}(n)$ is also a topological group. Transformations corresponding to orthogonal matrices are linear isometries of $\mathbb{R}^n$.

The simplest example of a C_0-semigroup is a translation semigroup on $\mathbb{R}$ or $\mathbb{R}_+$. The transformations are given by the formula $g_t(x) = x + t$, where $t \in \mathbb{R}$ if $X = \mathbb{R}$ and $t \ge 0$ if $X = \mathbb{R}_+$. Observe that if $X = \mathbb{R}$, then the translation semigroup $\{g_t\}$ is a flow and if $X = \mathbb{R}_+$ it is a semiflow. Another useful example of a C_0-semigroup is the rotation group on a unit circle S^1 on a complex space $\mathbb{C}$. This group consists of all transformations of the form $g_t(z) = ze^{it}$, $t \in \mathbb{R}$.

More general example of a C_0-semigroup can be defined by using systems of ordinary differential equations. Consider a moving particle in an open set $X \subset \mathbb{R}^d$. We assume that if a particle is at point x, then its velocity is $b(x)$, where $b\colon X \to \mathbb{R}^d$ is a differentiable function. It means that if $x(t)$ is its position at time t, then the function $x(t)$ satisfies the following equation:

$$x'(t) = b(x(t)). \tag{2}$$

We also assume that the particle does not leave the set X. For each $\bar{x} \in X$, we denote by $\pi_t \bar{x}$ the solution $x(t)$ of (2) with the initial condition $x(0) = \bar{x}$. Then, $\{\pi_t\}_{t \geq 0}$ is a semiflow on X. If additionally the solution $x(t)$ exists and $x(t) \in X$ for all $t \in \mathbb{R}$, then $\{\pi_t\}_{t \in \mathbb{R}}$ is a flow on X.

In Sect. 2.5 we introduce measure-preserving dynamical systems which are one-parameter semigroups but may not be C_0-semigroups. Such systems play an essential role in ergodic theory.

An important role in the study of properties of algebraic structures plays the notion of homomorphism. If $(\mathbf{G}_1, \cdot)$ and $(\mathbf{G}_2, \circ)$ are semigroups, then a *homomorphism* from $\mathbf{G}_1$ to $\mathbf{G}_2$ is a map $h \colon \mathbf{G}_1 \to \mathbf{G}_2$ such that

$$h(g_1 \cdot g_2) = h(g_1) \circ h(g_2) \quad \text{for all } g_1, g_2 \in \mathbf{G}_1. \tag{3}$$

If $\mathbf{G}_1$ and $\mathbf{G}_2$ are monoids, then we additionally assume that a homomorphism maps the identity element of $\mathbf{G}_1$ to the identity element of $\mathbf{G}_2$. If $\mathbf{G}_1$ and $\mathbf{G}_2$ are groups and h satisfies (3), then the homomorphism h maps the identity to the identity and $h(g^{-1}) = (h(g))^{-1}$ for all $g \in \mathbf{G}_1$.

We now give an example of a homomorphism from a semigroup without linear structure to a semigroup defined on a linear space. We will use the presented construction in Sect. 2.5. Let X be any set and let $(\mathbf{G}, \circ)$ be the semigroup which consists of all functions from X to X with composition as the binary operation: $(\varphi, \psi) \mapsto \varphi \circ \psi$. Let $V = F(X, \mathbb{R})$ be the set of all maps from X to $\mathbb{R}$ and let $\mathbf{L} = L(V, \mathbb{R})$ be the set of all linear operators on the space V. The space $\mathbf{L}$ is a linear space, but it is also a semigroup with the binary operation: $(P, Q) \mapsto Q \circ P$. Given a function $\varphi \colon X \to X$, we define a linear operator $U_\varphi \in \mathbf{L}$ by the formula $(U_\varphi f)(x) = f(\varphi(x))$ for $f \in V$. Then

$$U_{\varphi \circ \psi} = U_\psi \circ U_\varphi,$$

which means that the map $h \colon \mathbf{G} \to \mathbf{L}$ given by $h(\varphi) = U_\varphi$ is a homomorphism. The homomorphism h is a representation of the group $\mathbf{G}$. Generally, if $(\mathbf{G}, \cdot)$ is an arbitrary semigroup, V is a linear space and $\mathrm{End}(V)$ is the set of all linear transformations (endomorphisms) $V \to V$, and $h \colon \mathbf{G} \to \mathrm{End}(V)$ is a homomorphism, then h is called a *representation of the semigroup* $\mathbf{G}$ on the vector space V.

The operator U_φ is called the *Koopman operator* or the *composition operator* related to φ. Usually the set X has an additional structure. If, for example, (X, Σ) is a measurable space, then we can consider a subgroup of $\mathbf{G}$ consisting of all measurable transformations. Then, the space V can be the set of all measurable functions from X to $\mathbb{R}$.

If instead of the above semigroup $\mathbf{G}$ we consider the group $(\mathrm{S}(X), \circ)$ of all invertible functions from X onto X, then the homomorphism h transfers $\mathrm{S}(X)$ into the set of invertible linear transformations $\mathrm{GL}(V)$, $V = F(X, \mathbb{R})$, and the homomorphism h is a representation of the group $\mathrm{S}(X)$. In general, if $(\mathbf{G}, \cdot)$

is an arbitrary group, V is a linear space, $\mathrm{GL}(V)$ is the set of bijective linear transformations from V to V, and $h\colon \mathbf{G} \to \mathrm{GL}(V)$ is a homomorphism, then h is called a *representation of the group* $\mathbf{G}$ on the vector space V.

2.2 Transformation Groups in Geometry

As we have mentioned in Introduction, different geometries can be classified by transformation groups. Consider a given group $\mathbf{G}$ of transformations of the space $\mathbb{R}^3$. Klein proposed that a geometry related to the group $\mathbf{G}$ consists of the objects (properties or figures) which, as a result of the action of any transformation $S \in \mathbf{G}$, transfer to objects of the same type. *Euclidean geometry* corresponds to the *isometric transformations* (translations, rotations, reflections, and their compositions). We recall that any isometric transformation is given by the formula $Sx = Ax + b$, where $b \in \mathbb{R}^3$ and the matrix A is orthogonal. The objects of Euclidean geometry are, for example, the distance, perpendicularity and parallelity of straight lines, circles, and preservation of angles.

If we replace the group of isometries by the group of *similarities*, i.e., the transformations of the form $Sx = cAx + b$, where $c \neq 0$ is a real constant and A is an orthogonal matrix, then we obtain another geometry that Euclid actually dealt with. This geometry has the same aforementioned objects as Euclidean geometry with the exception of distance.

Affine geometry is the geometry of affine transformations, i.e., the maps of the form $Sx = Ax + b$, where A is any invertible linear transformation. The objects in affine geometry are parallelity of straight lines and ratios of lengths.

More advanced is the *projective geometry* which is induced by homographies. A point in the *projective space* X of dimension n is a one-dimensional subspace of the vector space $\mathbb{R}^{n+1}$. If $\mathbf{x} = [x_0, x_1, \dots, x_n]$ is a nonzero element of $\mathbb{R}^{n+1}$, then

$$[x_0 : x_1 : \cdots : x_n] = \{[\theta x_0, \theta x_1, \dots, \theta x_n]\colon \theta \in \mathbb{R}\}$$

is a one-dimensional subspace of $\mathbb{R}^{n+1}$ spanned by the vector $\mathbf{x}$. Let

$$\mathbb{R}_0^{n+1} = \mathbb{R}^{n+1} \setminus \{\mathbf{0}\}, \quad \mathbb{R}_{0,\bullet}^{n+1} = \mathbb{R}^{n+1} \setminus \{\mathbf{x} \in \mathbb{R}^{n+1}\colon x_0 = 0\}.$$

Then, we can consider a map $P\colon \mathbb{R}_0^{n+1} \to X$ given by

$$P([x_0, x_1, \dots, x_n]) = [x_0 : x_1 : \cdots : x_n].$$

Let $A\colon \mathbb{R}^{n+1} \to \mathbb{R}^{n+1}$ be an invertible linear map. A *homography*, called also *projective transformation*, induced by A is a mapping f_A from X to X given by

$$f_A([x_0 : x_1 : \cdots : x_n]) = [(A\mathbf{x})_0 : (A\mathbf{x})_1 : \cdots : (A\mathbf{x})_n].$$

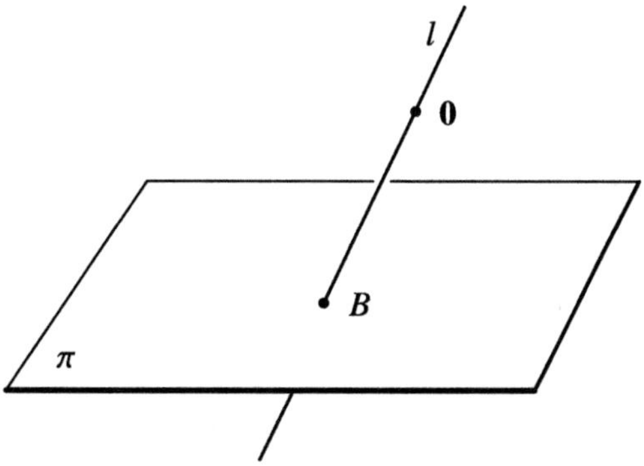

Fig. 1 The map Q transfers the line l to the point B

If $[a_{ij}]$ is a matrix of the map A, then it is called the *matrix of the homography* f_A. The following diagram commutes:

$$\begin{array}{ccc} \mathbb{R}_0^{n+1} & \xrightarrow{A} & \mathbb{R}_0^{n+1} \\ {\scriptstyle P}\downarrow & & \downarrow{\scriptstyle P} \\ X & \xrightarrow[f_A]{} & X, \end{array}$$

i.e., $f_A \circ P = P \circ A$. If V is a linear subspace of dimension $m+1 \leq n$ in $\mathbb{R}^{n+1}$, then $P(V \setminus \{\mathbf{0}\})$ is called a *flat* of dimension m in X. In particular the lines in X are images under P of two-dimensional linear subspaces. From the formula $f_A \circ P = P \circ A$, it follows that a homography transfers a flat on a flat of the same dimension.

We can also introduce a projective space in a way that is closer to geometrical intuition. Let π be the hyperplane $\{\mathbf{x} \in \mathbb{R}^{n+1} : x_0 = 1\}$. Fix a point $\mathbf{x} \in \mathbb{R}^{n+1}_{0,\bullet}$. Then, the vector $\overrightarrow{\mathbf{x0}}$ spans the line $l = [x_0 : x_1 : \cdots : x_n]$. Denote by B the intersection of l and π (see Fig. 1). In this way we obtain a bijective transformation of the set of all lines which are not parallel to the hyperplane π onto the space $\mathbb{R}^n$ (we omit 1—the first coordinate of the point B). The projective space Y is defined by adding the set Y_∞ of "points at infinity" to $\mathbb{R}^n$, which corresponds to lines parallel to the hyperplane π. Thus $Y = \mathbb{R}^n \cup Y_\infty$. In one-dimensional case Y_∞ is a single point, so we can write $Y = \mathbb{R} \cup \{\infty\}$. If $n = 2$, then $\mathbb{R}^2$ and Y_∞ can be identified with a complex plane $\mathbb{C}$ and with a semicircle at infinity, respectively, so we can write $Y_\infty = \{\infty \cdot e^{it} : t \in [0, \pi)\}$. For arbitrary n the set Y_∞ is an $(n-1)$-dimensional hemisphere at infinity, precisely, the hemisphere is formed by identification of antipodal points on the sphere.

Let $\mathbf{x} \in \mathbb{R}^{n+1}_{0,\bullet}$. Then $B = \left(1, \frac{x_1}{x_0}, \ldots, \frac{x_n}{x_0}\right)$, and we can define the map $Q \colon \mathbb{R}^{n+1}_{0,\bullet} \to \mathbb{R}^n$ by the formula $Q(x_0, x_1, \ldots, x_n) = \left(\frac{x_1}{x_0}, \ldots, \frac{x_n}{x_0}\right)$. It is not difficult to check that if V is an $(m+1)$-dimensional subspace of $\mathbb{R}^{n+1}$ such that $V \setminus \{\mathbf{0}\} \subset \mathbb{R}^{n+1}_{0,\bullet}$, then $Q(V \setminus \{\mathbf{0}\})$ is an m-dimensional flat in $\mathbb{R}^n$, i.e., $Q(V \setminus \{\mathbf{0}\}) = W + \mathbf{b}$, where W is an m-dimensional subspace of $\mathbb{R}^n$ and $\mathbf{b} \in \mathbb{R}^n$.

In this model the homography f_A restricted to the set $\mathbb{R}^n$ is given by the formula:

$$f_A(x_1, \ldots, x_n) = (y_1, \ldots, y_n), \quad \text{where } y_k = \frac{a_{k0} + a_{k1}x_1 + \cdots + a_{kn}x_n}{a_{00} + a_{01}x_1 + \cdots + a_{0n}x_n},$$

$k = 1, \ldots, n$, for all points $(x_1, \ldots, x_n)$ such that the last denominator differs from 0; otherwise, the point is transferred on a point at infinity. The group of all homographies contains affine maps and central perspectivities. It is interesting that though a homography is generally a nonlinear map, it transfers a flat on a flat, but does not preserve parallelity. Since the group of homographies is larger than the group of affine maps, all theorems of projective geometry remain valid in affine geometry.

It should be noted that n-dimensional projective space can be treated as a compact smooth manifold M and construct as the closed n-dimensional ball with identified antipodal points on the boundary. The map $h(\mathbf{x}) = \frac{\mathbf{x}}{1+\|\mathbf{x}\|}$ transfers part $\mathbb{R}^n$ of Y onto the open ball $B(\mathbf{0}, 1)$, and the infinite points are transferred on the boundary of M. When $n = 1$, then the set Y_∞ is a one-point set; thus, the projective space is topologically equivalent to a circle. In $n = 2$ the manifold M cannot be embedded in $\mathbb{R}^3$.

The idea of using transformation groups to define invariants has found applications in other branches of geometry, for example, in elliptic geometry. In topology that has separated from geometry, we consider objects invariant with respect to a group of homeomorphisms. Moreover, homotopy groups and cohomotopy groups play an important role in algebraic topology.

2.3 Symmetry Groups in Chemistry

Transformation groups are also applied to study symmetries of some geometrical or chemical objects. Planar or spatial figures are invariant with respect to some transformations called symmetries. All symmetries of a given figure X form a subgroup of $\mathrm{O}(n)$, $n = 2, 3$, called a *symmetry group* and denoted by $\mathrm{Sym}(X)$. For example, if X is a non-square rectangle, then this group consists of four symmetries: identity I, two reflections about the horizontal and vertical middle lines S_h, S_v, and the central symmetry S_O. The map $h\colon \mathrm{Sym}(X) \to \{0, 1, 2, 3\}$ given by the formulas, $h(I) = 0$, $h(S_h) = 1$, $h(S_v) = 3$, $h(S_O) = 2$, is the homomorphism between the groups $\mathrm{Sym}(X)$ and $(\mathbb{Z}_4, \oplus)$, where $\mathbb{Z}_4 = \{0, 1, 2, 3\}$ and $a \oplus b = (a + b) \mod 4$.

In chemistry symmetry groups describe molecular and crystal symmetries. Group theory enables simplification of analysis of chemical and physical properties of molecules and crystals. Molecular symmetries are operations performed on chemical molecules leading to an overlapping arrangement of the molecule's atoms before and after the operation.

As an example consider an ammonia molecule NH_3 (see Fig. 2). Let l be a straight line passing through the nitrogen atom and the center of the triangle defined by the hydrogen atoms. Then, the symmetry group of this molecule consists of six

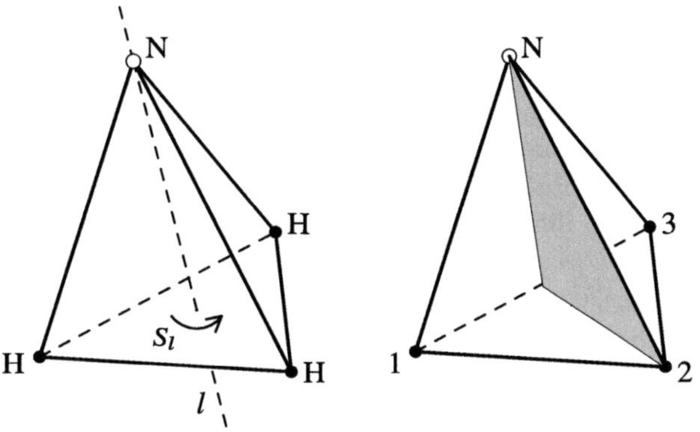

Fig. 2 Rotation S_l around the axis l by an angle of 120° does not change an ammonia molecule. The mirror plane σ_2 is marked on the right picture

Table 1 The results of the operation $g \circ f$ in the group C_{3v}

g \ f	I	S_l	S_l^2	S_{σ_1}	S_{σ_2}	S_{σ_3}
I	I	S_l	S_l^2	S_{σ_1}	S_{σ_2}	S_{σ_3}
S_l	S_l	S_l^2	I	S_{σ_3}	S_{σ_1}	S_{σ_2}
S_l^2	S_l^2	I	S_l	S_{σ_2}	S_{σ_3}	S_{σ_1}
S_{σ_1}	S_{σ_1}	S_{σ_2}	S_{σ_3}	I	S_l	S_l^2
S_{σ_2}	S_{σ_2}	S_{σ_3}	S_{σ_1}	S_l^2	I	S_l
S_{σ_3}	S_{σ_3}	S_{σ_1}	S_{σ_2}	S_l	S_l^2	I

isometries: identity I; two rotations $S_1 = S_l$, $S_2 = S_l^2$ around the axis l; and three reflections S_{σ_1}, S_{σ_2}, S_{σ_3}. Here S_l is the rotation around the axis l by an angle of 120°. The mirror planes σ_1, σ_2, σ_3 ran through the axis l and one of the hydrogen atoms. The symmetry group of NH_3 is denoted by C_{3v} (see Table 1). It is obvious that this group is isomorphic to the symmetry group of the equilateral triangle.

A group of transformations of the spaces $\mathbb{R}^n$ is called a *point group* if all transformations have a common fixed point. All isometries of a bounded object are point groups. Thus, a molecular symmetry group is also a point group. A *crystallographic group* is the group of all isometries that leave the crystal structure unchanged, i.e., the same kinds of atoms would be placed in similar positions as before the transformation. A crystallographic group includes some translations of the space. A *crystallographic point group* is a subgroup of the crystallographic group consisting of all transformations having a common fixed point. Though there are infinitely many 3-dimensional point groups, there are only 32 crystallographic point groups.

The crystallographic point group determines the directional variation of physical properties that arise from its structure, including optical and electro-optical features. The invariance of a crystal with respect to a translational defines crystallinity. Other applications of symmetry groups in quantum mechanical and in the spectroscopy can be found in [9, 18, 22].

2.4 Applications to Newtonian Mechanics

We will study the motion of material points in a three-dimensional Euclidean space. The position of points changes with time, so it is natural to assume that time is an element of space. Thus, physical space, called *world*, is a four-dimensional space. Since in this space we do not distinguish the origin of the coordinate system, the world is considered to be a four-dimensional *affine space* $A^4 = A^1 \times A^3$, in which A^1 is identified with the time axis, and at each fixed time t a set of points (events) of the form $\{t\} \times A^3$ is called *subspace of simultaneous events*. The difference of points from the space A^3 is a vector from $\mathbb{R}^3$, so in the subspace of simultaneous events, it is possible to introduce a distance:

$$\varrho((t, \mathbf{a}), (t, \mathbf{b})) = \|\mathbf{a} - \mathbf{b}\| = \sqrt{(\mathbf{a} - \mathbf{b}, \mathbf{a} - \mathbf{b})},$$

where $\| \cdot \|$, $(\cdot, \cdot)$ is respectively the norm and the scalar product in $\mathbb{R}^3$.

Many coordinate systems can be introduced in the space A^4, among which we distinguish *inertial systems*. These systems are characterized by *Galileo's principle of relativity*:

(1) all the laws of nature are the same at all moments of times in all inertial coordinate systems,
(2) each coordinate system moving in uniform rectilinear motion with respect to an inertial system is also an inertial system.

We also assume *Newton's principle of determinacy*, which states that the positions and velocities of the points of the system at the initial moment uniquely determine the entire motion of the system.

We introduce the coordinate system in the space $\mathbb{R} \times \mathbb{R}^3$. Then, the motion of a single point can be described by the function $\mathbf{x}\colon I \to \mathbb{R}^3$, where I is the interval on the time axis. The set $\mathbf{x}(I)$ is called the *trajectory* of a point, and the graph of the mapping, i.e., the set $\{(t, \mathbf{x}(t))\colon t \in I\}$, is called the *world line*.

The following transformations of the space $\mathbb{R} \times \mathbb{R}^3$ have the property that the world lines of any mechanical system subjected to the same transformation pass to other world lines of the same system:

(a) uniform motion with velocity $\mathbf{v} \in \mathbb{R}^3$

$$g_1(t, \mathbf{x}) = (t, \mathbf{x} + \mathbf{v}t),$$

(b) translation of the origin $(s, \mathbf{a}) \in \mathbb{R} \times \mathbb{R}^3$

$$g_2(t, \mathbf{x}) = (t + s, \mathbf{x} + \mathbf{a}),$$

(c) orthogonal transformation (rotation or plane symmetry)

$$g_3(t, \mathbf{x}) = (t, G\mathbf{x}),$$

where $G : \mathbb{R}^3 \to \mathbb{R}^3$ is an orthogonal transformation. The set of all possible compositions of transformations of types g_1, g_2, g_3 forms the *group of Galilean transformations*. It can be easily checked that any Galilean transformation is of the form $g_1 \circ g_2 \circ g_3$. Preserving the world line by the g_1 transformation follows from Galileo's principle of relativity, while by the transformations g_2 and g_3 from the geometrical structure of space: invariance to displacements (homogeneity) and rotations (isotropy).

It is possible to derive the general form of the equations of motion of material points from Newton's principle of determinacy. Since the motion of points is uniquely determined by their initial positions and velocities, so the acceleration at time t is a function of the position and velocity also at time t. Consider n material points. Let $\mathbf{x}_i(t)$ describe the position of the i-th point at time t. Then the function

$$\mathbf{x}(t) = (\mathbf{x}_1(t), \dots, \mathbf{x}_n(t))$$

describes the motion of a system of n points. Let $N = 3n$. Since acceleration is a function of position and velocity, so there exists a function $\mathbf{F} : \mathbb{R}^N \times \mathbb{R}^N \times \mathbb{R} \to \mathbb{R}^N$ such that

$$\mathbf{x}''(t) = \mathbf{F}(\mathbf{x}(t), \mathbf{x}'(t), t), \tag{4}$$

where $\mathbf{x}'(t) = (\mathbf{x}_1'(t), \dots, \mathbf{x}_n'(t))$.

We will now show how the invariance of the world line with respect to the Galilean group imposes certain conditions on the function $\mathbf{F}$. One of the Galilean transformations is a time shift. This means that if $\mathbf{x}(t)$ is a solution of Eq. (4), then also for any s the function $\mathbf{y}(t) = \mathbf{x}(t+s)$ is a solution of Eq. (4). From Eq. (4) we obtain:

$$\mathbf{x}''(t+s) = \mathbf{F}(\mathbf{x}(t+s), \mathbf{x}'(t+s), t+s). \tag{5}$$

The function $\mathbf{y}$ also satisfies Eq. (4), thus

$$\mathbf{y}''(t) = \mathbf{F}(\mathbf{y}(t), \mathbf{y}'(t), t) = \mathbf{F}(\mathbf{x}(t+s), \mathbf{x}'(t+s), t). \tag{6}$$

Since $\mathbf{x}''(t+s) = \mathbf{y}''(t)$, from Eqs. (5) and (6), it follows that

$$\mathbf{F}(\mathbf{x}(t+s), \mathbf{x}'(t+s), t+s) = \mathbf{F}(\mathbf{x}(t+s), \mathbf{x}'(t+s), t).$$

The initial positions and initial velocities of points can be chosen in any way and at any moment, so

$$\mathbf{F}(\mathbf{x}, \mathbf{y}, t+s) = \mathbf{F}(\mathbf{x}, \mathbf{y}, t)$$

for all $\mathbf{x} \in \mathbb{R}^N$, $\mathbf{y} \in \mathbb{R}^N$, $t \in \mathbb{R}$ and $s \in \mathbb{R}$. Thus, the function $\mathbf{F}$ does not depend on time and the equation of motion is of the form:

$$\mathbf{x}''(t) = \mathbf{F}(\mathbf{x}(t), \mathbf{x}'(t)). \tag{7}$$

Fix now a vector $\mathbf{a} \in \mathbb{R}^3$. If $\mathbf{x}(t)$ satisfies Eq. (4), then also the function $\mathbf{y}(t) = (\mathbf{x}_1(t) + \mathbf{a}, \dots, \mathbf{x}_n(t) + \mathbf{a})$ satisfies Eq. (4). Thus, as before, we conclude that

$$\mathbf{F}(\mathbf{x}_1 + \mathbf{a}, \dots, \mathbf{x}_n + \mathbf{a}, \mathbf{y}) = \mathbf{F}(\mathbf{x}_1, \dots, \mathbf{x}_n, \mathbf{y}),$$

where $\mathbf{x}_1 \in \mathbb{R}^3, \dots, \mathbf{x}_n \in \mathbb{R}^3$ and $\mathbf{y} \in \mathbb{R}^N$. From the invariance of the transformation $g_1(t, \mathbf{x}) = (t, \mathbf{x} + \mathbf{v}t)$, we conclude that

$$\mathbf{F}(\mathbf{x}, \mathbf{y}_1 + \mathbf{v}, \dots, \mathbf{y}_n + \mathbf{v}) = \mathbf{F}(\mathbf{x}, \mathbf{y}_1, \dots, \mathbf{y}_n)$$

for all $\mathbf{x} \in \mathbb{R}^N$ and $\mathbf{y}_1 \in \mathbb{R}^3, \dots, \mathbf{y}_n \in \mathbb{R}^3$. If $\mathbf{x}(t)$ is a solution of Eq. (4), then from invariance with respect to $g_3(t, \mathbf{x}) = (t, G\mathbf{x})$, the function $\mathbf{y}(t) = (G\mathbf{x}_1(t), \dots, G\mathbf{x}_n(t))$ is also a solution of Eq. (4). Thus

$$(G\mathbf{x}_1'', \dots, G\mathbf{x}_n'') = \mathbf{F}(G\mathbf{x}_1, \dots, G\mathbf{x}_n, G\mathbf{x}_1', \dots, G\mathbf{x}_n'). \tag{8}$$

Since

$$(G\mathbf{x}_1'', \dots, G\mathbf{x}_n'') = (G\mathbf{F}_1(\mathbf{x}, \mathbf{x}'), \dots, G\mathbf{F}_n(\mathbf{x}, \mathbf{x}')),$$

we obtain from Eq. (8) that

$$\mathbf{F}(G\mathbf{x}_1, \dots, G\mathbf{x}_n, G\mathbf{y}_1, \dots, G\mathbf{y}_n) = (G\mathbf{F}_1(\mathbf{x}, \mathbf{y}), \dots, G\mathbf{F}_n(\mathbf{x}, \mathbf{y}))$$

for all $\mathbf{x}_1 \in \mathbb{R}^3, \dots, \mathbf{x}_n \in \mathbb{R}^3$, $\mathbf{y}_1 \in \mathbb{R}^3, \dots, \mathbf{y}_n \in \mathbb{R}^3$. Let us introduce the following notations. If $\mathbf{a} \in \mathbb{R}^3$ and $\mathbf{x} = (\mathbf{x}_1, \dots, \mathbf{x}_n) \in \mathbb{R}^N$, then $\mathbf{x} + \mathbf{a} := (\mathbf{x}_1 + \mathbf{a}, \dots, \mathbf{x}_n + \mathbf{a})$ and if $G : \mathbb{R}^3 \to \mathbb{R}^3$, then $G\mathbf{x} := (G\mathbf{x}_1, \dots, G\mathbf{x}_n)$. Then, the function $\mathbf{F}$ satisfies the following conditions:

(i) $\mathbf{F}(\mathbf{x}, \mathbf{y}) = \mathbf{F}(\mathbf{x} + \mathbf{a}, \mathbf{y} + \mathbf{b})$,
(ii) $G\mathbf{F}(\mathbf{x}, \mathbf{y}) = \mathbf{F}(G\mathbf{x}, G\mathbf{y})$

for all vectors $\mathbf{a} \in \mathbb{R}^3$, $\mathbf{b} \in \mathbb{R}^3$, and for all orthogonal transformations $G : \mathbb{R}^3 \to \mathbb{R}^3$.

We now give an example of the application of the obtained result. Let us consider a system consisting of only one point. Then, it follows from equation (i) that there exists a vector $\mathbf{c} \in \mathbb{R}^3$ such that $\mathbf{F}(\mathbf{x}, \mathbf{y}) = \mathbf{c}$ for each $\mathbf{x} \in \mathbb{R}^3$ and $\mathbf{y} \in \mathbb{R}^3$. From (ii) it follows that we have $G\mathbf{c} = \mathbf{c}$ for any rotation G. Hence $\mathbf{c} = \mathbf{0}$. Thus, a point moves according to the equation:

$$\mathbf{x}''(t) = \mathbf{0}.$$

It implies that $\mathbf{x}(t) = \mathbf{a}t + \mathbf{b}$ for some vectors $\mathbf{a} \in \mathbb{R}^3$ and $\mathbf{b} \in \mathbb{R}^3$. Thus, a point moves in a straight line at constant velocity. We have proved *Newton's first law of motion*: if no force acts on a material point, then the point remains at rest or keeps moving in a straight line at constant speed.

2.5 Measure-Preserving Dynamical Systems

Let (X, Σ, μ) be a probability space, i.e., μ is a probability measure defined on a σ-algebra Σ of subsets of X. Consider a one-parameter semigroup $\{\pi_t\}_{t\in T}$. We assume that the function $(t, x) \mapsto \pi_t x$ is measurable as a function from the Cartesian product $T \times X$ to X and that the measure μ is *invariant* with respect to each transformation π_t, $t \in T$, i.e., $\mu(\pi_t^{-1}(A)) = \mu(A)$ for any set $A \in \Sigma$. Then, the quadruple (X, Σ, μ, π_t) is called a *measure-preserving dynamical system*. The triple (X, Σ, μ) is called a *phase space* and the set $\{\pi_t x : t \in T\}$ is called *trajectory of a point* $x \in X$. As in Sect. 2.1, we divide measure-preserving dynamical systems into endomorphisms, automorphisms, flows, and semiflows.

Our goal is to study the properties of measure-preserving dynamical systems: ergodicity, mixing, and exactness. A set A is called *invariant* with respect to a semigroup $\{\pi_t\}_{t\in T}$, if $\pi_t^{-1}(A) = A$ for any $A \in \Sigma$ and $t \in T$. The family of invariant sets forms a σ-algebra Σ_{inv}. If σ-algebra Σ_{inv} is trivial, i.e., it consists only of sets of measure μ zero or one, then the measure-preserving dynamical system is said to be *ergodic*.

A stronger property than ergodicity is mixing. A measure-preserving dynamical system (X, Σ, μ, π_t) is called *mixing* if

$$\lim_{t\to\infty} \mu(A \cap \pi_t^{-1}(B)) = \mu(A)\mu(B) \quad \text{for all } A, B \in \Sigma. \tag{9}$$

Identifying the measure μ with the probability P, one can formulate the condition (9) as follows:

$$\lim_{t\to\infty} \mathrm{P}(\pi_t(x) \in A | x \in B) = \mathrm{P}(A) \quad \text{for all } A, B \in \Sigma \text{ and } \mathrm{P}(B) > 0,$$

which means that the trajectories of almost all points enter a set A with asymptotic probability $\mu(A)$.

A stronger property than mixing is exactness. A system (X, Σ, μ, π_t) with double measurable transformations π_t, i.e., $\pi_t(A) \in \Sigma$ and $\pi_t^{-1}(A) \in \Sigma$ for all $A \in \Sigma$ and $t \in T$, and with an invariant probability measure μ is called *exact* if for every set $A \in \Sigma$ with $\mu(A) > 0$ we have $\lim_{t\to\infty} \mu(\pi_t(A)) = 1$. Exactness cannot hold for flows and automorphisms because in these cases

$$\mu(\pi_t(A)) = \mu(\pi_t^{-1}(\pi_t(A))) = \mu(A).$$

Exactness is equivalent to the following condition: σ-algebra $\bigcap_{t\geq 0}\pi_t^{-1}(\Sigma)$ contains only sets of measure μ zero or one. Here $\pi_t^{-1}(\Sigma)=\{\pi_t^{-1}(A)\colon A\in\Sigma\}$.

Directly checking whether a dynamical system is ergodic, mixing, or exact is not an easy issue. This problem can be reduced to the study of linear operators related to the representation of semigroups. In order to do it, we introduce the notion of Frobenius–Perron operators.

Let (X,Σ,m) be a σ-finite measure space. A measurable map $\varphi\colon X\to X$ is called *nonsingular* if it satisfies the following condition:

$$m(A)=0\Longrightarrow m(\varphi^{-1}(A))=0 \text{ for } A\in\Sigma. \tag{10}$$

Let $L^1=L^1(X,\Sigma,m)$ and let φ be a measurable nonsingular transformation of X. An operator $P_\varphi\colon L^1\to L^1$ which satisfies the following condition:

$$\int_A P_\varphi f(x)\,m(dx)=\int_{\varphi^{-1}(A)} f(x)\,m(dx) \text{ for } A\in\Sigma \text{ and } f\in L^1 \tag{11}$$

is called the *Frobenius–Perron operator* for the transformation φ. The operator P_φ is linear, is *positive* (if $f\geq 0$, then $P_\varphi f\geq 0$), and preserves the integral ($\int_X P_\varphi f\,dm=\int_X f\,dm$). Any operator having the above properties is called *stochastic (or Markov) operator.* We can equivalently define a stochastic operator as a linear operator on L^1 carrying out densities on densities. We recall that $f\in L^1$ is a *density* if $f\geq 0$ and $\|f\|=1$. By D we denote the set of all densities.

The adjoint of the Frobenius–Perron operator $P^*\colon L^\infty\to L^\infty$ is given by $P_\varphi^* g(x)=g(\varphi(x))$; thus, P_φ^* is the Koopman operator restricted to the space $L^\infty=L^\infty(X,\Sigma,m)$.

Example 1 Let X be a subset of $\mathbb{R}^d$ with non-empty interior and with the boundary of zero Lebesgue measure. Let $\varphi\colon X\to X$ be a measurable transformation. We assume that there exists at most countable family of pairwise disjoint open sets U_i, $i\in I$, contained in X having the following properties:

(a) the sets $X_0=X\setminus\bigcup_{i\in I}U_i$ and $\varphi(X_0)$ have zero Lebesgue measure,
(b) maps $\varphi_i=\varphi\Big|_{U_i}$ are diffeomorphisms from U_i onto $\varphi(U_i)$, i.e., φ_i are C^1 and invertible transformations and $\det\varphi_i'(x)\neq 0$ at each point $x\in U_i$.

Then transformations $\psi_i=\varphi_i^{-1}$ are also diffeomorphisms from $\varphi(U_i)$ onto U_i, and the Frobenius–Perron operator P_φ exists and is given by the formula:

$$P_\varphi f(x)=\sum_{i\in I_x} f(\psi_i(x))|\det\psi_i'(x)|, \tag{12}$$

where $I_x=\{i\colon x\in\varphi(U_i)\}$ (see [80, 83]).

Example 2 If we consider a flow $\{\pi_t\}_{t\in\mathbb{R}}$ on X related to Eq. (2), then according to formula (12) the group $\{P^t\}_{t\in\mathbb{R}}$ of Frobenius–Perron operators is given by the formula:

$$P^t f(x) = f(\pi_{-t}x)\det\left[\frac{d}{dx}\pi_{-t}x\right].$$

The set of all nonsingular transformations with composition as the binary operation forms a monoid $(\mathbf{G}, \circ)$. The set of all stochastic operators $\mathrm{StO}(X)$ is also a monoid with composition as the binary operation and $\mathrm{StO}(X)$ is a linear space. Since

$$P_{\varphi\circ\psi} = P_\varphi \circ P_\psi,$$

the map $h\colon \mathbf{G} \to \mathrm{StO}(X)$ given by $h(\varphi) = P_\varphi$ is a representation of the semigroup $\mathbf{G}$ on the space $L^1(X, \Sigma, m)$.

Frobenius–Perron operators can be successfully used to study ergodic properties of dynamical systems [49]. Let $\varphi\colon X \to X$ be a nonsingular transformation of some σ-finite measure space (X, Σ, m). First, we check when a probability measure μ absolutely continuous with respect to m is invariant with respect to φ. Let f^* be the density of μ with respect to m, i.e., $\mu(A) = \int_A f^*\,dm$ for $A \in \Sigma$. The measure μ is invariant with respect to φ if and only if

$$\int_{\varphi^{-1}(A)} f^*(x)\,m(dx) = \int_A f^*(x)\,m(dx) \text{ for } A \in \Sigma.$$

From (11) it follows immediately that the measure μ is invariant with respect to φ if and only if $P_\varphi f^* = f^*$. It means that if the map $\pi : T \times X \to X$ is measurable and $\{\pi_t\}_{t\in T}$ is a one-parameter semigroup of nonsingular transformations of (X, Σ, m), P^t denotes the Frobenius–Perron operator corresponding to π_t, then the quadruple (X, Σ, μ, π_t) is a measure-preserving dynamical system if and only if $P^t f^* = f^*$ for all $t \in T$.

It is also not difficult to check that the measure μ is ergodic if and only if f^* is a unique fixed point of all operators P^t in the set of densities; and the dynamical system is mixing if and only if for every $f \in D$, the density f^* is the weak limit of $P^t f$ as $t \to \infty$. By the *weak limit* $\lim_{t\to\infty} P^t f$, we understand a function $h \in L^1$ such that for every $g \in L^\infty$ we have:

$$\lim_{t\to\infty} \int_X P^t f(x) g(x)\,m(dx) = \int_X h(x) g(x)\,m(dx)$$

and we denote it by $\text{w-}\lim_{t\to\infty} P^t f$. More difficult is to prove that the dynamical system is exact if and only if $\lim_{t\to\infty} P^t f = f^*$ for every $f \in D$.

We collect the relations between ergodic properties of a dynamical system and the behavior of the Frobenius–Perron operator in Table 2.

Table 2 The relations between ergodic properties of the dynamical system (X, Σ, μ, π_t) and the Frobenius–Perron operators P^t

μ	f^*
Invariant	$P^t f^* = f^*$ for all $t \in T$
Ergodic	f^* is a unique fixed point in D of all P^t
Mixing	$\text{w-}\lim_{t\to\infty} P^t f = f^*$ for every $f \in D$
Exact	$\lim_{t\to\infty} P^t f = f^*$ for every $f \in D$

Example 3 Let $\varphi\colon [0,1] \to [0,1]$ be the transformation given by

$$\varphi(x) = \begin{cases} 2x & \text{for } x \in [0, 1/2], \\ 2 - 2x & \text{for } x \in (1/2, 1]. \end{cases} \tag{13}$$

The transformation φ is called the *tent map* because its graph looks like a tent. The Frobenius–Perron operator P_φ is of the form:

$$P_\varphi f(x) = \tfrac{1}{2} f(\tfrac{1}{2}x) + \tfrac{1}{2} f(1 - \tfrac{1}{2}x). \tag{14}$$

It is easy to see that the density $f^* = 1_{[0,1]}$ satisfies $P_\varphi f^* = f^*$, and consequently φ is an endomorphism on the phase space $([0,1], \mathcal{B}([0,1]), \ell)$, where $\mathcal{B}([0,1])$ is the σ-algebra of Borel subsets of $[0,1]$ and ℓ is the Lebesgue measure. We check that $\lim_{t\to\infty} P_\varphi^t f = f^*$ for any density f. It is sufficient to check this condition for densities which are Lipschitz continuous. Let L be the Lipschitz constant for f. Then

$$|P_\varphi f(x) - P_\varphi f(y)| \le \tfrac{1}{2}|f(\tfrac{x}{2}) - f(\tfrac{y}{2})| + \tfrac{1}{2}|f(1 - \tfrac{1}{2}x) + f(1 - \tfrac{1}{2}y)| \le \tfrac{L}{2}|x - y|.$$

Thus, $L/2$ is the Lipschitz constant for $P_\varphi f$, and by induction we conclude that $L/2^t$ is the Lipschitz constant for $P_\varphi^t f$. Hence, the sequence $(P_\varphi^t f)$ converges uniformly to a constant function. Since $P_\varphi^t f$ are densities, $(P_\varphi^t f)$ converges to f^* uniformly, which implies the convergence in L^1. The condition $\lim_{t\to\infty} P_\varphi^t f = f^*$ implies the exactness of the endomorphism φ.

Example 4 Now we check that the *logistic map* $\psi(x) = 4x(1-x)$ on $[0,1]$ is exact. Let φ be the tent map and $\alpha(x) = \frac{1}{2} - \frac{1}{2}\cos(\pi x)$. Then $\psi \circ \alpha = \alpha \circ \varphi$. This implies that $P_\psi P_\alpha = P_\alpha P_\varphi$. Hence $P_\psi = P_\alpha P_\varphi P_{\alpha^{-1}}$ and, by induction, $P_\psi^t = P_\alpha P_\varphi^t P_{\alpha^{-1}}$. Let $f \in D$. Then $P_{\alpha^{-1}} f \in D$ and from the previous example $\lim_{t\to\infty} P_\varphi^t P_{\alpha^{-1}} f = 1_{[0,1]}$. Therefore

$$\lim_{t\to\infty} P_\psi^t f = P_\alpha 1_{[0,1]} = g^*, \quad \text{where } g^*(x) = \left[\pi\sqrt{x(1-x)}\right]^{-1}.$$

Thus, the map ψ is an exact endomorphism on the space $([0,1], \mathcal{B}([0,1]), \mu)$, where $d\mu = g^*(x)\,dx$.

More advanced examples of applications of Frobenius–Perron operators to study ergodic properties of dynamical systems can be found in [12, 49, 50].

3 Stochastic Semigroups

3.1 Semigroups of Operators

Stochastic semigroups belong to a wider family of semigroups called C_0-semigroups of operators, so we first need to introduce the notion of a C_0-semigroup and to show its relation with evolution equations.

Let $(E, \|\cdot\|)$ be a Banach space and let $\{T(t)\}_{t\geq 0}$ be a family of linear bounded operators on E. The family $\{T(t)\}_{t\geq 0}$ is called a *semigroup of bounded operators* if it satisfies the following conditions:

(a) $T(0) = I$, i.e., $T(0)x = x$,
(b) $T(t+s) = T(t)T(s) \quad \text{for} \quad s, t \geq 0.$

If additionally

(c) for each $x \in E$ the function $t \mapsto T(t)x$ is continuous,

then $\{T(t)\}_{t\geq 0}$ is called a *C_0-semigroup* or *strongly continuous semigroup*. Let $\{T(t)\}_{t\geq 0}$ be a C_0-semigroup and let $\mathfrak{D}(A)$ be the set of such $x \subset E$ that there exists the limit:

$$Ax = \lim_{t\to 0^+} \frac{T(t)x - x}{t}. \tag{15}$$

Then, the set $\mathfrak{D}(A)$ is a linear subspace dense in E, and A is a linear operator from $\mathfrak{D}(A)$ to E. The operator A is called the *infinitesimal generator* (briefly the *generator*) of the semigroup $\{T(t)\}_{t\geq 0}$. We also say that the operator $A\colon \mathfrak{D}(A) \to E$ generates the semigroup $\{T(t)\}_{t\geq 0}$.

The notion of a C_0-semigroup and its generator is strictly connected with differential equations in the Banach space E. Let $\{T(t)\}_{t\geq 0}$ be a C_0-semigroup on E and let $A : \mathfrak{D}(A) \to E$ be its generator. Then, for every $x_0 \in \mathfrak{D}(A)$, the function $x\colon [0, \infty) \to E$ defined by the formula $x(t) = T(t)x_0$ is differentiable (in the sense of Fréchet) for $t \geq 0$ and satisfies the equation:

$$x'(t) = Ax(t) \quad \text{with initial condition } x(0) = x_0. \tag{16}$$

An equation of the form (16) is called an *evolution equation*. We also say that Eq. (16) *generates semigroup* $\{T(t)\}_{t\geq 0}$.

In the theory of semigroups of operators, the key issue is when a linear operator is a generator of a C_0-semigroup. If A is a bounded linear operator on E, then A generates a C_0-semigroup on E given by the formula:

$$T(t)x = e^{At}x = \sum_{k=0}^{\infty} \frac{t^k A^k x}{k!} \tag{17}$$

and the semigroup $\{T(t)\}_{t\geq 0}$ is *uniformly continuous*, i.e.,

$$\lim_{t\to t_0} \|T(t) - T(t_0)\| = 0 \text{ for } t_0 \geq 0. \tag{18}$$

We will study mainly semigroups of contractions, so we only characterize the generators of such semigroups. We recall that a linear operator T on E is a *contraction* if $\|Tx\| \leq \|x\|$ for $x \in E$.

Theorem 5 (Hille–Yosida) *Let E be a Banach space, and let A be a linear operator on E with domain $\mathfrak{D}(A)$. The operator A is a generator of a strongly continuous semigroup of contractions if and only if the following conditions are satisfied:*

(a) *the set $\mathfrak{D}(A)$ is dense in E;*
(b) *for each $\lambda > 0$ there exists an operator $(\lambda I - A)^{-1}$ defined everywhere on E;*
(c) $\|(\lambda I - A)^{-1}\| \leq \lambda^{-1}$ *for each $\lambda > 0$.*

The operator $(\lambda I - A)^{-1}$ is called the *resolvent* of A at λ and it is denoted by $R(\lambda, A)$ or R_λ. The resolvent is defined for λ from the *resolvent set* $\rho(A)$ which consists of all $\lambda \in \mathbb{C}$ such that the operator $\lambda I - A$ is invertible and the inverse operator is a bounded operator defined on the whole space E. The set $\sigma(A) = \mathbb{C} \setminus \rho(A)$ is called the *spectrum* of the operator A. If the operator A generates a semigroup $\{T(t)\}_{t\geq 0}$, then R_λ is called the *resolvent of the semigroup* $\{T(t)\}_{t\geq 0}$. The resolvent of a strongly continuous semigroup of contractions is also given by the formula:

$$R_\lambda x = \int_0^\infty e^{-\lambda t} T(t)x \, dt \qquad \text{for } x \in E. \tag{19}$$

Remark 12 We can also consider nonlinear strongly continuous semigroups. A family $\{T(t)\}_{t\geq 0}$ of nonlinear operators on a Banach space is called a *nonlinear strongly continuous semigroup* if it satisfies conditions (a), (b), and the map $(t, x) \mapsto T(t)x$ is continuous.

3.2 Stochastic and Substochastic Semigroups

Let (X, Σ, m) be a σ-finite measure space and let $L^1 = L^1(X, \Sigma, m)$. A C_0-semigroup $\{P(t)\}_{t\geq 0}$ of positive contractions on L^1 is called a *substochastic semigroup*. A C_0-semigroup of stochastic operators is called a *stochastic semigroup* or a *Markov semigroup*. We can also consider a *nonlinear stochastic semigroup* if $\{P(t)\}_{t\geq 0}$ is a nonlinear strongly continuous semigroup and $P(t)(D) \subset D$ for $t \geq 0$, where D is the set of all densities.

An example of a stochastic semigroup is a semiflow of Frobenius–Perron operators $\{P^t\}_{t\geq 0}$ considered in Sect. 2.5 if we assume that this semiflow satisfies condition (c) of the definition of a C_0-semigroup. We will now examine what conditions a linear operator A on a L^1 space should satisfy in order to be a generator of a stochastic semigroup.

First we consider the case when A is a bounded operator. A bounded linear operator A is a generator of a positive semigroup on L^1 (generally on a Banach lattice) if and only if $A + \|A\|I \geq 0$ (see, e.g., Theorem 1.11, page 255 [1]). We recall that a semigroup $\{P(t)\}_{t\geq 0}$ is *positive* if each operator $P(t)$ is positive. Let $\lambda = \|A\|$ and $B = A + \lambda I$, then B is a positive operator. Thus, the generator of the semigroup of positive operators $\{P(t)\}_{t\geq 0}$ on L^1 is of the form $A = -\lambda I + B$, where B is a positive and bounded operator. If A is a generator of a stochastic semigroup, then $P(t)$ preserves the integral. Since $P(t) = I + tA + o(t)$, we obtain:

$$\int_X Af(x)\, m(dx) = 0 \quad \text{for } f \in L^1. \tag{20}$$

Thus, $\int_X Bf(x)\, m(dx) = \lambda$ for $f \in D$. Assume that $\lambda > 0$ and let $P = B/\lambda$. Then $A = -\lambda I + \lambda P$, where P is a stochastic operator. Analogously if a linear and bounded operator A is a generator of a substochastic semigroup, then A is of the form $A = -\lambda I + \lambda P$, where P is a substochastic operator and $\lambda \geq 0$. The semigroup $\{P(t)\}_{t\geq 0}$ is given by the formula:

$$P(t)f = \sum_{k=0}^{\infty} \frac{(\lambda t)^k e^{-\lambda t}}{k!} P^k f. \tag{21}$$

Example 6 Let $X = \{1, 2, \ldots, n\}$, $\Sigma = 2^X$ and let m be the counting measure. In this case the space $L^1(X, \Sigma, m)$ is denoted by l_n^1. In the space l_n^1 a function $f : X \to \mathbb{R}$ is represented as a sequence $y = (y_1, \ldots, y_n)$, the integral of y over X is given by $\sum_{i=1}^n y_i$, and the space l_n^1 is isomorphic to the space $\mathbb{R}^n$ with the norm $\|y\| = |y_1| + \cdots + |y_n|$. The generator A of a stochastic semigroup on l_n^1 is of the form $Ay = yQ$, where the matrix $Q = [q_{ij}]$ satisfies conditions:

(i) $q_{ij} \geq 0$ for $i \neq j$,
(ii) $\sum_{j=1}^n q_{ij} = 0$ for $i = 1, \ldots, n$.

The matrix Q is the *intensity matrix* of a Markov chain $(\xi_t)_{t\geq 0}$. The entry $q_{ij}, i \neq j$, is the rate of jump from the state i to j, precisely

$$\mathrm{P}(\xi_{t+\Delta t} = j | \xi_t = i) = q_{ij}\Delta t + o(\Delta t). \tag{22}$$

Moreover, $|q_{ii}|$ is the rate of leaving the state i.

Remark 13 One can ask the question whether condition (20) is sufficient for a bounded linear operator to generate a stochastic semigroup. The answer is negative. Let $X = [0, 1]$ and m be the Lebesgue measure. Consider the operator $A = I - P$, where $Pf(x) = 2f(2x)$ for $x \in [0, 1/2]$ and $Pf(x) = 0$ otherwise. The operator A satisfies (20) and generates a semigroup $\{T(t)\}_{t\geq 0}$, but this semigroup is not positive. We have:

$$T(t)f = e^t e^{-tP} f = e^t \sum_{k=0}^{\infty} \frac{(-1)^k t^k P^k}{k!} f.$$

If $f = \mathbf{1}_{(1/2,1]}$, then $P^k f = 2^k \mathbf{1}_{(1/2^{k+1},1/2^k]}$ and $T(t)f(x) < 0$ for $x \in (1/2^{k+1}, 1/2^k]$ and k odd.

Theorem 5 can be applied to check whether an operator A is a generator of a stochastic or substochastic semigroup. From (19) it follows that if a semigroup is substochastic, then the operators R_λ are positive for each $\lambda > 0$. It turns out that vice versa, too, if A satisfies conditions (a)–(c) of Theorem 5 and

(d) $R_\lambda f \geq 0$ for all $\lambda > 0$ and $f \geq 0$,

then A is the generator of a substochastic semigroup.

If a semigroup $\{P(t)\}_{t\geq 0}$ is stochastic, then from (19) it follows that the operators λR_λ are stochastic. If the operator A satisfies conditions (a)–(d) and additionally

(e) for some $\lambda > 0$ the operator λR_λ is stochastic,

then A generates a stochastic semigroup. Condition (e) can be replaced by the following:

(e′) $\int_X Af(x)\, m(dx) = 0$ for $f \in \mathfrak{D}(A)$.

In applications, the following problem often arises. We know that an operator A is the generator of a stochastic or substochastic semigroup $\{P(t)\}_{t\geq 0}$ and B is a linear operator on L^1. When does the operator $A + B$ also generate a stochastic or a substochastic semigroup? This question is answered by perturbation theorems. Now we present one of them useful in applications.

Theorem 7 *Let A be a generator of a stochastic semigroup $\{S(t)\}_{t\geq 0}$, K be a stochastic operator and $\lambda > 0$. The semigroup $\{P(t)\}_{t\geq 0}$ generated by the operator $A + \lambda K - \lambda I$ is stochastic and it is given by the Dyson–Phillips expansion:*

$$P(t)f = e^{-\lambda t} \sum_{n=0}^{\infty} \lambda^n S_n(t) f, \tag{23}$$

where

$$S_0(t)f = S(t)f, \quad S_{n+1}(t)f = \int_0^t S_n(t-s)KS(s)f\,ds, \quad n \ge 0. \tag{24}$$

If $\{S(t)\}_{t\ge 0}$ is a substochastic semigroup and K is a substochastic operator, then $\{P(t)\}_{t\ge 0}$ is a substochastic semigroup.

3.3 Semigroups Related to Markov Chains

It can be expected that the construction of the stochastic semigroup presented in Example 6 can be extended to the case where X is a countable set. However, this case is more difficult.

Let $l^1 = L^1(\mathbb{N}, 2^{\mathbb{N}}, m)$, where m is the counting measure. Now elements of l^1 are infinite sequences $y = (y_0, y_1, \dots)$ and $\|y\| = \sum_{i=0}^{\infty} |y_i|$. We assume that an infinity dimensional matrix $Q = [q_{ij}]$ satisfies conditions:

(i) $q_{ij} \ge 0$ for $i \ne j$,
(ii) $\sum_{j=0}^{\infty} q_{ij} = 0$ for $i = 0, 1, \dots$.

If $\sup_{i\in\mathbb{N}} |q_{ii}| < \infty$, then the operator $Ay = yQ$ is bounded and $P(t)y = e^{At}y$, $t \ge 0$, is a stochastic semigroup on l^1.

The case when A is an unbounded operator is more complex. Two linear subspaces can be associated with the matrix Q:

$$\mathfrak{D}_0(Q) = \{x \in l^1\colon \sum_{i=0}^{\infty} |q_{ii}||x_i| < \infty\},$$

$$\mathfrak{D}(Q) = \Big\{x \in l^1\colon \text{ for each } j \in \mathbb{N} \text{ the series } \sum_{i=0}^{\infty} x_i q_{ij} \text{ is absolutely convergent and } \sum_{j=0}^{\infty}\Big|\sum_{i=0}^{\infty} x_i q_{ij}\Big| < \infty\Big\}.$$

Firstly, a mapping A with domain $\mathfrak{D}_0(Q)$ or $\mathfrak{D}(Q)$ need not be the generator of a strongly continuous semigroup on l^1. Secondly, even when A with an appropriately chosen domain generates a strongly continuous semigroup on l^1, this semigroup may not be a stochastic semigroup. Thirdly, the same matrix Q can be used to generate different positive semigroups (including stochastic semigroups) on l^1 depending on the choice of the domain of A.

We consider a family $\mathfrak{F}$ of all positive semigroups on l^1 with generators A such that $\mathfrak{D}_0(Q) \subseteq \mathfrak{D}(A)$ and $Ax = xQ$ for $x \in \mathfrak{D}_0(Q)$. A semigroup $\{P(t)\}_{t\ge 0}$ from $\mathfrak{F}$ is called *minimal* if for any other semigroup $\{T(t)\}_{t\ge 0}$ from $\mathfrak{F}$ we have $T(t)f \ge P(t)f$ for each nonnegative $f \in \mathfrak{D}_0(Q)$ and $t \ge 0$.

Theorem 8 *Assume that the matrix Q satisfies* (i)–(ii). *Then, there is the minimal semigroup $\{P(t)\}_{t\ge 0}$ and $\{P(t)\}_{t\ge 0}$ is a substochastic semigroup.*

The answer for the question when a minimal semigroup is stochastic is given by the following Kato theorem [43].

Theorem 9 *Let $\lambda > 0$ be a positive constant. The minimal semigroup related to Q is a stochastic semigroup if and only if the equation $Qx = \lambda x$ has no nonzero and nonnegative solution $x \in l^\infty$.*

If the minimal semigroup $\{P(t)\}_{t\ge 0}$ related to a matrix Q is a stochastic semigroup, then the matrix Q and the semigroup $\{P(t)\}_{t\ge 0}$ are called *nonexplosive*. If the matrix is nonexplosive, then Q is an intensity matrix of a Markov chain $(\xi_t)_{t\ge 0}$ on $\mathbb{N}$, i.e., the entries q_{ij} satisfy condition (22).

Remark 14 If $\mathfrak{D}(A)$ is the domain of the generator of a stochastic semigroup related to the intensity matrix Q, then

$$\mathfrak{D}_0(Q) \subseteq \mathfrak{D}(A) \subseteq \mathfrak{D}(Q)$$

and none of these inclusions can be replaced by equality.

3.4 Diffusion Semigroup

The theory of stochastic processes provides many interesting examples of semigroups. Semigroups related to diffusion processes play a fundamental role.

Consider the *Itô equation*:

$$dx_t = \sigma(x_t)\,dw_t + b(x_t)\,dt, \tag{25}$$

where w_t is an m-dimensional Brownian motion, $\sigma(x) = [\sigma^i_j(x)]$ is a $d \times m$ matrix and $b(x)$ is a vector in $\mathbb{R}^d$ for every $x \in \mathbb{R}^d$. We assume that for all $i = 1, ..., d$, $j = 1, ..., m$, the functions b_i, σ^i_j are sufficiently smooth and have bounded derivatives of all orders, and the functions σ^i_j are also bounded. A process x_t, $t \ge 0$, which satisfies (25) is called *diffusion process*.

Assume that the random variable x_0 has the density $f(x)$. Then solution x_t of (25) also has the density $u(t, x)$ and u satisfies the *Fokker–Planck equation* (also called the *Kolmogorov forward equation*):

$$\frac{\partial u}{\partial t} = \sum_{i,j=1}^{d} \frac{\partial^2 (a_{ij}(x)u)}{\partial x_i \partial x_j} - \sum_{i=1}^{d} \frac{\partial (b_i(x)u)}{\partial x_i}, \tag{26}$$

where $a_{ij}(x) = \frac{1}{2}\sum_{k=1}^{m} \sigma_k^i(x)\sigma_k^j(x)$. Let $P(t)f(x) = u(t, x)$. Then $\{P(t)\}_{t\geq 0}$ is a stochastic semigroup ([80]).

Consider the operator:

$$A_0 f = \sum_{i,j=1}^{d} \frac{\partial^2(a_{ij}(x)f)}{\partial x_i \partial x_j} - \sum_{i=1}^{d} \frac{\partial(b_i(x)f)}{\partial x_i} \tag{27}$$

on the set $E = \{f \in L^1(\mathbb{R}^d) \cap C_b^2(\mathbb{R}^d) \colon A_0 f \in L^1(\mathbb{R}^d)\}$, where $C_b^2(\mathbb{R}^d)$ denotes the set of all twice differentiable bounded functions whose derivatives of order ≤ 2 are continuous and bounded. The closure of the operator A_0 is the generator of a stochastic semigroup $\{P(t)\}_{t\geq 0}$ ([28, Proposition 1.3.3]). We recall that the *closure of the operator* A_0 is the operator A whose graph is the closure of the graph of A_0, as long as such an operator A exists.

If the functions a_{ij} satisfy the uniform elliptic condition:

$$\sum_{i,j=1}^{d} a_{ij}(x)\lambda_i\lambda_j \geq \alpha\|\lambda\|^2 \tag{28}$$

for some $\alpha > 0$ and every $\lambda \in \mathbb{R}^d$ and $x \in \mathbb{R}^d$, then the stochastic semigroup $\{P(t)\}_{t\geq 0}$ is an *integral semigroup*, i.e.,

$$P(t)f(y) = \int_{\mathbb{R}^d} q(t, x, y) f(x)\, dx, \quad t > 0,$$

and the *kernel* q is continuous and positive.

It should be mentioned that in the case when σ is a null matrix, then Eq. (25) reduces to Eq. (2). Consequently, the generator of the group of Frobenius–Perron operators from Example 2 is of the form:

$$Af = -\sum_{i=1}^{d} \frac{\partial(b_i(x)f)}{\partial x_i}. \tag{29}$$

3.5 Processes with Jumps

Let (X, Σ, m) be a σ-finite measure space and let $(x_t)_{t\geq 0}$ be a family of Markov processes such that if the distribution of the random variable x_0 has a density f, then the random variable x_t has a density $S(t)f$ and $\{S(t)\}_{t\geq 0}$ is a stochastic semigroup. Let $(A_0, \mathfrak{D}(A_0))$ be the generator of the semigroup $\{S(t)\}_{t\geq 0}$. For example, $\{S(t)\}_{t\geq 0}$ can be a diffusion semigroup or a semigroup of Frobenius–Perron operators.

Consider a family of Markov processes $(\xi_t)_{t\geq 0}$ describing the motion of points in the space X and defined in the following way. A point moves along the sample paths of the process $(x_t)_{t\geq 0}$, but it can randomly jump to a new position according to a jump function defined below, and then it continues to move along the sample paths of the process $(x_t)_{t\geq 0}$, etc.

We define the jump function by specifying the jump intensity from a point x to a set $B \in \Sigma$. For a fixed point $x \in X$, the *jump function* is a measure $\mathcal{K}(x, B)$ defined in such a way that the probability of jump from a point x to a set B during the time Δt is $\mathcal{K}(x, B)\Delta t + o(\Delta t)$. Usually the jump function $\mathcal{K}(x, B)$ is of the form $\mathcal{K}(x, B) = \psi(x)\mathcal{P}(x, B)$, where $\psi(x)$ is the rate of jump and $\mathcal{P}(x, B)$ is the probability of jump from x to a given set B. The function $\mathcal{P}$ is a *transition probability function* on (X, Σ), i.e., $\mathcal{P}(x, \cdot)$ is a probability measure on (X, Σ) for each $x \in X$ and the function $x \mapsto \mathcal{P}(x, B)$ is measurable for each $B \in \Sigma$.

Assume that $\mathcal{P}$ has the following property:

$$m(B) = 0 \Longrightarrow \mathcal{P}(x, B) = 0 \text{ for } m\text{-a.e. } x \text{ and } B \in \Sigma. \tag{30}$$

Then for every $f \in D$ the measure

$$\mu(B) = \int f(x)\mathcal{P}(x, B)\, m(dx)$$

is absolutely continuous with respect to the measure m. This fact is a simple consequence of the *Radon–Nikodym theorem*, which says that the measure ν is absolutely continuous with respect to the measure m if and only if the following implication $m(B) = 0 \Rightarrow \nu(B) = 0$ holds for all sets $B \in \Sigma$. Now, the formula $Pf = d\mu/dm$ defines a stochastic operator $P : L^1 \to L^1$ called a *jump operator*. Moreover, if $P^*\colon L^\infty \to L^\infty$ is the adjoint operator of P, then $P^*g(x) = \int g(y)\mathcal{P}(x, dy)$.

We consider two examples of the jump operator. One of them is Frobenius–Perron operator P_φ introduced in Sect. 2.5. The transition probability function corresponding to P_φ is of the form:

$$\mathcal{P}(x, B) = \begin{cases} 1, \text{if } \varphi(x) \in B, \\ 0, \text{if } \varphi(x) \notin B. \end{cases} \tag{31}$$

The second example is an integral stochastic operator. If (X, Σ, m) is a σ-finite measure space $q\colon X \times X \to [0, \infty)$ is a measurable function such that

$$\int_X q(x, y)\, m(dy) = 1$$

for almost all $x \in X$, then

$$Pf(y) = \int_X q(x,y) f(x)\, m(dx) \tag{32}$$

is a stochastic operator. The function q is called the *kernel* of the operator P and P is called an *integral* or *kernel* operator. The function $\mathcal{P}(x, B) = \int_B q(x,y)\, m(dy)$ is the transition probability function corresponding to P.

We assume that the function $\psi : X \to [0, \infty)$ describing the intensity of jumps is measurable and bounded. Then, the process $(\xi_t)_{t\ge 0}$ defines a semigroup $\{P(t)\}_{t\ge 0}$ on L^1 with generator:

$$Af = A_0 f - \psi f + P(\psi f). \tag{33}$$

We check that $\{P(t)\}_{t\ge 0}$ is a stochastic semigroup. Let $\lambda = \sup_X \psi(x)$ and $Kf = \lambda^{-1}[P(\psi f) + (\lambda - \psi) f]$. Then K is a stochastic operator and $A = A_0 - \lambda I + \lambda K$. According to Theorem 7 the operator $(A, \mathfrak{D}(A_0))$ is the generator of a stochastic semigroup defined by formulas (23) and (24).

If A_0 is of the form (29) and is therefore the generator of a semigroup of Frobenius–Perron operators, then $(\xi_t)_{t\ge 0}$ is an example of a piecewise deterministic process [23]. A stochastic process $(\xi_t)_{t\ge 0}$ is called a *piecewise deterministic processes* if there is an increasing sequence of random times (t_n), called jumps, such that sample paths of $(\xi_t)_{t\ge 0}$ are defined in a deterministic way in each interval $[t_n, t_{n+1})$; in our case they satisfy Eq. (2).

3.6 Diffusion Process with Random Switching

Consider the following stochastic equation:

$$dx_t = \sigma(x_t, y_t)\, dw_t + b(x_t, y_t)\, dt, \tag{34}$$

where the process $(y_t)_{t\ge 0}$ takes values in the set $I = \{0, 1, \dots, n\}$, $n \ge 1$. We assume that the matrices $\sigma(x, k)$ and the vectors $b(x, k)$ satisfy the same assumptions as $\sigma(x)$ and $b(x)$ from Sect. 3.4. We also assume that if $y_t = k$, then the process $(y_t)_{t\ge 0}$ jumps to the state l with rate $q_{kl}(x_t)$, i.e.,

$$\mathrm{P}(y_{t+\Delta t} = l \mid y_t = k) = q_{kl}(x_t)\Delta t + o(\Delta t). \tag{35}$$

We assume that the functions q_{kl} are bounded, continuous, and nonnegative. Then, $\xi_t = (x_t, y_t)$, $t \ge 0$ is a Markov process with values in the space $X = \mathbb{R}^d \times I$, and it is called *diffusion process with random switching* [94]. Figure 3 shows an example of the trajectory of such a process.

If $I = \{0, 1\}$ and $\sigma(x, 1) = 0$, then Eq. (34) describes the process which randomly jumps between stochastic and deterministic states. Such processes appear in transport phenomena in sponge–type structures [4, 8, 56].

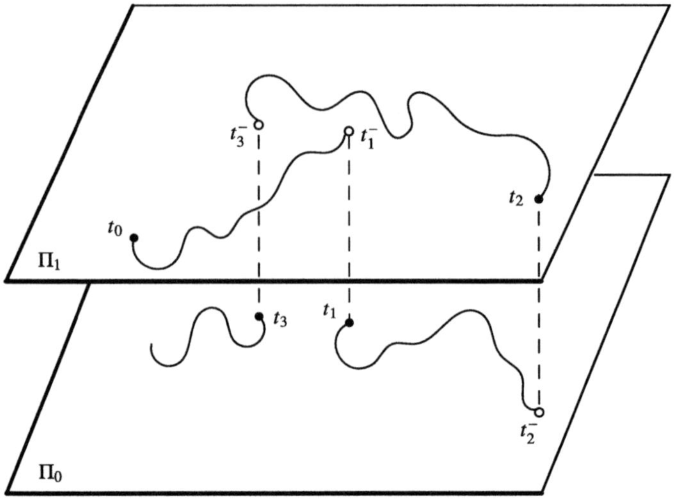

Fig. 3 Example of a trajectory of a diffusion process with random switches. In this case $X = \mathbb{R}^2 \times \{0, 1\}$ and t_0, t_1, t_2, t_3 are the successive switching times

We will now find the stochastic semigroup related to the process $(\xi_t)_{t\ge0}$. Let us denote by $\{S^k(t)\}_{t\ge0}$ the stochastic semigroup with diffusion matrix $\sigma(x, k)$ and with drift vector $b(x, k)$. It has the generator of the form:

$$A_k f = \sum_{i,j=1}^{d} \frac{\partial^2(a_{ij}(x,k)f)}{\partial x_i \partial x_j} - \sum_{i=1}^{d} \frac{\partial(b_i(x,k)f)}{\partial x_i}. \tag{36}$$

Let $q_{kk}(x) = -\sum_{l\neq k} q_{kl}(x)$. For a function f of the variables (x, k), we use the notation $f = [f_0, \dots, f_n]$, where $f_k(x) = f(x, k)$. Let $Af = [A_0 f_0, \dots, A_n f_n]$ and $Qf = [f_0, \dots, f_n]Q$. Then, the operator $A + Q$ is the generator of a stochastic semigroup $\{P(t)\}_{t\ge0}$ associated with the process $(\xi_t)_{t\ge0}$. This semigroup is defined on the space $E = L^1(X, \mathcal{B}(X), m)$, where m is the product of the Lebesgue measure on $\mathbb{R}^d$ and the counting measure on I. The operator A is the generator of a stochastic semigroup $\{S(t)\}_{t\ge0}$ on E given by the formula:

$$S(t)f = (S^0(t)f_0, \dots, S^n(t)f_n).$$

Let $\lambda = \sup_{x\in\mathbb{R}^d} \max\{-q_{00}(x), \dots, -q_{nn}(x)\}$ and $K = \lambda^{-1}Q + I$. Then K is a stochastic operator and $Q = -\lambda I + \lambda K$. Since the generator of the semigroup $\{P(t)\}_{t\ge0}$ is of the form $A - \lambda I + \lambda K$, this semigroup is given by the formulas (23) and (24).

In the case when $\sigma \equiv 0$, the Markov process $(\xi_t)_{t\ge0}$ is a piecewise deterministic. This process describes the action of several *dynamical systems with random switching* between them. Such dynamical systems have many applications in biology.

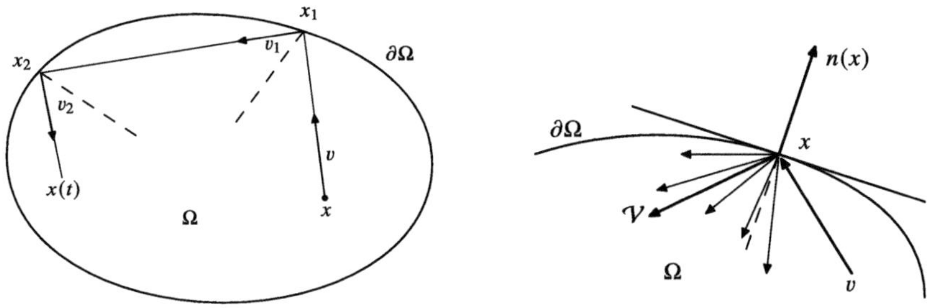

Fig. 4 Left: classical billiards—angles of incidence are equal to the angles of reflection. Right: stochastic billiards. Regular and diffuse reflection: $\mathcal{V}$, the specular reflection; thin vectors, diffuse reflection

They are natural models for describing gene regulatory networks. Depending on the state of activity of the genes present in the network, it is described by different systems of differential equations. A change in gene activity switches the system [10, 53, 82, 84, 95, 96].

3.7 Processes with Nonrandom Moments of Jumps

We will now present some processes in which a change in dynamics or a jump in phase space occurs when the process reaches the boundary of the domain or some fixed subset of the phase space. An example of such a process is stochastic billiards [20, 29, 54, 59]. Similar properties have biological models of cell cycle [52, 72, 73, 78], neuron activity [17, 71, 86, 87], gene regulatory networks [38, 46], and vesicular transport in the cell [13–15].

Let $\Omega \subset \mathbb{R}^d$ be a compact set with non-empty interior and with piecewise smooth boundary $\partial\Omega$. In *stochastic billiards* a moving point does not change its velocity in the interior of Ω, but when it collides with its boundary $\partial\Omega$. The new velocity is chosen randomly from among the velocities directed toward the interior of the set of Ω, and the motion continues until the next collision with the boundary (see Fig. 4).

Let x_t be the position of a moving particle and v_t be its velocity. Consider a stochastic process $\xi_t = (x_t, v_t)$, $t \geq 0$. This process takes values in the space $X = \Omega \times V$, $V = \mathbb{R}^d$. We can also consider stochastic billiards when a particle is moving with constant speed $\|v\| = 1$. Then $V = S^{d-1}$, where S^{d-1} is a unit sphere in the space $\mathbb{R}^d$. Assume that if the particle strikes the boundary at point x and with velocity v, then a new velocity $\bar{v}$ is chosen according to the density distribution $k(x, v, \bar{v})$. Assume that the distribution of the random variable ξ_0 has density $u_0 \in L^1(X, \mathcal{B}(X), \ell)$, where ℓ is the Lebesgue measure on X. Then, each random variable ξ_t has density $u(t, x, v)$, and the function u satisfies the following initial-boundary problem [54]:

$$\frac{\partial u}{\partial t} + v \cdot \nabla_x u(t, x, v) = 0, \quad \text{where } \nabla_x = \left[\frac{\partial}{\partial x_1}, \ldots, \frac{\partial}{\partial x_d}\right], \tag{37}$$

$$u(0, x, v) = u_0(x, v), \tag{38}$$

$$u_{|\Gamma_-} = H(u_{|\Gamma_+}), \tag{39}$$

where

$$\Gamma_\pm = \{(x, v) \in \partial\Omega \times V;\ \pm v \cdot n(x) > 0\},$$

$n(x)$ is the outward unit normal at $x \in \partial\Omega$ and H is a linear boundary operator relating the outgoing and incoming fluxes $u_{|\Gamma_+}$ and $u_{|\Gamma_-}$. The operator H is linear and bounded on the trace spaces:

$$L^1_\pm = L^1(\Gamma_\pm,\ |v \cdot n(x)|\pi(dx) \otimes m_V(dv)) = L^1(\Gamma_\pm, d\mu_\pm(x, v)),$$

where π denotes the Lebesgue surface measure on $\partial\Omega$ and m_V is the Lebesgue measure on the space V. The operator $H: L^1_+ \to L^1_-$ is given by the formula:

$$H\varphi(x, \bar{v}) = \int_{\Gamma_+(x)} k(x, v, \bar{v})\varphi(x, v)|v \cdot n(x)|\, m_V(dv),$$

where $\Gamma_+(x) = \{v \in V: (x, v) \in \Gamma_+\}$. Under some technical assumptions, the solutions of the system (37)–(39) generate a stochastic semigroup on the space $L^1(X, \mathcal{B}(X), \ell)$, and this semigroup has an interesting long-time behavior (see [54]).

One-dimensional version of stochastic billiards was studied in [59]. This version was applied in [73] to study Lebowitz and Rubinow model of the cell cycle [52]. Some models of the cell cycle will be presented in more detail in Sect. 5.8.

4 Long-Time Behavior of Semigroups of Operators

4.1 The Perron–Frobenius Theorem

A good starting point to study long-time behavior of semigroups of operators is a continuous version of the classical Perron–Frobenius theorem. To recall it, let E be the space $\mathbb{R}^n$ with the scalar product $\langle \cdot, \cdot \rangle$. The zero vector in E is denoted by $\mathbf{0}$. If $x \in E$, then we write $x > \mathbf{0}$ when $x_i > 0$ and $x \geq \mathbf{0}$ when $x_i \geq 0$ for $i = 1, 2, \ldots, n$. Let $Q = [q_{ij}]$ be an $n \times n$ real matrix. Then the formula $T(t)x = xe^{tQ}$ defines a semigroup $\{T(t)\}_{t\geq 0}$ on E. We assume that

(P) $q_{ij} \geq 0$ for $i \neq j$,

(I) for all i and j there exists a sequence $(i_1, i_2, \ldots, i_m)$ such that $i_1 = i$, $i_m = j$ and $q_{i_k i_{k+1}} > 0$ for $k = 1, \ldots, m-1$.

Condition (P) guarantees that the semigroup $\{T(t)\}_{t\geq 0}$ is positive, i.e. $T(t)x \geq \mathbf{0}$ for $x \geq \mathbf{0}$. From both conditions (P) and (I), it follows that there are a constant $r \in \mathbb{R}$ and vectors $x^* > \mathbf{0}$ and $y^* > \mathbf{0}$ such that

$$\lim_{t\to\infty} e^{-rt}T(t)x = x^*\langle y^*, x\rangle \quad \text{for } x \in E \tag{40}$$

and the convergence in (40) is exponential. The *irreducibility* assumption (I) implies that the limit set in (40) is one-dimensional. A semigroup $\{T(t)\}_{t\geq 0}$ satisfying condition (40) is said to have *asynchronous exponential growth.* It should be mentioned that one can find in literature, e.g. [92], a more general definition of asynchronous exponential growth, where it is merely assumed that $e^{-rt}T(t)x$ converges to a nonzero finite-rank operator.

This theorem is a consequence of quite nontrivial properties of the matrix Q. First of all, the number r is a *dominant eigenvalue* of Q, which means that any other eigenvalue λ satisfies $\operatorname{Re}\lambda < r$. Also, the general eigenspace associated with r is one-dimensional, and x^* and y^* are left and right eigenvectors corresponding to r, i.e., $x^*Q = rx^*$ and $Qy^* = ry^*$ (it follows that $T(t)x^* = e^{rt}x^*$). Finally, the functional α defined by $\alpha(x) = \langle y^*, x\rangle$ is positive, i.e., $\alpha(x) \geq 0$ for $x \geq 0$.

If assumption (I) is dropped, the dominant eigenvalue $r \in \mathbb{R}$ still exists, but (40) does not hold. More specifically, if the general eigenspace associated with r is one-dimensional, then (P) implies (40), but the inequalities $x^* > \mathbf{0}$ and $y^* > \mathbf{0}$ need not be fulfilled. If the general eigenspace associated with r is not one-dimensional, the behavior of the semigroup $\{T(t)\}_{t\geq 0}$ is more complex. For example, if Q is the 2×2 matrix with $q_{11} = q_{12} = q_{22} = 1$ and $q_{21} = 0$, then $r = 1$ is the dominant eigenvalue and $T(t)x = e^t(x_1, x_2 + x_1 t)$. This means that $\lim_{t\to\infty} e^{-t}t^{-1}T(t)x = xR$, where $xR = (0, x_1)$. Generally, if (P) holds, then the semigroup $\{T(t)\}_{t\geq 0}$ has polynomial exponential growth, i.e., there exists a nonnegative integer k and a $n \times n$ nonzero matrix R with nonnegative elements such that, for each x, $\lim_{t\to\infty} e^{-rt}t^{1-k}T(t)x = xR$. More precisely, $k \times k$ is the size of the largest Jordan block related to the eigenvalue r, and xR is a linear combination of certain eigenvectors related to these Jordan blocks.

If (P) is not satisfied, a dominant eigenvalue, defined as an eigenvalue with the largest real part, may be complex, and there may exist a number of eigenvalues with this property. These two new features result in two new types of asymptotic behavior: using the Jordan–Chevalley decomposition of Q, we show that (a) if imaginary parts of all dominant eigenvalues are commensurable, then the semigroup has a periodic polynomial exponential growth, and (b) if they are not, the semigroup has an almost periodic exponential growth. To summarize, asymptotic behavior of the semigroup $\{T(t)\}_{t\geq 0}$ is determined by its behavior on the invariant subspace E_d, spanned by generalized eigenspaces, with the largest dimension, related to dominant eigenvalues.

If we turn to semigroups in infinite-dimensional spaces, new problems appear because spectral properties of infinite-dimensional semigroups are more complex. For example, consider the intensity matrix Q indexed by integers, with elements $q_{i\,i+1} = q_{i\,i-1} = 1$, $q_{ii} = -2$ and $q_{ij} = 0$ in other cases. Since this Q is related to a bounded linear operator on the space l^1, we may think of the semigroup $\{T(t)\}_{t\geq 0}$ defined by $T(t) = e^{tQ}$. It can be shown that $\{T(t)\}_{t\geq 0}$ is a stochastic semigroup, i.e., $\|T(t)x\| = \|x\|$ for $x \geq 0$. The matrix Q also satisfies conditions (P) and (I), and thus one could expect that there exists $x^* > \mathbf{0}$ such that $T(t)x^* = x^*$, but that is not the case. In other words, for any t, $r = 1$ is not an eigenvalue of $T(t)$, even though the spectral radius of $T(t)$ is 1, and thus condition (40) fails to hold. The main reasons for this is non-compactness of trajectories $t \mapsto T(t)x$.

The second problem in infinite-dimensional spaces is related to the rate of convergence in (40). Let $\{T(t)\}_{t\geq 0}$ be a semigroup on a Banach space E. We say that this semigroup has *asynchronous* (or *balanced*) *exponential growth* if there exist $r \in \mathbb{C}$, nonzero $x^* \in E$, and a nonzero linear functional $\alpha\colon E \to \mathbb{C}$ such that

$$\lim_{t\to\infty} e^{-rt}T(t)x = x^*\alpha(x) \quad \text{for } x \in E. \tag{41}$$

This definition implies that r is a simple eigenvalue of the infinitesimal generator A of the semigroup $\{T(t)\}_{t\geq 0}$. In finite-dimensional cases, convergence in (41) is exponential because the spectrum of A is finite and all eigenvalues λ of A, other than r, satisfy the inequality $\operatorname{Re}\lambda < \operatorname{Re} r$. But in the infinite-dimensional case, condition (41) does not guarantee that r is an isolated element of the spectrum of A and the rate of convergence in (41) needs not be exponential.

Finally, note that if $E_0 = \operatorname{Ker}\alpha$, then E_0 is invariant with respect to the semigroup. By restricting $\{T(t)\}_{t\geq 0}$ to this subspace, the problem of the rate of convergence (41) is replaced by that of the rate of convergence of $\|e^{-rt}T(t)x\| \to 0$ for $x \in E_0$.

4.2 Asymptotic Decomposition via Spectral Properties

There are several results on asymptotic behavior of strongly continuous semigroups which are in the spirit of the previous section. These results are based on compactness properties of semigroups and are thoroughly discussed in monographs [27, 61].

Let A be the infinitesimal generator of a strongly continuous semigroup $\{T(t)\}_{t\geq 0}$ on a Banach space E and let $\mu \in \mathbb{C}$ be an isolated point of the spectrum of A. Since the resolvent R_λ is a holomorphic function of λ, it can be expanded as a Laurent series:

$$R_\lambda = \sum_{n=-\infty}^{\infty} (\lambda - \mu)^n U_n$$

for $0 < |\lambda - \mu| < \delta$ and some sufficiently small $\delta > 0$. The coefficients U_n of this series are bounded operators given by the formulas:

$$U_n = \frac{1}{2\pi i} \oint_\gamma \frac{R_\lambda}{(\lambda - \mu)^{n+1}}\, d\lambda, \quad n \in \mathbb{Z},$$

where γ is the positively oriented boundary of a disc with radius $< \delta$ and center at μ. The operator $P = U_{-1}$ is called the *residue* of R_λ at μ. We have $U_{-(n+1)} = (A - \mu I)^n P$. If there exists a $k > 0$ such that $U_{-k} \neq 0$ while $U_{-n} = 0$ for all $n > k$, then μ is called a pole of R_λ of order k.

Let $(L(E), \|\cdot\|)$ be the space of all bounded linear operators on E endowed with the operator norm. Denote by $K(E) \subseteq L(E)$ the set of all compact operators on E. A strongly continuous semigroup $\{T(t)\}_{t\geq 0}$ on E is called *eventually compact* if there exists $t_0 > 0$ such that the operator $T(t_0)$ is compact. It can be shown that an eventually compact semigroup is norm continuous for $t \geq t_0$. A strongly continuous semigroup $\{T(t)\}_{t\geq 0}$ on E is called *quasi-compact* if

$$\lim_{t\to\infty} \inf\{\|T(t) - K\| \colon K \in K(E)\} = 0.$$

It is clear that an eventually compact semigroup is quasi-compact but not vice versa. The asymptotic behavior of a quasi-compact semigroup is fully characterized by the following theorem (see [27, p. 333]).

Theorem 10 *Let $\{T(t)\}_{t\geq 0}$ be a quasi-compact semigroup on a Banach space E with generator A. Then the set $\{\lambda \in \sigma(A) \colon \operatorname{Re}\lambda \geq 0\}$ is finite (or empty) and consists of poles of R_λ of finite algebraic multiplicity. If we denote these poles by $\lambda_1, \ldots, \lambda_m$, and their residues by $P_1, \ldots, P_m$ with poles of order $k_1, \ldots, k_m$, we have $T(t) = T_1(t) + T_2(t) + \cdots + T_m(t) + R(t)$, where*

$$T_n(t) = e^{\lambda_n t} \sum_{j=0}^{k_n - 1} \frac{t^j}{j!}(A - \lambda_n I)^j P_n \quad \text{for } t \geq 0 \text{ and } 1 \leq n \leq m, \tag{42}$$

and

$$\|R(t)\| \leq M e^{-\varepsilon t} \quad \text{for some } \varepsilon > 0,\ M \geq 1 \text{ and all } t \geq 0. \tag{43}$$

For a particular instance of this theorem, consider the case when there exists a unique dominant eigenvalue $r \in \mathbb{C}$ being a first-order pole. Then, r is the *spectral bound* $s(A)$ of A, i.e.,

$$s(A) = \sup\{\operatorname{Re}\lambda \colon \lambda \in \sigma(A)\}$$

and $P = \lim_{\lambda\to r} (\lambda - r) R_\lambda$. From Theorem 10 it follows that

$$\|e^{-rt}T(t) - P\| \le Me^{-\varepsilon t} \quad \text{for some } \varepsilon > 0,\ M \ge 1 \text{ and all } t \ge 0. \tag{44}$$

Theorem 10 is also a proper background for a generalization of the Perron–Frobenius theorem to positive semigroups on Banach lattices. To obtain a version of the Perron–Frobenius theorem for such a semigroup, one typically checks, instead of (I), that the semigroup is *irreducible*; this means that for every $x \in E_+ \setminus \{\mathbf{0}\}$ and for every $x^* \in E_+^* \setminus \{\mathbf{0}\}$, there exists a $t > 0$ such that $x^*(T(t)x) > 0$ (here $E_+ = \{x \in E \colon x \ge 0\}$ and $E_+^* = \{x \in E^* \colon x \ge 0\}$). Indeed, the irreducibility rules out the possibility for the range of P to be more than one-dimensional, and together with quasi-compactness implies asynchronous exponential growth of the semigroup. A similar result can be obtained if we replace quasi-compactness by an assumption based on the notion of the *measure of non-compactness* of A; see, e.g., [92].

The asynchronous exponential growth (41) is a special case of the situation when there exists the limit:

$$\lim_{t\to\infty} e^{-rt}T(t) = P, \tag{45}$$

where $P\colon E \to E$ is an arbitrary nonzero operator and the convergence can be uniform, strong, or weak. The latter property was studied in [89, 90], and a comprehensive overview of applications of this theory to population dynamics can be found in [93]. If $\widetilde{T}(t) = e^{-rt}T(t)$ and $\widetilde{E} = \operatorname{Ker} P$, then condition (45) in the case of strong convergence is equivalent to *strong asymptotic stability* of the semigroup $\{\widetilde{T}(t)\}_{t\ge 0}$, i.e., to the condition

$$\lim_{t\to\infty} \|\widetilde{T}(t)x\| = 0 \quad \text{for } x \in \widetilde{E}. \tag{46}$$

In the case when the semigroup is bounded and the set $\sigma(A) \cap (i\mathbb{R})$ is at most countable, the strong asymptotic stability holds if and only if the adjoint operator A^* has no pure imaginary eigenvalues [2, 55]. Some versions of this result for unbounded operators are proved in [85].

The question of deriving the optimal rate of decay in (46) is highly nontrivial. This problem is extensively discussed in [6]. To give a taste of this theory, we recall one of the results of [11] (Theorem 2.4). Let $\{T(t)\}_{t\ge 0}$ be a strongly continuous semigroup on a Hilbert space E with generator A such that $\sigma(A) \cap (i\mathbb{R}) = \emptyset$. Then for a fixed $\alpha > 0$, the following conditions are equivalent:

(i) $R_{is} = O(|s|^\alpha), \quad s \to \infty,$
(ii) $\|T(t)A^{-1}\| = O(t^{-\alpha}), \quad t \to \infty,$
(iii) $\|T(t)A^{-1}x\| = o(t^{-\alpha}), \quad t \to \infty, x \in E.$

4.3 Asymptotics of Stochastic Semigroups: Introductory Results

We will consider stochastic semigroups on the space $L^1 = L^1(X, \Sigma, m)$ with a σ-finite measure m. The iterates of a stochastic operator form a *discrete-time* semigroup, and by writing $P(t) = P^t$ for these iterates, we may formulate most of definitions and results for both types of semigroups in a unified manner.

We will begin by studying the asymptotics of uniformly continuous stochastic semigroups. We recall that such a semigroup is given by the formula:

$$P(t)f = \sum_{k=0}^{\infty} \frac{(\lambda t)^k e^{-\lambda t}}{k!} P^k f, \tag{47}$$

where P is a stochastic operator and $\lambda \geq 0$.

Theorem 11 *If $Pf^* = f^*$ for some density $f^* > 0$ a.e., then for any density f, there exists an invariant density $\bar{f}$ such that $\bar{f} = \lim_{t\to\infty} P(t)f$ in L^1. If additionally, f^* is a unique invariant density of P, then $\lim_{t\to\infty} P(t)f = f^*$.*

The proof of this theorem is given in [75]. In the proof we define a new operator $\tilde{P}f = P(ff^*)/f^*$ on the space $L^1(X, \Sigma, \mu)$ with probability measure $\mu(dx) = f^*(x)\, m(dx)$. Then $\tilde{P}$ is a *double stochastic operator*, i.e., $\tilde{P}$ is a stochastic operator and $\tilde{P}\mathbf{1}_X = \mathbf{1}_X$. Next we show that Theorem 11 results from the following lemma.

Lemma 1 *If P is a double stochastic operator, then $P\,\mathrm{E}(f|\Sigma_{\mathrm{inv}}) = \mathrm{E}(f|\Sigma_{\mathrm{inv}})$ and*

$$\lim_{t\to\infty} \|e^{t(P-I)}f - \mathrm{E}(f|\Sigma_{\mathrm{inv}})\| = 0 \quad \text{for } f \in L^1(X, \Sigma, \mu).$$

The σ-algebra Σ_{inv} is given by $\Sigma_{\mathrm{inv}} = \{A \in \Sigma\colon P^*\mathbf{1}_A = \mathbf{1}_A\}$.

A stochastic semigroup $\{P(t)\}_{t\geq 0}$ is called *asymptotically stable* if there exists a density f^* such that

$$\lim_{t\to\infty} \|P(t)f - f^*\| = 0 \quad \text{for} \quad f \in D. \tag{48}$$

We recall that D denotes the set of all densities. From (48) it follows that f^* is an *invariant density*, i.e., $P(t)f^* = f^*$ for $t \geq 0$. If the semigroup $\{P(t)\}_{t\geq 0}$ is generated by an evolution equation $u'(t) = Au(t)$, then its asymptotic stability means that the stationary solution $u(t) = f^*$ is globally asymptotically stable in the sense of Lyapunov on the set of densities D. Observe that if there is an invariant density f^*, then the stochastic semigroup $\{P(t)\}_{t\geq 0}$ is asymptotically stable if and only if $\lim_{t\to\infty} \|P(t)f\| = 0$ for $f \in L^1_0 = \{f \in L^1\colon \int_X f(x)\, m(dx) = 0\}$.

There are several results concerning asymptotic stability of stochastic semigroups. We start with the *lower function theorem* of Lasota and Yorke [50]. Let $f^-(x) = \max(0, -f(x))$. A function $h \in L^1$, such that $h \geq 0$ and $h \neq 0$, is called a *lower function* for a stochastic semigroup $\{P(t)\}_{t\geq 0}$ if

$$\lim_{t\to\infty} \|(P(t)f - h)^-\| = 0 \quad \text{for every } f \in D. \tag{49}$$

Theorem 12 *A continuous or discrete-time stochastic semigroup is asymptotically stable if and only if it has a lower function h.*

This theorem was generalized in many directions. The interested reader is referred to the book [26], where one can find versions formulated in the language of von Neumann algebras and to the paper [31] which provides a comprehensive overview of earlier results and presents new theorems for semigroups on Banach lattices including a version with lower functions depending on f.

4.4 Partially Integral Semigroups

In next two sections, we present two theorems on asymptotic behavior of stochastic semigroups, which stem from the theory of Harris operators [30, 58, 62].

A stochastic semigroup $\{P(t)\}_{t\ge 0}$ is called *partially integral* if there exist a $t > 0$ and a measurable function $q(t, \cdot, \cdot): X \times X \to [0, \infty)$, called the *kernel of the semigroup*, such that

$$\int_X \int_X q(t, x, y)\, m(dx)\, m(dy) > 0$$

and

$$P(t)f(y) \ge \int_X q(t, x, y) f(x)\, m(dx) \quad \text{for } f \in D. \tag{50}$$

Theorem 13 *Let $\{P(t)\}_{t\ge 0}$ be a partially integral stochastic semigroup. We assume that this semigroup has invariant density f^* and there are no other periodic points in the set of densities. If $f_* > 0$ a.e., then the semigroup $\{P(t)\}_{t\ge 0}$ is asymptotically stable.*

The proof of Theorem 13 is given in [79]. If $\{P(t)\}_{t\ge 0}$ is a continuous-time stochastic semigroup, then Theorem 13 can be strengthened significantly [68].

Theorem 14 *Suppose that a continuous-time partially integral stochastic semigroup $\{P(t)\}_{t\ge 0}$ has a unique invariant density f^*. If $f^* > 0$ a.e., then the semigroup $\{P(t)\}_{t\ge 0}$ is asymptotically stable.*

Corollary 1 *If a continuous-time partially integral stochastic semigroup is irreducible and has an invariant density, then it is asymptotically stable.*

In order to formulate other corollaries to Theorem 13, we need two auxiliary definitions.

We say that a stochastic semigroup $\{P(t)\}_{t\ge 0}$ *spreads supports* if for every set $A \in \Sigma$ and for every $f \in D$, we have:

$$\lim_{t\to\infty} m(\operatorname{supp} P(t)f \cap A) = m(A)$$

and *overlaps supports,* if for every $f, g \in D$ there exists $t > 0$ such that

$$m(\operatorname{supp} P(t)f \cap \operatorname{supp} P(t)g) > 0.$$

Corollary 2 *A partially integral stochastic semigroup which spreads supports and has an invariant density is asymptotically stable.*

Corollary 3 *A partially integral stochastic semigroup which overlaps supports and has an invariant density $f_* > 0$ a.e. is asymptotically stable.*

We will now introduce the notion of sweeping, which, together with asymptotic stability, will allow us to give a characterization of the behavior of a rather wide class of substochastic semigroups.

A stochastic semigroup $\{P(t)\}_{t\geq 0}$ is called *sweeping* with respect to a set $B \in \Sigma$ if for every $f \in D$

$$\lim_{t\to\infty} \int_B P(t)f(x)\, m(dx) = 0.$$

The notion of sweeping was introduced in [45], but it is also known as *zero-type property*. In [45] the following theorem was proved.

Theorem 15 *Let $\{P(t)\}_{t\geq 0}$ be an integral stochastic semigroup. Assume that this semigroup has no invariant density, while for a certain set $A \in \Sigma$, the following condition holds:*
(KT) *there exists a measurable function f^* such that $0 < f^* < \infty$ a.e. and $P(t)f^* \leq f^*$ for $t \geq 0$, $f^* \notin L^1$ and $\int_A f^*(x)\, m(dx) < \infty$.*
Then the semigroup $\{P(t)\}_{t\geq 0}$ is sweeping with respect to A.

In paper [79] it was shown that Theorem 15 holds for a broader class of semigroups than integrals, for example, to overlapping semigroups.

Theorem 16 *Let $\{P(t)\}_{t\geq 0}$ be a stochastic semigroup which overlaps supports. Assume that the semigroup $\{P(t)\}_{t\geq 0}$ and a set $A \in \Sigma$ satisfy condition* (KT). *Then the semigroup $\{P(t)\}_{t\geq 0}$ is sweeping with respect to A.*

The advantage of Theorems 15 and 16 is that knowing the function f^*, we can determine the sets from which the semigroup is sweeping. The main problem in applying Theorems 15 and 16 is to prove that a stochastic semigroup satisfies condition (KT).

4.5 Asymptotic Decomposition via Kernel Minorants

Theorems 13 and 14 allow us to prove the asymptotic stability of Markov semigroups when we know that there is a positive invariant density. Unfortunately, in most applications it is difficult to determine the invariant density or even to prove its existence.

Our immediate goal will be to present results on the asymptotic behavior of substochastic semigroups, in which we do not directly assume that the semigroup has invariant density. One of the corollaries will be a criterion of sweeping from compact sets, not based on condition (KT) and convenient for applications.

We present a method developed in [69, 70] to study asymptotic properties of stochastic semigroups via their asymptotic decomposition. From now on we assume additionally that (X, ρ) is a separable metric space and $\Sigma = \mathcal{B}(X)$ is the σ-algebra of Borel subsets of X, and consider a partially integral substochastic semigroups $\{P(t)\}_{t\geq 0}$ with a kernel q which satisfies the following condition:

(K) for every $x_0 \in X$ there exist an $\varepsilon > 0$, a $t > 0$, and a measurable function $\eta \geq 0$ such that $\int \eta(x)\, m(dx) > 0$ and

$$q(t, x, y) \geq \eta(y) \quad \text{for } x \in B(x_0, \varepsilon),\ y \in X, \tag{51}$$

where $B(x_0, \varepsilon) = \{x \in X : \ \rho(x, x_0) < \varepsilon\}$.

It is clear that substochastic semigroups satisfying condition (K) are partially integral.

Now we present two theorems on asymptotic decomposition: the first for continuous-time substochastic semigroups and the second for substochastic operators (discrete-time semigroups) [69, 70].

Theorem 17 *Let $\{P(t)\}_{t\geq 0}$ be a continuous-time substochastic semigroup which satisfies* (K). *Then, there exist an at most countable set J, a family of invariant densities $\{f_j^*\}_{j\in J}$ with disjoint supports $\{A_j\}_{j\in J}$, and a family $\{\alpha_j\}_{j\in J}$ of positive linear functionals defined on L^1 such that*

(i) *for every $j \in J$ and for every $f \in L^1$, we have*

$$\lim_{t\to\infty} \|\mathbf{1}_{A_j} P(t) f - \alpha_j(f) f_j^*\| = 0, \tag{52}$$

(ii) *if $Y = X \setminus \bigcup_{j\in J} A_j$, then for every $f \in L^1$ and for every compact set F, we have:*

$$\lim_{t\to\infty} \int_{F\cap Y} P(t) f(x)\, m(dx) = 0. \tag{53}$$

We note that not only the sets A_j, $j \in J$, are disjoint but also their closures are disjoint, which means that the measures $\mu_j(dx) = f_j^*(x)\,m(dx)$ have disjoint topological supports [70].

We will precede the formulation of the asymptotic decomposition for substochastic operators with some auxiliary definitions.

Let E be a vector space. A linear mapping $T\colon E \to E$ satisfying the condition $T^2 = T$ is called a *projection*. If $y \in T(E)$, then $y = Tx$ for some $x \in E$, and from the definition of projection, we get $Ty = T^2x = Tx = y$, so T is the identity on $T(E)$.

Let (X, Σ, m) be a σ-finite measure space. A substochastic operator S is called *periodic* on the set $B \in \Sigma$, if there is a sequence of densities $h_1, \dots, h_k$ satisfying the following conditions:

$$h_i(x) = 0 \quad \text{for } i = 1, \dots, k \text{ and } x \notin B, \tag{54}$$

$$h_i h_j = 0 \quad \text{for } i \neq j, \qquad h_1(x) + \cdots + h_k(x) > 0 \quad \text{for } x \in B, \tag{55}$$

$$Sh_i = h_{i+1} \quad \text{for } i \leq k, \qquad \text{where} \quad h_{k+1} := h_1, \tag{56}$$

and $Sf = STf$ for $f \in L^1(X, \Sigma, m)$, where T is a positive projection from the space $L^1(X, \Sigma, m)$ onto its linear subspace spanned by functions $h_1, \dots, h_k$.

Theorem 18 *Suppose that the substochastic operator P has the property* (K). *Then there exist at most countable set J, a family $\{A_j\}_{j\in J}$ of pairwise disjoint measurable subsets of X and the family of substochastic operators $\{S_j\}_{j\in J}$, satisfying the conditions:*

(i) $P^*\mathbf{1}_{A_j} \geq \mathbf{1}_{A_j}$ *for* $j \in J$;
(ii) *for each $j \in J$ the operator S_j is periodic on the set A_j;*
(iii) *for each $j \in J$ and for any function $f \in L^1(X, \Sigma, m)$ we have*

$$\lim_{n\to\infty} \|\mathbf{1}_{A_j} P^n f - S_j^n f\| = 0; \tag{57}$$

(iv) *if $Y = X \setminus \bigcup_{j\in J} A_j$, then for any function $f \in L^1(X, \Sigma, m)$ and any compact set $F \subset X$, we have:*

$$\lim_{n\to\infty} \int\limits_{F\cap Y} P^n f(x)\,m(dx) = 0. \tag{58}$$

Remark 15 From (i) it follows that the operator P can be restricted to the space $L^1(A_j)$ and P is a stochastic operator on this space. Also the operator S_j restricted to the space $L^1(A_j)$ is a stochastic operator, and this restriction is given by the formula:

$$S_j f = \sum_{i=1}^{k} \alpha_i(f) h_{i+1}, \quad \alpha_i(f) = \int_{B_i} f(x)\, m(dx), \ B_i = \operatorname{supp} h_i, \tag{59}$$

where the functions $h_1, \ldots, h_k$ are the densities appearing in the definition of periodic operator. The operator P satisfies (56), i.e., $Ph_i = h_{i+1}$.

The following criterion of sweeping follows from Theorems 17 and 18.

Corollary 4 *Assume that a substochastic (continuous- or discrete-time) semigroup* $\{P(t)\}_{t\ge 0}$ *satisfies condition* (K) *and has no invariant densities. Then* $\{P(t)\}_{t\ge 0}$ *is sweeping from compact sets.*

Our next aim is to find conditions guaranteeing that a stochastic semigroup $\{P(t)\}_{t\ge 0}$ satisfies the *Foguel alternative*, i.e., it is asymptotically stable or sweeping from all compact sets [49]. The following result is a step in this direction.

Corollary 5 *If a continuous-time substochastic semigroup* $\{P(t)\}_{t\ge 0}$ *is irreducible and satisfies condition* (K)*, then the Foguel alternative holds.*

In the case of substochastic operator, we have a slightly weaker version of this corollary. A substochastic operator P is called *asymptotically periodic* if there exists a periodic operator S such that

$$\lim_{n\to\infty} \|P^n f - S^n f\| = 0 \quad \text{for } f \in D.$$

Corollary 6 *If a substochastic operator satisfies* (K) *and is irreducible, then it is either asymptotically periodic or sweeping from compact sets.*

Now we apply Corollary 5 to Markov chains. Let Q be the intensity matrix of a continuous-time Markov chain $(\xi_t)_{t\ge 0}$ on $\mathbb{N}$. Since the space $\mathbb{N}$ is discrete, the stochastic semigroup $\{P(t)\}_{t\ge 0}$ related to the process $(\xi_t)_{t\ge 0}$ satisfies condition (K).

Theorem 19 *Let* $\{P(t)\}_{t\ge 0}$ *be the stochastic semigroup on* l^1 *related to a continuous-time Markov chain* $(\xi_t)_{t\ge 0}$ *with intensity matrix* Q*. Let us assume that the entries of the matrix* Q *satisfy the following condition:*

(I) *for all* i *and* j *there exists a sequence* $(i_1, i_2, \ldots, i_m)$ *such that* $i_1 = i$, $i_m = j$ *and* $q_{i_k i_{k+1}} > 0$ *for* $k = 1, \ldots, m-1$.

Then the semigroup $\{P(t)\}_{t\ge 0}$ *satisfies the Foguel alternative:*

(a) *if the semigroup* $\{P(t)\}_{t\ge 0}$ *has an invariant density, then it is asymptotically stable,*
(b) *if the semigroup* $\{P(t)\}_{t\ge 0}$ *has no invariant density, then for every* $x \in l^1$ *and* $i \in \mathbb{N}$ *we have*

$$\lim_{t\to\infty} (P(t)x)_i = 0. \tag{60}$$

If a stochastic semigroup (continuous or discrete-time) overlaps supports, then the statements of Theorems 17 and 18 can be significantly strengthened (see [72] Proposition 1).

Corollary 7 *Assume that a stochastic semigroup* $\{P(t)\}_{t\geq 0}$ *satisfies* (K) *and overlaps supports. Then, this semigroup is sweeping or has an invariant density* f^* *with a support A, and there exists a positive linear functional* α *defined on* $L^1(X, \Sigma, m)$ *such that*

(i) *for every* $f \in L^1(X, \Sigma, m)$, *we have:*

$$\lim_{t\to\infty} \|\mathbf{1}_A P(t)f - \alpha(f)f^*\| = 0, \tag{61}$$

(ii) *if* $Y = X \setminus A$, *then for every* $f \in L^1(X, \Sigma, m)$ *and for every compact set* F, *we have:*

$$\lim_{t\to\infty} \int\limits_{F\cap Y} P(t)f(x)\, m(dx) = 0. \tag{62}$$

In particular, if $\{P(t)\}_{t\geq 0}$ *has an invariant density* f^* *with the support A and* $X \setminus A$ *is a subset of a compact set, then this semigroup is asymptotically stable.*

In order to formulate a further corollary, we need to introduce two additional conditions, which are weak versions of well-known concepts: irreducibility and tightness of measures.

(WI) There exists a point $x_0 \in X$ such that for each $\varepsilon > 0$ and for each density f we have

$$\int\limits_{B(x_0,\varepsilon)} P(t)f(x)\, m(dx) > 0 \quad \text{for some } t > 0. \tag{63}$$

(WT) There exists $\kappa > 0$ such that

$$\sup_{F\in\mathcal{F}} \limsup_{t\to\infty} \int_F P(t)f(x)\, m(dx) \geq \kappa \tag{64}$$

for $f \in D_0$, where D_0 is a dense subset of D and $\mathcal{F}$ is the family of all compact subsets of X.

Corollary 8 *Let* $\{P(t)\}_{t\geq 0}$ *be a continuous-time stochastic semigroup. Assume that* $\{P(t)\}_{t\geq 0}$ *satisfies conditions* (K), (WI), *and* (WT). *Then the semigroup* $\{P(t)\}_{t\geq 0}$ *is asymptotically stable.*

Corollary 9 *Suppose that X is a compact space. Then, a continuous-time stochastic semigroup which satisfies* (K) *and* (WI) *is asymptotically stable.*

The proof of Corollary 8 can be found in [70]. The second corollary follows directly from the first one. New results concerning positive operators on Banach lattices similar in spirit to Theorems 14 and 17 may be found in [32].

An advantage of the formulation of some results here in the form of the Foguel alternative is that in order to show asymptotic stability, we do not need to prove the existence of an invariant density. It is enough to check that the semigroup is not sweeping with respect to compact sets; then, automatically, the semigroup $\{P(t)\}_{t\geq 0}$ is asymptotically stable. We can eliminate the sweeping by means of some method similar to that of Lyapunov function called Hasminskiĭ function.

Theorem 20 *Let A be the generator of a stochastic semigroup $\{P(t)\}_{t\geq 0}$, $V: X \to [0, \infty)$ be a measurable function, Z be a measurable set, and $\varepsilon > 0$. Assume that there exists a measurable function, formally denoted by LV, satisfying conditions:*

$$\int_X LV(x)g(x)\,m(dx) = \int_X V(x)Ag(x)\,m(dx) \quad \text{for } g \in \mathfrak{D}(A) \cap D_V, \tag{65}$$

$$LV(x) \leq M - \varepsilon \quad \text{for } x \in Z, \qquad LV(x) \leq -\varepsilon \quad \text{for } x \in X \setminus Z, \tag{66}$$

where $D_V = \{f \in D\colon\ \int_X f(x)V(x)\,m(dx) < \infty\}$. Then the semigroup $\{P(t)\}_{t\geq 0}$ is not sweeping with respect to the set Z. If Z is compact, then the semigroup satisfies condition (WT).

The main difficulty in applying this method lies in correctly defining the action of the operator L on the function V, since V does not usually belong to the domain of the operator L. The Hasminskiĭ function method has been used to study Markov chains [77], multistate diffusion processes [66], diffusion with jumps [67], and dynamical systems with jumps [65].

4.6 Constrictive Operators

The next theorem we discuss here shows that convergence of trajectories of a stochastic semigroup to a compact set leads to interesting asymptotic properties. A stochastic semigroup $\{P(t)\}_{t\geq 0}$ is called strongly (weakly) *constrictive* [51] if there exists a strongly (weakly) compact subset F of L^1 such that

$$\lim_{t\to\infty} \inf\{\|P(t)f - g\|: g \in F\} = 0 \quad \text{for } f \in D.$$

Even though the definition of the constrictive operator and the property (K) are significantly differ, similar conclusions follow. One of them is the following theorem.

Theorem 21 *Suppose that a stochastic operator P defined on the space $L^1(X, \Sigma, m)$ is constrictive. Then there exist a finite set J, a family $\{A_j\}_{j\in J}$ of pairwise disjoint measurable subsets of X and the family $\{S_j\}_{j\in J}$ of substochastic operators satisfying the conditions:*

(i) $P^*\mathbf{1}_{A_j} \ge \mathbf{1}_{A_j}$ *for* $j \in J$*;*
(ii) for each $j \in J$ *the operator* S_j *is periodic on the set* A_j*;*
(iii) for each $j \in J$ *and for any function* $f \in L^1(X, \Sigma, m)$*, we have:*

$$\lim_{n\to\infty} \|\mathbf{1}_{A_j} P^n f - S_j^n f\| = 0; \tag{67}$$

(iv) if $Y = X \setminus \bigcup_{j\in J} A_j$*, then for any function* $f \in L^1(X, \Sigma, m)$ *we have*

$$\lim_{n\to\infty} \int_Y P^n f(x)\, m(dx) = 0. \tag{68}$$

Theorem 21 was first proved in [51] under assumption of strong constrictivity; 2 years later Komorník [44] showed that weak constrictivity suffices. The original formulation of Theorem 21 was different from that cited here. We have given its statement formulated in the spirit of Theorems 17 and 18.

If there exists a function $h \in L^1$ such that

$$\lim_{t\to\infty} \|(P(t)f - h)^+\| = 0 \quad \text{for every } f \in D, \tag{69}$$

then Theorem 21 holds. It is a simple consequence of the fact that the set $F = \{f \in D\colon f \le h\}$ is weakly compact. Theorem 21 can be strengthened considerably in the case when $\{P(t)\}_{t\ge 0}$ is a continuous-time stochastic semigroup. Then we replace condition (68) by condition (52).

The property of constrictiveness was generalized and studied in many papers (see [26] for a comprehensive overview of these results). Examples of applications of Theorem 21 and the lower function theorem are given in [49].

5 Applications to Biological Models

5.1 Model of DNA Evolution

The degree of relatedness between humans, species, genes, and proteins can be studied by comparing DNA chain sequences. DNA is inherited from parents, but mutations occur as a result of errors in DNA replication and repair. Mutations occur throughout the DNA chain at a constant level on average, and although the potential mutations may be numerous, as a result of natural selection, most of them are not conserved and mainly selection-neutral mutations are observed.

The simplest way to measure evolutionary distance is to check in how many places the aligned DNA sequences differ. More precisely, we first compare and align DNA sequences so that they match, and then check the differences between them. The initial issue here is modeling nucleotide substitutions at a fixed point in the DNA chain. Since we have four different nucleotides corresponding to nucleobases A, G, C, T, so models describing nucleotide evolution, called *models of substitution* of nucleotides in the DNA sequences, are Markov chains on a four-state space $S = \{A, G, C, T\}$. The intensity matrix in this chain is of the form:

$$Q = \begin{bmatrix} q_{AA} & q_{AG} & q_{AC} & q_{AT} \\ q_{GA} & q_{GG} & q_{GC} & q_{GT} \\ q_{CA} & q_{CG} & q_{CC} & q_{CT} \\ q_{TA} & q_{TG} & q_{TC} & q_{TT} \end{bmatrix}.$$

The simplest substitution model is *Jukes–Cantor model* [41]. In this model, we assume that probability of replacing any nucleotide by another nucleotide is the same. The intensity matrix Q in this model is of the form:

$$Q = \begin{bmatrix} -3\lambda & \lambda & \lambda & \lambda \\ \lambda & -3\lambda & \lambda & \lambda \\ \lambda & \lambda & -3\lambda & \lambda \\ \lambda & \lambda & \lambda & -3\lambda \end{bmatrix}.$$

Let $p_{ij}(t) = \mathrm{P}(\xi_t = j | \xi_0 = i)$. It is easy to check that

$$p_{ij}(t) = \frac{1}{4} - \frac{1}{4}e^{-4\lambda t} \quad \text{for } i \neq j, \qquad p_{ii}(t) = \frac{1}{4} + \frac{3}{4}e^{-4\lambda t} \quad \text{for } i \in S.$$

These formulas allow us to estimate the evolutionary distance of two different individuals or species based on differences in aligned DNA sequences. We will assume that both sequences are derived from a common ancestor and that, as a result of mutations, there have been independent changes in both sequences in successive generations. As a measure of evolutionary distance, we will take the time elapsed since the common ancestor, or the number of generations.

We assume that a mutation of a single nucleotide occurred on average every t_0 years. To begin, we compare the sequence of an ancestor that lived t years ago with that of a contemporary living individual. Since the intensity of nucleotide change is 3λ, so $3\lambda t_0 = 1$. The probability of changing a nucleotide at a fixed site is

$$p = 3\left(\tfrac{1}{4} - \tfrac{1}{4}e^{-4\lambda t}\right).$$

Since $3\lambda t_0 = 1$, we have:

$$\frac{t}{t_0} = -\tfrac{3}{4}\ln\left(1 - \tfrac{4}{3}p\right). \tag{70}$$

If the sequences are of length n and we have n_d differences, we assume that $\hat{p} = n_d/n$ is the estimated value of p.

Let us now consider two sequences of contemporary individuals. Suppose we want to determine how many years ago their common ancestor lived. The Markov chain in this model is reversible. Thanks to this symmetry, we can treat one of the individuals as the „ancestor" of their common ancestor. Thus, the time to the common ancestor will be twice as small as the determined by the formula (70).

5.2 *Growth of Erythrocyte Population*

Consider the following *model of erythrocyte population growth* (red blood cells). Erythrocytes are formed in the bone marrow by the division and differentiation of erythroid stem cells. Mature erythrocytes enter the bloodstream, where they serve mainly to transport oxygen to the cells and are destroyed after about 4 months. Assume that $b > 0$ is the intensity of erythrocyte production in the bone marrow, and $d > 0$ is the destruction rate of a single erythrocyte. We check that the number of erythrocytes in the bloodstream is described by a Markov chain (a certain birth–death process) on the space $\mathbb{N}$. The intensity matrix Q has entries: $q_{i\,i+1} = b$, $q_{i\,i-1} = di$, and $q_{ii} = -(b + di)$ for $i \in \mathbb{N}$ and $q_{ij} = 0$ otherwise.

First, using Theorem 9 we prove that the matrix Q generates a stochastic semigroup $\{P(t)\}_{t\ge 0}$. Let $\lambda > 0$ be a positive constant and assume a contrary that there exists a nonzero and nonnegative $x \in l^\infty$ which is a solution of the equation $Qx = \lambda x$. Then

$$dix_{i-1} - (b + di)x_i + bx_{i+1} = \lambda x_i \quad \text{for } i \in \mathbb{N}, \tag{71}$$

where we formally assume that $x_{-1} = 0$. Then

$$x_{i+1} - x_i = \frac{\lambda}{b}x_i + \frac{d}{b}(x_i - x_{i-1})i. \tag{72}$$

Hence, the sequence (x_i) is increasing and $x_{i+1} \ge \left(1+\frac{\lambda}{b}\right)x_i$. Consequently, $x \notin l^\infty$, and the matrix Q generates a stochastic semigroup.

We show that the semigroup $\{P(t)\}$ has an invariant density. Since the generator A of this semigroup is given by the formula $Ax = xQ$, we are looking for a positive solution of the equation $xQ = 0$. Then

$$bx_{i-1} - (b + di)x_i + d(i + 1)x_{i+1} = 0 \quad \text{for } i \in \mathbb{N}, \tag{73}$$

where $x_{-1} = 0$. The sequence $x_i = b^i/(d^i i!)$ satisfies (73). Since $\sum_{i=0}^{\infty} x_i = e^{b/d}$, in order to obtain a probabilistic distribution, we divide x_i by $e^{b/d}$. This will yield the invariant density (π_i), which is a Poisson distribution with the parameter $\lambda = b/d$:

$$\pi_i = \frac{\lambda^i e^{-\lambda}}{i!} \quad \text{for } i \in \mathbb{N}.$$

Since the matrix Q satisfies condition (I) of Theorem 19, the semigroup $\{P(t)\}_{t\geq 0}$ is asymptotically stable.

5.3 Population Growth with Stochastic Noise

Consider the following equation:

$$dx_t = b(x_t)\,dt + \sigma(x_t)\,dw_t. \tag{74}$$

Equation (74) is a stochastic version of the deterministic models describing the growth of a single population $x'(t) = b(x)$, where $x(t)$ is the size of the population at time t and the function b is the growth coefficient. We assume that $b\colon [0,\infty) \to \mathbb{R}$ is a C^1-function on the whole interval $[0,\infty)$ satisfying additionally the conditions $b(0) = 0$ and $b(x) \leq Kx$ for $x \geq 0$ and some $K > 0$.

We can consider two different stochastic perturbations. The first, called *environmental noise*, is used to describe small populations in which random fluctuations act in the same way on all individuals in the population, so $\sigma(x) = \sigma x$, $\sigma > 0$. The second is called *demographic noise*. In this case the stochastic perturbation acts independently on each part of population, and it has zero mean. According to the central limit theorem, such noise is proportional to $\sqrt{x}$, so $\sigma(x) = \sigma\sqrt{x}$. Models with demographic noise are more difficult to study because the trajectories of the process (x_t) are absorbed with positive probability at 0.

We only consider here models with environmental noise; thus, Eq. (74) has the form:

$$dx_t = b(x_t)\,dt + \sigma x_t\,dw_t. \tag{75}$$

The stochastic semigroup $\{P(t)\}_{t\geq 0}$ related to the process $(x_t)_{t\geq 0}$ is integral, and the kernel $q(t,x,y)$ is positive and continuous for $t, x, y > 0$. Thus, the semigroup $\{P(t)\}_{t\geq 0}$ is irreducible and satisfies condition (K). Then

$$\frac{\partial u}{\partial t} = \frac{1}{2}\frac{\partial^2}{\partial x^2}(\sigma^2 x^2 u) - \frac{\partial}{\partial x}(b(x)u) \tag{76}$$

is the Fokker–Planck equation corresponding to this semigroup. An invariant density f^*, if it exists, is a solution of the equation:

$$\frac{1}{2}\big(\sigma^2x^2f^*(x)\big)'' - \big(b(x)f^*(x)\big)' = 0. \tag{77}$$

We easily determine from (77) the invariant density:

$$f^*(x) = \frac{C}{x^2}\exp\left(\int_1^x \frac{2b(s)}{\sigma^2 s^2}\,ds\right), \tag{78}$$

as long as C can be chosen such that f^* is a density. Thus, an invariant density exists, and consequently the semigroup $\{P(t)\}_{t\geq 0}$ is asymptotically stable (see Corollary 1), if and only if

$$\int_0^\infty \frac{1}{x^2}\exp\left(\int_1^x \frac{2b(s)}{\sigma^2 s^2}\,ds\right)dx < \infty. \tag{79}$$

If (79) does not hold, then according to Corollary 4 the semigroup $\{P(t)\}_{t\geq 0}$ is sweeping from intervals $[\alpha, \beta]$, $0 < \alpha < \beta < \infty$, i.e.,

$$\lim_{t\to\infty}\int_\alpha^\beta P(t)f(x)\,dx = 0 \quad \text{for } f \in D. \tag{80}$$

Using Theorem 15, we can show more precise sweeping conditions and give their biological interpretation. If (79) does not hold, then f^* given by (78) with $C > 0$ is a non-integrable positive function such that $P(t)f^* = f^*$. Assume that there exists the limit $b'(\infty) = \lim_{x\to\infty} b(x)/x$. It is easy to check that

$$\int_0^\alpha f^*(x)\,dx < \infty \text{ if } \sigma^2 < 2b'(0), \qquad \int_0^\alpha f^*(x)\,dx = \infty \text{ if } \sigma^2 > 2b'(0), \tag{81}$$

$$\int_\beta^\infty f^*(x)\,dx < \infty \text{ if } \sigma^2 > 2b'(\infty), \qquad \int_\beta^\infty f^*(x)\,dx = \infty \text{ if } \sigma^2 < 2b'(\infty) \tag{82}$$

for $0 < \alpha < \infty$ and $0 < \beta < \infty$. From Theorem 15 it follows that:

(i) if $\sigma^2 < \min\{2b'(0), 2b'(\infty)\}$, then $\lim\limits_{t\to\infty} \mathrm{P}(x_t \geq \beta) = 1$,

(ii) if $\sigma^2 > \max\{2b'(0), 2b'(\infty)\}$, then $\lim\limits_{t\to\infty} \mathrm{P}(x_t \leq \alpha) = 1$,

(iii) if $2b'(0) < \sigma^2 < 2b'(\infty)$, then $\limsup\limits_{t\to\infty} \mathrm{P}(x_t \leq \alpha) > 0$ and $\limsup\limits_{t\to\infty} \mathrm{P}(x_t \geq \beta) > 0$.

It can be checked that the limits superior in (iii) can be replaced by limits, but their values depend on the distribution of x_0. Let $b_0 := b'(0) - \sigma^2/2$ and $b_\infty := b'(\infty) - \sigma^2/2$. Then b_0 and b_∞ are called the *actual growth rate* at 0 and ∞. Now conditions (i)–(iii) say: if both actual growth rates are positive, then the population grows indefinitely; if both actual growth rates are negative, then the population is becoming extinct; if $b_0 < 0$ and $b_\infty > 0$, then the population can with positive probability both grow indefinitely and become extinct.

5.4 Gene Expression

Gene expression is a process of production and decay of various biomolecules dependent on the activity of a number of genes. Changes in concentration of these biomolecules can be basically described by a system of differential equations, but this system depends on random parameters related to the genes' activity.

Assume that genes can form $k+1$ configurations depending on their state of activity and for configuration $i \in I = \{0, \dots, k\}$ concentration of d different biomolecules changes according to a system of differential equations of the form $x' = b^i(x)$, $x \in \mathbb{R}^d_+$. We assume that for each $x_0 \in \mathbb{R}^d_+$, there exists a unique solution $x\colon \mathbb{R}_+ \to \mathbb{R}^d_+$ of the last equation with the initial condition $x(0) = x_0$ and we denote this solution by $\pi^i_t x_0$. Assume that for $i, j \in I$ and $i \neq j$ and for concentration x, the rate of switching configuration from i to j is a nonnegative continuous and bounded function $x \mapsto q_{ij}(x)$.

We start with a simple model in which protein molecules are directly "produced" by a single gene. A gene can be in an active $i(t) = 1$ or inactive state $i(t) = 0$. When gene is active, then proteins are produced with a constant speed P. In both states μ is the rate of protein degradation. Let $x(t)$ be the concentration of protein. Then $x(t)$ satisfies the equation:

$$x'(t) = Pi(t) - \mu x(t). \tag{83}$$

Gene expression is described by dynamical systems with random switching $\xi_t = (x(t), i(t))$, $t \geq 0$, with values in the space $\mathbb{R}_+ \times \{0, 1\}$ and with random jumps from $(x, 0)$ to $(x, 1)$ and from $(x, 1)$ to $(x, 0)$ with rates $q_{01}(x)$ and $q_{10}(x)$, respectively. According to Sect. 3.6 this process generates a stochastic semigroup on the space $L^1(X, \mathcal{B}(X), m)$, where $X = \mathbb{R}_+ \times \{0, 1\}$ and m is the product of the Lebesgue measure on $\mathbb{R}_+$ and the counting measure on the set $\{0, 1\}$.

We now present a version of this model with nonrandom moments of jumps. We assume that the concentration of protein must not fall below a fixed level $\theta > 0$. We assume that the gene is inactivated with rate $q_{10}(x)$, but if the protein concentration reaches the level θ, the protein production is automatically activated (Fig. 5).

Now gene expression is described by a piecewise deterministic Markov process $\xi_t = (x(t), i(t))$, $t \geq 0$, with values in the space $\left[\theta, \frac{P}{\mu}\right] \times \{0, 1\}$ with random jumps from $(x, 1)$ to $(x, 0)$ and with nonrandom jumps from $(\theta, 0)$ to $(\theta, 1)$.

If x_0 is the initial number of protein in the active state, then the length of the active state has the distribution:

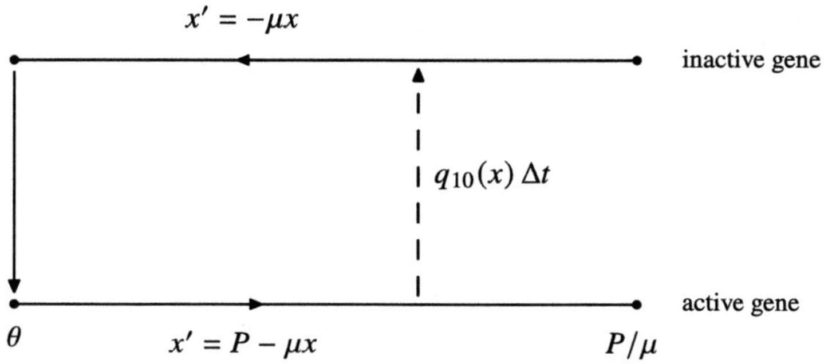

Fig. 5 Gene expression scheme with jumps on the boundary

$$F_1(t) = 1 - \exp\left\{-\int_0^t q_{10}(1 + (x_0 - 1)e^{-s})\,ds\right\}. \tag{84}$$

The duration of stay in the inactive state T satisfies the equation $x_0 e^{-\mu t} = \theta$; thus, $T = \frac{1}{\mu}\ln(x_0/\theta)$. The length distribution of the inactive state is therefore the function:

$$F_0(t) = \begin{cases} 0, & \text{if } t < T, \\ 1, & \text{if } t \geq T. \end{cases} \tag{85}$$

Now we consider more realistic model introduced in [53] and studied in [10]. A model consists of two types of molecules, mRNA, and protein, and we assume that variables x_1, x_2 denote their concentrations. If the gene is active, then mRNA transcript molecules are synthesized with a constant speed R. The protein translation proceeds with the rate $Px_1(t)$, where P is a constant. The mRNA and protein degradation rates are μ_R and μ_P, respectively. Thus, the process is described by the following system:

$$\begin{cases} x_1'(t) = Ri(t) - \mu_R x_1(t), \\ x_2'(t) = Px_1(t) - \mu_P x_2(t), \end{cases} \tag{86}$$

where $i(t) = 1$ if the gene is active and $i(t) = 0$ in the opposite case.

We assume that the rates q_{01} and q_{10} are continuous and bounded functions of $\mathbf{x} = (x_1, x_2)$. Then the process $\xi_t = (x_1(t), x_2(t), i(t))$, $t \geq 0$, is given by dynamical systems with random switching. The trajectories of the process enter the invariant set $X = \left[0, \frac{R}{\mu_R}\right] \times \left[0, \frac{PR}{\mu_P \mu_R}\right] \times \{0, 1\}$, so we can consider the process $(\xi_t)_{t\geq 0}$ with values in this set. It generates a stochastic semigroup $\{P(t)\}_{t\geq 0}$ on the space $L^1(X, \mathcal{B}(X), m)$.

In this model we have two dynamical systems, which correspond, respectively, to $i = 0$ and $i = 1$. Each system has a unique stationary point, either $\mathbf{x}_0 = (0, 0)$

or $\mathbf{x}_1 = \left(\frac{R}{\mu_R}, \frac{PR}{\mu_P\mu_R}\right)$, which is asymptotically stable. We assume that $q_{01}(\mathbf{x}_0) > 0$ and $q_{10}(\mathbf{x}_1) > 0$. This assumption implies that the trajectories of the process visit any neighborhood of the point $(\mathbf{x}_0, 0)$ and the point $(\mathbf{x}_1, 1)$ for infinitely many times. It means that condition (WI) holds. Moreover, if we assume additionally that $q_{01}(\mathbf{x}_1) > 0$ or $q_{10}(\mathbf{x}_0) > 0$, then we can check that condition (K) also holds (see [75] for the proof based on Hörmander's condition). The set X is compact. According to Corollary 9, the semigroup $\{P(t)\}_{t\geq 0}$ is asymptotically stable.

In some cases, the process of protein production has a pre-transcriptional step. Activation of the gene is followed by the production of RNA molecules called pre-mRNA, which cannot directly produce a protein. Then, while still in the cell nucleus, noncoding sequences (introns) are truncated to form mRNA molecules, which are then transported to the cytoplasm, where they serve as matrices in the translation process of protein production. Such a three-step model was introduced and investigated in [82].

Let us denote by $x_1(t), x_2(t), x_3(t)$ the concentration of pre-mRNA, mRNA, and protein at time t. The change in these quantities over time can be represented by a system of three equations:

$$\begin{cases} x_1'(t) = Ai(t) - (R + \mu_{pR})x_1(t), \\ x_2'(t) = Rx_1(t) - \mu_R x_2(t), \\ x_3'(t) = Px_2(t) - \mu_P x_3(t), \end{cases} \tag{87}$$

where A is the speed of pre-mRNA molecule synthesis when the gene is active, R is the conversion rate of pre-mRNA molecules into active mRNA, μ_{pR} is the degradation rate of pre-mRNA molecules, and the other constants are the same as in the previous model.

The model (86) can also be modified to a model with nonrandom jumps. As in the previous model, the gene is inactivated with rate $q_{10}(x_1, x_2)$, but it is activated when the protein concentration $x_2(t)$ reaches θ, where $0 < \theta \leq PR/\mu_R\mu_P$. Note that in this case, it may happen that $x_1(\bar{t}) < \mu_p\theta$ at the time $\bar{t}$ of activation, and then $x_2'(\bar{t}) < 0$ and the protein concentration $x_2(t)$ will decrease in a certain interval $(\bar{t}, \bar{t} + \varepsilon)$, in particular will fall below the level of θ. It is clear that the gene should remain active until the concentration of the protein $x_2(t)$ does not exceed at least the level of θ, we should therefore assume that the inactivation intensity $q_{10}(x_1, x_2)$ is zero when $x_2 \leq \theta$. The process $\xi_t = (x_1(t), x_2(t), i(t))$, $t \geq 0$ is defined on the space $X = \left[0, \frac{R}{\mu_R}\right] \times \left[0, \frac{PR}{\mu_P\mu_R}\right] \times \{0, 1\}$ and has nonrandom jumps on the set $S = \left[0, \frac{R}{\mu_R}\right] \times \{\theta\} \times \{0, 1\}$. In this case, additional jumps occur inside the phase space.

The papers [38, 46] present more advanced models describing the process of *subtiline* production by the bacterium *Bacillus subtilis*. Although the process plays an important role in biochemical technology, we will not discuss the model in detail, but only highlight its interesting mathematical aspects.

To survive, the bacterium produces a type of antibiotic, subtiline, which eliminates other microorganisms competing with it. The model considered in [38] consists of five differential equations describing the size x_1 of the *B. subtilis* population, the amount of food x_2, and the concentration x_3, x_4, x_5 of the three proteins involved in the production. Two of these equations are without switching, two with switching dependent on the state of two genes and one with switching dependent on the amount of food:

$$x_3'(t) = \begin{cases} -\mu x_3(t), & \text{if } x_2(t) \geq \theta, \\ P_3 - \mu x_3(t), & \text{if } x_2(t) < \theta, \end{cases}$$

where θ is a positive constant. The corresponding process is defined on the space $\mathbb{R}^5_+$ and has additional jumps on the subspace $S = \{x \in \mathbb{R}^5_+ : x_2 = \theta\}$.

5.5 Movement with Velocity Jumps

The pure jump-type processes considered in Sect. 5.7 are a far-reaching simplification of the description of the movement of biological objects. Much closer to reality models of the movement of cells or organisms are *velocity jump processes* [63]. These processes are also used to describe vesicular transport in cells. We will start with a general description and then give two examples of applications.

An individual is moving in the space $\mathbb{R}^d$ with a constant velocity, and at jump times (t_n) it chooses a new velocity. The intensity of the jump ψ depends on its location $x \in \mathbb{R}^d$ and velocity $v \in V$, where $V = \mathbb{R}^d$ or V is some subset of $\mathbb{R}^d$, e.g., V is the unit sphere S^{d-1} in $\mathbb{R}^d$. Assume that $\psi : \mathbb{R}^d \times V \to [0, \infty)$ is a bounded measurable function. Let x, v be the position and velocity at jump time. Denote by T the time until the next jump. Then

$$\mathrm{P}(t < T \leq t + \Delta t \mid T > t) = \psi(x + vt, v)\Delta t + o(\Delta t).$$

We can write this formula using the cumulative distribution function F of the random variable T:

$$\frac{F(t + \Delta t) - F(t)}{1 - F(t)} = \psi(x + vt, v)\Delta t + o(\Delta t).$$

Dividing both sides by Δt and going to the limit when $\Delta t \to 0$ we obtain

$$\frac{F'(t)}{1 - F(t)} = \psi(x + vt, v).$$

We write the last equation as

$$\frac{d}{dt}\ln(1-F(t)) = -\psi(x+vt, v),$$

and then we solve it and obtain

$$F(t) = 1 - e^{-\int_0^t \psi(x+vs,v)\,ds}.$$

Let $x(t)$ be the position and $v(t)$ be the velocity of an individual at time t. We assume that for every $x \in \mathbb{R}^d$, $v \in V$, there is a probability Borel measure $\mathcal{P}(x, v, B)$ on V which describes the change of the velocity after a jump, i.e.,

$$\mathrm{P}(v(t_n) \in B \mid x(t_n^-) = x,\ v(t_n^-) = v) = \mathcal{P}(x, v, B)$$

for every Borel subset B of V, where $x(t_n^-)$ and $v(t_n^-)$ are the left-hand side limits of $x(t)$ and $v(t)$, respectively, at the point t_n.

Knowing the distribution of the time to the next jump and the measure $\mathcal{P}(x, v, B)$ describing the change in velocity after the jump, we can construct a piecewise homogeneous deterministic Markov process $(\xi_t)_{t\geq 0}$ on the space $X = \mathbb{R}^d \times V$. Such processes have been used in [7, 63, 88] to describe the movement of bacteria, insects, and mammals.

Let m be a σ-finite measure on $(V, \mathcal{B}(V))$. We usually take it to be the Lebesgue measure when the set V is of positive Lebesgue measure, or the Lebesgue-Riemann surface measure when V is a manifold. Suppose that, for every point $x \in \mathbb{R}^d$, the transition probability function $\mathcal{P}(x, v, B)$ corresponds to some stochastic operator P_x on $L^1(V, \mathcal{B}(V), m)$ and ψ is a bounded function. Then the operator

$$Af(x,v) = -v \cdot \nabla_x f(x,v) + P_x(\psi f)(v) - \psi(x,v)f(x,v) \tag{88}$$

is the generator of a stochastic semigroup $\{P(t)\}_{t\geq 0}$ on $L^1(X, \mathcal{B}(X), dx \times m)$. As in Sect. 5.7, the operator A can be written in the form:

$$Af = -v \cdot \nabla_x f + \lambda \bar{P}_x f - \lambda f. \tag{89}$$

Example 22 Consider a symmetric movement on the real line $\mathbb{R}$. In this case we assume that an individual is moving with constant speed, say one, and at a jump time it changes the direction of movement to the opposite one. The process $(\xi_t)_{t\geq 0}$ has values in the space $X = \mathbb{R} \times \{-1, 1\}$ and $\mathcal{P}(x, v, \{-v\}) = 1$ for $v = -1, 1$. We assume that jump times are determined by a Poisson process $(N_t)_{t\geq 0}$ with intensity λ; it means that $\psi \equiv \lambda$. We can give explicit formulas for the process $(\xi_t)_{t\geq 0}$:

$$v_t = v_0(-1)^{N_t}, \quad x_t = x_0 + v_0 \int_0^t (-1)^{N_s}\,ds. \tag{90}$$

The stochastic semigroup related to this process is defined on the space $L^1(X, \mathcal{B}(X), m)$, where m is the product of the Lebesgue measure on $\mathbb{R}$ and the counting measure on the set $\{-1, 1\}$. If the random variable $\xi_0 = (x_0, v_0)$ has a density $u_0(x, i)$, then the random variable $\xi_t = (x_t, v_t)$ has the density $u(t, x, i)$ satisfying the following system of equations:

$$
\begin{aligned}
\frac{\partial u(t,x,1)}{\partial t} &= -\frac{\partial u(t,x,1)}{\partial x} - \lambda u(t,x,1) + \lambda u(t,x,-1),\\
\frac{\partial u(t,x,-1)}{\partial t} &= \frac{\partial u(t,x,-1)}{\partial x} - \lambda u(t,x,-1) + \lambda u(t,x,1),\\
u(0,x,i) &= u_0(x,i).
\end{aligned}
\tag{91}
$$

We recall that the function $u(t, x, i)$ is a density of ξ_t if

$$
\mathrm{P}(\xi_t \in A \times \{1\} \cup B \times \{-1\}) = \int_A u(t,x,1)\,dx + \int_B u(t,x,-1)\,dx,
$$

where A and B are any Borel subsets of $\mathbb{R}$. The system (91) has been studied by Goldstein [33] and Kac [42], among others. By suitable substitution it can be reduced to the telegraph equation.

There are many interesting examples of velocity jump processes with applications to aggregation and chemotaxis phenomena (see, e.g., [37]). Stochastic billiard introduced in Sect. 3.7 is also a velocity jump process, but it has jumps at the boundary, so its description is different. Also the cell cycle can be modeled by a velocity jump process [52, 78].

Other interesting velocity jump processes with jumps on certain subsets are used in [13, 14] to describe vesicular transport in cells. In these models, a molecule moves along the interval $[0, L]$ to some destination point x_0. A molecule can be in three states $i \in I = \{-1, 0, 1\}$; we assume that in the state i, it moves with a velocity i. The transition between states is controlled by a homogeneous Markov chain $\vartheta(t)$ on the set I. The particle starts from a point $x = 0$ and moves to the right. Assume that 0 is an elastic screen and L is an absorbing screen. We assume that if a particle is in a certain neighborhood U of the point x_0 and in the state $i = 0$, it is absorbed by the target point with intensity κ. If the molecule is captured by the target point or the point L, the process repeats itself. Between jumps the molecule moves according to a jump process $\xi_t = (x(t), i(t))$, $t \geq 0$, defined on the phase space $[0, L] \times I$, where the pair $(x(t), i(t))$ satisfies the system of equations:

$$
\begin{cases}
x'(t) = i(t),\\
i'(t) = 0.
\end{cases}
\tag{92}
$$

If $x \in (0, L)$, then the process ξ_t can jump from (x, i) to (x, j) with intensity q_{ij} for $i, j \in I$; if $(x, j) \in U \times \{0\}$, the process can jump from $(x, 0)$ to $(0, 1)$ with intensity κ. The process has additional jumps from $(0, -1)$ and from $(L, 1)$ to $(0, 1)$ with probability 1.

The literature on stochastic models of intracellular transport is quite extensive. Interested readers are referred to the review paper [15].

5.6 Structured Population Models

A population is usually heterogeneous, and it is important to divide it into homogeneous groups according to some significant parameters such as age, size, maturity, and trait in phenotypic models. Models which describe the time evolution of the distribution of the population according to such set of fixed parameters are called *structured*. Since living creatures reproduce from time to time, the evolution of these parameters has a "jumping nature."

Let $X \subset \mathbb{R}^d$ and let $x \in X$ be the vector of parameters that characterize any individual. We assume that the set X has non-empty interior and has the boundary of measure zero. We assume that x changes according to the equation:

$$x'(t) = b(x(t)), \tag{93}$$

so the function b and the set X have such properties that if $x_0 \in X$, then there exists a unique solution of (93) and $x(t) \in X$ for $t \geq 0$.

We assume that $\mu(x)$ is the death rate which includes both the real death and the loss of cells during the process of division in cellular population models. Let $\psi(x)$ be the birth rate, that is, $\psi(x)\Delta t$ is the probability that an individual with parameter x has a descendant in time interval $[t, t + \Delta t]$. We assume that μ and ψ are nonnegative and continuous functions and that ψ is bounded.

We also assume that the transition probability function $\mathcal{P}(x, dy)$ shows the distribution of parameters of a descendant at the birth. For example, if x is the age, then $\mathcal{P}(x, \{0\}) = 1$; if x is the size of a cell, then $\mathcal{P}(x, \{x/2\}) = 1$; and in a phenotypic model, the measure $\mathcal{P}(x, \cdot)$ is usually distributed near point x.

We are interested in finding the distribution of the parameter x at time t. This distribution is described by a density function $u(t, x)$; precisely

$$\int_A u(t, x)\, dx$$

is the number of individuals (or biomass) with the parameter x in the set A. It should be underlined that u is not a density in the probabilistic sense because the integral of $u(t, x)$ over the whole phase space X can be different from 1 and this integral can change with time.

First we consider the case when $\mathcal{P}$ satisfies (30), i.e., $\mathcal{P}$ corresponds to a stochastic operator P on the space $L^1(X, \mathcal{B}(X), \ell)$. Then, u satisfies an equation similar to (33):

$$\frac{\partial u}{\partial t} = -\nabla_x(bu) - \mu u + P(\psi u) \tag{94}$$

with initial condition $u(0,x) = u_0$. Moreover, if $P(t)u_0(x) = u(t,x)$, then $\{P(t)\}_{t\ge 0}$ is a positive C_0-semigroup on $L^1(X,\mathcal{B}(X),\ell)$.

The case when $\mathcal{P}$ does not satisfy (30) is generally more difficult. We restrict ourselves to an age-structured model, i.e., a model with $\mathcal{P}(a,\{0\}) = 1$, a denotes the age of an individual. Then, u satisfies the following boundary-initial problem:

$$\begin{aligned} &\frac{\partial u(t,a)}{\partial t} + \frac{\partial u(t,a)}{\partial a} = -\mu(a)u(t,a), \\ &u(t,0) = \int_0^\infty u(t,a)\psi(a)\,da, \quad u(0,a) = u_0. \end{aligned} \tag{95}$$

This problem also generates a positive C_0-semigroup.

The study of these models has a long history, and there are many papers on them and their generalizations; see, for example, [5, 24, 36, 57, 81, 91]. One of the key issues was asynchronous exponential growth (41). This property can be demonstrated using the spectral decomposition of quasi-compact positive semigroups (see Sect. 4.2). This property can also be proved using theorems concerning stochastic semigroups. In this case the proof goes as follows. Let A be an infinitesimal generator of a positive semigroup $\{T(t)\}_{t\ge 0}$ on $L^1(X,\mathcal{B}(X),\ell)$. We prove that there exist $\lambda \in \mathbb{R}$ and a bounded above and bounded away from zero function w such that $A^*w = \lambda w$. From this it follows that the semigroup $\{P(t)\}_{t\ge 0}$ given by $P(t) = e^{-\lambda t}T(t)$ is a stochastic semigroup on the space $L^1(X,\mathcal{B}(X),m)$, where m is a measure given by $m(B) = \int_B w(x)\,dx$. Next we prove asymptotic stability of the semigroup $\{P(t)\}_{t\ge 0}$. Since measures ℓ and m are equivalent, we have (41).

Dynamical systems with random jumps appear in many biological models. Examples of such processes are *ecological models with catastrophes* [16, 35, 40, 48]. The population size is described by a stochastic process $(\xi_t)_{t\ge 0}$, which between successive catastrophes changes according to a differential equation

$$\xi_t' = b(\xi_t). \tag{96}$$

We assume that $\psi(x)$ is the intensity of occurrence of a catastrophe when the population size is x and the transition probability function $\mathcal{P}(x,dy)$ shows the distribution of population size after the catastrophe. We assume that $\mathcal{P}$ corresponds to a stochastic operator P on the space $L^1(\mathbb{R}_+,\mathcal{B}(\mathbb{R}_+),\ell)$. If $u(t,x)$ is the density of the distribution of size of population, then u satisfies the equation:

$$\frac{\partial u}{\partial t} = -\frac{\partial (bu)}{\partial x} - \psi u + P(\psi u) \tag{97}$$

and generates a stochastic semigroup on $L^1(\mathbb{R}_+,\mathcal{B}(\mathbb{R}_+),\ell)$.

A similar model describes the *dynamics of immune status* [25, 74]. The immune status is the concentration of specific antibodies, which appear after infection with a pathogen and remain in serum, providing protection against future attacks by the same pathogen. Over time the number of antibodies decreases until the next infection. During an infection, the immunity is boosted and then it is gradually waning, until it is boosted at the next infection. Concentration of antibodies between subsequent infections is a decreasing stochastic process ξ_t, which satisfies Eq. (96) with $b(x) < 0$ for $x > 0$. The moments of infections are distributed according to a Poisson process with intensity $\lambda > 0$. If x is the concentration of antibodies at the moment of infection, then $\varphi(x) > x$ is the concentration of antibodies just after clearance of infection. In this case the jump operator is the Frobenius–Perron operator P_φ corresponding to φ and $Af = -(bf)' + \lambda P_\varphi f - \lambda f$ is a generator of a stochastic semigroup on $L^1(\mathbb{R}_+, \mathcal{B}(\mathbb{R}_+), \ell)$.

5.7 Semi–Markov Kangaroo Movement

Continuous-time Markov chains belong to a slightly larger class called pure jump-type Markov processes. A *pure jump-type Markov process* is a Markov process which remains constant between jumps. Such processes are used to describe a movement in which an individual makes a series of jumps and the time of a single jump is neglected. This process is also referred to as *kangaroo process* or *kangaroo movement* [63].

A pure jump-type homogeneous Markov process on a measurable space (X, Σ) can be defined in the following way. Let $\psi\colon X \to [0,\infty)$ be a given measurable function and let $\mathcal{P}$ be a given transition probability function on X. Let $t_0 = 0$ and let ξ_0 be an X-valued random variable. For each $n \ge 1$, one can choose the nth *jump time* t_n as a positive random variable satisfying

$$\mathrm{P}(t_n - t_{n-1} \le t \,|\, \xi_{t_{n-1}} = x) = 1 - e^{-\psi(x)t}, \quad t \ge 0,$$

and we assume that $\xi_t = \xi_{t_{n-1}}$ for $t_{n-1} \le t < t_n$ and ξ_{t_n} is a random X-valued variable satisfying condition

$$\mathrm{P}(\xi_{t_n} \in B \,|\, \xi_{t_{n-1}} = x) = \mathcal{P}(x, B), \quad B \in \Sigma.$$

If m is a σ-finite measure on the space (X, Σ) and the transition probability function $\mathcal{P}$ satisfies condition (30), then $\mathcal{P}$ corresponds to a jump operator P, which is a stochastic operator on the space $L^1(X, \Sigma, m)$. If ψ is a bounded function, then the operator

$$Af(x) = P(\psi f)(x) - \psi(x) f(x) \tag{98}$$

is a generator of uniformly continuous stochastic semigroup $\{P(t)\}_{t\ge 0}$ on $L^1(X, \Sigma, m)$. Indeed, if $\lambda \ge \sup\{\psi(x)\colon x \in X\}$, then we define the operator:

$$\bar{P} f(x) = \lambda^{-1}\big[P(\psi f)(x) + (\lambda - \psi(x)) f(x)\big]$$

and we represent the operator A in the form $Af = \lambda\bar{P}f - \lambda f$. The semigroup $\{P(t)\}_{t\geq 0}$ takes the form (47): $P(t)f = e^{\lambda t(\bar{P}-I)}f$.

The time between jumps in the kangaroo movement described above has an exponential distribution. Let us now consider the movement of a kangaroo with the same jump operator P as defined above, but with other distributions of time between jumps. Such a process has no Markov property and it belongs to the family of semi-Markov processes.

We assume that the *holding time*, i.e., the time a that the process spends at a given point x, has the probability density distribution $a \mapsto q(x,a)$. We assume that q is a measurable function and we will need further technical assumptions, which allow us to introduce some processes. In the case of previously introduced pure jump-type homogeneous Markov process, q is given by $q(x,a) = \psi(x)e^{-a\psi(x)}$.

Assume that the position of a jumping point is described by a stochastic process $(\xi_t)_{t\geq 0}$. The process $(\xi_t)_{t\geq 0}$ is generally non-Markovian, so it does not correspond to a stochastic semigroup. To overcome these obstacles, we define a Markov process $(\eta_t)_{t\geq 0}$ with the formula:

$$\eta_t = (\xi_{t_n}, t - t_n) \quad \text{for } t_n \leq t < t_{n+1},$$

where $t_0 < t_1 < t_2 < t_3 < \ldots$ are jump times. If $\xi_{t_n} = x$, then the jump rate at time $t_n + a$ is given by

$$p(x,a) = \lim_{\Delta t \downarrow 0} \frac{\mathrm{P}(\tau \in [a, a+\Delta t] \mid \tau \geq a)}{\Delta t}, \tag{99}$$

where the random variable τ is the holding time. Then

$$p(x,a) = \frac{q(x,a)}{\int_a^\infty q(x,r)\,dr} \tag{100}$$

and formally $p(x,a) = 0$ when the last integral is zero. The function p plays the same role in this model as the function ψ from the previous model except that it is a function of two variables (x,a).

Here we assume that $t_0 \leq 0$ and start observing the process from the moment $t = 0$, so the first jump occurs at a random time $t_1 > 0$. Assume that the distribution of the random variable η_0 has density $u_0(x,a)$. Then, the random variable η_t has some density $u(t,x,a)$. Since $p(x,a)$ is the rate of jump of the process $(\eta_t)_{t\geq 0}$ from (x,a) to a point $(y,0)$, where y is chosen according to the distribution $\mathcal{P}$, we have:

$$u(t,x,0) = \int_0^\infty \Big(P\big(p(\cdot,a)u(t,\cdot,a)\big)\Big)(x)\,da. \tag{101}$$

We will use the shortened notation $\mathfrak{P}u(t,x)$ for the expression on the right-hand side of (101). Thus, Eq. (101) takes the form $u(t,x,0) = \mathfrak{P}u(t,x)$.

Hence, the function u satisfies the following initial-boundary problem:

$$\frac{\partial u}{\partial t}(t,x,a) + \frac{\partial u}{\partial a}(t,x,a) = -p(x,a)u(t,x,a), \tag{102}$$

$$u(t,x,0) = \mathfrak{P}u(t,x), \tag{103}$$

$$u(0,x,a) = u_0(x,a). \tag{104}$$

The issue of when the resulting system of equations generates a stochastic semigroup is quite difficult and as we mentioned before we need additional assumptions. We assume that X is an open subset of $\mathbb{R}^d$ and the function q satisfies the following conditions:

(A1) $q\colon X \times [0,\infty) \to [0,\infty)$ is a continuous function and for each $x \in X$ the function $a \mapsto q(x,a)$ is a probability density,
(A2) there exists $\varepsilon > 0$ such that $\int_0^\varepsilon q(x,a)\,da < 1-\varepsilon$ for all $x \in X$.

We define $\overline{a}(x) \le \infty$ to be the minimum number such that

$$\int_0^{\overline{a}(x)} q(x,a)\,da = 1 \quad \text{for } x \in X.$$

Let

$$Y = \{(x,a)\colon x \in X,\ 0 \le a < \overline{a}(x)\}. \tag{105}$$

The process $(\eta_t)_{t\ge 0}$ has values in Y.

Let $E = L^1(Y, \mathcal{B}(Y), m)$, where $\mathcal{B}(Y)$ is the σ-algebra of Borel subsets of Y and m is the Lebesgue measure. We replace problem (102)–(104) by an evolution equation $u'(t) = Au(t)$ on the space E with operator A of the form:

$$Af = -\frac{\partial f}{\partial a} - pf. \tag{106}$$

and with domain

$$\mathfrak{D}(A) = \left\{ f \in E\colon\ \frac{\partial f}{\partial a} \in E,\ pf \in E,\ f(x,0) = \mathfrak{P}f(x) \right\}. \tag{107}$$

The operator A generates a stochastic semigroup $\{P(t)\}_{t\ge 0}$ on the space E. We omit the proof of this result here. The interested reader is referred to publication [76] where the proof of a similar result is given. This proof is based on perturbation method related to operators with boundary conditions developed in [34] and an extension of this method to unbounded perturbations in L^1 space in [39].

5.8 Cell Cycle Model

The cell cycle is a series of events that take place in a cell leading to its replication. It is regulated by a complex network of protein interactions [60]. One of the most important questions concerning the cell cycle is what factor determines when a cell divides. Such a factor may be the size of the cell and therefore different size-structured models are considered. Such models can be experimentally verified and allow us to understand processes inside single cells.

We assume that the population grows in steady-state conditions and neglected mortality of cells. Cells can be described by their age a and size x alone and reproduction occurs by fission into two equal parts. We assume that a cell grows with velocity $g(x)$, i.e., $x'(t) = g(x(t))$.

First, we assume that also the rate of division λ depends only on a cell size; then the evolution of size distribution of cells is described by the model considered in Sect. 5.6, where $\mathcal{P}(x, \{x/2\}) = 1$, $\mu \equiv 0$ and $\psi(x) = 2\lambda(x)$. In the definition of ψ, we have a factor 2 because division results in two cells. Moreover, $Pf(x) = 2f(2x)$ is the stochastic operator P corresponding to $\mathcal{P}$. Thus, the density function $u(t, x)$ satisfies a version of Eq. (94):

$$\frac{\partial u(t,x)}{\partial t} = -\frac{\partial\big(g(x)u(t,x)\big)}{\partial x} + 4\lambda(2x)u(t,2x). \tag{108}$$

Although models of type (108) are commonly used, they have one disadvantage in that it is difficult to determine the function λ experimentally. In [76] we propose a model that is based on two factors: the initial size of a cell x_b and the length of the cell cycle. Both factors are not difficult to measure, and there are quite a few results and models in the literature based on experiments on the distribution of cell cycle length and its dependence on initial cell size. However, a model based on the length of the cell cycle does not lead to an equation for $u(t, x)$, but using its properties we can investigate the long-time behavior of $u(t, x)$.

We now present this model and give its main properties. We assume that sizes of cells and cell cycle durations are bounded above and bounded away from zero. Let $\underline{x}_b$ and $\overline{x}_b$ be the minimum and maximum sizes of newborn cells. We also assume that cells age with unitary velocity and grow with a velocity $g(x)$, i.e., if a cell has the initial size x_b, then the size at age a satisfies the equation:

$$x'(a) = g(x(a)), \quad x(0) = x_b. \tag{109}$$

We denote by $\pi_a x_b$ the solution of (109). If the initial size of the mother cell is x_b and $\tau = a$ is the length of its cell cycle, then the initial size of the daughter cell is $S_a(x_b) = \frac{1}{2}\pi_a x_b$.

The length τ of the cell cycle is a random variable which depends on the initial cell size x_b; has values in some interval $[\underline{a}(x_b), \overline{a}(x_b)]$; and has the probability density distribution $q(x_b, \cdot)$, i.e., the integral $\int_0^A q(x_b, a)\, da$ is the probability that

$\tau \le A$. According to the definition of q, if a cell has the initial size x_b, then $\Phi(x_b, a) = \int_a^\infty q(x_b, r)\,dr$ is its *survival function*, i.e., $\Phi(x_b, a)$ is the probability that a cell will not split before age a.

The functions g and q satisfy the following assumptions:

(A1) $g\colon [\underline{x}_b, 2\overline{x}_b] \to (0, \infty)$ is a C^1-function,
(A2) $q\colon [\underline{x}_b, \overline{x}_b] \times [0, \infty) \to [0, \infty)$ is a continuous function and for each x_b the function $a \mapsto q(x_b, a)$ is a probability density,
(A3) $0 < \underline{a}(x_b) < \overline{a}(x_b) < \infty$, $q(x_b, a) > 0$ if $a \in (\underline{a}(x_b), \overline{a}(x_b))$, and $q(x_b, a) = 0$ if $a \notin (\underline{a}(x_b), \overline{a}(x_b))$ for each $x_b \in [\underline{x}_b, \overline{x}_b]$,
(A4) $x_b \mapsto \underline{a}(x_b)$ and $x_b \mapsto \overline{a}(x_b)$ are continuous functions,
(A5) $S_{\underline{a}(x_b)}(x_b) \ge \underline{x}_b$ and $S_{\overline{a}(x_b)}(x_b) \le \overline{x}_b$ for each $x_b \in [\underline{x}_b, \overline{x}_b]$,
(A6) $S_{\underline{a}(x_b)}(x_b) < x_b < S_{\overline{a}(x_b)}(x_b)$ for each $x_b \in (\underline{x}_b, \overline{x}_b)$.

Assumption (A6) means that the initial daughter cell size is distributed in some interval which contains the mother initial size.

Let the function p be given by (99). Then p satisfies (100) and $q = p\Phi$. Let P_a be the Frobenius–Perron operator corresponding to the transformation S_a. Denote by $u(t, x_b, a)$ the number of individuals in a population having initial size x_b and age a at time t. Then, according to our assumptions concerning the model, $p(x_b, a)u(t, x_b, a)$ is the number of cells of initial size x_b and age a which split in a unit time interval. It means that

$$\mathfrak{P}u(t, x_b) = 2\int_0^\infty \Big(P_a\big(p(\cdot, a)u(t, \cdot, a)\big)\Big)(x_b)\,da$$

is the number of new born cells in this time interval. Thus, the function u satisfies the initial-boundary problem (102)–(104).

As in Sect. 5.7 we replace problem (102)–(104) by an evolution equation $u'(t) = Au(t)$ on the space $E = L^1(Y, \mathcal{B}(Y), m)$ with generator A given by (106) and (107). Here the set Y is given by (105). The operator A generates a positive C_0-semigroup $\{T(t)\}_{t\ge 0}$ on the space E.

In order to formulate the main result of [76], we need an additional assumption:

(A7) there exists $x \in (\underline{x}_b, \overline{x}_b)$ such that $g(2x) \ne 2g(x)$.

Theorem 23 *If conditions* (A1)–(A7) *are fulfilled, then the semigroup* $\{T(t)\}_{t\ge 0}$ *has asynchronous exponential growth, i.e., there exist* $\lambda \in \mathbb{R}$*, a density* $f^* \in E$ *and a positive function* $v \in L^\infty(Y, \mathcal{B}(Y), m)$ *such that*

$$\lim_{t\to\infty} e^{-\lambda t}T(t)f = f^* \iint\limits_Y f(x_b, a)v(x_b, a)\,dx_b\,da \quad \text{for } f \in E. \tag{110}$$

The proof of Theorem 23 is based on the method presented in Sect. 5.6 of reducing the problem to the asymptotic stability of some stochastic semigroup, and then applying Theorem 14.

Now we consider a size-age-version of our model. Let $w(t,x,a)$ be the number of cells having size x and age a at time t. Then

$$u(t,x,a) = \frac{\partial(\pi_a x)}{\partial x} w(t,\pi_a x,a) = \frac{g(\pi_a x)}{g(x)} w(t,\pi_a x,a).$$

From the formula (110) it follows that

$$e^{-\lambda t} w(t,x,a) \to h^*(x,a) \iint\limits_Z v(\pi_{-a} y,a) w(0,y,a)\,dy\,da$$

in E as $t \to \infty$, where $h^*(x,a) = \dfrac{g(\pi_{-a}x)}{g(x)} f^*(\pi_{-a}x,a)$ and

$$Z = \{(\pi_a x,a) \colon x \in X,\ \ 0 \le a < \overline{a}(x)\}.$$

In particular, if $\bar{w}(t,x) = \int_0^\infty w(t,x,a)\,da$, then

$$e^{-\lambda t} \bar{w}(t,x) \to \bar{h}^*(x) \iint\limits_Z v(\pi_{-a} y,a) w(0,y,a)\,dy\,da, \ \ \bar{h}^*(x) = \int_0^\infty h^*(x,a)\,da.$$

Thus, we have obtained an asymptotic distribution of cell sizes, even though the function $\bar{w}$ is not determined by a semigroup of operators.

Acknowledgments This research was partially supported by the National Science Centre (Poland) Grant No. 2017/27/B/ST1/00100.

References

1. Arendt, W., et al.: In: Nagel, R. (Ed.) One-Parameter Semigroups of Positive Operators. Lecture Notes in Mathematics, vol. 1184. Springer, Berlin (1986)
2. Arendt, W., Batty, C.J.K.: Tauberian theorems and stability of one parameter semigroups. Trans. Amer. Math. Soc. **306**, 837–852 (1988)
3. Arnold, V. I.: Mathematical Methods of Classical Mechanics. Springer, New York (1989)
4. Balakrishnan, V., Van den Broeck, C., Hanggi, P.: First-passage times of non-Markovian processes: the case of a reflecting boundary. Phys. Rev. A **38**, 4213–4222 (1988)
5. Banasiak, J., Pichór, K., Rudnicki, R.: Asynchronous exponential growth of a general structured population model. Acta Appl. Math. **119**, 149–166 (2012)
6. Batty, C.J.K., Chill, R., Tomilov, Y.: Fine scales of decay of operator semigroups. J. Eur. Math. Soc. **18**, 853–929 (2016)
7. Berg, H., Brown, D.: Chemotaxis in *Eschericha Coli*: analysis by three dimensional tracking. Nature **239**, 500–504 (1972)
8. Bezak, V.: A modification of the Wiener process due to a Poisson random train of diffusion-enhancing pulses. J. Phys. A Math. Gen. **25**, 6027–6041 (1992)

9. Blinder, S.M.: Quantum Chemistry. LibreTexts (2023). https://chem.libretexts.org/Bookshelves/Physical_and_Theoretical_Chemistry_Textbook_Maps/Quantum_Chemistry_(Blinder)
10. Bobrowski, A., Lipniacki, T., Pichór, K., Rudnicki, R.: Asymptotic behavior of distributions of mRNA and protein levels in a model of stochastic gene expression. J. Math. Anal. Appl. **333**, 753–769 (2007)
11. Borichev, A., Tomilov, Y.: Optimal polynomial decay of functions and operator semigroups. Math. Ann. **347**, 455–478 (2010)
12. Boyarsky, A., Góra, P.: Laws of Chaos: Invariant Measures and Dynamical Systems in One Dimension. Birkhäuser, Basel (1997)
13. Bressloff, P., Newby, J.: Directed intermittent search for hidden targets. New J. Phys. **11**, 023033 (2009)
14. Bressloff, P., Newby, J.: Quasi-steady state reduction of molecular motor-based models of directed intermittent search. Bull. Math. Biol. **72**, 1840–1866 (2010)
15. Bressloff, P., Newby, J.: Stochastic models of intracellular transport. Rev. Mod. Phys. **85**, 135–196 (2013)
16. Brockwell, P. J., Gani, J., Resnick, S.I.: Catastrophe processes with continuous state space. Austral. J. Stat. **83**, 208–226 (1983)
17. Burkitt, A.N.: A review of the integrate-and-fire neuron model: I. Homogeneous synaptic input. Biol. Cybern. **95**, 1–19 (2006)
18. Carter, R.L.: Molecular Symmetry and Group Theory. Wiley, New York (1997)
19. Cayley, A.: On the theory of groups, as depending on the symbolic equation $\theta^n = 1$. Philosoph. Mag. **7**, 40–47 (1854)
20. Comets, F., Popov, S., Schütz, G.M., Vachkovskaia, M.: Billiards in a general domain with random reflections. Arch. Ration. Mech. Anal. **191**, 497–537 (2009)
21. Cornfeld, I.P., Fomin, S.V., Sinai, Ya.G.: Ergodic Theory, Grundlehren der Mathematischen Wissenschaften, vol. 245. Springer, New York (1982)
22. Cotton, A.: Chemical Applications of Group Theory. Wiley, New York (1990)
23. Davis, M.H.A.: Piecewise-deterministic Markov processes: a general class of nondiffusion stochastic models. J. Roy. Statist. Soc. Ser. B **46**, 353–388 (1984)
24. Diekmann, O., Heijmans, H.J.A.M., Thieme, H.R.: On the stability of the cell size distribution. J. Math. Biol. **19**, 227–248 (1984)
25. Diekmann, O., de Graaf, W.F., Kretzschmar, M.E.E., Teunis, P.F.M.: Waning and boosting: on the dynamics of immune status. J. Math. Biol. **77**, 2023–2048 (2018)
26. Emel'yanov, E.Y.: Non-spectral Asymptotic Analysis of One-Parameter Operator Semigroups. Birkhäuser, Basel (2007)
27. Engel, K-J., Nagel, R.: One-Parameter Semigroups for Linear Evolution Equations. Springer Graduate Texts in Mathematics, vol. 194. Springer, New York (2000)
28. Ethier, S., Kurtz, T.: Markov Processes: Characterization and Convergence. Wiley, New York (1986)
29. Evans, S.N.: Stochastic billiards on general tables. Ann. Appl. Probab. **11**, 419–437 (2001)
30. Foguel, S.R.: The Ergodic Theory of Markov Processes. Van Nostrand Reinhold Company Publisher, New York (1969)
31. Gerlach, M., Glück, J.: Lower bounds and the asymptotic behaviour of positive operator semigroups. Ergod. Theor. Dyn. Syst. **38**, 3012–3041 (2018)
32. Gerlach, M., Glück, J.: Convergence of positive operator semigroups. Trans. Amer. Math. Soc. **372**, 6603–6627 (2019)
33. Goldstein, S.: On diffusion by discontinuous movements and on the telegraph equation. Quart. J. Mech. Appl. Math. **4**, 129–156 (1951)
34. Greiner, G.: Perturbing the boundary conditions of a generator. Houston J. Math. **13**, 213–229 (1987)
35. Gripenberg, G.: A Stationary distribution for the growth of a population subject to random catastrophes. J. Math. Biol. **17**, 371–379 (1983)

36. Gyllenberg, M., Heijmans, H.J.A.M.: An abstract delay-differential equation modelling size dependent cell growth and division. SIAM J. Math. Anal. **18**, 74–88 (1987)
37. Hillen, T., Hadeler, K.P.: Hyperbolic systems and transport equations in mathematical biology. In: Warnecke, G. (Ed.) Analysis and Numerics for Conservation Laws, pp. 257–279. Springer, Berlin (2005)
38. Hu, J., Wu, W.C., Sastry, S.S.: Modeling subtilin production in bacillus subtilis using stochastic hybrid systems. In: Alur, R., Pappas, G.J. (Eds.) Hybrid Systems: Computation and Control. LNCS, vol. 2993, pp. 417–431. Springer, Berlin (2004)
39. Gwiżdż, P., Tyran-Kamińska, M.: Positive semigroups and perturbations of boundary conditions. Positivity **23**, 921–939 (2019)
40. Hanson, F.B., Tuckwell, H.C.: Logistic growth with random density dependent disasters. Theor. Popul. Biol. **19**, 1–18 (1981)
41. Jukes, T.H., Cantor, C.R.: Evolution of protein molecules. In: Munro, H.N. (Ed.) Mammalian Protein Metabolism, pp. 21–132. Academic, New York (1969)
42. Kac, M.: A stochastic model related to the telegrapher's equation. Rocky Mountain J. Math. **4**, 497–510 (1974)
43. Kato, T.: On the semi-groups generated by Kolmogoroff's differential equations. J. Math. Soc. Japan **6**, 1–15 (1954)
44. Komorník, J.: Asymptotic periodicity of the iterates of Markov operators. Tôhoku Math. J. **38**, 15–27 (1986)
45. Komorowski, T., Tyrcha, J.: Asymptotic properties of some Markov operators. Bull. Pol. Ac. Math. **37**, 221–228 (1989)
46. Kouretas, P., Koutroumpas, K., Lygeros, J., Lygerou, Z.: Stochastic hybrid modeling of biochemical processes. In: Cassandras, C.G., Lygeros, J. (Eds.) Stochastic Hybrid Systems, Control Engineering Series, vol. 24, pp. 221–248. CRC Press, Boca Raton (2007)
47. Krengel, U.: Ergodic Theorems. de Gruyter Studies in Mathematics, vol. 6. Walter de Gruyter, Berlin (1985)
48. Lande, R.: Risks of population extinction from demographic and environmental stochasticity and random catastrophes. Am. Nat. **142**, 911–927 (1993)
49. Lasota, A., Mackey, M.C.: Chaos, Fractals and Noise. Stochastic Aspects of Dynamics, Springer Applied Mathematical Sciences, vol 97, II edn. Springer, New York (1994)
50. Lasota, A., Yorke, J.A.: Exact dynamical systems and the Frobenius–Perron operator. Trans. AMS **273**, 375–384 (1982)
51. Lasota, A., Li, T.Y., Yorke, J.A.: Asymptotic periodicity of the iterates of Markov operators. Trans. Am. Math. Soc. **286**, 751–764 (1984)
52. Lebowitz, J.L., Rubinow, S.L.: A theory for the age and generation time distribution of microbial population. J. Math. Biol. **1**, 17–36 (1974)
53. Lipniacki, T., Paszek, P., Marciniak-Czochra, A., Brasier, A.R., Kimmel, M.: Transcriptional stochasticity in gene expression. J. Theor. Biol. **238**, 348–367 (2006)
54. Lods, B., Mokhtar-Kharroubi, M., Rudnicki, R.: Invariant density and time asymptotics for collisionless kinetic equations with partly diffuse boundary operators. Ann. I. H. Poincaré - AN **37**, 877–923 (2020)
55. Lyubich, Yu.I., Phóng, Vũ.Q.: Asymptotic stability of linear differential equation in Banach space. Studia Math. **88**, 37–42 (1988)
56. Łuczka, J., Rudnicki, R.: Randomly flashing diffusion: asymptotic properties. J. Statist. Phys. **83**, 1149–1164 (1996)
57. Metz, J.A.J., Diekmann, O. (Eds.): The Dynamics of Physiologically Structured Populations. Springer Lecture Notes in Biomathematics, vol. 68. Springer, New York (1986)
58. Meyn, S., Tweedie, R.: Markov Chains and Stochastic Stability. Springer, London (1993)
59. Mokhtar-Kharroubi, M., Rudnicki, R.: On asymptotic stability and sweeping of collisionless kinetic equations. Acta Appl. Math. **147**, 19–38 (2017)
60. Morgan, D.O.: The Cell Cycle: Principles of Control. New Science Press, London (2007)
61. Nagel, R. (Ed.): One-parameter Semigroups of Positive Operators, Lecture Notes in Mathematics, vol. 1184. Springer, Berlin (1986)

62. Nummelin, E.: General Irreducible Markov Chains and Non-negative Operators. Cambridge Tracts in Mathematics, vol. 83. Cambridge University Press, Cambridge (1984)
63. Othmer, H.G., Dunbar, S.R., Alt, W.: Models of dispersal in biological systems. J. Math. Biol. **26**, 263–298 (1988)
64. Pollicott, M., Yuri, M.: Dynamical Systems and Ergodic Theory. London Mathematical Society Student Texts, vol. 40. Cambridge University Press, Cambridge (1998)
65. Pichór, K.: Asymptotic stability of a partial differential equation with an integral perturbation. Ann. Pol. Math. **68**, 83–96 (1998)
66. Pichór, K., Rudnicki, R.: Stability of Markov semigroups and applications to parabolic systems. J. Math. Anal. Appl. **215**, 56–74 (1997)
67. Pichór, K., Rudnicki, R.: Asymptotic behaviour of Markov semigroups and applications to transport equations. Bull. Pol. Ac. Math. **45**, 379–397 (1997)
68. Pichór, K., Rudnicki, R.: Continuous Markov semigroups and stability of transport equations. J. Math. Anal. Appl. **249**, 668–685 (2000)
69. Pichór, K., Rudnicki, R.: Asymptotic decomposition of substochastic operators and semigroups. J. Math. Anal. Appl. **436**, 305–321 (2016)
70. Pichór, K., Rudnicki, R.: Asymptotic decomposition of substochastic semigroups and applications. Stoch. Dynam. **18**, 1850001 (18 pages) (2018)
71. Pichór, K., Rudnicki, R.: Stability of stochastic semigroups and applications to Stein's neuronal model. Discrete Contin. Dyn. Syst. Ser. B **23**, 377–385 (2018)
72. Pichór, K., Rudnicki, R.: Applications of stochastic semigroups to cell cycle models. Discrete Contin. Dyn. Syst. Ser. B **24**, 2365–2381 (2019)
73. Pichór, K., Rudnicki, R.: One and two-phase cell cycle models. Biomath **8**, 1905261 (2019)
74. Pichór, K., Rudnicki, R.: Dynamics of antibody levels: asymptotic properties. Math. Method Appl. Sci. **43**, 10490–10499 (2020)
75. Pichór, K., Rudnicki, R.: Asymptotic properties of stochastic semigroups with applications to piecewise deterministic Markov processes. In: Banasiak, J., Bobrowski, A., Lachowicz, M., Tomilov, Y. (Eds.) Semigroups of Operators: Theory and Applications SOTA-2018. Springer Proceedings in Mathematics & Statistics, vol. 325, pp. 329–347. Springer, Cham (2020)
76. Pichór, K., Rudnicki, R.: Cell cycle length and long-time behavior of an age-size model. Math. Meth. Appl. Sci. **45**, 5797–5820 (2022)
77. Pichór, K., Rudnicki, R., Tyran-Kamińska, M.: Stochastic semigroups and their applications to biological models. Demonstratio Math. **45**, 463–495 (2012)
78. Rotenberg, M.: Transport theory for growing cell populations. J. Theor. Biol. **103**, 181–199 (1983)
79. Rudnicki, R.: On asymptotic stability and sweeping for Markov operators. Bull. Pol. Ac.: Math. **43**, 245–262 (1995)
80. Rudnicki, R.: Stochastic operators and semigroups and their applications in physics and biology. In: Banasiak, J., Mokhtar-Kharroubi, M. (Eds.) Evolutionary Equations with Applications in Natural Sciences. Lecture Notes in Mathematics, vol. 2126, pp. 255–318. Springer, Heidelberg (2015)
81. Rudnicki, R., Pichór, K.: Markov semigroups and stability of the cell maturation distribution. J. Biol. Syst. **8**, 69–94 (2000)
82. Rudnicki, R., Tomski, A.: On a stochastic gene expression with pre-mRNA, mRNA and protein contribution. J. Theor. Biol. **387**, 54–67 (2015)
83. Rudnicki, R., Tyran-Kamińska, M.: Piecewise Deterministic Processes in Biological Models. Springer Briefs in Applied Sciences and Technology, Mathematical Methods. Springer, Cham (2017)
84. Singh, A., Hespanha, J.P.: Stochastic hybrid systems for studying biochemical processes. Phil. Trans. R. Soc. A **368**, 4995–5011 (2010)
85. Sklyar, G.M., Polak, P.: Notes on the asymptotic properties of some class of unbounded strongly continuous semigroups. J. Math. Phys. Anal. Geo. **15**, 412–424 (2019)
86. Stein, R.B.: Some models of neuronal variability. Biophys. J. **7**, 37–68 (1967)

87. Stein, R.B., Gossen, E.R., Jones, K.E.: Neuronal variability: noise or part of the signal? Nat. Rev. Neurosci. **6**, 389–397 (2005)
88. Stroock, D.W.: Some stochastic processes which arise from a model of the motion of a bacterium. Z. Wahrscheinlichkeitstheorie verw. Gebiete **28**, 305–315 (1974)
89. Thieme, H.R.: Positive perturbation of operator semigroups: Growth bounds, essential compactness, and asynchronous exponential growth. Disc. Cont. Dyn. Sys. **4**, 735–764 (1998)
90. Thieme, H.R.: Balanced exponential growth of operator semigroups. J. Math. Anal. Appl. **223**, 30–49 (1998)
91. Webb, G.W.: Theory of Nonlinear Age-Dependent Population Dynamics. Dekker, New York (1985)
92. Webb, G.W.: An operator-theoretic formulation of asynchronous exponential growth. Trans. Amer. Math. Soc. **303**, 751–763 (1987)
93. Webb, G.W.: Structured population dynamics. In: Rudnicki, R. (Ed.) Mathematical Modelling of Population Dynamics, vol. 63, pp. 123–163. Banach Center Publication IMPAN, Warszawa (2004)
94. Yin, G.G., Zhu, C.: Hybrid Switching Diffusions: Properties and Applications. Stochastic Modelling and Applied Probability, vol. 63. Springer, New York (2010)
95. Zeiser, S., Franz, U., Müller, J., Liebscher, V.: Hybrid modeling of noise reduction by a negatively autoregulated system. Bull. Math. Biol. **71**, 1006–1024 (2009)
96. Zeiser, S., Franz, U., Liebscher, V.: Autocatalytic genetic networks modeled by piecewise-deterministic Markov processes. J. Math. Biol. **60**, 207–246 (2010)

Markov Semigroup Approach to Evolution Equations

Andrzej Tomski

Abstract This overview is devoted to investigation of stochastic evolution in which a family of evolution equations in L^1 is driven by continuous-time Markov processes. The results are illustrated by selected amount of physical and biological examples. Here we also emphasize the role of Markov semigroups in describing these processes. We investigate long-time behavior of this phenomenon. We discuss some future perspectives to investigate in this field.

Keywords Stochastic semigroup · Piecewise deterministic Markov processes · Evolution equations

1 Evolution Equations

The main topic of this paper is evolution equation. An evolution equation can be defined as a partial differential equation that describes the time evolution of a physical system starting from some given initial data (states). We want to know something about the behavior of this system in the future especially in terms of the qualitative nature of its action. It is generally very difficult to solve evolutionary equations directly because they usually involve higher-order partial derivatives, nonlinear elements, stochastic perturbations, and so on. However, such systems can be analyzed both qualitatively with functional analysis methods and using numerical algorithms using a computer, especially programming environments. Despite that fact, nowadays evolution equations arise from many areas of both natural and engineering sciences. For example, one can consider Schrödinger equations from quantum mechanics, Navier–Stokes equations from fluid mechanics, and nonlinear reaction-diffusion equations from heat transfers. These cases are special examples

A. Tomski (✉)
Institute of Mathematics, University of Silesia in Katowice, Katowice, Poland
e-mail: andrzej.tomski@us.edu.pl

M. Mitrović, M. N. Hounkonnou (eds.), *Algebra Without Borders*, Mathematics in Mind, https://doi.org/10.1007/978-3-031-86477-3_8

of evolution equations; however, it is a much more general notion. To define such equation, first we need the notion of closed operators.

Definition 1 Let $\mathbb{X}$ be a Banach space. A linear operator A having domain $D(A) \subset \mathbb{X}$ and values in $\mathbb{X}$ is called closed iff for any converging sequence $(x_n)_{n=1}^{\infty}$ of elements of $\mathbb{X}$ such that Ax_n converges, the limit $x = \lim_{n\to\infty} x_n$ is an element of $D(A)$ and $Ax = \lim_{n\to\infty} Ax_n$. It is a known fact all continuous operators are closed. However, one can give many examples of linear operators that are closed but not continuous as well.

Example 1 We consider $C[0, 1]$, which is the space of continuous functions on the interval [0, 1] equipped with standard supremum norm. We define the operator A on the algebraic subspace $D(A)$ of continuously differentiable functions by the formula $Ax = x'$. We claim that operator A is not bounded. To justify this, we note that it is not defined on the entire space. Despite the fact the domain of this operator is dense in $C[0, 1]$ maybe it is not possible to extend A to a bounded linear operator. It would be possible if there exists a constant K such that $||Ax|| \leq K||x||$ [6]. However, if we define the functions $x_n(i) = i^n, n \leq 1$, we can see that $||x_n|| = 1$ and $Axn = nx_{n-1}$, so $Axn = n$. Conversely, if for a converging sequence $(x_n)_{n=1}^{\infty}$ of $D(A)$ elements, there also exists the limit $\lim_{n\to\infty} x_n'$, then x is continuously differentiable and this limit is equal to x'. Now since the sequence $(x_n)_{n=1}^{\infty}$ converges to some continuous value y , the integral $\int_0^i x_n'(i)di$ converges to $\int_0^i y(i)di$ while $x_n(i)$ converges to $x(i)$. This ends the proof.

The basic definition of evolution equation given here is as follows.

Definition 2 Let A be a closed operator on the set X and $x \in D(A)$. We define the Cauchy problem or evolution equation in the following form:

$$u'(t) = Au(t), t \geq 0, u(0) = x. \tag{1.1}$$

As we can see, on the right-hand side of such an equation, we have a closed operator. We recall that a classical solution of this equation is a function $u \in C^1(R_{\geq 0}, X)$ taking values in $D(A)$ and satisfying it for all values $t \geq 0$. We conclude here that in such case, Au belongs to $C(R_{\geq 0}, X)$ and thus u belongs to $C(R_{\geq 0}, X)$. The values $u(t) \in X$ are being interpreted in a way that they describe some physical system whose special characteristics are encoded in the operator A along with its domain. There is a very clear reason for doing so. Our goal is to be able to predict the future behavior of the system by means of considered evolution equation. To achieve this, we need its solutions for various initial values x. Moreover, the solutions should be uniquely determined by the initial point x since otherwise we are not really able to predict the behavior of the process.

In addition, one will know the initial value only approximately, so that for a reasonable prediction, the solutions should not vary too much under small changes of the data. Natural question appears about well-posedness of this problem. We have then the following theorem.

Theorem 1 *We assume that the operator A is closed. The Cauchy problem is well-posed if the set $D(A)$ is dense in X, if for each $x \in D(A)$, there exists a unique solution u of evolution equation and if its solutions depend continuously on the data; i.e.,*

$$\forall b > 0 : (D(A), || \cdot ||_X) \to C([0, b], X); x \to u(\cdot; x),$$

is continuous function.

We note that more general definition of the evolution equation can be formulated. Briefly speaking, evolution equation is of the form:

$$\frac{du(t)}{dt} = A(t)u(t) + f(t), 0 < t \leq T. \tag{1.2}$$

Here the function $A(t)$ denotes the infinitesimal generator of a C_0 semigroup for each $t \in [0, T]$ and some given function f, which is usually a strongly continuous function with values in a Banach space X. We assume that the following properties are satisfied:

- (i) the domain $D(A(t))$ of a function $A(t)$ is a dense set in X and it is independent of t;
- (ii) $A(t)$ is a generator of a contraction semigroup for all $t \in [0, T]$;
- (iii) the bounded function $t \to (I - A(t))(I - A(0)) - 1$ is continuously differentiable.

In the work of Kato [26], the fundamental solution $U(t, s), 0 \leq s \leq t \leq T$ was constructed as a bounded operator-valued function which satisfies the following properties:

- (a) a function $U(t, s)$ is strongly continuous for $0 \leq s \leq t \leq T$;
- (b) $U(s, s) = I$ for $s \in [0, T]$;
- (c) $U(t, r)U(r, s) = U(t, s)$ for $0 \leq s \leq t \leq T$;

A solution of evolution equation which satisfies the initial condition $u(0) = u_0$ if it exists, then it can be expressed in the following form:

$$u(t) = U(t, 0)u_0 + \int_0^t U(t, s)f(s)ds. \tag{1.3}$$

We note that evolution equations formulated in this way addressed deterministic description of a phenomenon. The operator A which is present in such type of equations is nonrandom. New question arises: what happens if we investigate a situation when these operators switch randomly? This leads us first to the notion of random evolutions. Random evolutions or stochastic evolutions were successfully introduced by R. Hersh and R.J. Griego in [22]. Random evolutions are a family of models used to investigate some dynamical systems in random media (environment). Here, Poisson process represents the random medium. However, from a mathematical side, random evolutions are considered as operator-valued stochastic processes defined in a Banach space. They represent in a quite abstract form the behavior of stochastic evolutionary systems in random media. Hence, there is a special need to study such systems of equations. To introduce such systems was possible due to observation that various abstract Cauchy problems can be extensively related to some generators of Markov processes. To this day, the subject of debate is how to define precisely what random evolution is. We can use this term to describe a situation when an evolutionary system changes its mode of evolution following a stochastic mode. Such situations are quite natural in nature. For example, consider a molecule moving on the real line with constant velocity V. However, this value changes at a random time point, given by the arrival of Poisson process to $-V$, when the velocity of the molecule at time t is equal to $(-1)^{N(t)}V$, where by $N(t), t \geq 0$, we denote a Poisson process on the interval $[0, \infty)$. Other examples of scientifical problems which lead to random evolutions include:

- a molecule which moves on a straight line with constant speed, until it takes a part in a random collision; after that accident it changes its velocity, and moves again with a new constant speed,
- a radio signal which propagates through some turbulent medium, in which the index of refraction is changing randomly,
- a population of bacteria which evolve in randomly fluctuated environment.

Applications of random evolutions processes are present in such areas as population dynamics in random environment [13], insurance [45], or mathematical physics [24]. However, discrete-time random evolutions were investigated in a link with Markov chains, and as embedded random evolution in continuous-time Markov processes [27].

2 Piecewise Deterministic Markov Processes

In this section we present shortly piecewise deterministic Markov processes (PDMPs) which have been first introduced in the work of M.H. Davis [14]. Briefly speaking PDMP is a broad family of stochastic processes covering virtually most of the applications, omitting mainly diffusion-related phenomena. A book [42] is devoted to the description of biological models based on this class of stochastic processes. PDMP is a continuous-time Markov process taking values in some metric space. It can be defined in a way that there is an increasing sequence of random times $(t_n)_{n=1}^{\infty}$, called jump times, such that trajectories $\xi(t)$ of this process are defined deterministically in each interval of the form (t_n, t_{n+1}). Generally deterministic evolution is described by ordinary differential equations (ODEs) which induce dynamical systems. For PDMPs two types of behavior of the process at jump times are possible: the process can jump to some new point or it can change its dynamics which defines its trajectories.

Existing models usually describe the underlying phenomena for some population from the point of view of a single individual. In physics this is often known as a molecule perspective [8]. That means that the dynamics of every single individual is driven by separate stochastic laws depending on a variety of factors, e.g., its mass and energy. However, there are alternative situations in which the entire population is touched by randomly switching environmental conditions, e.g., molecules driven by a common environmental noise [8] and the response of a metabolic or gene regulatory system to an environmental stimulus [3]. This is known as a population perspective [8], and is the approach that we use in this paper since we consider one common source of randomness which affects all individuals in a population.

Here we treat the evolution of the density of a population distribution in the situation where every individual has its own deterministic dynamics, but the whole population is touched by some continuous-time Markov process with finite state space that changes the current state of all individuals. In this approach, a single state is represented by a population density which is an element of an infinite-dimensional space. The general approach to study the evolution of such densities is not known, and this is the main reason behind investigation of their moments of various orders. However, we must stress here that these processes are dual processes to the class of PDMPs called random evolutions introduced by Griego and Hersh [18], but on the basis of the work done by such authors like Goldstein [16] and Kac [25], [40]. Such processes are examples of models governed by switching partial differential equations (PDEs) that recently became quite popular, and there is plenty of papers devoted to this topic [5, 9, 10, 33, 34]. We note that these papers focus mainly on applications in biological sciences. These models based on diffusion processes and parabolic PDEs. In the papers [9, 32], the authors provide the moment and correlation equations in the case of diffusion processes. Here we will recall our generalization of such case by obtaining moment and correlation equations for much broader class of stochastic processes. Among the obtained results, we claim that the mean of a process given by randomly switching PDEs can be recognized as special stochastic semigroup (Theorem 2 and Corollary 3). In effect, the mean of random density in the population perspective is identical to a density from the individual perspective. We also investigate the mean of the process at large time for some biophysical processes. It allows us to investigate the asymptotic behavior for the mean of the process in the cases of various types of bifurcations. It appeared during our work that a computer was useful to provide numerical simulations for the mean of the process (i.e., Python-written FiPy [21] was used as an object-oriented PDE solver, which is based on a standard finite-volume approach).

3 Basic Definitions

Now we provide some basic definitions concerning Markov semigroups along with examples of such structures. It is these semigroups that will be used in the equations of evolution. Markov operators were introduced to investigate classical dynamical systems and stochastically perturbed dynamical systems. In general, such systems refer to a movement of points. Statistical look at such systems leads us to observation that we investigate the evolution of a distribution of points on the phase space. This is the way in which we obtain a transformation P which is defined on the space of probability measures.

Assume that P is defined by a transition probability function, i.e., the transformation of Dirac measures δ_x determines P. If there is some measure m on X, then we can only consider measures being absolutely continuous with respect to this measure. In that case, instead of the transformation of measures, we consider the transformation of densities of these measures. In this way we can obtain a linear transformation of the space of integrable functions which preserves the set of densities. Such a transformation is called a Markov operator which will be defined precisely below.

Let $(E, \mathcal{E}, m)$ be a triple with σ-finite measure space and we set $L^1 = L^1(E, \mathcal{E}, m)$. The set of densities $D \subset L^1$ is defined by the formula:

$$D = \{f \in L^1 : f \geq 0, \|f\| = 1\}.$$

A *stochastic (Markov) operator* is linear mapping $P \colon L^1 \to L^1$ such that it preserves D, i.e., $P(D) \subset D$ [31]. A family of linear operators defined $\{P(t)\}_{t\geq 0}$ on L^1 is called a *stochastic semigroup* if each operator $P(t)$ is stochastic and $\{P(t)\}_{t\geq 0}$ is a C_0-*semigroup*, i.e., the following three conditions are satisfied:

(i) $P(0) = \text{Id}$ (identity condition),
(ii) $P(t + s) = P(t)P(s)$ for all $t, s \geq 0$ (semigroup condition),
(iii) for every f the function $t \to P(t)f$ is continuous (continuity condition).

There are plenty of specific Markov semigroups used commonly in the literature. Moreover, one can assign them clear physical interpretation. We give examples of a few of them. A more comprehensive description of Markov semigroups can be found in the review paper [43].

Example 2 Frobenius–Perron operator. This is the operator which describes statistical features of point-to-point transformations. Let (X, Σ, m) be a σ-finite measure space and then let S be a measurable transformation of the set X. If a measure μ gives the distribution of points in X, then the measure υ defined by the formula $\upsilon(A) = \mu(S^{-1}(A))$ gives the distribution of points after transformation given by S. We assume here that the transformation S is non-singular, which means that if $m(A) = 0$, then $m(S^{-1}(A)) = 0$. Note that if the measure μ is absolutely continuous with respect to the measure m, then υ is also absolutely continuous. If f is the density of μ and if g is the density of υ, then we define the operator P_S by $P_S f = g$. This operator can be extended to a linear operator $P_S : L_1 \to L_1$. A Markov operator which is defined in this way is called the Frobenius–Perron operator for the transformation. The range of applications of this operator is very large, both in mathematics and physics. Frobenius–Perron operators can be also effectively applied to study ergodic properties of transformations.

Example 3 Iterated function system is another important example orientated on Markov semigroups. We assume that $S_1, ..., S_n$ are non-singular transformations of the space X. Moreover, $P_1, ..., P_n$ are the Frobenius–Perron operators corresponding to the transformations $S1, ..., S_n$. Moreover, $p_1(x), ..., p_n(x)$ are nonnegative measurable functions on X such that $p_1(x) + \cdots + pn_(x) = 1$ for all $x \in X$. We take some arbitrary point x. Next, we choose S_i with the probability $p_i(x)$. $S_i(x)$ constitutes the position of point x under action of the system. The evolution of distribution is the Markov operator given by the following formula:

$$P_f = \sum_{i=1}^{n} P_i(p_i f). \tag{3.1}$$

Biological application of iterated function system was given IN the paper [] where additionally the probability of choice of any iterated map depended on the state of the phase space X. More precisely, random jump times of the process set activation times of the gene when pre-mRNA molecules were produced before consecutive phases occurred. It was shown there that in such case, there exists a unique invariant limit measure. Probabilistic description of the process was provided along with a comparison to other biological models.

Example 4 Integral operator. If function $k : X \times X \to [0, \infty)$ is a measurable function such that $\int_X k(x, y)m(dx) = 1$ for almost all points $y \in X$, then

$$Pf(x) = \int_X k(x, y) f(y)m(dy)$$

s a Markov operator. Transformation k is called a kernel of the operator P. Many natural processes in science can be modeled by randomly perturbed hybrid dynamical systems whose distribution can be described by some integral Markov operators. Such systems adopt the form:

$$X_{n+1} = S(X_n, \xi_{n+1}),$$

where $(\xi_n)_{n=1}^{\infty}$ is a sequence of independent identically distributed random variables and the initial value of the system X_0 does not depend on $(\xi_n)_{n=1}^{\infty}$. Describing such processes leads to investigating behavior of the sequence of the measures (μ_n) defined as $\mu_n(A) = Prob(X_n \in A)$. Time evolution of this kind of measures can be represented by a Markov operator P such that $\mu_{n+1} = P\mu_n$. The operator P has its domain on the space of probability measures. If for almost all y the distribution μ_y of the random variable $S(y, \xi_n)$ is absolutely continuous with respect to m and $k(x, y)$ is the density of μ_y, then operator P represents the evolution of the system. Moreover, integral operators are used in a two-phase Tyrcha's model of cell cycle which is a generalization of Lasota–Mackey model.

Example 5 Fokker–Planck equation. We investigate stochastic differential equation in Stratonovitch formulation, i.e.,

$$dX_t = \sigma(X_t) \circ dW_t + \sigma_0(X_t)dt, \tag{3.2}$$

where W_t denotes a m-dimensional Wiener process, $\sigma(x) = [\sigma^i_j(x)]$ constitutes $d \times m$ matrix and $\sigma_0(x)$ is a vector in $\mathbb{R}^d$. For all $i = 1, ..., d,\ j = 0, ..., m$, we take sufficiently smooth functions σ^i_j with bounded derivatives of all orders and bounded coefficients of the matrix σ. We can express similar equation in Ito's formulation. If we assume that random variable X_t is a solution of this equation such that the distribution of random variable X_0 is absolutely continuous with the density function $v(x)$, in such case, $X(t)$ is also absolutely continuous with the density function $u(x, t)$, and it satisfies the general Fokker–Planck equation which has the form:

$$\frac{du}{dt} = \Sigma^d_{i=1} \frac{\partial}{\partial x_i}\left(\Sigma^d_{j=1} a_{ij}(x)\frac{\partial u}{\partial x_j}\right) - \Sigma^d_{i=1}\frac{\partial(\sigma^i_0(x)u)}{\partial x_i}, \tag{3.3}$$

with $a_{ij}(x) = \frac{1}{2}\Sigma^m_{k=1}\sigma^i_k(x)\sigma^j_k(x)$. Now we consider the operator:

$$Af = \Sigma^d_{i=1} \frac{\partial}{\partial x_i}\left(\Sigma^d_{j=1} a_{ij}(x)\frac{\partial f}{\partial x_j}\right) - \Sigma^d_{i=1}\frac{\partial(\sigma^i_0(x)f)}{\partial x_i}, \tag{3.4}$$

defined on $E = \{f \in L^1(\mathbb{R}^d) \cap C^2_b(\mathbb{R}^d) : Af \in L^1(\mathbb{R}^d)\}$ where $C^2_b(\mathbb{R}^d)$ is the set of all twice differentiable and bounded functions with derivatives of order ≤ 2 being continuous and bounded. If $v \in C^2_b(\mathbb{R}^d)$, then Fokker–Planck equation has a unique solution u satisfying the condition $u(x, 0) = v(x)$. Moreover, this solution and its spatial derivatives with the order ≤ 2 are uniformly bounded functions with a domain $[0, T] \times \mathbb{R}^d$.

Example 6 Semigroup for semiflow. A particularly important example, especially in terms of applications, concerns the construction of a semigroup for dynamic semiflows. We assume now that (X, Σ, m) is a σ−finite measure space. Moreover, we have a measurable semiflow on X , defined on $\pi : [0, \infty) \times X \to X$ such that the mapping $\pi(t, x) \to \pi(t, x)$ is measurable and the function $\pi_t : X \to X$, such that $\pi_t(x) = \pi(t, x)$ for $x \in X$ and $t \geq 0$, with properties:

$$\pi_0(x) = x \text{ and } \pi_{s+t}(x) = \pi_t(\pi_s(x)), t, s \geq 0, x \in X.$$

We also assume that for each $t > 0$, the function π_t is non-singular with respect to the measure m. We set $P_0(t)$ as the Frobenius–Perron operator on the space L^1 corresponding to π_t. One can check that it satisfies the following property:

$$\int_X \mathbf{1}_B(\pi_t(x)) f(x)m(dx) = \int_B P_0(t) f(x)m(dx);\ B \in \Sigma,\ f \in L^1. \tag{3.5}$$

Since $\{\pi_t\}_{t\geq 0}$ constitutes a semigroup on the set X, it follows that $P_0(0) = I$ and $P_0(t+s) = P_0(t)P_0(s)$ for all $s, t \leq 0$. In effect, the family $P_0(t)_{t\geq 0}$ is a semigroup on the space L^1.

There are plenty of examples of some special Markov semigroups used for very strict applications in physics and biology [43]. Modern science uses methods of Markov semigroups theory to describe and study environmental and molecular processes.

Example 7 Transport equation. It is a situation when we are able to generate new Markov semigroups if we have already some. It is known fact that if the differential equation $\frac{du}{dt} = Au$ generates stochastic semigroup $\{S(t)\}_{t\geq 0}$ and we take some Markov operator K and a positive constant λ, then the newly created equation of the following form:

$$\frac{du}{dt} = Au - \lambda u + Ku \tag{3.6}$$

generates Markov semigroup as well. Moreover, on the basis of the well-known Dyson–Phillips expansion, we are able to put a formula of such semigroup:

$$P(t)f = \exp(-\lambda t)\sum_{n=0}^{\infty} \lambda^n S_n(t)f, \tag{3.7}$$

with $S_0(t) = S(t)$ and

$$S_{n+1}(t)f = \int_0^t S(t-s)KS_n(s)f ds, n \geq 0. \tag{3.8}$$

We can find such type of equations in population dynamics [37] as well as in astrophysics where such equation is used to describe the fluctuations in the brightness of galaxies [11]. In reality, A is the operator from the transport equation, and the Markov operator denoted by K represents some transition probability function $P(x, E)$. The perturbed equation (3.8) has a meaning in a probabilistic sense. We investigate a collection of molecules moving under the action of the flow $x' = b(x)$. The motion of these molecules changes in the following way. In every time interval $[t, t + \Delta t]$, a single molecule changes its position from x to a point from the set E with the probability equal to $P(x, E)\Delta t + o(\Delta t)$. It appears that a solution of transport equation plays the role of probability density function of the position of the molecule estimated at some given time t.

Moreover, time- and size-dependent models of various populations can be expressed in the form of a transport equation (3.8), i.e.,

$$\frac{du}{dt} + \frac{dV(x)u}{dx} = -u(x,t) + Ku(x,t). \tag{3.9}$$

In such case, the function $V(x)$ can be interpreted as the velocity of the growth of the size of a cell, while K is Markov operator which describes the replication. If one assumes that the size of a descendant cell is just a half of the size of the parent cell, then we have $Kf(x) = 2f(2x)$. However, if one investigates unequal cell division, then K is an integral operator. We note that much more advanced models of population dynamics lead to equations of the form (3.10), but typically, instead of the operator on the right-hand side of this equation, we have some non-bounded linear operator which can be represented as Markov semigroup. Moreover, Eq. (3.8) can also describe the distribution of the solutions of a Poisson-driven stochastic differential equation of the following form:

$$dX_t = b(X_t)dt + f(X_t)dN_t \tag{3.10}$$

with the function N_t being the Poisson process. The operator A is given by the formula $A_u = -\sum_{i=1}^{d} \frac{d}{dx_i}(b_i(x)u)$, while K is a Frobenius–Perron operator connected to the function $T(x) = x + f(x)$.

Example 8 Randomly flashing diffusion [36]. We consider the stochastic equation of the following form:

$$dX_t = (Y_t\sigma(X_t))dW_t + b(X_t)dt, \tag{3.11}$$

with Y_t being a homogeneous, binary-valued Markov process independent of W_t and X_0. This equation represents randomly diffusing process which randomly jumps between stochastic and deterministic states. One can show that this process generates a Markov semigroup on the space $L_1(\mathbb{R} \times \{0, 1\})$. The density functions of the distribution of this process satisfy the following system of equations:

$$\begin{cases} \frac{du_1}{dt} = -pu_1 + qu_0 + \frac{d^2}{dx^2}(a(x)u_1) - \frac{d}{dx}(b(x)u_1), \\ \frac{du_0}{dt} = pu_1 - qu_0 - \frac{d}{dx}(b(x)u_0). \end{cases} \tag{3.12}$$

Such description allows us to introduce a multistate diffusion process on the space R^d, and one can check that it generates a Markov semigroup as well. Now, let the process Y_t to be a continuous time Markov chain defined on the phase space $\Gamma = \{1, ..., k\}$, having the property that the transition probability from the state j to the state $i = j$ in time interval Δt is equal to $p_{ij}\Delta t + o(\Delta t)$. Moreover, we can assume that $p_{ij} > 0$ for all $i = j$. Now, we define b as a $d-$dimensional vector function on the set $R^d \times \Gamma$. Let $d-$dimensional random variable X_0 be independent of the process Y_t. We investigate stochastic differential equation in the following form:

$$dXt = \sigma(X_t, Y_t)dW_t + b(X_t, Y_t)dt. \tag{3.13}$$

Two random variables (X_t, Y_t) give a Markov process on the space $\mathbb{R}^d \times \Gamma$. If the random variable X_0 has an absolutely continuous distribution, then the random variable X_t has it as well for all $t > 0$.

Now we recall one of the most important concepts in the theory of continuous semigroups. The infinitesimal generator of the semigroups is defined as its right derivative at the origin point where the sense in which the derivative is taken depends on the continuity hypothesis. For example, the generator of a semigroup is defined as the strong derivative. The infinitesimal *generator* of the semigroup $\{P(t)\}_{t\geq 0}$ is the operator A with domain $\mathcal{D}(A) \subset L^1$ given by

$$\mathcal{D}(A) = \{f \in L^1 : \lim_{t\downarrow 0} \frac{1}{t}(P(t)f - f) \text{ exists}\},$$

$$Af = \lim_{t\downarrow 0} \frac{1}{t}(P(t)f - f), \quad f \in \mathcal{D}(A).$$

It is obvious that if x and y belong to D(A) then $\alpha x + \beta y$ belong to this set as well for all real values α and β, which means that $A(\alpha x + \beta y) = \alpha Ax + \beta Ay$. We conclude that A is a linear operator. $D(A)$ is non-empty, and moreover, it is dense in the space L^1.

We can also provide some other properties of the generator of a stronly continuous (not necessarily Markov) semigroup.

Lemma 1 ([42]) *Let $S(t)_{t\geq 0}$ be a strongly continuous semigroup with generator $(A, D(A))$.*
Then, the following properties hold:

- *There exist constants $\omega \in \mathbb{R}$ as well $M \geq 1$ satisfying the inequality:*

$$||S(t)|| \leq Me^{\omega t}, t \geq 0. \tag{3.14}$$

- *For every $f \in X$, the function $[0, \infty] \ni t \to S(t)f \in X$ is continuous.*
- *If $f \in X$, then:*

$$\int_0^t S(s)fds \in D(A) \tag{3.15}$$

and

$$S(t)f - f = A\int_0^t S(s)fd(s), t > 0. \tag{3.16}$$

- *If $f \in D(A)$, then $S(t)f \in D(A)$,*

$$\frac{d}{dt}S(t)f = S(t)Af \tag{3.17}$$

and

$$S(t)f - f = \int_0^t S(s)Afd(s), t > 0. \quad (3.18)$$

- *The pair (A,D(A)) constitutes an operator which is closed and densely defined.*

An important property of the generator is that it uniquely determines a strongly continuous semi group. Define A_1, A_2 as two linear operators with domains $D(A_1), D(A_2)$, respectively. A_2 is an extension of A_1 or $A1 \subset A2$ iff $D(A_1) \subset D(A_2)$ and $A_2 f = A_1 f$ for $f \in D(A_1)$.

The idea is to use stochastic semigroups as a representation of some evolution equations.

A simple example of evolution equations is the deterministic Liouville equation [43].

Consider the movement of molecules in the phase space $\mathbb{R}^d$, $d \geq 1$, described by a differential equation:

$$x'(t) = b(x(t)), \quad (3.19)$$

where $b(x)$ is a d-dimensional vector. Then, the Liouville equation describes the evolution of the density of the distribution of molecules, i.e., if $x(t)$ has a density $u(t, x)$, then u is the solution of the following equation:

$$\frac{\partial u(t, x)}{\partial t} = -\operatorname{div}(b(x)u(t, x)). \quad (3.20)$$

Let, for any $x_0 \in \mathbb{R}^d$, Eq. (3.19) with initial condition $x(0) = x_0$ have a solution for all t, which we denote by $\pi(t, x_0)$, and let the mapping $x_0 \mapsto \pi(t, x_0)$ be *non-singular* with respect to the Lebesgue measure Leb on $\mathbb{R}^d$, i.e., $\text{Leb}(\{x \in \mathbb{R}^d : \pi(t, x) \in B\}) = 0$ for all Borel sets B with $\text{Leb}(B) = 0$. If $f : \mathbb{R}^d \to [0, +\infty)$ is the density of the $\mathbb{R}^d$-valued random vector ξ_0, then the density of $\pi(t, \xi_0)$ is given by

$$P(t)f(x) = f(\pi(-t, x)) \det[\frac{d}{dx}\pi(-t, x)].$$

The family of operators $\{P(t)\}_{t \geq 0}$ forms a stochastic semigroup on the space $L^1(\mathbb{R}^d)$ and $u(t, x) = P(t)f(x)$ is the solution of (3.20) with initial condition $u(0, x) = f(x)$.

We will also use later a notation from the theory of tensor products [44]. We will define two main types of a tensor product. Assume we have two Banach spaces of functions $\mathcal{X}_1$ and $\mathcal{X}_2$, either $\mathcal{X}_i$ is an L^1 space or it is a subspace of the set of all bounded and measurable functions with the supremum norm. For $f_1 \in \mathcal{X}_1$ and $f_2 \in \mathcal{X}_2$, we just identify the function $(x_1, x_2) \mapsto f_1(x_1)f_2(x_2)$ with the tensor $f_1 \otimes f_2$. The tensor product space $\mathcal{X}_1 \otimes \mathcal{X}_2$ is defined as the set of all possible linear combinations of these tensors. The completion of the space $\mathcal{X}_1 \otimes \mathcal{X}_2$ with the projective norm

$$\|h\|_\pi = \inf\left\{\sum_{k=1}^{n} \|f_k\|\|g_k\| : f_k \in \mathcal{X}_1, g_k \in \mathcal{X}_2, h = \sum_{k=1}^{n} f_k \otimes g_k\right\}$$

is called then the *projective tensor product* of the spaces $\mathcal{X}_1$ and $\mathcal{X}_2$ and is denoted by $\mathcal{X}_1 \hat{\otimes} \mathcal{X}_2$.

It is well-known fact [44, Chapter 2] that $L^1(E_1, \mathcal{E}_1, m_1)\hat{\otimes}L^1(E_2, \mathcal{E}_2, m_2)$ is isometrically isomorphic with the space $L^1(E_1 \times E_2, \mathcal{E}_1 \times \mathcal{E}_2, m_1 \times m_2)$.

However, if we consider now $\mathcal{X}_1 \otimes \mathcal{X}_2$ with the injective norm

$$\|h\|_\varepsilon = \sup\{|(\gamma_1 \otimes \gamma_2)(h)| : \gamma_i \in \mathcal{X}_i^*, \|\gamma_i\| \le 1\},$$

with $\mathcal{X}_i^*$ being the dual of $\mathcal{X}_i$ and

$$(\gamma_1 \otimes \gamma_2)(h) = \sum_{k=1}^{n} \gamma_1(f_k)\gamma_2(g_k) \quad \text{for } h = \sum_{k=1}^{n} f_k \otimes g_k,$$

then the completion of $\mathcal{X}_1 \otimes \mathcal{X}_2$ is called the *injective tensor product*, and it is denoted by $\mathcal{X}_1 \widehat{\otimes} \mathcal{X}_2$. Specifically, if $C(E_i)$ is the space of continuous functions defined on a compact space E_i, then $C(E_1)\widehat{\otimes}C(E_2)$ is the space $C(E_1 \times E_2)$, by [44].

Assume we have two linear and bounded operators $S_i : \mathcal{X}_i \to \mathcal{X}_i$. Then, the linear mapping $S_1 \otimes S_2 : \mathcal{X}_1 \otimes \mathcal{X}_2 \to \mathcal{X}_1 \otimes \mathcal{X}_2$ defined by the following formula:

$$(S_1 \otimes S_2)(f_1 \otimes f_2) = S_1(f_1) \otimes S_2(f_2)$$

has its continuous extension to tensor product spaces. The result from [1, Section A-I.3, Proposition] is known.

Proposition 1 *Let $\{S_1(t)\}_{t\ge0}$ and $\{S_2(t)\}_{t\ge0}$ be C_0-semigroups on Banach spaces $\mathcal{X}_1$, $\mathcal{X}_2$ and the operators $(A_1, \mathcal{D}(A_1))$, $(A_2, \mathcal{D}(A_2))$ are their generators. In such case, the family*

$$\{S_1(t) \otimes S_2(t)\}_{t\ge0} \tag{3.21}$$

is a C_0-semigroup on both of the tensor products of $\mathcal{X}_1$ and $\mathcal{X}_2$. Moreover, the closure of

$$A_1 \otimes \mathrm{Id} + \mathrm{Id} \otimes A_2, \tag{3.22}$$

being defined on the core $\mathcal{D}(A_1) \otimes \mathcal{D}(A_2)$, is its infinitesimal generator.

From this theorem, we can obtain the following corollary.

Corollary 1 *If both* $\{S_1(t)\}_{t\geq 0}$ *and* $\{S_2(t)\}_{t\geq 0}$ *are stochastic semigroups on* $L^1(E_1, \mathcal{E}_1, m_1)$ *and* $L^1(E_2, \mathcal{E}_2, m_2)$*, respectively, then also* $\{S_1(t) \otimes S_2(t)\}_{t\geq 0}$ *is a stochastic semigroup on* $L^1(E_1 \times E_2, \mathcal{E}_1 \times \mathcal{E}_2, m_1 \times m_2)$.

Mainly due to its simplicity, we present a short proof of this theorem.

Proof For $f_i \in L^1(E_i, \mathcal{E}_i, m_i)$

$$\int_{E_1\times E_2} (S_1(t) \otimes S_2(t))(f_1 \otimes f_2) d(m_1 \times m_2) = \int_{E_1} S_1(t) f_1 dm_1 \int_{E_2} S_2(t) f_2 dm_2.$$

This equality means that the operator $S_1(t) \otimes S_2(t)$ has the property of preserving the integral. As $S_1(t) \otimes S_2(t)$ is positive, this ends the proof. □

We will use tensor products later to derive equations for the correlations of the stochastic evolution equation. However, we will start with the description of randomly switching dynamics, i.e., a regime when we have a single particle driven by a stochastic Markov process.

4 Stochastic Evolution of Dynamics

Now we recall a classical setting of PDMP-based models seen from the perspective of individual units and taking place in a finite-dimensional space.

Here we consider sufficiently smooth family of vector fields b_i, $i \in I = \{0, 1, \ldots, k\}$, $k \in \mathbb{N}$, determined on an open subset G of $\mathbb{R}^d$. Now we let $E \subset G$ be a Borel set with non-empty interior and with its boundary having Lebesgue measure equal to zero. For each $i \in I$ and $x_0 \in E$, we consider the equation:

$$x'(t) = b_i(x(t)), \tag{4.1}$$

with initial condition $x(0) = x_0$, assuming it has some solution $\pi_i(t, x_0)$ for all $t > 0$ in the set E. The mapping $(t, x_0) \mapsto \pi_i(t, x_0)$ is continuous. Moreover, we assume that each transformation of the form $\pi_i(t, \cdot)$ is non-singular with respect to the Lebesgue measure on the set E. Let $f : E \to [0, +\infty)$ be a density of a random vector ξ_0 with values in the space E. Then, the density of the process $\pi_i(t, \xi_0)$ is given by

$$P_i(t) f(x) = \mathbf{1}_E(\pi_i(-t, x)) f(\pi_i(-t, x)) \det\left[\frac{d}{dx}\pi_i(-t, x)\right]. \tag{4.2}$$

We know that for all values $i \in I$, the family of operators $\{P_i(t)\}_{t\geq 0}$ forms a stochastic semigroup on the space $L^1(E)$ which is called a Frobenius–Perron semigroup [31, Section 7.4].

Now we can define a randomly switching dynamics as a Markov process $\xi(t) = (x(t), i(t))$ on the state space $E \times I$ such that $x(t)$ is given by the solution of the differential equation having the form:

$$x'(t) = b_{i(t)}\big(x(t)\big), \tag{4.3}$$

where $i(t)$ is a continuous-time Markov chain with values in I and intensity matrix $[q_{ij}]$. We note that if the system is at time t_0 at the state (x_0, i), then $x(t)$ behaves according to Eq. (4.1) with $i(t) = i$ and $\xi(t) = (\pi_i(t - t_0, x_0), i)$. If $i(t)$ switches its value to some j with some given intensity q_{ij}, i.e., the chance of switching from i to j after time Δt is $q_{ij}\Delta t + o(\Delta t)$, $j \in I, j \neq i$, then we set the new vector field as b_j and the process restarts. Let t_1 be the moment of a switch from i to j and $x_1 = \pi_i(t_1 - t_0, x_0)$. Then, we have $\xi(t) = (\pi_j(t - t_1, x_1), j)$ until the next switch of the state of the process $\{i(t)\}_{t\geq 0}$. This kind of action repeats interminably.

We define:

$$q_i = \sum_{j\neq i} q_{ij} \quad \text{and} \quad q_{ii} = -q_i, \quad i \in I. \tag{4.4}$$

We note that if $k = 1$, then we have $q_{10} = -q_{11} = q_1$ and $q_{01} = -q_{00} = q_0$.

Moreover, for every $i \in I$, let $\{P_i(t)\}_{t\geq 0}$ be the stochastic semigroup as in (4.2) and let $(A_i, \mathcal{D}(A_i))$ be its infinitesimal generator. If $f = (f_i)_{i\in I}$ is a column vector with functions f_i such that $f_i \in \mathcal{D}(A_i)$, we define then a vector $Af = (A_i f_i)_{i\in I}$. Here we also denote the matrix $[q_{ij}]$ by Q and its transpose $[q_{ji}]$ by Q^T. In such case, the operator $A + Q^T$ is the infinitesimal generator of a stochastic semigroup $\{P(t)\}_{t\geq 0}$ on the space $L^1(E\times I) = L^1(E\times I, \mathcal{B}(E\times I), \mu)$. In our case, $\mathcal{B}(E\times I)$ is the σ-algebra of Borel subsets of $E \times I$ and μ is the product of the d-dimensional Lebesgue measure and the counting measure on I.

If we define the function $u(t) = P(t)f$, then u satisfies the following evolution equationL

$$\begin{cases} u'(t) = Au(t) + Q^T u(t), \\ u(0) = f. \end{cases} \tag{4.5}$$

Using the notation $f = (f_i)_{i\in I}$ and $u(t) = (u_i(t))_{i\in I}$, we can rewrite again (4.5) as

$$\begin{cases} u_i'(t) = A_i u_i(t) + \sum_{j\in I} q_{ji} u_j(t), \\ u_i(0) = f_i, \quad i \in I. \end{cases} \tag{4.6}$$

To conclude, the process $\xi(t) = (x(t), i(t))$ induces the stochastic semigroup $\{P(t)\}_{t\geq 0}$ (see [42, Section 4.2]), i.e., if f is the density function of the process $\xi(0)$, then $P(t)f$ is the density function of the process $\xi(t)$ and

$$\mathbb{P}(x(t) \in B, i(t) = i) = \int_{B\times\{i\}} P(t) f \, d\mu = \int_B u_i(t, x)\, dx, \quad i \in I.$$

The whole regime can be called piecewise deterministic dynamics.
Piecewise deterministic processes are particularly useful in natural sciences when it comes to model natural phenomena. Now we give some examples of such situations.

Example 9 Stochastic gene expression [7, 38]. Gene expression is widely known as a very complex process. Genes can change its state randomly between their active or inactive status. When gene is in its active state, transcripts are produced in the so-called bursts. After gene is activated, the DNA sequence is transformed into pre-mRNA form. Next, the noncoding sequences are cut and then the process goes on. In such case, we obtain a functional form of mRNA, which is moved then into the cytoplasm, where translation phase transforms mRNA into a protein. Hence, we investigate here a model of gene expression with three positive variables x_1, x_2, x_3, representing the dynamics of pre-mRNA, mRNA, and protein molecules, respectively. We assume here that pre-mRNA molecules are produced with the speed $A_1\gamma(t)$, where A_1 is a constant and it is driven by a stochastic process $\gamma(t) \in \{0, 1\}$ which states, at $t \geq 0$, if the gene is in an active ($\gamma(t) = 1$) or inactive ($\gamma(t) = 0$) state. The speed of mRNA production is equal to $A_2x_1(t)$, where A_2 is a constant and $x_1(t)$ denotes the level of pre-mRNA molecules at t. Finally, the protein translation is done at the rate $A_2x_1(t)$, where $x_2(t)$ denotes the level of mRNA molecules at t. Pre-mRNA molecules are lost with the rate $d_1x_1 = d_1'x_1 + A_2x_1$, where d_1' is the speed of degradation of pre-mRNA molecules and the constant A_2 is the speed of pre-mRNA conversion into mRNA. It appears that $d_1 = d_1' + A_2$ can be interpreted as the total speed of degradation of pre-mRNA molecules. We define now two continuous and integrable functions: the activation and inactivation rate function $q_0(x_1, x_2, x_3)$ and $q_1(x_1, x_2, x_3)$, respectively. Typically, the gene is activated with the rate $q_0(x_1, x_2, x_3)$ and it is inactivated with the rate $q_1(x_1, x_2, x_3)$ as well. Such model leads to following system of equations:

$$\begin{cases} 0 \xrightarrow{q_0(x_1,x_2,x_3)} 1, \; 0 \xleftarrow{q_1(x_1,x_2,x_3)} 1 \\ \dfrac{dx_1}{dt} = A_1\gamma(t) - d_1x_1 \\ \dfrac{dx_2}{dt} = A_2x_1 - d_2x_2 \\ \dfrac{dx_3}{dt} = A_3x_2 - d_3x_3, \end{cases} \tag{4.7}$$

where A_i and d_i are positive constants.

We assume that $q_0(x_1, x_2, x_3)$ and $q_1(x_1, x_2, x_3)$ are two continuous and nonnegative functions on $\mathbb{R}^3$ such that

$$q_0\,(0, 0, 0) \neq 0 \text{and}\; q_1\,(1, 1, 1) \neq 0.$$

Let $i_0 \in \{0, 1\}$, $T_0 = 0$, $\mathbf{x}_0 \in \mathbb{R}^3_+$ and we define then
$\gamma \colon [0, \infty) \to \{0, 1\}$ such that $\gamma(0) = i_0$ and

$$\gamma(t) := \begin{cases} i, & \text{if } T_{n-1} \leqslant t < T_n, \\ 1 - i, & \text{if } t = T_n, \end{cases} \tag{4.8}$$

where for $n \geqslant 1$, T_n is a random function such that

$$\begin{aligned} F_{\mathbf{x}_s}(t) &= \text{Prob}(T_n - T_{n-1} \leqslant t | \, \gamma(T_{n-1}) = i) \\ &= 1 - \exp\left(- \int_0^t q_i(\pi_i(s, \mathbf{x}_s)) ds \right), \end{aligned} \tag{4.9}$$

$$\mathbf{x}_s := \begin{cases} \pi_i^{s - T_{n-1}}(\mathbf{x}_{n-1}), & \text{if } T_{n-1} \leqslant s < T_n, \\ \mathbf{x}_n, & \text{if } s = T_n, \end{cases} \tag{4.10}$$

where

$$\mathbf{x}_n = \pi_i^{T_n - T_{n-1}}(\mathbf{x}_{n-1}). \tag{4.11}$$

After a few simplifications, we consider the following system:

$$\begin{cases} \dfrac{dx_1}{dt} = \gamma(t) - x_1 \\ \dfrac{dx_2}{dt} = a(x_1 - x_2) \\ \dfrac{dx_3}{dt} = b(x_2 - x_3), \end{cases} \tag{4.12}$$

$$\zeta(t) := (x_1(t), x_2(t), x_3(t), \gamma(t)) = (\mathbf{x}(t), \gamma(t)). \tag{4.13}$$

The state space of this process is $\mathbb{X} = X \times \{0, 1\}$ along with the jump rates q_i and the jump distribution $\mathbb{J}((\mathbf{x}, i), \cdot)$ being the Dirac measure $\delta_{(\mathbf{x}, 1-i)}$ such that

$$\mathbb{J}((x, i), \mathbb{X}) = 1. \tag{4.14}$$

A random variable T_n is called *time of the n-th jump of the process.* Here $\Delta_k = T_k - T_{k-1} > 0$, with $k \geqslant 1$, $\Delta_k < \infty$ and

$$\lim_{k \to \infty} T_k = \infty, \tag{4.15}$$

which completes the definition of the process for all $t \geq 0$.

Example 10 Goodwin model [29]. In 1965, Goodwin [17] provided a model of gene expression, where the protein can repress the transcription phase of its own gene. Such a description turns this model into a system of nonlinear ordinary differential equations. In this model we deal with two kinds of variables which are the concentration levels of mRNA and the corresponding protein, respectively. Within years, several papers investigating Goodwin model appeared. Here, we consider two crucial stages for this process: mRNA processing in transcription phase and translation phase which aims to produce functional proteins. The feedback loop in this case is induced by the repression mechanism which is exerted by the protein to the mRNA production and it is given by a classical Hill function, which is nonlinear. The feedback term represents protein-regulated speed of transcription when gene is in an active state. Likewise the previous case, q_0 denotes switching intensity function from an inactive state of a gene to an active state, and q_1 denotes switching intensity function from an active state of a gene to an inactive state. We have also two positive constants k_1 and k_2 corresponding to the speed of mRNA and protein production, respectively, while positive constants d_1 and d_2 correspond to their degradation rates. In the case of mRNA, the production term consists of a nonlinear regulation function f, while in the protein case the concentration level just increases proportionally to the level of mRNA molecules. In such case, switching mechanism with place-dependent switching intensity functions is another example of PDMP. Goodwin describes the process using the following system of differential equations:

$$\begin{cases} \dfrac{dx_1}{dt} = k_1 f(x_n) - d_1 x_1 \\ \dfrac{dx_2}{dt} = k_2 x_1 - d_2 x_2 \\ \vdots \\ \dfrac{dx_m}{dt} = k_m x_{m-1} - d_m x_m \end{cases}, \tag{4.16}$$

with x_i being continuous, nonnegative functions for $i \in \{1, \ldots, m\}$ and f is given in a way such that this system has a unique solution for the initial condition $x(0) = x_0 \in R+^m$ and for all $t \geq 0$. The function f is a regulation function, x_i represents the level of concentration of i-th molecule, the constants k_1, ..., k_m are called production rates while the constants d_1, ..., d_m are called degradation rates.

The regulation function is considered first to be the *decreasing Hill function* of the form:

$$f^-(x) = \frac{\theta^n}{\theta^n + x^n}, \tag{4.17}$$

where $x \geq 0$, $\theta > 0$ is a threshold value and $n > 0$.

Here we note that the function (4.17) is decreasing in its domain, $f^-(0) = 1$ and

$$\lim_{x \to \infty} f^-(x) = 0.$$

The symbol f^{-} is used here to distinguish that function from the increasing Hill function f^{+}.

Now, we will consider two-variable negative feedback loop:

$$\begin{cases} \dfrac{dx}{dt} = k_1 \dfrac{\theta^n}{\theta^n + y^n} - d_1 x \\ \dfrac{dy}{dt} = k_2 x - d_2 y \end{cases} \tag{4.18}$$

with $x(t)$ and $y(t)$ being the levels of concentration of mRNA and proteins, respectively. Constants k_1, k_2, d_1, d_2 are greater than zero. After a few simplifications, we obtain the following equations:

$$\begin{cases} \dfrac{dx}{dt} = \dfrac{1}{1 + y^n} - ax \\ \dfrac{dy}{dt} = bx - cy \end{cases} \tag{4.19}$$

with four constants $a, b, c, n > 0$ along with initial conditions $x(0) = x_0 \geq 0$, $y(0) = y_0 \geq 0$. Here $b = \frac{k_1 k_2}{\theta}$ is a translation rate and a, c are degradation rates as the corresponding coefficients in the system (4.18), respectively. It is known that the system (4.19) has no analytical solutions for any given values of the coefficients $a, b, c,$ and n. However, one can observe that the system (4.19) has a unique solution, which is denoted by $\pi^{-}(t, x_0)$. The system has only a single steady-state $x^* = (x^*, y^*) \neq (0, 0)$ (see the paper [19]), satisfying the following equations.

For all values of the coefficients $a, b, c,$ and n, it is known that the only steady state of this system is asymptotically stable [15]. However, as the process of gene expression is regulated stochastically, i.e., only when the gene is activated is x being produced at the speed regulated by the Hill function $\dfrac{1}{1 + y^n}$. Otherwise, when the gene is not active, no production of new mRNA molecules takes place. In this case, the system (4.16) takes the following form:

$$\begin{cases} \dfrac{dx}{dt} = -ax \\ \dfrac{dy}{dt} = bx - cy \end{cases} \tag{4.20}$$

Now we investigate the system with the regulation function $f^{+}(x)$:

$$f^{+}(x) = \frac{x^n}{\theta^n + x^n}, \tag{4.21}$$

where x and the rest of the parameters n and θ have the same meaning as in the previous case.

Hence, now we investigate the following system:

$$\begin{cases} IA \xrightarrow{q_0(x,y)} A, \quad A \xrightarrow{q_1(x,y)} IA, \\ \dfrac{dx}{dt} = \gamma(t)\dfrac{y^n}{1+y^n} - ax \\ \dfrac{dy}{dt} = bx - cy \end{cases} \tag{4.22}$$

with $x(t)$, $y(t)$, IA, A and real coefficients a, b, c, n. In this case, when the stochastic function $\gamma(t) \equiv 0$, then this system has one asymptotically stable stationary point $(0, 0)$. However, when $\gamma(t) \equiv 1$, the number and character of steady points depend on the coefficients. If we have that $x^* = (x^*, y^*)$ is a steady state, then it satisfies the following system of equations [20]:

$$\begin{cases} \dfrac{(y^*)^n}{1+(y^*)^n} = ax^* \\ cy^* \quad = bx^* \end{cases} \tag{4.23}$$

Here we define the stochastic process $\xi^+(t) = (x(t), y(t), \gamma(t))$ with $t \geq 0$, adopting a form of PDMP. Hence, we obtain two models. The first one is a model with a negative feedback loop, while the second one is a model with a positive feedback loop. Using the notions above, one can show that distribution functions of the process are given by the following system of PDEs:

$$\begin{cases} \dfrac{\partial f_0}{\partial t} + \dfrac{\partial}{\partial x}(-axf_0) + \dfrac{\partial}{\partial y}((bx-cy)f_0) = q_1 f_1 - q_0 f_0 \\ \dfrac{\partial f_1}{\partial t} + \dfrac{\partial}{\partial x}((f^\star(y) - ax)f_1) + \dfrac{\partial}{\partial y}((bx-cy)f_1) = q_0 f_0 - q_1 f_1 \end{cases} \tag{4.24}$$

with $f^\star(y) = f^-(y)$ for a negative feedback loop (4.17) and $f^\star(y) = f^+(y)$ for a positive feedback loop (4.21) as well.

5 Stochastic Evolution of Distributions

Now we provide the idea of the role of stochasticity in explaining scientific phenomena from the point of view of the population in an environment disturbed by some random perturbations. Our considerations in this and the following section are based on the results obtained in paper [28]. This well-known situation will be contrasted in the following sections with a population perspective approach; thus, our analysis takes its place in infinite-dimensional space. The work of [8] provides a pictorial distinction between the individual and populational perspectives. Briefly speaking, multiple realizations of a single molecule driven by binary-valued stochastic process generate the density distribution function. However, when the population of molecules is large (i.e., a number of the individuals converge to infinity), we observe how it evolves in a single realization of a common stochastic noise source and generates the density. In effect, the stochastic evolution equation describes the evolution of the population density with respect to realizations of the common stochastic noise We use here examples of a population model with different birth rates [41] and a Goodwin model [17] with positive feedback on gene expression process [20].

Example 11 (Population Model with Births and Deaths) We investigate here a population which has a size equal to some $x \geq 0$, with a death rate μ and birth rate $\beta - cx$ where a parameter β is changing its value in time between β_0 and β_1 . The reason of such change can be some environmental perturbation, i.e., climate change or seasonal change. Moreover, the growth of the population in such model is determined by the basic differential equation of the form (4.1) with two alternative right-hand sides:

$$b_0(x) = (\beta_0 - cx)x - \mu x \quad \text{and} \quad b_1(x) = (\beta_1 - cx)x - \mu x. \tag{5.1}$$

We know that for both values of i, equation such as (4.1) defines a stochastic semigroup $\{P_i(t)\}_{t\geq 0}$ given by the Eq. (4.2). Analogically, if the population follows birth rate value related to $i \in \{0, 1\}$ and has some distribution density function g, then after the passage of time $t > 0$, the population distribution density function is given then by $P_i(t)g$. The idea we consider here is the switching between the semigroups $\{P_0(t)\}_{t\geq 0}$ and $\{P_1(t)\}_{t\geq 0}$ according to the function $i(t) \in \{0, 1\}$. Hence, we get a stochastic evolution equation of the following form [8]:

$$\frac{\partial u(t,x)}{\partial t} = -\frac{\partial \big(b_{i(t)}(x)u(t,x)\big)}{\partial x}, \tag{5.2}$$

with $u(t, x) = P_i(t)g(x)$ for $i(t) = i$ being the population density. We can consider this model as an infinite-dimensional version of the model (4.3). In this way we can make a simple and justified translation from a finite-dimensional space as a subset of R^d space to an infinite-dimensional space like L^1.

We will illustrate this approach using a reference to previously mentioned example of a gene expression model with a Hill function, a Goodwin model [17].

Example 12 (Goodwin Model with a Positive Feedback) Now we investigate the model [20] which describes gene expression driven by positive feedback control loop function of gene transcription. Briefly speaking, it describes a mechanism by which a protein can maintain the expression of its own gene as well as it changes its state between two levels of expression. Here we denote the concentration level at some given time t by $x(t) \geq 0$. Moreover, we assume the degradation rate is switching with constant intensities between two positive values γ_0 and γ_1, which is driven by some environmental condition.

For a given value $i \in \{0, 1\}$, the concentration level of the protein changes its values according to the following equation:

$$x'(t) = \frac{x^n(t)}{1 + x^n(t)} - \gamma_i x(t), \tag{5.3}$$

where $n \in \mathbb{N}$. We know that this equation induces a stochastic semigroup $\{P_i(t)\}_{t\geq 0}$ as in (4.2). If the gene cluster initially has some degradation rate γ_i and a distribution density g, then after the passage of time $t > 0$, the population distribution density function is given by $P_i(t)g$. Using $b_i(x)$ equal to the right-hand side of Eq. (5.3), we obtain again the stochastic evolution equation (5.2).

We see that the concept of randomly switching distributions naturally appears in many biophysical models. We will now formalize this theory, i.e., we investigate the case of a whole population of individuals living in an environment with random perturbations. The situation where a population of molecules is touched by a common environmental stochastic noise was considered by Bressloff in [8]. The deterministic dynamics of a population is given by Eq. (4.1) with a state space equal to a Borel subset of $\mathbb{R}^d$ with i from a finite set I. Here the environmental perturbations are represented by switching process between different values of i. Hence, the total population density $u(t, x)$ is given by the following equation:

$$\frac{\partial u(t,x)}{\partial t} = -\operatorname{div}(b_{i(t)}(x)u(t,x)), \tag{5.4}$$

with $i(t)$ being a Markov chain taking values from a discrete set I. Obviously, this equation has an initial condition $u(0, x) = g(x)$. In this way we consider equation (5.4) as a generalization of the stochastic Liouville equation (5.2) mentioned earlier. We have to note that this is not the only regime in which randomly switching environments have been considered. They have been investigated also in the case of diffusion processes [32] but as a kind of a parabolic equation. However, our approach is able to generalize such cases into a more general scheme for randomly switching stochastic semigroups.

We set $L^1(E) = L^1(E, \mathcal{B}(E), m)$, with (E, ρ) being a separable metric space and m being a σ-finite measure. We define then $I = \{0, 1, \ldots, k\}$, $k \in \mathbb{N}$. For every $i \in I$, we investigate a generator $(A_i, \mathcal{D}(A_i))$ of a stochastic semigroup $\{P_i(t)\}_{t\geq 0}$ on the set $L^1(E)$. As in the previous cases, we have the stochastic process $\{i(t)\}_{t\geq 0}$ which is a continuous-time Markov chain taking values (states) in I and having constant intensities q_{ij}, $i, j \in I$. We investigate the following evolution equation:

$$\begin{cases} u'(t) = A_{i(t)}u(t), \\ u(0) = g. \end{cases} \tag{5.5}$$

which defines a PDMP [42, p. 32] given by the equation:

$$X(t) = (u(t), i(t)), \quad t \geq 0, \tag{5.6}$$

having values in the product space $L^1(E) \times I$.

Moreover, we will obtain the formula for the solution $u(t)$ of (5.5). We set $t_0 = 0$ and for each $n \in \mathbb{N}$, we define t_n as the nth jump time of the Markov chain $i(t)$ so that

$$\mathbb{P}(t_{n+1} - t_n > t | i(t_n) = i) = e^{-q_i t}, \quad n \geq 0.$$

Now we define $\mathbb{P}_j$ as the probability measure defined on trajectories ω of the Markov chain $\{i(t)\}_{t\geq 0}$ with property $i(0) = j$. We denote integration with respect to the measure $\mathbb{P}_j$ by $\mathbb{E}_j$, set $U(t_0) = U(0) = \mathrm{Id}$ and

$$U(t) = P_{i(t_n)}(t - t_n) \circ U(t_n) \quad \text{for } t \in [t_n, t_{n+1}), n \geq 0,$$

with

$$U(t_n) = P_{i(t_{n-1})}(t_n - t_{n-1}) \circ \cdots \circ P_{i(t_1)}(t_2 - t_1) \circ P_{i(0)}(t_1).$$

If we set

$$N(t) = \max\{n \geq 0 : t_n \leq t\} \tag{5.7}$$

as the number of jumps of the process $\{i(t)\}_{t\geq 0}$ up to some given time moment t. Using $N(t)$ we denote:

$$U(t) = P_{i(t_{N(t)})}(t - t_{N(t)})U(t_{N(t)}), \quad t \geq 0. \tag{5.8}$$

In this way $u(t) = U(t)g$ for $g \in L^1(E)$ is the solution of (5.5). We note that for every $t > 0$, $g \in L^1(E)$, and ω, if $t \in [t_n(\omega), t_{n+1}(\omega))$ for $n = n(\omega)$, then for $i = i(t_n(\omega))$, we have equality:

$$U(t,\omega)g = P_i(t - t_n(\omega))U(t_n(\omega))g,$$

where $U(t,\omega)$ is a composition of stochastic operators. Hence

$$\|U(t,\omega)g\| = \int_E |U(t,\omega)g(x)|m(dx) \leq \int_E |g(x)|m(dx) = \|g\|$$

and we can treat $U(t)g$ as a random variable with values in the set $L^1(E)$. Moreover, for every $j \in I$, $\mathbb{E}_j(U(t)g)$ is given by the almost everywhere finite function $\mathbb{E}_j(u(t,x))$ with $u(t,x) = U(t)g(x)$, i.e., for any $h \in L^\infty(E, \mathcal{B}(E), m)$

$$\int_E \mathbb{E}_j(U(t)g)(x)h(x)m(dx) = \int_E \mathbb{E}_j(u(t,x))h(x)m(dx).$$

This constitutes a general characterization of stochastic evolution equation and its solutions. As we can see, we have expressed the solutions of the stochastic evolution equation in the language of semigroups. However, the question arises whether they can be described in a more concrete way and whether they can somehow be related to the solutions of the equations of evolution in the case of classical randomly switching systems of equations. The answer to this question turns out to be positive. In the next section, we will characterize the moments of solutions of the stochastic evolution equation, i.e., its first- and second-order moment respectively, according to the paper [28].

6 Moment Characterization

Here we will obtain equations for the moments of the solution $u(t, x)$ and find a link to the evolution equation (4.5) from Sect. 4.

Using the setting from Sect. 5, we investigate the first moment of the solution $u(t)$ of stochastic evolution equation (5.5). We know that $u(t, x) = U(t)g(x)$,with $U(t)$ given by (5.8).

Just from the definition of V, the first moment of $u(t, x)$ is equal to

$$V(t,x) = \mathbb{E}(u(t,x)) = \sum_{j\in I} \mathbb{E}_j(u(t,x)) = \sum_{j\in I}\sum_{i\in I} \mathbb{E}_j(\mathbf{1}_{\{i(t)=i\}}u(t,x)),$$
$$x \in E, t \geq 0,$$

Here, by $\mathbb{E}$ we denote the expectation and by $\mathbf{1}_F$ we denote the indicator function of an event F. Taking

$$V_i(t,x) = \sum_{j\in I} \mathbb{E}_j(\mathbf{1}_{\{i(t)=i\}}u(t,x)), \tag{6.1}$$

and

$$V(t,x) = \sum_{i\in I} V_i(t,x), \quad x \in E, t \geq 0. \tag{6.2}$$

it appears that:

$$\frac{\partial}{\partial t} V_i = A_i V_i + \sum_j q_{ji} V_j, \quad i \in I. \tag{6.3}$$

Moreover, $(V_i)_{i\in I}$ can be expressed in terms of a stochastic semigroup.

We set $L^1(E \times I) = L^1(E \times I, \mathcal{B}(E \times I), \mu)$, with μ being the Cartesian product of the measure m on E and the counting measure on I. We also denote elements of the set $L^1(E \times I)$ by $f := (f_i)_{i\in I} \in L^1(E \times I)$, with $f_i \in L^1(E), i \in I$, and define the family of operators $\{P(t)\}_{t\geq 0}$ on $L^1(E \times I)$ by the following formula:

$$(P(t)f)_i = \sum_{j\in I} \mathbb{E}_j(\mathbf{1}_{\{i(t)=i\}}U(t)f_j), \quad i \in I. \tag{6.4}$$

We formulate two following conditions:

(i) One can find a Banach space B, a set of Borel measurable functions $\mathcal{H} \subseteq B$, and a family C of Borel subsets of E that is closed under intersection operation and generates the Borel σ-algebra $\mathcal{B}(E)$ with the property for every set $F \in C$ there exist a nonincreasing sequence of functions $h_n \in \mathcal{H}$ such that

$$\lim_{n\to\infty} h_n(x) = \mathbf{1}_F(x), \quad x \in E. \tag{6.5}$$

(ii) Let $\{P_i(t)\}_{t\geq 0}$ be a stochastic semigroup on the space $L^1(E)$ with infinitesimal generator $(A_i, \mathcal{D}(A_i))$, $i \in I$. For every $i \in I$ there exist a C_0-semigroup $\{T_i(t)\}_{t\geq 0}$ on B such that

$$\langle P_i(t)g, h\rangle = \langle g, T_i(t)h\rangle, \quad g \in L^1(E), h \in B, t \geq 0. \tag{6.6}$$

We define the scalar product of two functions g, h by $\langle g, h\rangle := \int_E g(x)h(x)m(dx)$.

Under these two conditions, the following theorem is true.

Theorem 2 *We assume conditions* (6) *and* (6) *hold. In such case, the family of operators* $\{P(t)\}_{t\geq 0}$ *defined in* (6.4) *forms a stochastic semigroup on the space* $L^1(E \times I)$. *The infinitesimal generator of the semigroup* $\{P(t)\}_{t\geq 0}$ *is the operator* $A + Q^T$ *with* $Af = (A_i f_i)_{i\in I}$ *and* $Q^T f = (\sum_j q_{ji} f_j)_{i\in I}$ *for* $f = (f_i)_{i\in I}$ *with* $f_i \in \mathcal{D}(A_i)$, $i \in I$.

For a proof of this theorem, we refer the reader to the paper [28].

Theorem formulated above shows that (6.4) constitutes a formula for a stochastic semigroup in a general case. However, it can be applied also to specific cases. This Theorem 2 holds also for the semigroups of Frobenius–Perron operators given by the Eq. (4.2).

Corollary 2 *If* E *is a Borel subset of* $\mathbb{R}^d$ *and for each* $i \in I$ $\{P_i(t)\}_{t\geq 0}$ *is given by* (4.2), *then both of the conditions* (6) *and* (6) *are true. Hence, Theorem 2 holds for* (4.2) *as well.*

Another consequence of Theorem 2 is the following.

Corollary 3 *We assume that conditions* (6) *and* (6) *hold. If* $\{P(t)\}_{t\geq 0}$ *is given by* (6.4) *and* $u(t)$ *by* (5.5), *then for every* $g \in L^1(E)$ *and* $l \in I$ *such that* $u(0) = g$ *and* $i(0) = l$ *we have* $V_i(t, x) = (P(t)f)_i(x)$, *with* V_i *is as in* (6.1) *and* $f = (f_j)_{j\in I}$ *is equal to*

$$f_j = \begin{cases} g, & j = l, \\ 0, & j \neq l. \end{cases}$$

Above all, the mean of the process in (5.6) *is given by the following formula:*

$$V(t,x) = \mathbb{E}(u(t,x)) = \sum_{i \in I} (P(t)f)_i(x). \tag{6.7}$$

We obtained general the characterization of the mean of the process by (6.4).

Now we will investigate the link between Fokker–Planck systems (4.5) describing a distribution of stochastic processes in Euclidean $\mathbb{R}^d$ space and the first moment equation (6.3). We have proved yet that it has the same form as (4.6) and the semigroup $P(t)f$ is the solution of the evolution equation (4.5) equipped with initial condition f. Thus, if $(u_0, u_1, u_2, \dots, u_k)$ is a solution of (4.6) and for every $i \in I$, there is $f_i^* \in L^1(E)$ satisfying the following equality:

$$\lim_{t\to\infty} \int_E \left|u_i(t,x) - f_i^*(x)\right| m(dx) = 0, \tag{6.8}$$

then, just by Corollary 3, we obtain:

$$\lim_{t\to\infty} \int_E \left|V_i(t,x) - f_i^*(x)\right| m(dx) = 0$$

and, by (6.2),

$$\lim_{t\to\infty} \int_E \left|V(t,x) - V^*(x)\right| m(dx) = 0 \tag{6.9}$$

with $V^*(x) = \sum_{i \in I} f_i^*(x)$. The function V^* can be considered as mean of the process at large time. Moreover, condition (6.8) is satisfied if the stochastic semigroup $\{P(t)\}_{t\geq 0}$ is asymptotically stable, i.e., there exists $f^* \in L^1(E \times I)$ such that for each density $f \in L^1(E \times I)$

$$\lim_{t\to\infty} \|P(t)f - f^*\| = 0 \tag{6.10}$$

Here we also note that f^* in (6.10) is an invariant density for $\{P(t)\}_{t\geq 0}$, which equivalently means that $P(t)f^* = f^*$ for all $t \geq 0$.

However, if the semigroup $\{P(t)\}_{t\geq 0}$ is sweeping from compact subsets of $E \times I$, i.e., for each compact subset F of E, any $f \in L^1(E \times I)$ and $i \in I$ we have:

$$\lim_{t\to\infty} \int_F (P(t)f)_i(x) m(dx) = 0,$$

then the mean of the process in (5.6) at large time vanishes, as for any compact subset F of the space E we obtain

$$\lim_{t\to\infty}\int_F V(t,x)m(dx) = 0. \tag{6.11}$$

Sufficient conditions for asymptotic behavior of the stochastic semigroup $\{P(t)\}_{t\geq 0}$ connected with the randomly switching dynamics $\xi(t) = (x(t), i(t))$ with $x(t)$ satisfying the Eq. (4.3) as in Sect. 4 with $E \subseteq \mathbb{R}^d$ are well known [2, 4] and [41, 42]. For that purpose will use here the notion of Lie brackets.

Lie bracket of two sufficiently smooth vector fields b_i and b_j is given by the formula:

$$[b_i, b_j](x) = Db_j(x)b_i(x) - Db_i(x)b_j(x)$$

with $Db(x)$ being the derivative of the vector field b at point x. For a family of vector fields $b_0, \ldots, b_k$, sufficiently smooth in a neighborhood of x we say that *Hörmander's condition is satisfied at* x if the vectors

$$b_1(x) - b_0(x), \ldots, b_k(x) - b_0(x),\ [b_i, b_j](x)_{0\leq i,j\leq k},\ [b_i,[b_j,b_l]](x)_{0\leq i,j,l\leq k}, \ldots$$

span the space $\mathbb{R}^d$.

From [2, Theorem 2] (see also [4, Theorem 4.4]) and [42, Corollary 5.3], we conclude that if Hörmander's condition is satisfied at every point $x \in E$ and a stochastic semigroup $\{P(t)\}_{t\geq 0}$ has no invariant density, then it is sweeping from all compact subsets of $E \times I$.

A point $x \in E$ is called reachable from y if one can find $n \in \mathbb{N}$, indices $i_1, \ldots, i_n \in I$ and times $s_1, \ldots s_n$ satisfying $x = \pi_{i_n}(s_n, \ldots, \pi_{i_1}(s_1, y))$, where for every $i \in I$ transformation $t \mapsto \pi_i(t, x_0)$ is the solution of (4.1) with initial condition $x(0) = x_0$. x is called *accessible from* y if each neighborhood of x contains a point which is reachable from y. Using [2, Theorem 1] with [4, Theorem 4.6], we conclude that if the semigroup $\{P(t)\}_{t\geq 0}$ has an invariant density and if Hörmander's condition is satisfied at a point $x \in E$ that is accessible from any point in , E then the semigroup $\{P(t)\}_{t\geq 0}$ is asymptotically stable.

The mean of the process at large time can be illustrated by some simple examples in trajectory dynamics [30].

Let $E \subset (0, \infty)$ and $I = \{0, 1\}$ so undoubtly there is switching between two dynamical systems b_0 and b_1 which constitutes a Markov process $\xi(t) = (x(t), i(t))$, $t \geq 0$ on $E \times \{0, 1\}$. Here $q_i = q_{i(1-i)}$, $i = 0, 1$, (see (4.4)) denote intensities of switching between these systems from b_i to b_{1-i}. We also assume that $b_0(0) = b_1(0) = 0$ and that one of b_0 or b_1 has one more fixed point a which is accessible from any point in the set E. Then, Hörmander's condition is satisfied at a point x if $b_1(x) - b_0(x) \neq 0$. We define:

$$r(x) = \frac{q_0}{b_0(x)} + \frac{q_1}{b_1(x)} \quad \text{and} \quad R(x) = \int_{x_0}^{x} r(s)ds,$$

with $x_0 \in (0, a)$. Two functions

$$f_0(x) = \frac{e^{-R(x)}}{|b_0(x)|} \quad \text{and} \quad f_1(x) = \frac{e^{-R(x)}}{|b_1(x)|}$$

constitute stationary solutions of the Fokker–Planck equation (4.6). However, if

$$\kappa = \int_0^a \big(f_0(x) + f_1(x)\big)dx < \infty, \tag{6.12}$$

then the stochastic semigroup $\{P(t)\}_{t\geq 0}$ is asymptotically stable. Moreover, the mean at large time is given by the formula:

$$V^*(x) = \kappa^{-1}(f_0(x) + f_1(x))\mathbf{1}_{(0,a)}(x). \tag{6.13}$$

If the product $b_0'(0)b_1'(0) \neq 0$, then condition (6.12) is satisfied for $\lambda > 0$ depending on the shape of b_0, b_1, and we can define it in the following way:

$$\lambda = p_0 b_0'(0) + p_1 b_1'(0) \tag{6.14}$$

with

$$p_0 = \frac{q_1}{q_0 + q_1} \quad \text{and} \quad p_1 = \frac{q_0}{q_0 + q_1}$$

being the chance of choosing b_0 and b_1, respectively. When $\lambda < 0$, it turns out that $\{P(t)\}_{t\geq 0}$ is sweeping from the family of all compact subsets of our state space. This implies that the mean at large time is equal to zero [41]. Numerical simulations of the corresponding Fokker-Planck-type equation performed by the author of this paper confirmed obtained results in FiPy [21].

We will illustrate the results using the example of a supercritical Hopf bifurcation [23, 28].

Example 13 (Supercritical Hopf Bifurcation) The standard form for a supercritical Hopf bifurcation, after performing a transformation to polar coordinates (θ, r), is given by the system of equations:

$$\begin{cases} \theta'(t) = \omega + br^2(t), \\ r'(t) = \mu r(t) - r^3(t). \end{cases} \tag{6.15}$$

We note that asymptotic behavior of this system depends on the value of μ. For $\mu < 0$ we have a single stationary state $(r_*, \theta_*) = (0, 0)$, but for $\mu > 0$ we have an unstable stationary state at a point $(0, 0)$ and there is also a limit cycle with $r = \sqrt{\mu}$. Now we take $\omega = \omega_i$ and $\mu = \mu_i$, $i = 0, 1$, in the system (6.15) with $\mu_0 < 0$ and $\mu_1 > 0$. We define $E = \mathbb{S}^1 \times (0, \infty)$, with $\mathbb{S}^1$ being the unit circle in the set $\mathbb{R}^2$.

Without losing general consideration, we set $b = 0$. If $\omega_0 \neq \omega_1$, then the Hörmander condition is satisfied at $(\theta, r) \in E$. Moreover, $(0, \sqrt{\mu_1})$ is reachable from all points in the set E. The long-time behavior of the mean given by the Eq. (6.7) is dependent on the parameter λ in (6.14) with $b_i'(0) = \mu_i$, $i = 0, 1$. If $\lambda > 0$, then the mean at large time is

$$V^*(\theta, r) = \frac{1}{2\pi\kappa}\big(f_0(r) + f_1(r)\big)\mathbf{1}_{\mathbb{S}^1}(\theta).$$

so it is positive on the unit circle. When $\lambda < 0$, the mean of the process at large time is equal to zero.

However, if $\omega = \omega_0 = \omega_1$, then θ is independent of r and we have $\theta'(t) = \omega$ for each i. Hence, the process $(\theta(t), r(t), i(t))$ can be split into two independent parts: deterministic $\theta(t)$ and $(r(t), i(t))$.

Now we investigate long-time behavior of the stochastic process (5.6) when $\lambda > 0$. If we consider g in (5.6) as the product of marginal densities $g(\theta, r) = g_1(\theta)g_2(r)$, then we have the following formula:

$$\lim_{t\to\infty} \int_E |V(t, \theta, r) - \frac{1}{\kappa} g_1(\theta - \omega t)\big(f_0(r) + f_1(r)\big)| d\theta dr = 0,$$

with f_0 and f_1 being defined by the same formula as in the previous example.

Now we will investigate further the stochastic process (5.6) by deriving equations for the correlations. We treat this topic as a natural extension of the first moment equation problem investigated in Sect. 6. We will provide then the theory for second-order correlations; however, equations for higher-order moments can be obtained in a similar way.

Second-order correlations are defined by the following formula:

$$C_i(t, x, y) = \mathbb{E}(\mathbf{1}_{\{i(t)=i\}} u(t, x)u(t, y)), \quad x, y \in E, \ i \in I, \ t \geq 0. \tag{6.16}$$

In such case, the following differential equation is satisfied:

$$\frac{\partial}{\partial t} C_i = (A_i \otimes \mathrm{Id})C_i + (\mathrm{Id} \otimes A_i)C_i + \sum_{j\in I} q_{ji} C_j, \quad i \in I, \tag{6.17}$$

with $A_i \otimes \mathrm{Id}$ and $\mathrm{Id} \otimes A_i$ having its domain as the set of functions $(x, y) \mapsto f(x, y)$ where $f \in L^1(E^2)$. In such case, if $A_i f(x) = -\operatorname{div}(b_i(x) f(x))$, then

$$(A_i \otimes \mathrm{Id}) f(x, y) = -\operatorname{div}(b_i(x) f(x, y)), \tag{6.18}$$

$$(\mathrm{Id} \otimes A_i) f(x, y) = -\operatorname{div}(b_i(y) f(x, y)). \tag{6.19}$$

Equation (6.17) is defined on the space $L^1(E^2 \times I) = L^1(E^2 \times I, \mathcal{B}(E^2 \times I), \mu^2)$, with μ^2 being the product of a double of the measure m on E and the counting measure on the set I as it often comes for PDMP. Now, if we define the operators on the set $L^1(E^2 \times I)$ by the formula

$$(S(t)f)_i = \sum_{j \in I} \mathbb{E}_j(\mathbf{1}_{\{i(t)=i\}} U(t) \otimes U(t) f_j), \quad i \in I, \tag{6.20}$$

for $f = (f_j)_{j \in I} \in L^1(E^2 \times I)$, $f_j \in L^1(E^2)$, $j \in I$, and $U(t)$ as in (5.8), then the following theorem holds [28].

Theorem 3 *We assume that both of the conditions* (6) *and* (6) *are satisfied. In such case, the family of operators* $\{S(t)\}_{t \geq 0}$ *defined in* (6.20) *constitutes Markov semigroup on the space* $L^1(E^2 \times I)$*. Moreover, the infinitesimal generator of this semigroup is equal to* $A + Q^T$*, defined as*

$$A(f_i)_{i \in I} = \big((A_i \otimes Id) f_i + (Id \otimes A_i) f_i\big)_{i \in I} \text{ and } Q^T(f_i)_{i \in I} = (\sum_{j \in I} q_{ji} f_j)_{i \in I}, \tag{6.21}$$

where $f_i \in \mathcal{D}(A_i) \otimes \mathcal{D}(A_i), i \in I$.

7 Conclusion

Here we introduced the idea behind evolution equations and the concept of Markov semigroups used to its construction. We can use this regime to describe randomly switching dynamics, where gene can change its state from active to inactive and vice versa. Such equations describe the evolution of many biological systems. However, in a large part of this article, we have provided the notion of stochastic evolution equation in L^1 space. Such equation describes the evolution of some physical population driven by a common environmental source. We investigated the moments of the stochastic evolution equation solutions and found the link between this moment equation and a deterministic system of Fokker–Planck-type equations for the distributions of the process. It appeared that the mean of the process at large time can be expressed by the stationary solutions of a Fokker-Planck-type system providing that they exist. We gave then an example of the application of our results to scientific model in which the underlying dynamics display Hopf bifurcation. We also studied second-order correlations of solutions of the stochastic evolution equation. There are some promising perspectives for the future. One of the ideas would be to show convergence in distribution of the infinite-dimensional process $(u(t), i(t))$ to some stationary distribution. Many examples linked to diffusion processes with such convergence are given; see [12, 33]. Possible future extension of this topic is connected with addition of switching to stochastic PDEs driven by various types of stochastic noises; see [39] and complex hybrid networks appearing in many important modeling problems as in the case of [35]. Moreover, numerous computer packages, like Python package FiPy [21], raise hopes for the possibility of simulating solutions of various evolution equations, and hence they give the chance to make interesting hypothesis about them.

Acknowledgments The author is grateful to professors Ryszard Rudnicki and Marta Tyran-Kamińska for a guidance and numerous discussions on evolutionary equations and its connections with semigroup theory.

References

1. Arendt, W., Grabosch, A., Greiner, G., Groh, U., Lotz, H.P., Moustakas, U., Nagel, R., Neubrander, F., Schlotterbeck, U.: One-Parameter Semigroups of Positive Operators. Lecture Notes in Mathematics, vol. 1184. Springer, Berlin (1986)
2. Bakhtin, Y., Hurth, T.: Invariant densities for dynamical systems with random switching. Nonlinearity **25**(10), 2937–2952 (2012)
3. Belle, A., Tanay, A., Bitincka, L., Shamir, R., O'Shea, E.K.: Quantification of protein half-lives in the budding yeast proteome. Proc. Natl. Acad. Sci. USA **103**(35), 13004–13009 (2006)
4. Benaïm, M., Le Borgne, S., Malrieu, F., Zitt, P.-A.: Qualitative properties of certain piecewise deterministic Markov processes. Ann. Inst. Henri Poincaré Probab. Stat. **51**(3), 1040–1075 (2015)
5. Berezhkovskii, A.M., Shvartsman, S.Y.: Diffusive flux in a model of stochastically gated oxygen transport in insect respiration. J. Chem. Phys. **144**(20), 204101 (2016)
6. Bobrowski, A.: Functional Analysis for Probability and Stochastic Processes: An Introduction. Cambridge University Press, Cambridge (2005). https://doi.org/10.1017/CBO9780511614583
7. Bobrowski, A., Lipniacki, T., Pichór, K., Rudnicki, R.: Asymptotic behavior of distributions of mRNA and protein levels in a model of stochastic gene expression. J. Math. Anal. Appl. **333**(2), 753–769 (2007)
8. Bressloff, P.C.: Stochastic Liouville equation for particles driven by dichotomous environmental noise. Phys. Rev. E **95**(1), 012124 (2017)
9. Bressloff, P.C., Lawley, S.D.: Moment equations for a piecewise deterministic PDE. J. Phys. A **48**(10), 105001 (2015)
10. Bressloff, P.C., Lawley, S.D., Murphy, P.: Effective permeability of a gap junction with age-structured switching. SIAM J. Appl. Math. **80**(1), 312–337 (2020)
11. Chandrasekhar, S., M¨unch, G.: The theory of fluctuations in brightness of the Milky-Way. Astrophys. J. **125**, 94–123 (1952)
12. Cloez, B., Hairer, M.: Exponential ergodicity for Markov processes with random switching. Bernoulli **21**(1), 505–536 (2015)
13. Cohen, J.E.: Eignevalue inequalities for random evolutions: origins and open problems. Ineq. Statist. Probab. (IMS) **5**, 41–53 (1984)
14. Davis, M.H.A.: Piecewise-deterministic Markov processes: a general class of nondiffusion stochastic models. J. Roy. Statist. Soc. Ser. B **46**(3), 353–388 (1984)
15. De Jong, H.: Modeling and simulation of genetic regulatory systems: a literature review. J. Comput. Biol. **9**(1), 67–103 (2002). https://doi.org/10.1089/10665270252833208
16. Goldstein, S.: On diffusion by discontinuous movements, and on the telegraph equation. Quart. J. Mech. Appl. Math. **4**, 129–156 (1951)
17. Goodwin, B.: Oscillatory behavior in enzymatic control processes. Adv. Enzyme Regul. **3**, 425–428 (1965). https://doi.org/10.1016/0065-2571(65)90067-1
18. Griego, R., Hersh, R.: Theory of random evolutions with applications to partial differential equations. Trans. Amer. Math. Soc **156**, 405–418 (1971)
19. Griffith, J.: Mathematics of cellular control processes: I. Negative feedback to one gene., J. Theoret. Biol. **20**, 202–208 (1968). http://doi.org/10.1016/0022-5193(68)90189-6

20. Griffith, J.: Mathematics of cellular control processes II. Positive feedback to one gene. J. Theoret. Biol. **20**(2), 209–216 (1968)
21. Guyer, J.E., Wheeler, D., Warren, J.A.: FiPy: partial differential equations with Python, Comput. Sci. Eng. **11**(3), 6–15 (2009). http://www.ctcms.nist.gov/fipy
22. Hersh, R., Griego, R.J.: Brownian motion and potential theory. Sci. Am. **220**(3), 66–77 (1969)
23. Hurth, T., Kuehn, C.: Random switching near bifurcations, Stochast. Dyn. **20**, 2050008 (2019)
24. Jefferies, B.: Evolution Processes and the Feynman-Kac Formula. Kluwer, Dordrecht (1996)
25. Kac, M.: A stochastic model related to the telegrapher's equation. Rocky Mountain J. Math. **4**, 497–509 (1974)
26. Kato, T.: Perturbation Theory for Linear Operators, Reprint of the 1980 edn. Classics in Mathematics Series. Springer, Berlin (1995)
27. Korolyuk, V.S., Limnios, N.: Stochastic Systems in Merging Phase Space. World Scientific, Singapore (2005)
28. Klimasara, P., Mackey, M.C., Tomski, A., Tyran-Kamińska, M.: Randomly switching evolution equations. Nonlinear Analy. Hybrid Syst. **39**, 100948 (2021)
29. Kozdeba, A., Tomski, A.: Application of the Goodwin model to autoregulatory feedback for stochastic gene expression. Math. Biosci. **327**, 1–12 (2020)
30. Kuznetsov, Y.A.: Elements of Applied Bifurcation Theory, vol. 112. Springer Science & Business Media, Cham (2013)
31. Lasota, A., Mackey, M.C.: Chaos, Fractals, and Noise. Applied Mathematical Sciences, vol. 97. Springer, New York (1994)
32. Lawley, S.D.: Boundary value problem for statistcs of diffusion in randomly switching environment: PDE and SDE perspectives. SIAM J. Appl. Dyn. Syst. **15**, 1410–1433 (2016)
33. Lawley, S.D., Mattingly, J.C., Reed, M.C.: Stochastic switching in infinite dimensions with applications to random parabolic PDE. SIAM J. Math. Anal. **47**(4), 3035–3063 (2015)
34. Lawley, S.D., Best, J.A., Reed, M.C.: Neurotransmitter concentrations in the presence of neural switching in one dimension. Discrete Contin. Dyn. Syst. Ser. B **21**(7), 2255–2273 (2016)
35. Liu, Q., Wang, Z., He, X., Zhou, D.: Event-based distributed filtering over Markovian switching topologies. IEEE Trans. Automatic Control **64**(4), 1595–1602 (2018)
36. Luczka, J., Rudnicki, R.: Randomly flashing diffusion: asymptotic properties. J. Statist. Phys. **83**, 1149–1164 (1996)
37. Mackey, M.C., Rudnicki, R.: Global stability in a delayed partial differential equation describing cellular replication. J. Math. Biol. **33**, 89–109 (1994)
38. Rudnicki, R., Tomski, A.: On a stochastic gene expression with pre-mRNA, mRNA and protein contribution. J. Theor. Biol. **387**, 54–67 (2015)
39. Peszat, S., Zabczyk, J.L.: Stochastic Partial Differential Equations with Lévy Noise: An Evolution Equation Approach. Encyclopedia of Mathematics and Its Applications, vol. 113. Cambridge University Press, Cambridge (2007)
40. Pinsky, M.A.: Lectures on Random Evolution. World Scientific Publishing, River Edge (1991)
41. Rudnicki, R., Tyran-Kamińska, M.: Piecewise deterministic Markov processes in biological models. In: Semigroups of Operators - Theory and Applications. Springer Proceedings in Mathematics & Statistics, vol. 113. Springer, New York (2015), pp. 235–255
42. Rudnicki, R., Tyran-Kamińska, M.: Piecewise Deterministic Processes in Biological Models. SpringerBriefs in Applied Sciences and Technology. SpringerBriefs in Mathematical Methods. Springer, Cham (2017)
43. Rudnicki, R., Pichór, K., Tyran-Kamińska, M.: Markov semigroups and their applications. In Dynamics of Dissipation. Lectures Notes in Physics, vol. 597. Springer, Berlin (2002), pp. 215–238
44. Ryan, R.A.: Introduction to Tensor Products of Banach Spaces. Springer Monographs in Mathematics. Springer, London (2002)
45. Swishchuk, A.: Stochastic stability and optimal control of semi-Markov risk processes in insurance mathematics. In: Janssen, J., Limnios, N. (eds.) Semi-Markov Models and Applications, pp 313–323. Kluwer, Dordrecht (1999)

Inverse Semigroups and Omaha Kinship

John P. Boyd

Abstract Many kinship terminologies can be modeled as inverse semigroups. This is especially true of Omaha-type kinship systems and is illustrated here with the Fox consanguineal (or "blood") relatives. The semigroup operation comes not from the composition of relations, but from a more symbolic, or linguistic, operation. For example, a man's brother's brother reduces symbolically to his brother, even though in the composition of relations a man's brother composed with brother would also include the man himself. The semigroup for the Fox system can be further specified to be an inverse semigroup, which is characterized in terms of their idempotent elements and as a web of the even more specific bicyclic semigroups.

Keywords Inverse semigroup · Kinship · Consanguineal · Ideal · Fox

1 Introduction

This chapter applies some of the basic concepts of the theory of semigroups [1, 2], to certain aspects of kinship terminology. The main kinship example will be that found among the Fox[1] [9] tribe, originally found in the Great Lakes region of North America. However, pressure from the French, British, and Americans forced the Fox southwestward to Iowa and Oklahoma, where they reside today.

Kinship is of the utmost importance in all societies, especially for children. However, in modern societies, adults also have many interactions with unrelated individuals, such as co-workers, civil officials, friends, or strangers. When kinship terms are used in modern societies, their meaning may be metaphorical. For example, "father" can mean a founder, so that George Washington is called the

[1] The word "Fox" is an exonym for the indigenous people who call themselves the "Meskwaki."

J. P. Boyd (✉)
University of California, Irvine, CA, USA
e-mail: jpboyd@uci.edu

M. Mitrović, M. N. Hounkonnou (eds.), *Algebra Without Borders*, Mathematics in Mind, https://doi.org/10.1007/978-3-031-86477-3_9

"father" of the USA. When capitalized, "Father" may refer to priest or a supernatural being. The social implication of calling someone a "father" is to give respect or even deference. Similarly, "brother" can be extended to mean any male that you are socially close to. Sometimes it can be a plea for empathy as when said to the police, "Don't tase me, bro!" Finally, to call someone "son" may be a term of either affection or belittlement, depending on the context.

Kinship is much more important in "primitive" societies, such as hunting and gathering, horticultural, or pastoral tribes. In many of these societies, everyone is assigned some kinship term, even those with no known genealogical relationship, such as traders or anthropologists.

Finally, the semigroup theory necessary to describe the Fox kinship system in terms of inverse semigroups includes the concepts of ideals, commuting idempotents, and bicyclic semigroups.

2 Kinship Terminology

2.1 Classification of Relatives

An enlightening aspect of some kinship systems around the world is that they may make distinctions that your own language does not, or merge together relatives that you distinguish. One of the most common systems found in smaller societies is the *bifurcate merging* method of naming parents and their siblings, defined by "merging" father and his brother under the same term, and similarly merging mother and her sister with another term. Meanwhile, father's sister has a separate term to herself, as does mother's brother. If we use the obvious abbreviations of "MoBr" for "mother's brother," etc., then bifurcate merging and the English *collateral lineal* kinship terminology are defined by the following two partitions on the set of parents and their siblings:

$$\{\{\text{FaSi}\}, \{\text{Fa}, \text{FaBr}\}, \{\text{Mo}, \text{MoSi}\}, \{\text{MoBr}\}\} \qquad \text{(bifurcate merging)}, \tag{1}$$

$$\{\{\text{FaSi}, \text{MoSi}\}, \{\text{FaBr}, \text{MoBr}\}, \{\text{Mo}\}, \{\text{Fa}\}\} \qquad \text{(collateral lineal)}. \tag{2}$$

These two systems are "orthogonal" in the sense that the intersection of these two partitions is the identity partition, with each term in its own class. On the other hand, their join in the lattice of partitions results in just two classes, the males and the females. The collateral lineal system (formerly called "Eskimo," an offensive term now avoided) makes the distinction between lineal relatives (in this case father and mother) and collaterals (in this case siblings of parents). While this may sound absurdly logical to speakers of English and other languages with the same partition, it is not inevitable. The bifurcate merging system distinguishes, or "bifurcates," relatives on the mother's side from those on the father's side. Simultaneously, it merges same-sex siblings of the parental generation, so we get father and father's brother both being called by the same term, "father." Yes, these people know the difference between the two senses of "father," just as the English are capable of distinguishing, say, the two kinds of "aunts" with the adjectives "maternal" and "paternal" or by using the possessive construction, "mother's sister" and "father's sister"[6, 10].

The two partitions of parents and their siblings in Eq. (1) have implications for the terminology of their children [10]. The concept of "cousin" in English has to be refined into two categories, *cross cousin*, the child of a parent's opposite-sex sibling, and *parallel cousin*, the child of a parent's same-sex sibling. In terms of the standard kinship abbreviations above, these two types of cousin are defined to be:

$$\{\text{FaSiCh}, \text{MoBrCh}\} \qquad \text{(cross cousins)}, \tag{3}$$

$$\{\text{FaBrCh}, \text{MoSiCh}\} \qquad \text{(parallel cousins)}. \tag{4}$$

The distinction between cross and parallel cousins occurs most commonly in societies with the bifurcate merging terminology of Eq. (1). One rationale for this is that given that you call your father's brother a "father," it is somehow logical to call his children "brother" and "sister." The same reasoning applies to the parallel cousins on your mother's side. This is associated with an extension of the incest and marriage ban between siblings to parallel cousin, while in some cases allowing marriage between cross cousins. In fact, one of the most marriageable relatives in many societies, including some of the biblical patriarchs of Israel, is between a man and his mother's brother's daughter. One of these patriarchs, Jacob, was simultaneously married to two of his mother's brother's daughters, Leah and Rachael. Note that this relative may be distinguished from the other three "first cousins," both terminologically and in terms of who can marry whom.

Malinowski [5] pointed out an amusing dilemma facing a hypothetical Christian missionary Bible translator in the Trobriand Islands, which has a matrilineal society with the bifurcate merging system. A literal translation of the English phrase "God, the Father" would be inappropriate because the Trobriand father is a loving figure with no real authority, land being inherited from the mother's brother. Malinowski suggested that "God, the Mother's Brother" might be a better translation, carrying more the force of authority. However, the disadvantage would be that Trobriand mother's brothers typically do not show the love consistently expressed by a Trobriand father and occasionally bestowed by the Biblical God.

The models of kinship in this paper do not address any of the connotative, psychological, or social implications of kin terminology. However, relatives that are merged under the same kinship term tend to be more similar socially than those who are distinguished. Furthermore, there is no attempt to consider the nonbinary gender roles and pronouns that play such a prominent role in liberal democracies.

The other disclaimer for kinship terminology discussed here is that it is only loosely associated with the composition of relations. For example, the relations, "father" and "son," on the set of males in a community can be combined using the standard composition of relations to form another relation, "a man's mother's son."

However, this relation would include some ordered pairs that are members of the identity, or equality, relation as well as actual "brothers." What is addressed in this paper are the purely algebraic aspects of kinship terminology considered as a subset of the culture's language, not its actual social relations.

With this introduction to kinship terminology, one can see the outlines of a "kinship algebra." The operation of concatenation, indicated by "'s," is a binary operation. The merging in classes is like a congruence relation on the set of strings of kinship terms. Finally, the fact that the definitions of cross and parallel cousins require no parentheses implies that the binary operation in a kinship system is associative, resulting in a semigroup, discussed in Sect. 3.1.

2.2 Kinship Notation

The basic set of kinship symbols is $B := \{\triangle, \bigcirc, +, -\}$. The first two members of B are standard symbols for "male" and "female," respectively [4, 7, 8][2]. The last two members of B, the plus and minus signs, represent the "parent" and "child" relations, respectively. In addition, let us introduce one derived symbol, $\sqcap$, the "sibling" relation.[3]

These basic symbols can be composed (from left to right) to form other kinship relations such as $+\triangle$, meaning "father." Conversely, $\triangle+$ would mean "a male's parent." The sibling relation can be defined in terms of the basic kinship primitive B by an equation, defining a sibling as a parent's child (the half-sibling rule):

$$\sqcap := +\bigcirc- = +\triangle-, \qquad \text{(half-sibling rule).} \tag{5}$$

This almost universal rule may suggest that societies try to encourage half-siblings to be accepted as a full sibling. Since we have to specify the sex of both the speaker and the referent, this system gives us four types of siblings: $\triangle \sqcap \triangle$, $\triangle \sqcap \bigcirc$, $\bigcirc \sqcap \triangle$, and $\bigcirc \sqcap \bigcirc$. Unfortunately, it does not distinguish between elder and younger siblings. See [10] for a model of possible sibling terminologies

[2] Other kinship notations exist. Romney and D'Andrade [8] used "m" and "f" for "male" and "female," which could be produced on a manual typewriter. However, these letters have the opposite association with the first letters of "mother" and "father" in the kinship context. Boyd et al. [1] use the Mars and Venus symbols, which require a special package in LaTeX. The notation in this paper has the advantage of closely following the notation in traditional "kinship diagrams," but its disadvantage is the possible confusion between the female symbol, $\bigcirc$, and the mathematical zero, 0. Also, the parent $+$ and child $-$ symbols might be mistaken for the usual binary operations on the integers.

[3] One could also define a "spouse" symbol, $\sqcup := a - b + c$, where $a \neq c$, replacing the 0 in Eq. (7). As with the sibling symbol $\sqcap$, the spouse symbol $\sqcup$ is consistent with some kinship diagrams. However, the spouse concept would take you into the realm of *affinal* relatives (related by marriage), whereas we are confining ourselves to *consanguineal* relative (related by blood) in this paper.

corresponding to "conjunctive" partitions of the eight sibling terms that include the older/younger distinction. Sibling terminologies can be found that include such interesting concepts as "older sibling of the opposite sex."

Let $\mathcal{F}_K$ be the free semigroup on the eight terms, $K = \{a \pm b\}$, where $a, b \in \{\triangle, \bigcirc\}$. The term "free" here refers to the unique semigroup with generators K but which satisfies no additional equations [2]. We want to impose some equations for all $a, b, c \in \{\triangle, \bigcirc\}$:

$$a - b + c = ac, \qquad \text{(unique parenthood rule)}. \tag{6}$$

Next we must decide how to multiply the ac in the right-hand side above:

$$ac = \begin{cases} a, & \text{if } a = c \quad \text{(idempotent rule)}, \\ 0, & \text{if } a \neq c \quad \text{(no affinal rule)}, \end{cases} \tag{7}$$

where the 0 is a distinct element adjoined to $\mathcal{F}_K$ satisfying the usual equations for a zero element: $x0 = 0x = 0$ for all $x \in \mathcal{F}_K \cup \{0\}$. With these equations, a unique system of representatives for *consanguineal* terms, denoted by **Con**, is given by strings of the forms:

$$a_1 + a_2 + \cdots + a_m, \qquad \text{(ancestors)}, \tag{8}$$

$$b_1 - b_2 - \cdots - b_n, \qquad \text{(descendants)}, \tag{9}$$

$$a_1 + a_2 + \cdots + a_m \sqcap b_1 - b_2 - \cdots - b_n, \qquad \text{(collaterals)}, \tag{10}$$

for $a_i, b_j \in \{\triangle, \bigcirc\}$ and positive integers, m, n. Note that the semigroup **Con** includes zero, 0.

If an English speaker has a relative called an "aunt," then which kinship term does this aunt use for the speaker? This question cannot be answered without knowing the sex of the speaker: for a male, the answer is "nephew," while for a female, "niece." This is why the sex of the speaker is always specified in our notation. In normal speech, of course, the sex of the speaker is obvious. Furthermore, if a male calls two different relatives by the term "aunt," then one would expect those two women to call him, "nephew." This is called the *principle of reciprocity* and justifies the use of *inverse semigroups*, defined below. More formally, if x is any consanguineal relative, then its (*generalized*) *inverse* (or *reciprocal*) is the consanguineal x^{-1}, defined by Eqs. (11) and (12):

$$x = a_1 + \cdots + a_m \; b_1 - \cdots - b_n, \qquad \text{(consanguineal)}, \tag{11}$$

$$x^{-1} = b_n + \cdots + b_1 \; a_m - \cdots - a_1, \qquad \text{(inverse)}, \tag{12}$$

$$x = y \quad \Rightarrow \quad x^{-1} = y^{-1}, \qquad \text{(principle of reciprocity)}, \tag{13}$$

for all $a_i, b_j \in \{\triangle, \bigcirc\}$, with $a_m = b_1$ and $m, n > 0$.

2.3 Fox Consanguineal Kinship

The Fox kinship system is classified as Omaha[4] Type I by Lounsbury [4], requiring two additional equations for all $a \in \{\triangle, \bigcirc\}$:

$$a + \triangle \sqcap \triangle = a + \triangle, \qquad \text{(father merging)}, \qquad (14)$$

$$a + \bigcirc \sqcap \bigcirc - b = a \sqcap b, \qquad \text{(mother merging)}, \qquad (15)$$

$$a + \bigcirc \sqcap \triangle - b = a + \bigcirc \sqcap b, \qquad \text{(skewing)}. \qquad (16)$$

Note that the Fox merging rule (indicated by the equals sign) is split into two subtypes: father merging (14) and mother merging (15). In *father merging*, the father is merged with father's brother, as in the usual bifurcate merging rule. On the other hand, in *mother merging*, the mother is distinguished from mother's sister, while the merging is delayed until the next generation, resulting in mother's sister's children being merged with siblings. These two types of merging can be observed in the Fox kinship diagram in Fig. 1. If we can allow ourselves one peek beyond consanguineal relatives into the realm of affinal relatives, we can find another piece of evidence for a tendency toward merging: "mother's sister's husband" is equated with "father," which makes sense if "mother's sister" were similar to "mother."

While there is only partial merging in the Fox kinship system, the "bifurcate" part of bifurcate merging holds completely. That is, "father's sister" is distinguished from "mother's sister" and "mother's brother" is distinguished from "father's brother."

As for the first cousins among the Fox, there is a clear distinction between cross and parallel cousins. In fact, the two types of parallel cousins are equated with siblings, while the two cross cousins are distinguished from the parallel cousins and from each other. So overall, the Fox closely resemble a bifurcate merging system.

What is rather amazing about the Fox cross cousins, however, is the skewing rule (16), which equates MoBrSo with MoBr and FaSiSo with SiSo. That is, MoBrSo is "promoted" to MoBr, while FaSiSo is "demoted" to a lower generation and equated with SiSo, or Ne ("nephew") in Fig. 1. This suggests that the meaning of MoBr is not at all captured by the English word "uncle" and that a better translation for this term would be "male of my mother's patrilineage."

3 Binary Operations

A *binary operation* on a set S is a function from the set of all ordered pairs from S into S. A binary operation is also called a *groupoid* and we shall adopt this terminology. It is convenient to use infix notation for groupoids, so if $*$ is a groupoid

[4] The Omaha tribe is also found today in the Great Plains of the USA but has a distinct language and history from the Fox.

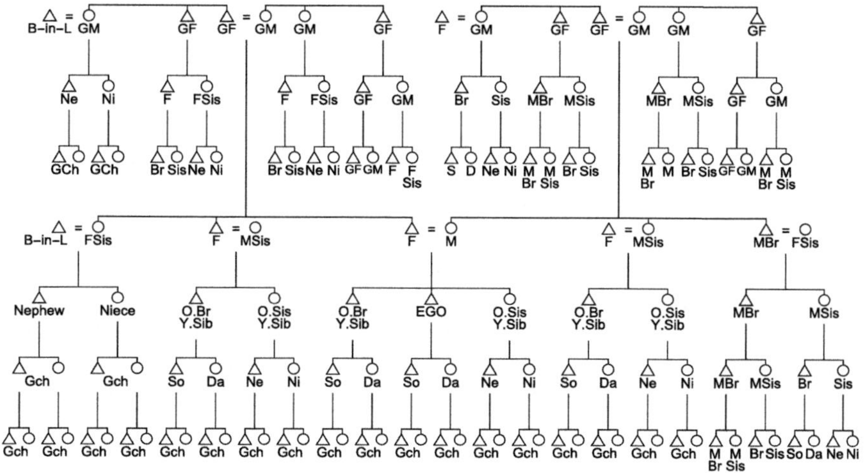

Fig. 1 Chart of the Fox kinship system, taken from Tax [9]. The point of view is for a male ego. The chart for a female ego has significant differences that are predictable from the equations

on S, then the $*$-image of the ordered pair (x, y) is denoted by $x * y$. That is,

$$* : S^2 \to S;\ (x, y) \mapsto x * y.$$

Note that a groupoid $*$ on a set S is determined by the ordered pair $(S, *)$, where S is the *underlying set* and $*$ is the function. If only one groupoid is under consideration, then the underlying set S is called the groupoid without confusion. An example of a groupoid on the set of nonzero real numbers is the usual division, $\div$. However, division is not a groupoid on all the real numbers, because division by zero, $x \div 0$, is undefined. Other common symbols for binary operations are $+$, $\times$, $\circ$, $\oplus$, or especially by no symbol at all, so the result of the binary operation would be written as xy.

A *homomorphism* from one groupoid, $(S, *)$, to another, $(T, \circ)$, is a function, $h : S \to T$, such that for all $x, y \in S$, multiplication is *preserved*, in the sense that $h(x * y) = h(x) \circ h(y)$. If the homomorphism h is also a *bijection* (a function such that for every $y \in T$, there is a unique $x \in S$ such that $y = h(x)$), then h is said to be an *isomorphism* from S to T. Indeed, this isomorphism h determines another isomorphism $h^{-1} : T \to S$ called the *inverse* of h. For example, the logarithm is an isomorphism $\log : \mathrm{Re}_+ \to \mathrm{Re}$, from the positive reals with multiplication, $(\mathrm{Re}_+, \times)$ to the reals with addition. That is, $\log(x \times y) = \log(x) + \log(y)$. The inverse of the log function is the exponential function, satisfying $\exp(x + y) = \exp(x) \times \exp(y)$. An isomorphism $h : S \to S$ from a groupoid to itself is called an *automorphism*.

A *congruence* on a groupoid, S, is an equivalence relation $\equiv$ on S such that for all $x, y, z \in S$, we have that $x \equiv y$ implies both $xz \equiv yz$ (*right compatibility*) and $zx \equiv zy$ (*left compatibility*). As with any algebraic system, every homomorphism defines a unique congruence and vice versa. That is, given a homomorphism $S \to T$ of groupoids, then the equivalence relation on S, defined by $x \equiv_f y$ if and only if $f(x) = f(y)$, can be shown to be a congruence. See the "Main Homomorphism Theorem" in [2].

Conversely, if $\equiv$ is a congruence on a groupoid S, then the *equivalence class* for each $x \in S$ is denoted by $[x]_\equiv$, or if the context is clear by $[x]$, and is defined by $[x] := \{ y \in S \mid y \equiv x \}$. The induced multiplication between equivalence classes is defined by $[x][y] = [xy]$ for all $x, y \in S$. The set of all equivalence classes, $S/\equiv$, together with the induced multiplication, is called the *quotient groupoid* for the congruence. See the "Induced Homomorphism Theorem" in [2].

The set $C(S)$ of all congruence on a groupoid S can be ordered by set inclusion, forming a *complete lattice* [2], meaning that if C is a set of congruences on S, then there is a unique *greatest lower bound* for C, denoted by $\bigwedge C$ (or by $\inf C$), defined by

$$\bigwedge C := \bigcap C. \tag{17}$$

That is, the greatest lower bound for C is the intersection of members of C. If there are only two congruences, say ρ and σ, then the greatest lower bound is denoted by the infix notation $\rho \wedge \sigma$.

Of course, it has to be proven that the intersection of congruences is again a congruence, but this has been shown for all algebras, which includes groupoids. Note that the congruence lattice on S has a least element, the *identity relation* on S, defined by $\iota_S := \{ (x, x) \mid x \in S \}$, and a greatest element, the *universal relation* on S, which equals the Cartesian product, $S^2 := \{ (x, y) \mid x, y \in S \}$. If the context is clear, the identity element may be denoted simply by ι.

A lattice must also have a *least upper bound*, which in the case of congruences is less straightforward. By analogy with the greatest lower bound in Eq. (17), one might guess that the union would be the least upper bound. However, the union of congruences, which may seem to be the natural choice for the least upper bound, may not be a congruence or even an equivalence relation. For example, if $S = \{ a, b, c \}$, and

$$\rho = \{ (a, b), (b, a) \} \cup \iota_S \quad \text{and} \tag{18}$$

$$\sigma = \{ (b, c), (c, b) \} \cup \iota_S, \quad \text{then} \tag{19}$$

$$\rho \cup \sigma = \{ (a, b), (b, a), (b, c), (c, b) \} \cup \iota_S, \tag{20}$$

which is not transitive and hence not even an equivalence, much less a congruence. The (correct) least upper bound for a set C of congruences on a groupoid S is the intersection of all congruences in $C(S)$ that contain every member of C:

$$\bigvee C := \bigwedge \left\{ \rho \in C(S) \mid \rho \supseteq \bigcup C \right\}. \tag{21}$$

The join is always defined since the universal relation, S^2, is a congruence that contains any union of congruences. As with the meet in Eq. (17), the join of just two congruences, ρ and σ, may be written using infix notation, $\rho \vee \sigma$.

Following the musical theme suggested by ordering, we introduce two congruences determined by an arbitrary relation R and an equivalence relation E, both defined on S:

$$R^{\sharp} := \bigwedge\{\, \rho \in C(C) \mid \rho \supseteq R \,\} \qquad \text{(closure)}, \tag{22}$$

$$E^{\flat} := \bigvee\{\, \rho \in C(C) \mid \rho \subseteq R \,\} \qquad \text{(interior)}. \tag{23}$$

Warning: the interior operation in the above equation only works for equivalence relations. To see how a nonequivalence can fail to have an interior congruence, consider the example of Eq. (18) where the union, $R := \rho \cup \sigma$, of two equivalences is not an equivalence. If ρ and σ were both congruences, then they would both be lower bounds for R, but their join is S^2, which properly contains R.

The typical use of the closure operation is when R is a set of generating equations for the congruence $R^{\sharp}$ and the quotient groupoid $S/R^{\sharp}$. A typical use of the interior operation is when the equivalence E is rather large, with relatively few equivalence classes, and one wants a finer congruence $E^{\flat}$ such that each equivalence class of E is a union of $E^{\flat}$ congruence classes. In the extreme case where E has exactly two equivalence classes, then the groupoid $S/E^{\flat}$ is called the *syntactic monoid* for either one of the classes, noting that if one of two equivalence classes is a union of congruence classes, then so is the other equivalence class.

Two common properties of a groupoid S are the *commutative* property, where $xy = yx$ for all $x, y \in S$, and the *associative* property, where $(xy)z = x(yz)$ for all $x, y, z \in S$. Associative operations are called *semigroups*, the title concept of this paper. The classic book by Clifford and Preston [2] requires that the underlying set S of a semigroup be nonempty, while newer sources allow S to be empty.

Notice that the example above of the usual division operation on the nonzero reals is neither commutative nor associative. That division is not commutative is obvious, while the associative property fails because $(2 \div 3) \div 4 = 1/6$, which is not equal to $2 \div (3 \div 4) = 8/3$. However, the usual multiplication on the reals is both commutative and associative.

A minimal example of a groupoid that is neither associative nor commutative is the two-element set, $G = \{a, b\}$, with the multiplication $ab = a$, $ba = b$, and $aa = b$. The group G fails to be associative because $a(ba) = ab = a$, while $(ab)a = aa = b$. Actually, this gives *two* examples of nonassociative groupoids, since bb, which was unspecified above, could be set equal to either a or b.

Two examples of groupoids that satisfy one of the above properties and not the other also exist. First, the *mean* operation $\oplus$ defined on the reals by $x \oplus y := (x+y)/2$ is commutative but not associative. Second, the usual matrix multiplication on the set $\mathbb{R}^{n\times n}$ of n by n real-valued matrices (for a fixed positive integer n) is associative but not commutative.

If the groupoid S is on a finite set with n elements, then S can be displayed as a *Cayley multiplication table*, an n by n square matrix. The rows and columns are labeled by elements of S, and the entry for row a and column b is the product ab. For example, the Cayley multiplication in Table 1 is a minimal inverse semigroup, defined later in Sect. 3.2, which is not commutative: i.e., $a * b = e$, while $b * a = f$.

How many groupoids is it possible to have on a finite set with n elements? The answer in terms of labeled function is n^{n^2} if you think of how many ways can you fill in the Cayley table. So, for example, if $n = 3$, then there are $3^{3^2} = 19{,}683$ such tables. However, this counts isomorphic groupoids separately. For example, if $S = \{a, b, c\}$, then there are three isomorphic constant groupoids that map every ordered pair onto the same element. That is, the Cayley table is filled with all a's, all b's, or all c's. Counting isomorphism classes is not simple and usually involves group theory. Fortunately, the free On-Line Encyclopedia of Integer Sequences (OEIS) h has solutions to many of the known enumerative problems. In addition to the exact enumerations in Table 2, the OEIS gives an asymptotic formula, $n^{n^2}/n!$, for the number of non-isomorphic groupoids. This formula converges extremely rapidly: for $n = 2$, 3, and 4, the relative errors are 20, 1.5, and 0.014 percent, respectively.

3.1 Semigroups

Associativity implies that parentheses are not necessary for products in a semigroup. For example, in the multiplicative semigroup of natural numbers, $(\mathbb{N}, \times)$, we can write $2 \times 3 \times 4$ without parentheses.

Unlike groups, a semigroup S may not have an *identity*, defined as an element $1 \in S$ such that for all $x \in S$, $1x = x1 = x$. Another difference is that a semigroup may have a *zero* element $0 \in S$ defined by $0x = x0 = 0$ for all $x \in S$, whereas a nontrivial group cannot have a zero. Note that if S is a singleton set $\{x\}$, then x is both an identity and a zero.

If a semigroup S fails to have an identity, then a derived semigroup can be constructed by adding a distinct new element to S and imposing the equations for an identity. This new semigroup has just one more element and the same multiplication table as the old one, except for the row and column belonging to the identity. Similar

Table 1 The Cayley multiplication table for a minimal noncommutative, inverse semigroup. The elements a and b are inverses of each other, e and f are idempotents, as is the zero element, 0

*	a	b	e	f	0
a	0	e	0	a	0
b	f	0	b	0	0
e	a	0	e	0	0
f	0	b	0	f	0
0	0	0	0	0	0

Table 2 Number of non-isomorphic groupoids of sizes 1–4 that satisfy a given property. The second column is the OEIS A-number and the last row counts commutative semigroups

Property	OEIS #	1	2	3	4
None	A001329	1	10	3,330	178,981,952
Commutative	A001425	1	4	129	43,968
Associative	A027851	1	5	24	188
Both	A001426	1	1	3	12

remarks apply, mutatis mutandis, to semigroups that fail to have a zero element. The notations S^0 and S^1 are taken to mean the semigroup formed by the adjunction of a zero or an identity, respectively, if S does not already have one. That is,

$$S^0 := \begin{cases} S, & \text{if } S \text{ has a zero element,} \\ S \cup \{0\}, & \text{otherwise.} \end{cases} \tag{24}$$

$$S^1 := \begin{cases} S, & \text{if } S \text{ has an identity element,} \\ S \cup \{1\}, & \text{otherwise.} \end{cases} \tag{25}$$

These two definitions follow [2], but one could add new 0s or 1s unconditionally.

A semigroup S may also contain an *idempotent* element e, defined by the property that $e = e^2$. Every zero and identity element is an idempotent, but not conversely. Also, a semigroup can contain at most one zero and at most one identity, but may have multiple idempotents. There is a *natural partial order* on the set E of all idempotents in S, defined by $e \leq f$ if and only if $ef = fe = e$. If the semigroup has an identity 1, then 1 is the greatest elements of E, since $e\,1 = 1\,e = e$ for all $e \in E$. Conversely, if the semigroup has a zero, 0, then 0 is the least element of E. An idempotent e of a semigroup is said to be *primitive* if it is a minimal nonzero element in the natural partial order.

A classic example of a semigroup is the set $\mathcal{B}_X = 2^{X^2}$ of all binary relations on a set X, where the binary operation, $(\circ)$, is the composition of relations. Relations are often symbolized by lower-case Greek letters, so if $\rho \in \mathcal{B}_X$, then ρ is a set of ordered pairs of members of X and $(x, y) \in \rho$ is written as $x\;\rho\;y$. The *composition* of two relations $\rho, \sigma \in \mathcal{B}_X$ is defined by $x\;(\rho \circ \sigma)\;z$ if and only if there exists $y \in X$ such that $x\;\rho\;y$ and $y\;\sigma\;z$. While it is easy to prove that this operation is associative, other properties, such as the commutative law and the existence of inverses, fail to hold in general. Despite our disavowal of kinship relations in Sect. 2.1, we cannot help using them for an example of a noncommutative relation: a "mother's brother" is not in general equal to a "brother's mother."

The semigroup of relations on X does contain an identity element, the *identity relation*, $\iota_X := \{\,(x, x) \mid x \in X\,\}$. It is easy to verify that this is an identity element. That is, for all $\rho \in \mathcal{B}_X$, one has $\iota_X \circ \rho = \rho \circ \iota_X = \rho$. In addition, $\mathcal{B}_X$ contains a zero element $\emptyset_X$, defined to be the empty set.

The semigroup of binary relations $\mathcal{B}_X$ also contains many other idempotents, such as equivalence relations, partial orders, any subset of the identity relation, and the tiny relations of the form $\tau_{a,b} := \{ (a, b), (b, b) \}$, where $a, b \in X$ and $a \neq b$. It is easy to verify that each primitive idempotent of $\mathcal{B}_X$ is a singleton ordered pair of the form $\{ (x, x) \}$ for $x \in X$.

Note that the natural partial order on idempotents in the semigroup of relations does not correspond to the inclusion order of relations, considered as sets. For example, in the natural partial order, ι_X is greater than any other idempotent relation in $\mathcal{B}_X$, but in the set inclusion order, ι_X contains the identity relation on any subset of X and yet is contained by the *universal* relation X^2. Note that it is X^2, not ι_X, which is the greatest element in the inclusion ordering. In addition, ι_X neither contains nor is contained by the tiny $\tau_{a,b}$ relations defined above.

One may define a bijection on any semigroup of relations:

$$* : \mathcal{B}_X \to \mathcal{B}_X; \rho \mapsto \rho^*, \tag{26}$$

where $x\,\rho\,y \;\Leftrightarrow\; y\,\rho^*\,x$. The mapping of Eq. (26) is called the *reciprocal* (or *converse*) and has the property of being an *involuntary anti-automorphism*:

$$(\rho \circ \sigma)^* = \sigma^* \circ \rho^*, \tag{27}$$

for all $\rho, \sigma \in \mathcal{B}_X$.

A second example of a semigroup, which is more closely related to kinship terminology, is the *bicyclic* [2] semigroup, $C(p, q)$, which is defined by a single generating equation $pq = 1$. The elements of $C(p, q)$ can be arranged in a rectangular array:

$$\begin{array}{cccc} 1 & p & p^2 & \cdots \\ q & qp & qp^2 & \cdots \\ q^2 & q^2p & q^2p^2 & \cdots \\ \vdots & \vdots & \vdots & \ddots \end{array} \tag{28}$$

Two elements $q^k p^l$ and $q^m p^n$, where the exponents are nonnegative integers, multiply as follows:

$$(q^k p^l)(q^m p^n) = q^i p^j, \quad \text{where,} \tag{29}$$

$$i = k + m - \min(l, m), \tag{30}$$

$$j = l + n - \min(l, m). \tag{31}$$

Figure 2 gives a graphical representation of multiplication for $C(p, q)$ for the two cases: $l \leq m$ (in this example, $l = 1$ and $m = 2$) and $l > m$ (in the second case, $l = 2$ and $m = 1$). These two cases follow from Eq. (28), which implies that $p^l q^m$ equals q^{m-l} if $l \leq m$, while $p^l q^m$ equals p^{l-m} if $l > m$.

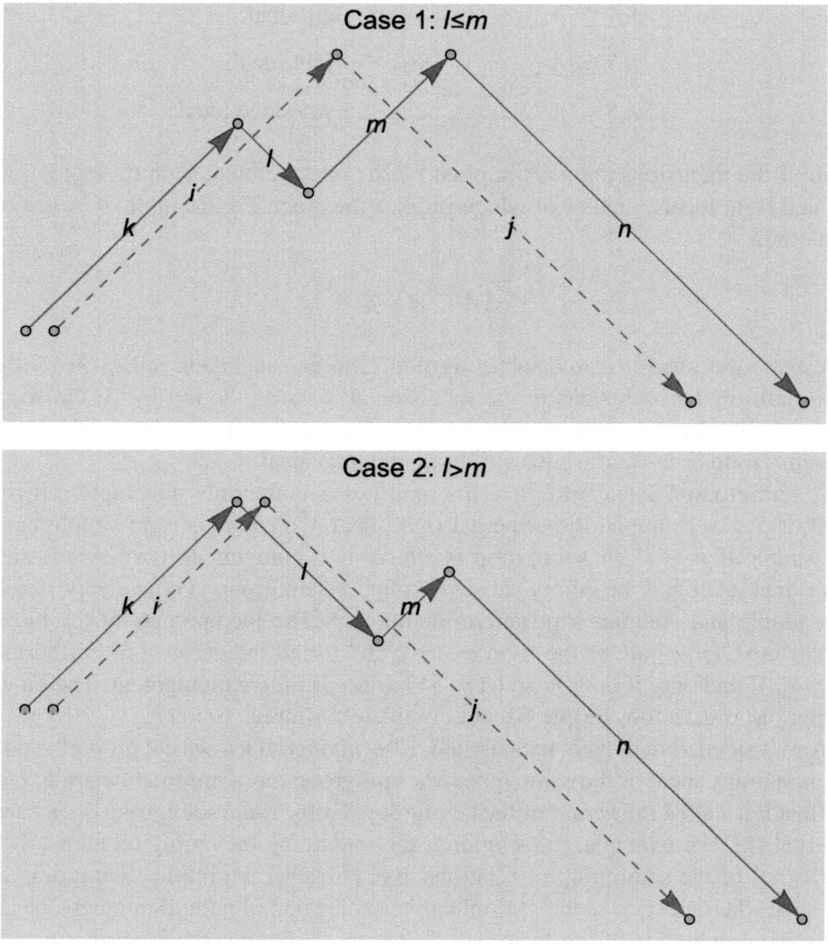

Fig. 2 The two cases for multiplication in the bicyclic semigroup depend on which of the two adjoining "legs" is longer

The idempotents of $C(p, q)$ are elements of the form $q^n p^n$, found along the main diagonal of Eq. (28).

If A, B are subsets of a semigroup S, then the set of all products from these subsets is defined by $AB := \{ ab \mid a, b \in S \}$. This set product respects set inclusion in the sense that if A, B, X, and Y are subsets of a semigroup such that $A \subseteq B$ and $X \subseteq Y$, then $AX \subseteq BY$. If A is a singleton set, $\{ a \}$, then the set products, $\{ a \}B$ and $B\{ a \}$, are denoted by aB and Ba, respectively. This notation is useful in defining some of the most basic properties of subsets of semigroups. Below is a list of properties (and their names) for a subset A of a semigroup S:

$$AA \subseteq A \qquad \text{subsemigroup,} \tag{32}$$

$$SA \subseteq A \qquad \text{left ideal,} \tag{33}$$

$$AS \subseteq A \qquad \text{right ideal,} \tag{34}$$

$$SAS \subseteq A \qquad \text{two-sided ideal.} \tag{35}$$

Each of the inclusions above is implied by the ones below it with the exception of left and right ideals, neither of which implies the other. For example, if A is a right ideal, then

$$AA \subseteq AS \subseteq A. \tag{36}$$

So every right ideal is also a subsemigroup. The first inclusion in Eq. (36) follows from multiplying both sides of the inclusion $A \subseteq S$ on the left by A, the second, by the definition of a right ideal in Eq. (34). Thus, every (two-sided) ideal is also a subsemigroup because it is also a right (and a left) ideal.

A semigroup S is called *left*[*right*] *simple* if S is the only nonempty left[right] ideal of S. For example, the principal right ideal $q^k C(p, q)$ is right simple but not left simple if $k > 1$. A semigroup is *simple* if it contains no (two-sided) ideals, other than itself and the empty subsemigroup. A semigroup is *completely simple* if it is simple and contains a primitive idempotent. The idempotents of the bicyclic semigroup $C(p, q)$ are of the form $e_n := q^n p^n$ for all integer $n \geq 0$. Furthermore, $e_m \leq e_n$ if and only if $m \geq n$, so $C(p, q)$ has no primitive idempotent. The bicyclic semigroup is therefore simple but not completely simple.

A two-sided, left, or right ideal is said to be *minimal* if it does not properly contain any nonempty ideal of the same type. If a semigroup has a minimal two-sided ideal K, then it is called the *kernel* of the semigroup. Every finite semigroup does contain a kernel [2]. For example, the singleton set containing the empty relation $\{ \emptyset_X \}$ is the kernel of the semigroup of relations $\mathcal{B}_X$. However, an infinite semigroup may not have a kernel, such as the multiplicative semigroup of natural numbers, because every ideal, which is of the form $a \times \mathbb{N}$, properly contains the ideal $b \times a \times \mathbb{N}$ if $b > 1$.

Note that the concept of minimal ideals is trivial for semigroups with zero, since a zero is the unique minimal left, right, or two-sided ideal. Therefore, it is useful to introduce the concept of 0-minimal ideals: a two-sided [left, right] ideal M of a semigroup S is 0-*minimal* if $M \neq 0$ and 0 is the only nonempty two-sided [left, right] ideal of S properly contained in M.

There is a corresponding concept of 0-simplicity. A semigroup S with a zero element is said to be 0-*simple* [*left* 0-*simple*,*right* 0-*simple*] if $S \neq 0$ and 0 is the only nonempty two-sided [left, right] ideal of S.

An example of a subsemigroup, defined by Eq. (32), is, $\mathcal{T}_X = X^X$, the semigroup of functions on a set X, as a subsemigroup of the semigroup of relations, $\mathcal{B}_X$. This is possible because a function $f : X \to X$ can be defined as a relation on X such that for every $x \in X$, there exists a unique $y \in X$ such that xfy holds, written $f(x) = y$. It is easy to verify that the composition of two functions on set X is again a function on X, so that these functions do form a semigroup, $\mathcal{T}_X$, which is a subsemigroup of $\mathcal{B}_X$. If A is a subset of a semigroup S, then the intersection of all semigroups of S containing A is evidently the smallest subsemigroup of S containing A and is called the *subsemigroup generated by* A, denoted by $\langle A \rangle$. In more concrete terms, $\langle A \rangle$ consists of all finite products of elements of A. Similar definitions hold for right, left, or two-sided ideals generated by a subset of a semigroup.

A more constructive result [2], using Eq. (25), shows that the ideal generated by a subset A of a semigroup is given by $S^1 A S^1$. If A is a singleton element, $\{ a \}$, then the (left, right, or two-sided) ideal generated by $\{ a \}$ is called a *principal* left, right, or two-sided ideal, and is given by $S^1 a$, $a S^1$, or $S^1 a S^1$, respectively. If the semigroup S already has an identity element, as is the case in the examples below, the 1s can be omitted from the exponents since $S = S^1$ holds.

Examples of principal left, right, and two-sided ideals, defined by Eqs. (33), (34), and (35), can be found in $\mathcal{T}_X$, the semigroup of functions on a set X. These ideals of a function f are related to its *image* (or *range*) and *equivalence* (or *partition*):

$$\mathrm{im}(f) = \{ y \in X \mid \exists x \in X : y = f(x) \} \qquad \text{(image of } f\text{)}, \tag{37}$$

$$\mathrm{eq}(f) = \{ (x, y) \in X^2 \mid f(x) = f(y) \} \qquad \text{(equivalence for } f\text{)}. \tag{38}$$

The principal left-, right-, and two-sided ideals for f can be found using the results [2] below:

$$\mathcal{T}_3 f = \{ g \in \mathcal{T}_3 \mid \mathrm{im}(g) \subseteq \mathrm{im}(f) \}, \tag{39}$$

$$f \mathcal{T}_3 = \{ g \in \mathcal{T}_3 \mid \mathrm{eq}(g) \supseteq \mathrm{eq}(f) \}, \tag{40}$$

$$\mathcal{T}_3 f \mathcal{T}_3 = \{ g \in \mathcal{T}_3 \mid |\mathrm{im}(g)| \leq |\mathrm{im}(f)| \text{ and } |\mathrm{eq}(g)| \geq |\mathrm{eq}(f)| \}. \tag{41}$$

This result assumes we keep the practice, inherited from the semigroup of relations, of composition from left to right. If we had reverted to the frequent practice of composing functions from right to left, then the left and right ideals would be switched. Also note that the ordering of equivalences in Eq. (39) is the usual one for partitions: $\pi_1 \subseteq \pi_2$ if every equivalence class of π_1 is a subset of some equivalence class of π_2. In this case, π_1 is said to be a *refinement* of π_2, or equivalently, π_2 is said to be *coarser* than π_1.

Specifically, consider the semigroup of functions on three elements, $\mathcal{T}_3$, which has exactly $3^3 = 27$ functions. An example of a function in $\mathcal{T}_X$ can be specified by listing its arguments and values, e.g., $f(1) = 2$, $f(2) = 2, f(3) = 1$, or by listing the values in a row vector, $f = (2, 2, 1)$, where the arguments are implicit in the ordering. The various principal ideals generated by this one function are listed in Fig. 3, with the further abbreviation for functions of omitting the parentheses and the commas. Note that all the ideals in $\mathcal{T}_X$ are principal if X is finite [2].

The principal left ideal, $221\,\mathcal{T}_3$, has eight elements: 212, 221, ..., 111, as shown in Fig. 3. Functions in this ideal are characterized by having images that are a subset of $\{ 1, 2 \}$, as required by Eq. (39). Conversely, the principal right ideal, $\mathcal{T}_3\,221$, has

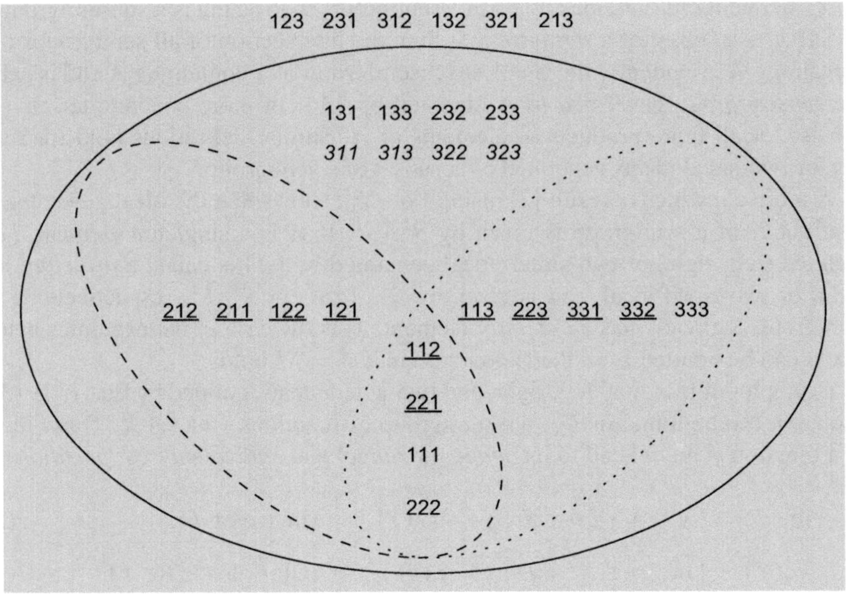

Fig. 3 Principal ideals in $\mathcal{T}_3$ generated by the function 221. The lines for the left ideal are dashed; the right, dotted; and the two-sided, solid. The underlined functions—such as 212, 113, and 112—are generators equivalent to 221 of the left or right ideals containing them. Note that the complement of the two-sided idea generated by (2, 2, 1) is the group of six permutations of $\{1, 2, 3\}$. Finally, the four generalized inverses of 211 are italicized

nine elements: $113, 221, \ldots, 111$. Functions of this right ideal are characterized by having equivalences (partitions) that are greater than (coarser than) or equal to $\{\{1, 2\}, \{3\}\}$. Finally, since these two left and right ideals are also subsemigroups, their intersection, $\{112, 221, 111, 222\}$, is a subsemigroup because the intersection of subsemigroups is again a subsemigroup.

Finally, the kernel of $\mathcal{T}_3$ (the minimal ideal) is the set of constant functions:

$$K = \{(1, 1, 1), (2, 2, 2), (3, 3, 3)\}.$$

The bicyclic semigroup, $C(p, q)$, has principal ideals that can be read off from the display in Eq. (28). For example, the principal right ideal $qC = qpC = qp^2C$ is the set of elements from the second row and below. In fact, all the elements in the second row generate the same principal right ideal. Similarly, all the elements from a given column generate the same principal left ideal.

Two examples of a bicyclic semigroup can be constructed from consanguineal kinship strings defined in Eq. (10). First, consider the set $\mathrm{CR}_\triangle$ of all consanguineal relatives such that $a_i = b_j = \triangle$, i.e., consanguineal relatives with no female symbols $\bigcirc$. This is isomorphic to the bicyclic semigroup $C(p, q)$ with the mapping:

$$\triangle + \triangle \mapsto p \qquad \text{(map of first generator)}, \tag{42}$$

$$\triangle - \triangle \mapsto q \qquad \text{(map of second generator)}, \tag{43}$$

$$\triangle + (+\triangle)^m(-\triangle)^n \mapsto p^m q^n \qquad \text{(derived map)}, \tag{44}$$

for all $m, n \geq 0$. The second example of a bicyclic semigroup is $\text{CR}_{\bigcirc}$, defined as all consanguineal relatives with no *male* symbols. The derived map (44) is an isomorphism because it is a bijection and preserves the generating equations, where the unique parenthood rule (6) corresponds to the bicyclic equation (28). The male symbol $\triangle$ is the identity of this semigroup.

The importance of bicyclic semigroups is given by this theorem found in Clifford and Preston [2, pp. 81–82].

Theorem 1 (Web of Bicyclic Semigroups) *If e is any nonzero idempotent of a simple semigroup S that is not completely simple, then S contains a bicyclic subsemigroup having e as identity element.*

3.2 Inverse Semigroups

An *inverse semigroup* [3] is a semigroup S such that each $x \in S$ has a unique *inverse*, $x^{-1} \in S$, such that

$$xx^{-1}x = x \quad \text{and} \tag{45}$$

$$x^{-1}xx^{-1} = x^{-1}. \tag{46}$$

Elements of an inverse semigroup of the form xx^{-1} or $x^{-1}x$ are idempotents since

$$(xx^{-1})(xx^{-1}) = (xx^{-1}x)x^{-1} = xx^{-1}, \tag{47}$$

where the first equation in (47) is due to associativity and the second to the definition of inverses. A similar argument shows that $x^{-1}x$ is also an idempotent.

An inverse semigroup need not be commutative, as can be seen in the five-element example in Table 1. In that table, a and b are inverses of each other, while the other three elements are inverses of themselves since they are idempotents. The most general possible example of an inverse semigroup is the set $\mathcal{I}_X$ of all one-to-one partial functions on a set X, called the *symmetric inverse semigroup* on X. The sense in which the symmetric inverse semigroup is the "most general example" is given in the following theorem [2].

Theorem 2 (Representation Theorem) *Every inverse semigroup is isomorphic to an inverse subsemigroup of the symmetric inverse* $\mathcal{I}_S$.

Proof *(Proof Outline)* The proof is much more difficult than the analogous representation of a semigroup by its right translations. The main idea is to define the one-to-one partial translations: $\rho_a : Sa^{-1} \to Sa; x \mapsto xa$. □

The next theorem, also found in [2], gives three necessary and sufficient conditions for a semigroup to be an inverse semigroup, any one of which could be its definition.

Theorem 3 (Equivalent Conditions for Inverse Semigroups) *The following three conditions on a semigroup are equivalent:*

(i) *S is* regular *(i.e., for all $x \in S$, there is a $y \in S$ such that $xyx = x$), and any two idempotents commute;*
(ii) *S is regular and there exists an involutionary anti-automorphism $x \mapsto x^*$ such that $e = e^*$ for every idempotent in S [2, page 62];*
(iii) *every principal right ideal and every principal left ideal has a unique idempotent generator;*
(iv) *S is an inverse semigroup.*

The finite transformation semigroups $\mathcal{T}_n$ do have generalized inverses, but they are not inverse semigroups for $n > 1$, since each function may have more than one generalized inverse. For example, the function $f = 221$, shown in Fig. 3 depicting its ideals, has four inverses: 311, 313, 322, and 323. Similarly, by the second condition in Theorem 3, $\mathcal{T}_3$ is not an inverse semigroup because the principal left ideal generated by 221 has two idempotent generators, 122 and 121.

The iconic example of an inverse semigroup is the set of all n by n square matrices $\mathrm{Re}^{n\times n}$ with the usual matrix multiplication. Here the generalized inverse of a matrix A is called the *Moore-Penrose inverse* [2] and is usually denoted by $A^\dagger$ to distinguish it from the usual matrix inverse, which only exists for non-singular matrices. The Moore-Penrose inverse can be computed with the singular value decomposition of linear algebra and has many applications in probability and data analysis.

The homomorphic image of an inverse semigroup is also an inverse semigroup [2, page 57], but a subsemigroup of an inverse semigroup may not be an inverse semigroup [2]. An *inverse subsemigroup* of an inverse semigroup S is a subsemigroup T of S such that $x \in S$ implies $x^{-1} \in S$. If $\langle A \rangle$ is the subsemigroup generated by a subset A of an inverse semigroup S, then $\langle A \rangle$ may not contain all inverses. For example, the subsemigroup generated by q^2 in the bicyclic semigroup $C(p, q)$ is $\langle\{ p^2 \}\rangle = \{ p^2, p^4, \ldots\}$, all even powers of p, which does not contain an inverse for any of its elements. However, if we let A^{-1} denote all the inverses of elements of A, then the inverse subsemigroup generated by A is the subsemigroup generated by their union, $\langle A \cup A^{-1} \rangle$. In the bicyclic example above, the inverse subsemigroup generated by q^2 is $\langle\{ q^2, p^2 \}\rangle$, which is the set of all elements of the form $q^{2m} p^{2n}$ for all $m, n \geq 0$.

The two important properties of inverse semigroups in the following theorem [2] show that the mapping from a member of an inverse semigroup to its inverse, $x \mapsto x^{-1}$, is an involutionary anti-automorphism.

Theorem 4 *Any two elements x, y of an inverse semigroup S satisfy:*

$$(x^{-1})^{-1} = x \quad \textit{and} \quad (xy)^{-1} = y^{-1}x^{-1}. \tag{48}$$

Proof Since both x and $(x^{-1})^{-1}$ are inverses of x^{-1}, the unique of inverses shows that they are equal, proving the first equation.

The proof of the second equation uses the fact that for all $x \in S$, both xx^{-1} and $x^{-1}x$ are idempotent and thereby commute with each other. So to show that $y^{-1}x^{-1}$ is the inverse of xy, note, in brutal detail, that

$$\begin{aligned}(xy)(y^{-1}x^{-1})(xy) &= x(yy^{-1})(x^{-1}x)y \\ &= x(x^{-1}x)(yy^{-1})y \\ &= (xx^{-1}x)(yy^{-1}y) \\ &= xy.\end{aligned}$$

A similar argument shows that xy is the inverse of $y^{-1}x^{-1}$. □

3.3 The Fox Inverse Semigroups

In this section we shall show, in a proof omitted from [1], that the Fox consanguineal kinship system forms an inverse semigroup. First, the following lemmas show that the semigroup of consanguineals is regular.

Lemma 1 *If $u = u_1 + \cdots + u_m$, where $m \geq 1$, is an ancestor, then its reciprocal $u^{-1} = u_m - \cdots - u_1$, satisfies $u^{-1}u = u_m$. Similarly, if $v = v_1 - \cdots - v_n$, where $n \geq 1$, is a descendant (Eq. (9)), then the reciprocal satisfies $vv^{-1} = v_1$.*

Proof The proof is by mathematical induction on the number of sex terms. Let u be an ancestor of length m. If $m = 1$, then $u = u_1 = u^{-1}$, and by the idempotent rule, Eq. (7), $u^{-1}u = u_1u_1 = u_1 = u$. If the lemma is true for $m > 1$, then let u be an ancestor of length $m + 1$. Then

$$\begin{aligned}u^{-1}u &= u_{m+1} - \cdots - u_2 - u_1u_1 + u_2 + \cdots + u_{m+1} \\ &= u_{m+1} - \cdots - u_2u_2 + \cdots + u_{m+1} \\ &= u_{m+1},\end{aligned}$$

where the first equation follows by definition; the second, by the unique parenthood principle (Eq. (6)); and the last, by the induction hypothesis. By a similar argument, every descendant term v satisfies $vv^{-1}v = v_1$. □

The next lemma shows that the semigroup of consanguineals **Con** is regular, but stops short of finding that it is an inverse semigroup.

Lemma 2 *The semigroup of consanguineal kin terms* **Con** *is regular.*

Proof First note that the inverse of 0 is 0, resulting in the surprising expression $0^{-1} = 0$. Next, let $x = uv$ be any nonzero consanguineal term, as in Eq. (11), where $u = u_1 + \cdots + u_m$ and $v = v_1 - \cdots - v_n$ for $m, n \geq 1$. An inverse for x is $x^{-1} = v^{-1}u^{-1}$, as in Eq. (12), and where $u_m = v_1$. □

This theorem shows that **Con** is in fact an inverse semigroup.

Theorem 5 *The semigroup of consanguineal terms* **Con** *is an inverse semigroup.*

Proof Lemma 2 shows that **Con** is regular. Equation (13) shows that reciprocity is an involutionary anti-automorphism, defined in Eq. (27). The semigroup **Con** is an inverse semigroup from the second condition from Theorem 3. □

The Fox kinship system is formed by imposing additional equations specifying the merging rules (14) and (15) plus skewing (16). The resulting semigroup is an inverse semigroup since the homomorphic image of an inverse semigroup, in this case the consanguineals **Con**, is again an inverse semigroup.

4 Conclusions

At least two questions must be asked of any scientific model: first, how well does it fit the data and, second, how parsimonious and even beautiful is the model. In the case of categorical data analysis, as described statistically, these two questions have been systematically condensed into two functions relating the data and the model: the goodness of fit, measured by the log-likelihood, and the degree of parsimony, measured by number of free parameters. Furthermore, these two numbers are elegantly combined into the eponymous Akaike information criterion (AIC), which minimizes the expression:

$$\mathrm{AIC} = 2k - 2\ln \hat{L}, \tag{49}$$

where k is the number of free parameters and $\hat{L}$ is the maximized likelihood. This is equivalent to adding $2k$ to the likelihood as a penalty for having too many parameters. That is, the fewer the number of free parameters, the more parsimonious and "beautiful" is the model. The AIC does balance goodness of fit with beauty, and its apparently arbitrary form is based on information theory, which gives the model a theoretical justification and relates it to other fields.

Although statistical modeling for algebraic theories is advancing apace, expanding the results and techniques of categorical data analysis, there are difficulties in applying statistics to the model presented here for the Fox consanguineal kinship system. These difficulties spring from the fact that there is no sample space or sampling method: instead the "data" is a single kinship diagram from a book written by an anthropologist after long interviews with an unknown set of "informants." Similar remarks apply to any grammatical description of a language. So what, if anything, is to be done?

The question of goodness of fit can be naively answered by noting the errors of classification of the semigroup congruence compared to the Fox kinship diagram of Fig. 1. The most glaring example is the older/younger distinction among siblings. This is glossed over in our Fox semigroup model and cannot be corrected by any added equation. However, it would be easy to incorporate this distinction by adding two more generators, say, E and Y, for "elder" and "younger," respectively. Then, a typical example of a consanguineal relative could distinguish between two kinds of a man's mother's brother's daughter:

$$\triangle + \bigcirc E \triangle - \bigcirc \qquad \text{(MoElBrDa)}$$

$$\triangle + \bigcirc Y \triangle - \bigcirc \qquad \text{(MoYoBrDa),}$$

but which would be equated by additional equations in the Fox case. Thus, the "fit" would be better, but the number of generators and equations, playing the role of "free parameters," would be increased. However, there is no clear analogue of the AIC to help decide whether or not the elder/younger distinction is really an improvement to the model. For the purposes of this paper, it would probably make everything even more complicated, so it is better to not introduce it.

Another type of "error" in our analysis is the fact that there are several different congruence classes contained in the equivalence class of Fox relatives who are called "grandfather." Looking from left to right at the four brothers of parents' parents in Fig. 1, note that their sons are labeled F, GF, MBr, and GF, respectively, indicating that there are at least three congruence classes. If we were just trying to fit these four brothers of grandparents, together with the two original grandfathers, then the equation

$$\triangle + a_1 + \triangle = \triangle + a_2 + a_3 \sqcap \triangle, \tag{50}$$

where $a_i \in \{ \triangle, \bigcirc \}$, would put all the grandfathers in Fig. 1 into a single class in the syntactic semigroup produced by the interior operation in Eq. (22). Similar remarks apply to the relatives lumped with grandchildren (Gch).

There are at least two problems with the solution proposed by Eq. (50). First, it does not address the possibility that there might be even more relatives addressed as "GF." The Fox kinship diagram of Fig. 1 is only a five-generational subset of a potentially infinite set of consanguineal relatives. For example, the Fox word for "grandfather" might really mean "a male relative two or more generations above me." This could easily be described by another equation equating higher generations of relatives. This ambiguity could easily be resolved by asking a Fox person, "What do you call your grandfather's father?" The second problem with the proposed solution is that while it may describe the GF class perfectly, it is not at all consistent with the other kinship terms. For example, it would separate the MBr class into generational subclasses.

A more serious problem with the inverse semigroup model presented here is with one of the members of the GF class: a man's mother's father's brother, which we shall show, is congruent to a man's mother's brother. If we multiply mother's brother on the right by father, we get:

$$\begin{aligned} \text{MoBrFa} &= (\triangle + \bigcirc \sqcap \triangle) + \triangle && (51)\\ &= \triangle + \bigcirc + \triangle - \triangle + \triangle && \text{definition of } \sqcap \quad (52)\\ &= \triangle + \bigcirc + \triangle && \text{unique parenthood} \quad (53)\\ &= \text{MoFa}. && (54) \end{aligned}$$

Since MoBrSo is congruent to MoBr (by the skewing rule), then if we also multiply MoBrSo by Fa, we end up with MoBr. This means that MoFa really should be called MoBr instead of GF. This is actually done in other societies that have a related kinship system called Omaha Type II [4]. This problem led the authors [1] to reject the unique parenthood principle. Upon reflection, however, it seems better to retain this important social principle and settle for describing GF as meaning something like "elderly man," independent from the detailed kinship system.

The audacious model presented here attempts to efficiently describe not just one set of strings called by the same name, but all of them at once, satisfying the same axioms of inverse semigroups (with the exception of GF and Gch, discussed previously). The advantage of inverse semigroups over general semigroups is that the inverse elements correspond to the nearly universally observed principle of reciprocity defined in Eq. (13).

How can the model presented here be generalized? The most obvious extension is to include affinal relatives such as brother-in-law, indicated by "B-in-L" in Fig. 1. It appears from the examples in this Fox kinship diagram that "B-in-L" might mean "a man who marries into my patrilineage," whereas mother's sister has been merged with father's wife. This extension to affinals would mean removing the no-affinal rule in Eq. (7) and replacing it with rules consistent with the Fox kinship diagram and a careful reading of the ethnography [9].

Another generalization would be to relate our linguistic model **Con** to an empirical kinship subsemigroup S_X of the semigroup $\mathcal{B}_X$ all relations on a set X of people. Any consideration of the semigroup S_X was disavowed in Sect. 2.1, because it was not obvious how to relate the two semigroups. For example, it is easy to construct a set of kinship relations that do not form an inverse semigroup with respect to the usual composition of relations. Specifically, suppose we have just four people, $X = \{e, f, b, s\}$, corresponding to an "ego," his father, brother, and son, respectively. This implies two binary relations, "son" and its inverse "father," denoted by σ and ϕ, respectively. That is, omitting $\circ$ in favor of juxtaposition,

$$\phi = \{(f, e), (f, b), (e, s)\} \qquad \text{(son relation)}, \tag{55}$$

$$\phi = \{(e, f), (b, f), (s, e)\} \quad (\sigma \text{ inverse}), \tag{56}$$

$$\sigma\phi = \{(f, f), (e, e)\} \quad (\text{idempotent } 1), \tag{57}$$

$$\phi\sigma = \{(e, e), (b, b), (e, b), (b, e), (s, s)\} \quad (\text{idempotent } 2), \tag{58}$$

$$\sigma\phi\ \phi\sigma = \{(e, e), (e, b)\} \quad (1 \text{ times } 2), \tag{59}$$

$$\phi\sigma\ \sigma\phi = \{(e, e), (b, e)\} \quad (2 \text{ times } 1), \tag{60}$$

and since the last two lines are unequal, idempotents do not commute, which shows by Theorem 3 that this is not an inverse semigroup. Since the homomorphism of an inverse semigroup is again an inverse semigroup, the small example of 21 relations in the semigroup S_X generated by σ and ϕ cannot be a homomorphic image of the Fox inverse semigroup.

However, the concept of a homomorphism has several generalizations that might connect the semigroup of relations over a group of people with their Fox kinship system. Two such generalizations are *partial homomorphisms*, which preserve every nonzero product, and *relational morphisms*.

If we go beyond the realm of kinship into more general social relations, then the algebraic adjustments alluded to in the previous paragraph will have to be supplemented by a statistical theory of relations. My enemy's enemy is often a friend, but not always. Is the number of such "errors" more than can be expected by chance? The answer requires a well-defined sample space and probably much computing power. If the answer is yes, an enemy's enemy is statistically a friend, then the question is: "how much does this carry over to further compositions of these relations?" Perhaps other types of relational compositions need to be considered, such as the "demonic" composition, which originated in computer science.

Finally, the focus on inverse semigroups seems justified in the realm of kinship, but it may be too restrictive for other types of social relations. Perhaps other varieties (equationally defined classes) of semigroups, such as orthodox semigroups, which are more general than inverse semigroups but less general than regular semigroups, would be more useful. Also, ordered semigroups, which have an ordering that is compatible with their multiplication, might capture the inclusion ordering of semigroups of relations.

References

1. Boyd, J.P., Haehl, J.H., Sailer, L.D.: Kinship systems and inverse semigroups. J. Math. Sociol. **2**(1), 37–61 (1972)
2. Clifford, A.H., Preston, G.B.: The Algebraic Theory of Semigroups, Volumes I and II. American Mathematical Society, Providence (1961, 1967)
3. Lipscomb, S.: Symmetric Inverse Semigroups. American Mathematical Society, Providence (1996)
4. Lounsbury, F.G.: A formal account of the Crow- and Omaha-type kinship terminologies. In: Goodenough, W.H. (ed.) Explorations in Cultural Anthropology: Essays in Honor of George Peter Murdock, pp. 351–393. McGraw-Hill, New York (1964)

5. Malinowski, B.: Sex and Repression in Savage Society. Meridian Books, New York (1955)
6. Rivers, W.H.R.: The genealogical method of anthropological inquiry. Sociol. Rev. **a3**(1), 1–12 (1910)
7. Romney, A.K.: Kalmuk Mongol and the classification of lineal kinship terminologies. In: Hammel, E.A. (ed.) Formal Semantic Analysis. American Anthropologist, vol. 67(5), Part 2, pp. 127–141. American Anthropological Association, Menasha (1965)
8. Romney, A.K., D'Andrade, R.G.: Cognitive aspects of English kin terms. Am. Anthropol. **66**(3), Part 2: 146–170 (1964)
9. Tax, S.: The social organization of the Fox Indians. In: Eggan, F. (ed.) Social Anthropology of North American Tribes, pp. 243–284. The University of Chicago Press, Chicago (1937)
10. Wilson, R.A.: Kinship past, kinship present: bio-essentialism in the study of kinship. Am. Anthropol. **118**(3), 570–584 (2016)

Part IV
Generalizations in Algebra

Set-theoretical Considerations and Zero Division in Hom-Associative Algebras

Germán García Butenegro and Sergei Silvestrov

Abstract The purpose of this article is to examine zero division relations between elements and ideals of general algebras using set-theoretical tools. By splitting an algebra into subsets according to zero division relations, it is possible to establish certain zero division rules on linear combinations and thus arbitrary elements of a given algebra. We investigate in detail these subsets, properties of zero division relation, subsets of zero divisors, and annihilators in hom-algebras and in particular hom-associative algebras. For hom-associative algebras, we investigate properties of zero division relation subsets of the kernel of the twisting map, annihilators. Furthermore, we investigate kernels of right and left multiplication operators, ideals, hom-ideals, simplicity, hom-simplicity and domain properties for general hom-algebras and hom-associative algebras, and low-dimensional hom-associative algebras of nonassociative type using the kernels of twisting maps, annihilators, and their zero division relation subsets.

Keywords Hom-algebra · Zero divisor · Hom-ideal · Hom-subalgebra

1 Introduction

Zero division relation in algebras is one of the fundamental relations throughout Mathematics and its applications. This article focuses on zero division relations in hom-algebras. The following notations are used unless specified otherwise:

$\mathcal{A}$: hom-associative algebra.
A_{ij} : subsets of $\mathcal{A}$.
$[\cdot, \cdot, \cdot]_\alpha$: hom-associator.
$[\cdot, \cdot]$: commutator.
$\mathcal{L}_0(S)$: set of left 0-divisors of $S \subseteq \mathcal{A}$.
$\mathcal{R}_0(S)$: set of right 0-divisors of $S \subseteq \mathcal{A}$.
$Ann^l_{\mathcal{A}}(S)$: left annihilator of $S \subseteq \mathcal{A}$.
$Ann^r_{\mathcal{A}}(S)$: right annihilator of $S \subseteq \mathcal{A}$.
L_a : left multiplication by a
R_a : right multiplication by a

G. García Butenegro (✉) · S. Silvestrov
Division of Mathematics and Physics, School of Education, Culture and Communication, Mälardalen University, Västerås, Sweden
e-mail: german.garcia@mdu.se; sergei.silvestrov@mdu.se

M. Mitrović, M. N. Hounkonnou (eds.), *Algebra Without Borders*, Mathematics in Mind, https://doi.org/10.1007/978-3-031-86477-3_10

An algebra over a field $\mathbb{F}$ is a linear space $\mathcal{A}$ over $\mathbb{F}$ with a bilinear multiplication operation $\mu : \mathcal{A} \times \mathcal{A} \to \mathcal{A}$, $(x, y) \mapsto xy$ denoted by juxtaposition. The algebra is *associative* if $x(yz) = (xy)z$ for all $x, y, z \in \mathcal{A}$, and it is *commutative* if $xy = yx$ for all $x, y \in \mathcal{A}$. If no confusion is possible, we identify the algebra with the associated linear space $\mathcal{A}$.

Definition 1 A hom-algebra is a triple $(\mathcal{A}, \mu, \alpha)$ where $\mathcal{A}$ is a linear space, $\mu : \mathcal{A} \times \mathcal{A} \to \mathcal{A}$ is a bilinear product operation, denoted by the dot or juxtaposition, and $\alpha : \mathcal{A} \to \mathcal{A}$ is a linear map.

The subject of hom-algebra structures has been started by Hartwig, Larsson, and Silvestrov in 2003 in [6], where hom-Lie algebras and more general quasi-hom-Lie algebras were introduced while developing a general method for construction of quasi-deformations and discretizations of Lie algebras of vector fields with more general twisted derivations in place of derivatives leading to deformations of Witt- and Virasoro-type algebras based on twisted derivations. During 2003–2005, the general abstract quasi-Lie algebras and the subclasses of quasi-hom-Lie algebras and hom-Lie algebras and their more general color (graded) counterparts including color and super quasi-Lie algebras, color and super quasi-hom-Lie algebras, and color and super hom-Lie algebras were introduced by Silvestrov, Larsson, and Sigurdsson in [6–8, 11]. Subsequently, in 2006, various classes of hom-Lie-admissible algebras have been considered by Makhlouf and Silvestrov in [9], where the hom-associative algebras have been introduced and shown to be hom-Lie-admissible, that is, leading to hom-Lie algebras using commutator map as new product, thus constituting a natural generalization of associative algebras since the associative algebras are Lie-admissible as the commutator product yields a Lie algebra for any associative algebra. In [9], moreover, G-hom-associative algebras including along with hom-associative algebras other interesting classes of hom-Lie-admissible algebras such as hom-pre-Lie and hom-left symmetric algebras, generalizing the corresponding classes of nonassociative algebras, were introduced, as well as first examples of low-dimensional hom-Lie algebras have been described. hom-algebra structures are very useful since hom-algebra structures of a given type include their classical counterparts and open more possibilities for deformations, extensions of cohomological structures, and representations.

Definition 2 ([9], Definition 1.1) A hom-associative algebra is a triple $(\mathcal{A}, \mu, \alpha)$ where $\mathcal{A}$ is a linear space, $\mu : \mathcal{A} \times \mathcal{A} \to \mathcal{A}$ is a bilinear product operation denoted by the dot or juxtaposition, and $\alpha : \mathcal{A} \to \mathcal{A}$ is a linear map such that

$$\alpha(x) \cdot (y \cdot z) = (x \cdot y) \cdot \alpha(z)$$

for all $x, y, z \in \mathcal{A}$.

If no confusion arise, we identify this triple with the pair $(\mathcal{A}, \alpha)$.

Example 1 (Yau Twist [12], Corollary 2.6) Given an associative algebra $(\mathcal{A}, \cdot)$ and an endomorphism α, the Yau twist is the hom-associative algebra $(\mathcal{A}, \mu, \alpha)$ with product given by $\mu(x, y) = \alpha(x \cdot y)$.

This twisting is one of the standard procedures for constructing hom-associative algebras. Many known examples are in fact twists of certain associative algebras. This is established in [12]. Moreover, for all classes of G-associative algebras, one can twist into their hom-counterparts.

Example 2 (Hom-assocification) Given an associative algebra $\mathcal{A}$, a non-associative algebra B and a homomorphism $f : B \to \mathcal{A}$, the algebra $(B \oplus \mathcal{A}, \mu, \alpha)$, where

$$\mu((b_1, a_1), (b_2, a_2)) = (b_1 b_2, f(b_1)a_2 + a_1 f(b_2) + a_1 a_2),$$
$$\alpha(b, a) = (0, f(b) + a)$$

is a multiplicative hom-associative algebra, known as *hom-assocification*. This construction was proposed by Lars Hellström in private communication. A variation of this construction was elaborated by Bäck and Richter [3, Propositions 6.1 and 6.2] in the particular case where B is an R-algebra over a commutative unital ring R and $\mathcal{A} = R$, under the name *weak unitalization*.

Hom-algebraic identities are the result of twisting the corresponding algebraic identity with linear map α. Similarly to the associative case, we can study these relations using the corresponding twisted *hom-associator*.

Definition 3 The hom-associator $[\cdot, \cdot, \cdot]^{\alpha}_{as} : \mathcal{A} \times \mathcal{A} \times \mathcal{A} \to \mathcal{A}$ is the trilinear operator defined by $[x, y, z]^{\alpha}_{as} = (xy)\alpha(z) - \alpha(x)(yz)$.

With this definition in mind, hom-associative algebras are those Hom-algebras where the hom-associator is always equal to 0.

Definition 4 A unity in an algebra $(\mathcal{A}, \cdot)$ is an element $1 \in \mathcal{A}$ such that, for all $q \in \mathcal{A}$, $1 \cdot q = q$ (left unity) or $q \cdot 1 = q$ (right unity). If both properties hold, 1 is a two-sided unity. An algebra equipped with a unity is known as unital.

Many hom-associative algebras become associative when equipped with a two-sided unity [5]. It is necessary to introduce a generalized notion of unity that accounts for the action of the twisting map α.

Definition 5 A hom-unity (also called weak unity, or weak identity element) of a hom-algebra $(\mathcal{A}, \alpha)$ is an element $w \in \mathcal{A}$ such that, for all $q \in \mathcal{A}$, $w \cdot q = \alpha(q)$ (left hom-unity) or $q \cdot w = \alpha(q)$ (right hom-unity). If both properties hold, w is a two-sided hom-unity. A hom-algebra equipped with a hom-unity is said to be hom-unital (or weakly unital).

Remark 1 We use the term *hom-unities* (instead of *weak unities* or *weak identity elements*) to emphasize the action of the twisting map and to avoid terminological mix-up with notions of weak ideals and weak subalgebras of a hom-algebra.

As particular as hom-unitality may look at first glance, many examples and constructions of hom-associative algebras are hom-unital.

Example 3 (Hom-unital Hom-associative Algebras) The following constructions provide hom-unital hom-associative algebras:

(i) Let $(\mathcal{A}, \cdot, 1)$ be unital. The Yau twist [12], given by $(\mathcal{A}, \mu, \alpha)$, where α is multiplicative and $\mu(x, y) = \alpha(x \cdot y)$, is hom-unital with hom-unity $\alpha(1)$.
(ii) Consider a left-unital algebra $(\mathcal{A}, \cdot, 1_l)$. The product $\cdot$ is hom-associative with twisting map L_a and hom-unity a if a satisfies:

1. a is a central element of $\mathcal{A}$.
2. $\mathcal{A}a \subseteq N(\mathcal{A})$ is an ideal.

Indeed, for any $x, y, z \in \mathcal{A}$,

$$\begin{aligned} L_a(x)(yz) &= (ax)(yz) \overset{1}{=} (xa)(yz) = (x(1_l a))(yz) \overset{2}{=} x((1_l a)(yz)) \\ &= x(a(yz)) \overset{1}{=} x((yz)a) = x((yz)(1_l a)) \overset{2}{=} x(y(z(1_l a))) \\ &= x(y(za)) \overset{2}{=} (xy)(za) \overset{1}{=} (xy)(az) = (xy)L_a(z) \end{aligned}$$

This construction is inspired by [4, Section 1.3], where stronger two-sided unitality condition is used.

(iii) The hom-assocification in Example 2 of two unital algebras B and $\mathcal{A}$, where u is the unity of B and $f(u)$ is the unity of $\mathcal{A}$, is two-sided unital with unity $(u, 0)$ and two-sided hom-unital with hom-unity $(0, f(u))$.
(iv) The weak unitalization [3] of a multiplicative hom-associative R-algebra B over a commutative associative unital ring R is two-sided hom-unital hom-associative with hom-unity $(0, 1_r)$.

Definition 6 Let $I \subseteq \mathcal{A}$ be a linear subspace. Then I is a:

- *Left ideal* if $a \cdot s \in I$ for all $a \in \mathcal{A}$ and $s \in I$.
- *Right ideal* if $s \cdot a \in I$ for all $a \in \mathcal{A}$ and $s \in I$.
- *Left hom-ideal* if $a \cdot s \in I$ and $\alpha(s) \in I$ for all $a \in \mathcal{A}$ and $s \in I$.
- *Right hom-ideal* if $s \cdot a \in I$ and $\alpha(s) \in I$ for all $a \in \mathcal{A}$ and $s \in I$.

Definition 7 A hom-algebra $(\mathcal{A}, \alpha)$ is (hom-)simple if it does not contain any proper (hom-)ideals, left (hom-)simple if it does not contain any proper left (hom-)ideals, and right (hom-)simple if it does not contain any proper right (hom-)ideals.

In the presence of hom-unities, ideals become hom-ideals: for any left hom-unity w and any left ideal $I \leq \mathcal{A}$, we have $\alpha(I) = wI \subseteq \mathcal{A}I \subseteq I$, and thus I is a hom-ideal. The case with right hom-unities is similar. This is established in [3].

Example 4 (Hom-simple Hom-associative Algebras) The first hom-associative Weyl algebras are simple [3, Proposition 5.14]. They are also hom-unital, and thus hom-simple.

Example 5 (Non-hom-simple Hom-associative Algebras) The following hom-algebras are neither simple nor hom-simple:

(i) The Yau twist of a non-simple associative algebra.
(ii) Any two-sided unital algebra which is also hom-associative with twisting map L_a (see Example 3). Here $\mathcal{A}a$ is an ideal, and $L_a(\mathcal{A}a) \subseteq \mathcal{A}a$, so it is a hom-ideal.
(iii) Hom-associficiations (see Example 2) have $(0, \mathcal{A})$ as a two-sided hom-ideal.

An essential aspect of the structure of algebras and hom-algebras is the presence of nonzero elements that multiply to 0. These elements can be described in terms of linear maps, as the kernels of left and right multiplication operators, explicitly as linear subspaces depending on the elements by which one multiplies, or as sets when a distinction is made between elements which cancel an entire set or *some element of a certain set*. We introduce notations to all three.

Definition 8 Let $S \subseteq \mathcal{A}$. The left annihilator of S is the subspace of elements of $\mathcal{A}$ that cancel every element in S by left multiplication:

$$Ann^l_{\mathcal{A}}(S) = \{a \in \mathcal{A} \mid \forall s \in S,\ a \cdot s = 0\} = \bigcap_{s \in S} \ker(R_s).$$

The right annihilator of S is the subspace of elements of $\mathcal{A}$ that cancel every element in S by right multiplication:

$$Ann^r_{\mathcal{A}}(S) = \{a \in \mathcal{A} \mid \forall s \in S,\ s \cdot a = 0\} = \bigcap_{s \in S} \ker(L_s).$$

Definition 9 The set of left zero divisors of S is the subset of elements of $\mathcal{A}$ that cancel at least one nonzero element in S by left multiplication:

$$\mathcal{L}_0(S) = \{a \in \mathcal{A} \mid \exists\, s \in S \backslash \{0\},\ a \cdot s = 0\}.$$

The set of right zero divisors of S is the subset of elements of $\mathcal{A}$ that cancel at least one nonzero element in S by right multiplication:

$$\mathcal{R}_0(S) = \{a \in \mathcal{A} \mid \exists\, s \in S \backslash \{0\},\ s \cdot a = 0\}.$$

The set of zero divisors is given by the union of these sets $\mathcal{L}_0(S) \cup \mathcal{R}_0(S)$. The set of two-sided zero divisors is given by the intersection $\mathcal{L}_0(S) \cap \mathcal{R}_0(S)$.

Any element in $\mathcal{L}_0(S) \cup \mathcal{R}_0(S)$ is a *zero divisor of* S.

It is particularly interesting to consider zero divisors of the entire algebra.

Definition 10 A zero divisor of $\mathcal{A}$ is *one-sided global* if it cancels all elements in $\mathcal{A}$ on one side, and *two-sided global* if it cancels all elements in $\mathcal{A}$ on both sides. Any one-sided global zero divisor is automatically nilpotent of order 2:

(i) if x is left global, for any $y \in \mathcal{A}$, $x \cdot y = 0 \Rightarrow x^2 = 0$.
(ii) if x is right global, for any $y \in \mathcal{A}$, $y \cdot x = 0 \Rightarrow x^2 = 0$.

The sets of global zero divisors coincide with the annihilators of the algebra:

- The subspace of global left zero divisors is $Ann^l_{\mathcal{A}}(\mathcal{A})$.
- The subspace of global right zero divisors is $Ann^r_{\mathcal{A}}(\mathcal{A})$.
- The subspace of global zero divisors is $Ann^l_{\mathcal{A}}(\mathcal{A}) \cup Ann^r_{\mathcal{A}}(\mathcal{A})$.
- The subspace of global two-sided zero divisors is

$$Ann_{\mathcal{A}}(\mathcal{A}) = Ann^l_{\mathcal{A}}(\mathcal{A}) \cap Ann^r_{\mathcal{A}}(\mathcal{A})$$

 and will be denoted by $\mathcal{A}_{00}$.

For any $S \subseteq \mathcal{A}$, the sets of zero divisors $\mathcal{L}_0(S)$ and $\mathcal{R}_0(S)$ can be expressed as unions of annihilators:

$$\mathcal{L}_0(S) = \bigcup_{s \in S \setminus \{0\}} Ann^l_{\mathcal{A}}(s), \qquad \mathcal{R}_0(S) = \bigcup_{s \in S \setminus \{0\}} Ann^r_{\mathcal{A}}(s)$$

or in terms of left and right multiplication operators:

$$\mathcal{L}_0(S) = \bigcup_{s \in S \setminus \{0\}} \ker(R_s), \qquad \mathcal{R}_0(S) = \bigcup_{s \in S \setminus \{0\}} \ker(L_s)$$

and they contain all elements that cancel *at least* one nonzero element of S. These are unions of linear subspaces, but not linear in general. Elements in $\mathcal{L}_0(S)$ and $\mathcal{R}_0(S)$ are referred to, respectively, as left and right zero divisors of S. Elements not in $\mathcal{L}_0(\mathcal{A})$ or $\mathcal{R}_0(\mathcal{A})$ are referred to as *regular*.

Definition 11 An element $a \in \mathcal{A}$ is:

- *Left-regular* if $a \notin \mathcal{L}_0(\mathcal{A})$, that is, $a \cdot x \neq 0 \; \forall x \in \mathcal{A}, x \neq 0$,
- *Right-regular* if $a \notin \mathcal{R}_0(\mathcal{A})$, that is, $x \cdot a \neq 0 \; \forall x \in \mathcal{A}, x \neq 0$,
- *Two-sided regular* if $a \notin \mathcal{L}_0(\mathcal{A}) \cup \mathcal{R}_0(\mathcal{A})$, that is, $a \cdot x \neq 0$, $x \cdot a \neq 0$, for all $x \in \mathcal{A}, x \neq 0$,
- *Regular* if $a \notin \mathcal{L}_0(\mathcal{A}) \cap \mathcal{R}_0(\mathcal{A})$, that is, it is either left- or right-regular or both.

Definition 12 A hom-associative algebra $(\mathcal{A}, \alpha)$ where all elements are two-sided regular is called *hom-associative domain.*

Example 6 (Hom-associative Domains) The first hom-associative Weyl algebras [2, 3] are hom-associative domains.

Definition 13 For a subset $S \subseteq \mathcal{A}$, define the following subsets:

$$S^l_0 = \{s \in S \mid \forall x \in \mathcal{A}, \; s \cdot x = 0\}, \qquad S^r_0 = \{s \in S \mid \forall x \in \mathcal{A}, \; x \cdot s = 0\},$$

$$S^l_1 = \{s \in S \mid \exists x \in \mathcal{A} : \; s \cdot x \neq 0\}, \qquad S^r_1 = \{s \in S \mid \exists x \in \mathcal{A} : \; x \cdot s \neq 0\}.$$

As a set, $S = S_0^l \sqcup S_1^l = S_0^r \sqcup S_1^r$. In addition, let

$$\begin{aligned}
S_{00} &= \{s \in S \mid \forall x, y \in \mathcal{A},\ s \cdot x = 0,\ y \cdot s = 0\},\\
S_{10} &= \{s \in S \mid \forall x \in \mathcal{A},\ x \cdot s = 0,\ \exists y \in \mathcal{A} :\ s \cdot y \neq 0\},\\
S_{01} &= \{s \in S \mid \forall x \in \mathcal{A},\ s \cdot x = 0,\ \exists y \in \mathcal{A} :\ y \cdot s \neq 0\},\\
S_{11} &= \{s \in S \mid \exists x, y \in \mathcal{A} :\ x \cdot s \neq 0,\ s \cdot y \neq 0\}.
\end{aligned}$$

As a set, $S = S_{00} \sqcup S_{01} \sqcup S_{10} \sqcup S_{11}$. Also $S_0^l = S_{00} \sqcup S_{01}$ and $S_0^r = S_{00} \sqcup S_{10}$.

Proposition 1 *If $S \subseteq \mathcal{A}$ is a linear subspace, then S_{00}, S_0^l and S_0^r are linear subspaces of S.*

2 Zero Division on a Linear Subspace

On this section we look into an arbitrary linear subspace $V \subseteq \mathcal{A}$. We can separate elements in V by zero division according to Definition 13. That is, denote:

$$\begin{aligned}
V_0^l &= \{v \in V \mid \forall x \in \mathcal{A},\ v \cdot x = 0\}, & V_0^r &= \{v \in V \mid \forall x \in \mathcal{A},\ x \cdot v = 0\},\\
V_1^l &= \{v \in V \mid \exists x \in \mathcal{A} :\ v \cdot x \neq 0\}, & V_1^r &= \{v \in V \mid \exists x \in \mathcal{A} :\ x \cdot v \neq 0\},
\end{aligned}$$

so that $V = V_0^l \sqcup V_1^l = V_0^r \sqcup V_1^r$. In addition, denote:

$$\begin{aligned}
V_{00} &= \{v \in V \mid \forall x, y \in \mathcal{A},\ v \cdot x = 0,\ y \cdot v = 0\},\\
V_{10} &= \{v \in V \mid \forall x \in \mathcal{A},\ x \cdot v = 0,\ \exists y \in \mathcal{A} :\ v \cdot y \neq 0\},\\
V_{01} &= \{v \in V \mid \forall x \in \mathcal{A},\ v \cdot x = 0,\ \exists y \in \mathcal{A} :\ y \cdot v \neq 0\},\\
V_{11} &= \{v \in V \mid \exists x, y \in \mathcal{A} :\ x \cdot v \neq 0,\ v \cdot y \neq 0\},
\end{aligned}$$

so that, as a set, $V = V_{00} \sqcup V_{01} \sqcup V_{10} \sqcup V_{11}$. Also $V_0^l = V_{00} \sqcup V_{01}$ and $V_0^r = V_{00} \sqcup V_{10}$. We can express these subsets in terms of annihilators:

$$\begin{aligned}
V_0^l &= V \cap Ann_{\mathcal{A}}^l(\mathcal{A}), & V_1^l &= V \cap (\mathcal{A} \backslash Ann_{\mathcal{A}}^l(\mathcal{A})),\\
V_0^r &= V \cap Ann_{\mathcal{A}}^r(\mathcal{A}), & V_1^r &= V \cap (\mathcal{A} \backslash Ann_{\mathcal{A}}^r(\mathcal{A})),
\end{aligned}$$

as well as the V_{ij}, which can be described as

$$\begin{aligned}
V_{00} &= V \cap Ann_{\mathcal{A}}(\mathcal{A}) = V \cap (Ann_{\mathcal{A}}^l(\mathcal{A}) \cap Ann_{\mathcal{A}}^r(\mathcal{A})),\\
V_{01} &= V \cap (Ann_{\mathcal{A}}^l(\mathcal{A}) \cap (\mathcal{A} \backslash Ann_{\mathcal{A}}^r(\mathcal{A}))),
\end{aligned}$$

$$V_{10} = V \cap ((\mathcal{A} \backslash Ann^l_{\mathcal{A}}(\mathcal{A})) \cap Ann^r_{\mathcal{A}}(\mathcal{A})),$$

$$V_{11} = V \cap ((\mathcal{A} \backslash Ann^l_{\mathcal{A}}(\mathcal{A})) \cap (\mathcal{A} \backslash Ann^r_{\mathcal{A}}(\mathcal{A}))).$$

Remark 2 While this form will not be used explicitly, it highlights the zero division relations inside.

2.1 Coordinate-Based Summation Rules

Across this section we establish zero division rules over sums of elements of an arbitrary linear subspace $V \subseteq \mathcal{A}$ based on zero division relations on the summands. We intend to answer the question: *when is a sum of two elements of V a global zero divisor of $\mathcal{A}$?*

2.1.1 Over an Arbitrary Basis

The sets of zero divisors of a given algebra $\mathcal{A}$ are in general not linear, but they are unions of linear spaces. Across this section we aim to establish zero division rules for sums of elements given whether or not they are global zero divisors on $\mathcal{A}$. In order to do so, we consider projections of the type $P_U : V \to U \subseteq V$ given by assigning to each element in V its corresponding (finite) linear combination of basis elements of U, together with the following general principle.

Lemma 1 *Let V be a linear space, $\mathcal{B}_V$ a basis for V, $U \leq V$ such that there is a sub-basis $\mathcal{B}_U$ of $\mathcal{B}_V$ that spans U, and denote $W = V \backslash U$. Then the set $W \cup \{0\}$ contains a linear subspace generated by all basis elements of $\mathcal{B}_V$ that do not belong to $\mathcal{B}_U$. Call said family of basis elements $\mathcal{B}_W$, then it holds that $\mathcal{B}_V = \mathcal{B}_U \sqcup \mathcal{B}_W$.*

One can immediately expand this split to a finite union of sub-bases. If we split $V \subseteq \mathcal{A}$ according to Definition 13, that is, $V = V^l_0 \sqcup V^l_1 = V_{00} \sqcup V_{01} \sqcup V_{10} \sqcup V_{11} = V^r_0 \sqcup V^r_1$, only V_{00}, $V_{00} \sqcup V_{01} = V^l_0$ and $V_{00} \sqcup V_{10} = V^r_0$ are linear subspaces. We can project onto them, but it is not clear where that projection lands in terms of the V_{ij}.

Remark 3 Subsets V_{ij}, $(i, j) \neq (0, 0)$ are stable under product with scalars but not linear, as the element 0 is in V_{00}. $V_{ij} \cup \{0\}$ is still not linear, but contains a linear subspace according to the lemma above.

Assume that we can choose a basis $\mathcal{B}_{\mathcal{A}}$ of $\mathcal{A}$ verifying the following properties:

(i) There is a sub-basis $\mathcal{B}$ which generates V.
(ii) There is a sub-basis $\mathcal{B}_{0l}$ of $\mathcal{B}$ which generates V^l_0.
(iii) All remaining basis elements of $\mathcal{B}$ are contained in V^l_1. Name this family of elements $\mathcal{B}_{1l}$

(iv) There is a sub-basis $\mathcal{B}_{00}$ of $\mathcal{B}_{0l}$ which generates V_{00}.
(v) All remaining basis elements of $\mathcal{B}_{0l}$ are contained in V_{01}. Name this family $\mathcal{B}_{01}$.
(vi) There is a subfamily $\mathcal{B}_{10}$ of $\mathcal{B}_{1l}$ which is contained in V_{10}.
(vii) All remaining basis elements of $\mathcal{B}_{1l}$ are contained in V_{11}. Name this family $\mathcal{B}_{11}$.

We can use this basis to describe the V_i^r as well. As sets, $V_0^r = V_{00} \sqcup V_{10}$ and $V_1^r = V_{01} \sqcup V_{11}$. Now, V_0^r is a linear subspace, and we have a sub-basis given by $\mathcal{B}_{0r} = \mathcal{B}_{00} \sqcup \mathcal{B}_{10}$. The remaining elements on $\mathcal{B}$ are given by $\mathcal{B}_{1r} = \mathcal{B}_{01} \sqcup \mathcal{B}_{11}$ and are all contained in V_1^r. For each (i, j), define the following subspaces:

(i) $\overset{\star}{V}_{ij}$, the linear span of all elements in $\mathcal{B}_{ij}$,
(ii) $\overset{\star}{V}{}_i^l$, the linear span of all elements in $\mathcal{B}_{il}$,
(iii) $\overset{\star}{V}{}_i^r$ as the linear span of all elements in $\mathcal{B}_{ir}$.

For each (i, j) it holds that $\overset{\star}{V}_{ij} \subseteq (V_{ij} \cup \{0\})$. The intersection of two such $\overset{\star}{V}_{ij}$ is always trivial, as the V_{ij} are pairwise disjoint. It also holds that $\mathcal{B} = \mathcal{B}_{0l} \sqcup \mathcal{B}_{1l} = \mathcal{B}_{00} \sqcup \mathcal{B}_{01} \sqcup \mathcal{B}_{10} \sqcup \mathcal{B}_{11}$, and thus $V = V_0^l \oplus \overset{\star}{V}{}_1^l = \bigoplus_{i,j=0}^{1} \overset{\star}{V}_{ij}$.

From now on, we denote by $P_{ij}(v)$ the projection of element v onto $\overset{\star}{V}_{ij}$ given by assigning v to the corresponding linear combination of elements in $\mathcal{B}_{ij}$. Also denote by $P_{0l}(v)$, $P_{1l}(v)$, $P_{0r}(v)$ and $P_{1r}(v)$ the projections over V_0^l, $\overset{\star}{V}{}_1^l$, V_0^r and $\overset{\star}{V}{}_1^r$. For all $v \in V$, we have $v = P_{00}(v) + P_{01}(v) + P_{10}(v) + P_{11}(v)$. $v = P_{0l}(v) + P_{1l}(v)$ and $v = P_{0r}(v) + P_{1r}(v)$.

In order to establish summation rules, we use these general principles.

Lemma 2 *Let $V \subseteq \mathcal{A}$ be a linear space, $W \subseteq V$ be a linear subspace, $S \subseteq V$ be a subset such that $V = W \sqcup S$. Also, let $w \in W$, $s \in S$, and let $\overset{\star}{S}$ be the linear span of a family of elements $\mathcal{B}_S$ such that $\mathcal{B}_V = \mathcal{B}_W \sqcup \mathcal{B}_S$ is a basis for V and $\overset{\star}{S}\backslash\{0\} \subseteq S$. Then the following properties hold:*

(i) $w = P_W(w)$ *and* $s = P_W(s) + P_S(s)$, *with* $P_S(s) \neq 0$,
(ii) $w + s \notin W$,
(iii) *If* $P_W(w) = -P_W(s)$, *then* $w + s \in \overset{\star}{S}$,
(iv) *Otherwise,* $w + s \in S\backslash\overset{\star}{S}$.

Let now $w \in W$, $s \in \overset{\star}{S}$. Then, the following properties hold:

(v) $w = P_W(w)$ *and* $s = P_S(s)$,
(vi) *If* $s = 0$, *then* $w + s \in W$.
(vii) *If* $w = 0$, *then* $w + s \in \overset{\star}{S}$.
(viii) *If* $w \neq 0, s \neq 0$, *then* $w + s \in S\backslash\overset{\star}{S}$.

Observe that $w + s \notin W$ is immediate for any $w \in W$ and $s \in S$. Indeed, $w + s = P_W(w) + P_W(s) + P_S(s) \notin W$, as $P_S(s) \neq 0$.

Consider the projections of each element. For all $v \in V$, we have:

$$v \in V_0^l \Leftrightarrow v = P_{0l}(v),$$
$$v \in V_1^l \Leftrightarrow v = P_{0l}(v) + P_{1l}(v),\ \ P_{1l}(v) \neq 0,$$
$$v \in \overset{\star}{V_1^l} \Leftrightarrow v = P_{1l}(v),$$
$$v \in V_0^r \Leftrightarrow v = P_{0r}(v),$$
$$v \in V_1^r \Leftrightarrow v = P_{0r}(v) + P_{1r}(v),\ \ P_{1r}(v) \neq 0,$$
$$v \in \overset{\star}{V_1^r} \Leftrightarrow v = P_{1r}(v).$$

Using these properties, we can identify sums of elements in V with their corresponding projections and obtain explicit summation rules (Tables 1, 2, and 3).

Theorem 1 *Let $V = V_0^l \oplus \overset{\star}{V_1^l}$. Elements in V obey the summation rules given by Table 4 in (the remaining ones are covered by commutativity; conditions marked with # are impossible).*

Any element in V can be expressed uniquely as $v + a$, where $v \in V_0^l$ and $a \in \overset{\star}{V_1^l}$. If $a \neq 0$, one has $v + a \in V_1^l$ and $v - a \in V_1^l$, but $(v + a) + (v - a) = 2v \in V_0^l$. Thus, it is sufficient to study the case where $v_1, v_2 \in V_0^l$, $a_1, a_2 \in \overset{\star}{V_1^l}$: indeed, if $a \in V_1^l$, then $a \in V$, and we can express it uniquely as $w + b$, with $w \in V_0^l$, $b \in \overset{\star}{V_1^l}$. Then, $v + a = (v + w) + b$, with $v + w \in V_0^l$, $b \in \overset{\star}{V_1^l}$. This idea can be interpreted algebraically: any element in V can be expressed as $v + a$, where v is a global left zero divisor and a is linear combination of elements which are not global zero divisors. It can also be interpreted geometrically: V can be seen as a family of affine

Table 1 Subsets of $\ker\alpha$ according to zero division

	K_{00}	K_{01}	K_{10}	K_{11}	$\ker\alpha$
A_1^3	$\langle e_3 - e_2\rangle$	$\emptyset$	$\emptyset$	$\langle e_2, e_3\rangle \setminus \langle e_3 - e_2\rangle$	$\langle e_2, e_3\rangle$
A_2^3	$\{0\}$	$\emptyset$	$\emptyset$	$\langle e_3\rangle \setminus \{0\}$	$\langle e_3\rangle$
A_4^3	$\{0\}$	$\emptyset$	$\emptyset$	$\langle e_1, e_3\rangle \setminus \{0\}$	$\langle e_1, e_3\rangle$
A_5^3	$\{0\}$	$\emptyset$	$\emptyset$	$\langle e_3\rangle \setminus \{0\}$	$\langle e_3\rangle$
A_6^3	$\{0\}$	$\emptyset$	$\langle e_1\rangle \setminus \{0\}$	$\emptyset$	$\langle e_1\rangle$
A_{11}^3	$\{0\}$	$\emptyset$	$\langle e_1\rangle \setminus \{0\}$	$\emptyset$	$\langle e_1\rangle$

Table 2 Subsets of $\ker\alpha \cdot \mathcal{A}$ and $\mathcal{A} \cdot \ker\alpha$

	K_1^l	$K_1^l \cdot \mathcal{A}$	K_1^r	$\mathcal{A} \cdot K_1^r$
A_1^3	$\langle e_2, e_3\rangle \backslash \langle e_3 - e_2\rangle$	$\langle e_1, e_3 - e_2\rangle$	$\langle e_2, e_3\rangle \backslash \langle e_3 - e_2\rangle$	$\langle e_2 + e_3\rangle$
A_2^3	$\langle e_3\rangle \backslash \{0\}$	$\langle e_3\rangle$	$\langle e_3\rangle \backslash \{0\}$	$\langle e_3\rangle$
A_4^3	$\langle e_1, e_3\rangle \backslash \{0\}$	$\langle e_1\rangle$	$\langle e_1, e_3\rangle \backslash \{0\}$	$\langle e_1\rangle$
A_5^3	$\langle e_3\rangle \backslash \{0\}$	$\langle e_3\rangle$	$\langle e_3\rangle \backslash \{0\}$	$\langle e_3\rangle$
A_6^3	$\langle e_1\rangle \backslash \{0\}$	$\langle e_1\rangle$	$\emptyset$	$\emptyset$
A_{11}^3	$\langle e_1\rangle \backslash \{0\}$	$\langle e_1\rangle$	$\emptyset$	$\emptyset$

Table 3 Annihilators of $\ker\alpha \cdot \mathcal{A}$ and $\mathcal{A} \cdot \ker\alpha$

	$Ann^r_{\mathcal{A}}(K_1^l \cdot \mathcal{A})$	$Ann^l_{\mathcal{A}}(\mathcal{A} \cdot K_1^r)$	$Ann^r_{\mathcal{A}}(K_1^l \cdot \mathcal{A}) \cap Ann^l_{\mathcal{A}}(\mathcal{A} \cdot K_1^r)$	$\alpha(\mathcal{A})$
A_1^3	$\langle e_1, e_3 - e_2\rangle$	$\langle e_1, e_3 - e_2\rangle$	$\langle e_1, e_3 - e_2\rangle$	$\langle e_1\rangle$
A_2^3	$\langle e_1, e_2\rangle$	$\langle e_1, e_2\rangle$	$\langle e_1, e_2\rangle$	$\langle e_1, e_2\rangle$
A_4^3	$\langle e_1, p_1e_3 - p_2e_2\rangle$	$\langle e_1, e_2\rangle$	$\langle e_1\rangle$	$\langle e_1\rangle$
A_5^3	$\langle e_1, e_2\rangle$	$\langle e_1, e_2\rangle$	$\langle e_1, e_2\rangle$	$\langle e_1, e_2\rangle$
A_6^3	$\langle e_1, e_3\rangle$	$\mathcal{A}$	$\langle e_1, e_3\rangle$	$\langle e_1, e_3\rangle$
A_{11}^3	$\langle e_1, e_2\rangle$	$\mathcal{A}$	$\langle e_1, e_2\rangle$	$\langle e_1, e_2\rangle$

Table 4 Summation rules for Theorem 1

v_1 in	v_2 in	$v_1 + v_2$ in	If and only if
V_0^l	$\overset{\star}{V}{}_1^l$	V_0^l	$v_2 = 0$
		$\overset{\star}{V}{}_1^l$	$v_1 = 0$
		V_1^l	$v_2 \neq 0$
V_0^l	V_1^l	V_0^l	$\#(\Leftarrow v_2 \in V_0^l \cap V_1^l = \emptyset)$
		$\overset{\star}{V}{}_1^l$	$P_{0l}(v_1) = -P_{0l}(v_2)$
		$V_1^l \backslash \overset{\star}{V}{}_1^l$	otherwise
V_1^l	V_1^l	V_0^l	$P_{1l}(v_1) = -P_{1l}(v_2)$
		$\overset{\star}{V}{}_1^l$	$P_{0l}(v_1) = -P_{0l}(v_2)$
		$V_1^l \backslash \overset{\star}{V}{}_1^l$	otherwise
V_1^l	$\overset{\star}{V}{}_1^l$	V_0^l	$P_{1l}(v_1) = -v_2$
		$\overset{\star}{V}{}_1^l$	$P_{0l}(v_1) = 0$
		$V_1^l \backslash \overset{\star}{V}{}_1^l$	otherwise

spaces $\{a + V_0^l \mid a \in \overset{\star}{V_1^l}\}$. If one sums two elements $v_1 + a_1, v_2 + a_2$ in such affine spaces, the only case where we land in the original V_0^l is when $a_1 + a_2$ is in V_0^l, that is, $a_1 + a_2 = 0$ since $V_0^l \cap \overset{\star}{V_1^l} = \{0\}$. Similarly, the only sum which lands in $\overset{\star}{V_1^l}$ is that where $v_1 + v_2 = 0$.

Corollary 1 *Let $v_1, v_2 \in V_0^l$, $a_1, a_2 \in V_1^l$. Then*

$$v_1 + v_2 + a_1 + a_2 \in V_0^l \Leftrightarrow a_1 + a_2 \in V_0^l,$$

$$v_1 + v_2 + a_1 + a_2 \in \overset{\star}{V_1^l} \Leftrightarrow v_1 + v_2 + P_{1l}(a_1 + a_2) = 0.$$

Otherwise, $v_1 + v_2 + a_1 + a_2 \in V_1^l \backslash \overset{\star}{V_1^l}$.

Remark 4 Although it suffices to study the case where $a_1, a_2 \in \overset{\star}{V_1^l}$, decomposing a_1 and a_2 may not be straightforward – it is, thus, interesting to consider these sums as close to the division in sets as possible.

Remark 5 These results apply to the split $V = V_0^r \sqcup V_1^r$ naturally, substituting the corresponding projections.

Consider now $V = V_{00} \oplus \overset{\star}{V_{01}} \oplus \overset{\star}{V_{10}} \oplus \overset{\star}{V_{11}}$. Elements in the subsets V_{ij} can be described using projections in this way:

$$v \in V_{00} \Leftrightarrow v = P_{00}(v),$$

$$v \in V_{01} \Leftrightarrow v = P_{00}(v) + P_{01}(v),\ \ P_{01}(v) \neq 0,$$

$$v \in \overset{\star}{V_{01}} \Leftrightarrow v = P_{01}(v),$$

$$v \in V_{10} \Leftrightarrow v = P_{00}(v) + P_{10}(v),\ \ P_{10}(v) \neq 0,$$

$$v \in \overset{\star}{V_{10}} \Leftrightarrow v = P_{10}(v),$$

$$v \in V_{11} \Leftrightarrow v = P_{00}(v) + P_{01}(v) + P_{10}(v) + P_{11}(v),\ \ P_{11}(v) \neq 0,$$

$$v \in \overset{\star}{V_{11}} \Leftrightarrow v = P_{11}(v).$$

Theorem 2 *Let $V = V_{00} \oplus \overset{\star}{V_{01}} \oplus \overset{\star}{V_{10}} \oplus \overset{\star}{V_{11}}$, and denote $v_k = P_{00}(v_k) + P_{01}(v_k) + P_{10}(v_k) + P_{11}(v_k)$. Elements in V obey the summation rules given by Table 5 in (the remaining ones are covered by commutativity; rules marked with # are contradictory).*

Table 5 Summation rules for Theorem 2

v_1 in	v_2 in	$v_1 + v_2$ in	If and only if
V_{00}	$\overset{\star}{V}_{01}$	V_{00}	$v_2 = 0$
		$\overset{\star}{V}_{01}$	$v_1 = 0$
		$V_{01} \backslash \overset{\star}{V}_{01}$	otherwise
V_{00}	V_{01}	V_{00}	$\#(\Leftarrow v_2 \in V_{00} \cap V_{01} = \emptyset)$
		$\overset{\star}{V}_{01}$	$P_{00}(v_1) = -P_{00}(v_2)$
		$V_{01} \backslash \overset{\star}{V}_{01}$	otherwise
V_{00}	$\overset{\star}{V}_{10}$	V_{00}	$v_2 = 0$
		$\overset{\star}{V}_{10}$	$v_1 = 0$
		$V_{10} \backslash \overset{\star}{V}_{10}$	otherwise
V_{00}	V_{10}	V_{00}	$\#(\Leftarrow v_2 \in V_{00} \cap V_{10} = \emptyset)$
		$\overset{\star}{V}_{10}$	$P_{00}(v_1) = -P_{00}(v_2)$
		$V_{10} \backslash \overset{\star}{V}_{10}$	otherwise
V_{00}	$\overset{\star}{V}_{11}$	V_{00}	$v_2 = 0$
		$\overset{\star}{V}_{11}$	$v_1 = 0$
		$V_{11} \backslash \overset{\star}{V}_{11}$	otherwise
V_{00}	V_{11}	V_{00}	$v_2 = 0$
		V_{01}	$\#(\Leftarrow v_2 \in V_{10} \cap V_{11} = \emptyset)$
		V_{10}	$\#(\Leftarrow v_2 \in V_{01} \cap V_{11} = \emptyset)$
		$\overset{\star}{V}_{11}$	$P_{00}(v_1 + v_2) = P_{01}(v_2) = P_{10}(v_2) = 0$
		V_{11}	otherwise
V_{01}	V_{01}	V_{00}	$P_{01}(v_1) = -P_{01}(v_2)$
		$\overset{\star}{V}_{01}$	$P_{00}(v_1) = -P_{00}(v_2)$
		$V_{01} \backslash \overset{\star}{V}_{01}$	otherwise
V_{01}	$\overset{\star}{V}_{01}$	V_{00}	$P_{01}(v_1) = -P_{01}(v_2)$
		$\overset{\star}{V}_{01}$	$P_{00}(v_1) = 0$
		$V_{01} \backslash \overset{\star}{V}_{01}$	otherwise
V_{01}	V_{10}	V_{00}	$\#(\Leftarrow v_2 \in V_{00} \cap V_{10} = \emptyset)$
		V_{01}	$\#(\Leftarrow v_2 \in V_{00} \cap V_{10} = \emptyset)$
		V_{10}	$\#(\Leftarrow v_1 \in V_{00} \cap V_{01} = \emptyset)$
		V_{11}	otherwise
V_{01}	$\overset{\star}{V}_{10}$	V_{00}	$\#(\Leftarrow v_1, v_2 \in V_{00})$
		V_{01}	$P_{10}(v_2) = 0$
		$\overset{\star}{V}_{01}$	$P_{00}(v_1) = P_{10}(v_2) = 0$
		V_{10}	$\#(\Leftarrow v_1 \in V_{01} \cap V_{00} = \emptyset)$
		V_{11}	otherwise

(continued)

Table 5 (continued)

v_1 in	v_2 in	$v_1 + v_2$ in	If and only if
V_{01}	$\overset{\star}{V}_{11}$	V_{00}	$\#(\Leftarrow v_1 \in V_{00}, v_2 = 0)$
		V_{01}	$v_2 = 0$
		$\overset{\star}{V}_{11}$	$\#(\Leftarrow v_1 \in V_{01} \cap \overset{\star}{V}_{11} = \emptyset)$
		$V_{11} \setminus \overset{\star}{V}_{11}$	otherwise
V_{01}	V_{11}	V_{00}	$\#(\Leftarrow v_2 \in V_{01} \cap V_{11} = \emptyset)$
		V_{01}	$\#(\Leftarrow v_2 \in V_{01} \cap V_{11} = \emptyset)$
		V_{10}	$\#(\Leftarrow v_2 \in V_{10} \cap V_{11} = \emptyset)$
		$\overset{\star}{V}_{11}$	$P_{00}(v_1 + v_2) = P_{01}(v_2) = P_{10}(v_2) = 0$
		V_{11}	otherwise
V_{10}	V_{10}	V_{00}	$P_{10}(v_1) = -P_{10}(v_2)$
		$\overset{\star}{V}_{10}$	$P_{00}(v_1) = -P_{00}(v_2)$
		$V_{10} \setminus \overset{\star}{V}_{10}$	otherwise
V_{10}	$\overset{\star}{V}_{10}$	V_{00}	$P_{10}(v_1) = -P_{10}(v_2)$
		$\overset{\star}{V}_{10}$	$P_{00}(v_1) = 0$
		$V_{10} \setminus \overset{\star}{V}_{10}$	otherwise
V_{10}	$\overset{\star}{V}_{01}$	V_{00}	$\#(\Leftarrow v_1 \in V_{10} \cap V_{00} = \emptyset)$
		V_{10}	$P_{01}(v_2) = 0$
		$\overset{\star}{V}_{10}$	$P_{00}(v_1) = P_{01}(v_2) = 0$
		V_{01}	$\#(\Leftarrow v_1 \in V_{10} \cap V_{00} = \emptyset)$
		V_{11}	otherwise
V_{10}	$\overset{\star}{V}_{11}$	V_{00}	$\#(\Leftarrow v_1 \in V_{10} \cap V_{00} = \emptyset)$
		V_{10}	$v_2 = 0$
		$\overset{\star}{V}_{11}$	$\#(\Leftarrow v_1 \in V_{10} \cap V_{00} = \emptyset)$
		$V_{11} \setminus \overset{\star}{V}_{11}$	otherwise
V_{10}	V_{11}	V_{00}	$\#(\Leftarrow v_2 \in V_{11} \cap V_{10} = \emptyset)$
		V_{10}	$\#(\Leftarrow v_2 \in V_{11} \cap V_{10} = \emptyset)$
		V_{01}	$\#(\Leftarrow v_2 \in V_{11} \cap V_{01} = \emptyset)$
		$\overset{\star}{V}_{11}$	$P_{00}(v_1 + v_2) = P_{01}(v_2) = P_{10}(v_2) = 0$
		V_{11}	otherwise
V_{11}	$\overset{\star}{V}_{11}$	V_{00}	$P_{01}(v_1) = P_{10}(v_1) = P_{11}(v_1 + v_2) = 0$
		V_{01}	$P_{10}(v_2) = P_{11}(v_1 + v_2) = 0,\ P_{01}(v_1) \neq 0$
		$\overset{\star}{V}_{01}$	$P_{00}(v_1) = P_{10}(v_1) = P_{11}(v_1 + v_2) = 0$
		V_{10}	$P_{01}(v_2) = P_{11}(v_1 + v_2) = 0,\ P_{10}(v_1) \neq 0$
		$\overset{\star}{V}_{10}$	$P_{00}(v_1) = P_{01}(v_1) = P_{11}(v_1 + v_2) = 0$
		$\overset{\star}{V}_{11}$	$P_{00}(v_1) = P_{01}(v_1) = P_{10}(v_1) = 0$
		V_{11}	$P_{11}(v_1 + v_2) \neq 0$

(continued)

Table 5 (continued)

v_1 in	v_2 in	$v_1 + v_2$ in	If and only if
V_{11}	V_{11}	V_{00}	$P_{01}(v_1+v_2) = P_{10}(v_1+v_2) = P_{11}(v_1+v_2) = 0$
		V_{01}	$P_{10}(v_1+v_2) = P_{11}(v_1+v_2) = 0,\ P_{01}(v_1+v_2) \neq 0$
		$\overset{\star}{V}_{01}$	$P_{00}(v_1+v_2) = P_{10}(v_1+v_2) = P_{11}(v_1+v_2) = 0$
		V_{10}	$P_{01}(v_1+v_2) = P_{11}(v_1+v_2) = 0,\ P_{10}(v_1+v_2) \neq 0$
		$\overset{\star}{V}_{10}$	$P_{00}(v_1+v_2) = P_{01}(v_1+v_2) = P_{11}(v_1+v_2) = 0$
		$\overset{\star}{V}_{11}$	$P_{00}(v_1+v_2) = P_{01}(v_1+v_2) = P_{10}(v_1+v_2) = 0$
		$V_{11}\backslash\overset{\star}{V}_{11}$	$P_{11}(v_1+v_2) \neq 0$

2.1.2 Over a Finite Basis

Across this section, letters $\lambda, \mu, \eta, \gamma$ represent coefficients from the field $\mathbb{F}$. Consider V as a finitely generated linear space with a certain basis $\mathcal{B}$ indexed by $\mathcal{I}$. Any element $v \in V$ can be expressed as $k = \sum_{i\in\mathcal{I}} \lambda_i e_i$.

According to previous definition, we choose a basis $\mathcal{B}$ for V in such a way that V_0^l is generated by a sub-basis $\mathcal{B}_{0l}$ with indices in $\mathcal{I}_{0l}$ and the remaining elements in $\mathcal{B}$ are contained in V_1^l and indexed by $\mathcal{I}_{1l} = \mathcal{I}\backslash\mathcal{I}_{0l}$. We have $V = V_0^l \oplus \overset{\star}{V}{}_1^l$, and thus any element $v \in V$ can be expressed uniquely as $k = \sum_{i\in\mathcal{I}_0} \lambda_i e_i + \sum_{j\in\mathcal{I}_1} \mu_j e_j$. From this it is immediate that $v \in V_0^l \Leftrightarrow \mu_j = 0\,\forall j \in \mathcal{I}_1$, $v \in \overset{\star}{V}{}_1^l \Leftrightarrow \lambda_j = 0\,\forall j \in \mathcal{I}_1$ and $v \in V_1^l\backslash\overset{\star}{V}{}_1^l$ otherwise. We express these and other summation relations between elements in V in terms of the coefficients λ_i and μ_j.

Theorem 3 *Let $V = V_0^l \oplus \overset{\star}{V}{}_1^l$, and denote $v_k = \sum_{i\in\mathcal{I}_0} \lambda_{ki} e_i + \sum_{j\in\mathcal{I}_1} \mu_{kj} e_j$. Elements in V obey the summation rules given by Table 6 in (the remaining ones are covered by commutativity; rules marked with # are contradictory).*

Similarly, by construction we choose a basis $\mathcal{B}$ such that $\mathcal{B} = \mathcal{B}_{00} \sqcup \mathcal{B}_{01} \sqcup \mathcal{B}_{10} \sqcup \mathcal{B}_{11}$, and let $\overset{\star}{V}_{ij}$ be the linear space generated by the elements in $\mathcal{B}_{ij}$. Here $V = V_{10} \oplus \overset{\star}{V}_{01} \oplus \overset{\star}{V}_{10} \oplus \overset{\star}{V}_{11}$ as a linear space and $V = V_{00} \sqcup V_{01} \sqcup V_{10} \sqcup V_{11}$ as a set. The sub-basis $\mathcal{B}_{ij}$ has elements indexed by $\mathcal{I}_{ij}$, and thus we can express any element in V as $v_k = \sum_{i\in\mathcal{I}_{00}} \lambda_{ki} e_i + \sum_{j\in\mathcal{I}_{01}} \mu_{kj} e_j + \sum_{l\in\mathcal{I}_0} \eta_{kl} e_l + \sum_{t\in\mathcal{I}_{11}} \gamma_{kt} e_t$.

Theorem 4 *Let $V = V_{00} \oplus \overset{\star}{V}_{01} \oplus \overset{\star}{V}_{10} \oplus \overset{\star}{V}_{11}$, and denote $v_k = \sum_{i\in\mathcal{I}_{00}} \lambda_{ki} e_i + \sum_{j\in\mathcal{I}_{01}} \mu_{kj} e_j + \sum_{l\in\mathcal{I}_{10}} \eta_{kl} e_l + \sum_{t\in\mathcal{I}_{11}} \gamma_{kt} e_t$. Elements in V obey the summation rules given*

Table 6 Summation rules for Theorem 3

v_1 in	v_2 in	$v_1 + v_2$ in	If and only if
V_0^l	$\overset{\star}{V}{}_1^l$	V_0^l	$\mu_{2j} = 0\ \forall j$
		$\overset{\star}{V}{}_1^l$	$\lambda_{1i} = 0\ \forall i$
		$V_1^l \backslash \overset{\star}{V}{}_1^l$	otherwise
V_0^l	V_1^l	V_0^l	#($\Leftarrow \lambda_{2i} = \mu_{2j} = 0\ \forall i, j$)
		$\overset{\star}{V}{}_1^l$	$\lambda_{1i} = -\lambda_{2i}\ \forall i$
		$V_1^l \backslash \overset{\star}{V}{}_1^l$	otherwise
V_1^l	V_1^l	V_0^l	$\mu_{1j} = -\mu_{2j}\ \forall j$
		$\overset{\star}{V}{}_1^l$	$\lambda_{1i} = -\lambda_{2i}\ \forall i$
		$V_1^l \backslash \overset{\star}{V}{}_1^l$	otherwise
V_1^l	$\overset{\star}{V}{}_1^l$	V_0^l	$\mu_{1j} = -\mu_{2j}\ \forall j$
		$\overset{\star}{V}{}_1^l$	$\lambda_{1i} = 0\ \forall i$
		$V_1^l \backslash \overset{\star}{V}{}_1^l$	otherwise

by Table 7 (the remaining ones are covered by commutativity. Rules marked with # are contradictory).

The particular case $v + a, v - a$ studied before can be considered in the case of a finite basis. Let $v_1 \in V_{00}$, that is, $v_1 = \sum_{i \in \mathcal{I}_{00}} \eta_{1i} v_i$, and $a_1 \in V_{01}$, that is, $a_1 = \sum_{i \in \mathcal{I}_{00}} \lambda_{1i} v_i + \sum_{j \in \mathcal{I}_{01}} \mu_{1j} v_j$, and a similar construction for v_2 and a_2. Here,

$$v_1 + v_2 = \sum_{i \in \mathcal{I}_{00}} (\eta_{1i} + \eta_{2i}) v_i,$$

$$a_1 + a_2 = \sum_{i \in \mathcal{I}_{00}} (\lambda_{1i} + \lambda_{2i}) v_i + \sum_{j \in \mathcal{I}_{01}} (\mu_{1j} + \mu_{2j}) v_j,$$

$$v_1 + v_2 + a_1 + a_2 = \sum_{i \in \mathcal{I}_{00}} (\eta_{1i} + \eta_{2i} + \lambda_{1i} + \lambda_{2i}) v_i + \sum_{j \in \mathcal{I}_{01}} (\mu_{1j} + \mu_{2j}) v_j.$$

From this decomposition, it is immediate that:

1. $v_1 + v_2 + a_1 + a_2 \in V_{00}$ if and only if $\mu_{1j} + \mu_{2j} = 0\ \forall j \in \mathcal{I}_{01}$,
2. $v_1 + v_2 + a_1 + a_2 \in \overset{\star}{V}_{01}$ if and only if $\eta_{1i} + \eta_{2i} + \lambda_{1i} + \lambda_{2i} = 0\ \forall i \in \mathcal{I}_{00}$,
3. $v_1 + v_2 + a_1 + a_2 \in V_{01} \backslash \overset{\star}{V}_{01}$ otherwise.

Table 7 Summation rules for Theorem 4

v_1 in	v_2 in	$v_1 + v_2$ in	If and only if
V_{00}	$\overset{\star}{V}_{01}$	V_{00}	$v_2 = 0$
		$\overset{\star}{V}_{01}$	$v_1 = 0$
		$V_{01} \backslash \overset{\star}{V}_{01}$	otherwise
V_{00}	V_{01}	V_{00}	$\#(\Leftarrow v_2 = 0)$
		$\overset{\star}{V}_{01}$	$\lambda_{1i} = -\lambda_{2i}\ \forall i$
		$V_{01} \backslash \overset{\star}{V}_{01}$	otherwise
V_{00}	$\overset{\star}{V}_{10}$	V_{00}	$v_2 = 0$
		$\overset{\star}{V}_{10}$	$v_1 = 0$
		$V_{10} \backslash \overset{\star}{V}_{10}$	otherwise
V_{00}	V_{10}	V_{00}	$\#(\Leftarrow v_2 = 0)$
		$\overset{\star}{V}_{10}$	$\lambda_{1i} = -\lambda_{2i}\ \forall i$
		$V_{10} \backslash \overset{\star}{V}_{10}$	otherwise
V_{00}	$\overset{\star}{V}_{11}$	V_{00}	$v_2 = 0$
		$\overset{\star}{V}_{11}$	$v_1 = 0$
		$V_{11} \backslash \overset{\star}{V}_{11}$	otherwise
V_{00}	V_{11}	V_{00}	$v_2 = 0$
		V_{01}	$\#(\Leftarrow \eta_{2l} e_l = \gamma_{2t} = 0\ \forall l, t)$
		V_{10}	$\#(\Leftarrow \mu_{2j} = \gamma_{2t} = 0\ \forall j, t)$
		$\overset{\star}{V}_{11}$	$\lambda_{1i} + \lambda_{2i} = \mu_{2j} = \eta_{2l} = 0\ \forall i, j, l$
		V_{11}	otherwise
V_{01}	V_{01}	V_{00}	$\mu_{1j} = -\mu_{2j}\ \forall j$
		$\overset{\star}{V}_{01}$	$\lambda_{1i} = -\lambda_{2i}\ \forall i$
		$V_{01} \backslash \overset{\star}{V}_{01}$	otherwise
V_{01}	$\overset{\star}{V}_{01}$	V_{00}	$\mu_{1j} = -\mu_{2j}\ \forall j$
		$\overset{\star}{V}_{01}$	$\lambda_{1i} = 0\ \forall i$
		$V_{01} \backslash \overset{\star}{V}_{01}$	otherwise
V_{01}	V_{10}	V_{00}	$\#(\Leftarrow \mu_{1j} = \eta_{2l} = 0\ \forall j, l)$
		V_{01}	$\#(\Leftarrow \eta_{2l} = 0\ \forall l)$
		V_{10}	$\#(\Leftarrow \eta_{1l} = 0\ \forall l)$
		V_{11}	otherwise
V_{01}	$\overset{\star}{V}_{10}$	V_{00}	$\#(\Leftarrow \mu_{1j} = \eta_{2l} = 0\ \forall j, l)$
		V_{01}	$\eta_{2l} = 0\ \forall l$
		$\overset{\star}{V}_{01}$	$\lambda_{1i} = \eta_{2l} = 0\ \forall i, l$
		V_{10}	$\#(\Leftarrow \mu_{1j} = 0\ \forall j)$
		V_{11}	otherwise

(continued)

Table 7 (continued)

v_1 in	v_2 in	$v_1 + v_2$ in	If and only if
V_{01}	$\overset{\star}{V}_{11}$	V_{00}	$\#(\Leftarrow \mu_{1j} = \gamma_{2t} = 0\ \forall j, t)$
		V_{01}	$v_2 = 0$
		$\overset{\star}{V}_{11}$	$\#(\Leftarrow v_1 = 0)$
		$V_{11} \backslash \overset{\star}{V}_{11}$	otherwise
V_{01}	V_{11}	V_{00}	$\#(\Leftarrow \mu_{1j} + \mu_{2j} = \eta_{2l} = \gamma_{2t} = 0\ \forall j, l, t)$
		V_{01}	$\#(\Leftarrow \eta_{2l} = \gamma_{2t} = 0\ \forall l, t)$
		V_{10}	$\#(\Leftarrow \mu_{2j} = \gamma_{2t} = 0\ \forall j, t)$
		$\overset{\star}{V}_{11}$	$\lambda_{1i} + \lambda_{2i} = \mu_{2j} = \eta_{2l} = 0\ \forall i, j, l$
		V_{11}	otherwise
V_{10}	V_{10}	V_{00}	$\eta_{1l} = -\eta_{2l}\ \forall l$
		$\overset{\star}{V}_{10}$	$\lambda_{1i} = -\lambda_{2i}\ \forall i$
		$V_{10} \backslash \overset{\star}{V}_{10}$	otherwise
V_{10}	$\overset{\star}{V}_{10}$	V_{00}	$\eta_{1l} = -\eta_{2l}\ \forall l$
		$\overset{\star}{V}_{10}$	$\lambda_{1i} = 0\ \forall i$
		$V_{10} \backslash \overset{\star}{V}_{10}$	otherwise
V_{10}	$\overset{\star}{V}_{01}$	V_{00}	$\#(\Leftarrow \mu_{2j} = \eta_{1l} = 0\ \forall j, l)$
		V_{10}	$\mu_{2j} = 0\ \forall j$
		$\overset{\star}{V}_{10}$	$\lambda_{1i} = \mu_{2j} = 0\ \forall i, j$
		V_{01}	$\#(\Leftarrow \eta_{1l} = 0\ \forall l)$
		V_{11}	otherwise
V_{10}	$\overset{\star}{V}_{11}$	V_{00}	$\#(\Leftarrow \eta_{1l} = \gamma_{2t} = 0\ \forall l, t)$
		V_{10}	$v_2 = 0$
		$\overset{\star}{V}_{11}$	$\#(\Leftarrow v_1 = 0)$
		$V_{11} \backslash \overset{\star}{V}_{11}$	otherwise
V_{10}	V_{11}	V_{00}	$\#(\Leftarrow \lambda_{1i} + \lambda_{2i} = \mu_{2j} = \gamma_{2t} e_t = 0\ \forall i, j, t)$
		V_{10}	$\#(\Leftarrow \mu_{2j} = \gamma_{2t} = 0\ \forall j, t)$
		V_{01}	$\#(\Leftarrow \eta_{2l} = \gamma_{2t} = 0\ \forall l, t)$
		$\overset{\star}{V}_{11}$	$\lambda_{1i} + \lambda_{2i} = \mu_{2j} = \eta_{2l} = 0\ \forall i, j, l$
		V_{11}	otherwise
V_{11}	$\overset{\star}{V}_{11}$	V_{00}	$\mu_{1j} = \eta_{1l} = \gamma_{1t} + \gamma_{2t} = 0\ \forall j, l, t$
		V_{01}	$\eta_{2l} = \gamma_{1t} + \gamma_{2t} = 0\ \forall l, t,\ \sum_{j \in \mathcal{I}_{01}} \mu_{1j} e_j \neq 0$
		$\overset{\star}{V}_{01}$	$\lambda_{1i} = \eta_{1l} = \gamma_{1t} + \gamma_{2t} = 0\ \forall i, l, t$
		V_{10}	$\mu_{2j} = \gamma_{1t} + \gamma_{2t} = 0\ \forall j, t,\ \sum_{l \in \mathcal{I}_{10}} \eta_{1l} e_l \neq 0$
		$\overset{\star}{V}_{10}$	$\lambda_{1i} = \mu_{1j} = \gamma_{1t} + \gamma_{2t} = 0\ \forall i, j, t$
		$\overset{\star}{V}_{11}$	$\lambda_{1i} = \mu_{1j} = \eta_{1l} = 0\ \forall i, j, l$
		V_{11}	$\sum_{t \in \mathcal{I}_{11}} (\gamma_{1t} + \gamma_{2t}) e_t \neq 0$

(continued)

Table 7 (continued)

v_1 in	v_2 in	v_1+v_2 in	If and only if
V_{11}	V_{11}	V_{00}	$\mu_{1j}+\mu_{2j}=\eta_{1l}+\eta_{2l}=\gamma_{1t}+\gamma_{2t}=0\ \forall j,l,t$
		V_{01}	$\eta_{1l}+\eta_{2l}=\gamma_{1t}+\gamma_{2t}=0\ \forall l,t,$ $\sum_{j\in\mathcal{I}_{01}}(\mu_{1j}+\mu_{2j})e_j\neq 0$
		$\overset{\star}{V}_{01}$	$\lambda_{1i}+\lambda_{2i}=\eta_{1l}+\eta_{2l}=\gamma_{1t}+\gamma_{2t}=0\ \forall i,l,t$
		V_{10}	$\mu_{1j}+\mu_{2j}=\gamma_{1t}+\gamma_{2t}=0\ \forall i,j,t,$ $\sum_{l\in\mathcal{I}_{10}}(\eta_{1l}+\eta_{2l})e_l\neq 0$
		$\overset{\star}{V}_{10}$	$\lambda_{1i}+\lambda_{2i}=\mu_{1j}+\mu_{2j}=\gamma_{1t}+\gamma_{2t}=0\ \forall i,j,t$
		$\overset{\star}{V}_{11}$	$\lambda_{1i}+\lambda_{2i}=\mu_{1j}+\mu_{2j}=\eta_{1l}+\eta_{2l}=0\ \forall i,j,l$
		$V_{11}\backslash\overset{\star}{V}_{11}$	$\sum_{t\in\mathcal{I}_{11}}(\gamma_{1t}+\gamma_{2t})e_t\neq 0$

3 Set-theoretical Results on Hom-Algebras

Hom-deformed identities invoke multiple relations between products of elements. In particular, hom-associativity with a non-injective twisting map provides an entire array of zero division relations involving elements in $\ker\alpha$ (see [1]). It is natural to consider the influence of those identities on the set-theoretical split we have constructed for $\mathcal{A}$ and $\ker\alpha$. We construct several arguments for (hom-)simplicity and zero division relations in hom-associative algebras based on ideal condition on certain annihilators. We originally aim to answer the question: *when can we find regular elements in* $\ker\alpha$?

Let $(\mathcal{A},\alpha)$ be any hom-algebra. Set-theoretically, we can open it using Definition 13 as $\mathcal{A}=\mathcal{A}_0^l\sqcup\mathcal{A}_1^l=\mathcal{A}_0^r\sqcup\mathcal{A}_1^r=\mathcal{A}_{00}\sqcup\mathcal{A}_{01}\sqcup\mathcal{A}_{10}\sqcup\mathcal{A}_{11}$. It is immediate that $\mathcal{A}_0^l$ and $\mathcal{A}_0^r$ are linear subspaces, but $\mathcal{A}_1^l$ and $\mathcal{A}_1^r$ are not. Indeed, consider $v\in\mathcal{A}_0^l$, $a\in\mathcal{A}_1^l$. According to summation rules, $v+a$, $v-a\in\mathcal{A}_1^l$, but $(v+a)+(v-a)=2v\in\mathcal{A}_0^l$.

Let us open $K=\ker\alpha$ according to Definition 13. We use notation K (for kernel) instead of S for the corresponding subsets K_0^l, K_1^l, K_0^r, K_1^r and K_{ij} for each i, j. With these notations, $\ker\alpha$ opens as $K_0^l\sqcup K_1^l=K_0^r\sqcup K_1^r=K_{00}\sqcup K_{01}\sqcup K_{10}\sqcup K_{11}$. Immediately K_0^l and K_0^r are linear subspaces, but K_1^l and K_1^r are not.

Proposition 2 *Let* $(\mathcal{A},\alpha)$ *be a hom-algebra. Then:*

(i) $\mathcal{A}_0^l$ *and* $\mathcal{A}_0^r$ *are, respectively, right and left ideals of* $\mathcal{A}$.
(ii) K_0^l *and* K_0^r *are, respectively, right and left hom-ideals of* $\mathcal{A}$.

Proof

(i) For all $x,a\in\mathcal{A}$ and $v\in\mathcal{A}_0^l$, we have $(v\cdot a)\cdot x=0\cdot x=0\Rightarrow v\cdot a\in\mathcal{A}_0^l\Rightarrow\mathcal{A}_0^l$ is a right ideal of $\mathcal{A}$. Similarly, for all $x,a\in\mathcal{A}$ and $v\in\mathcal{A}_0^r$, we have $x\cdot(a\cdot v)=x\cdot 0=0\Rightarrow a\cdot v\in\mathcal{A}_0^r\Rightarrow\mathcal{A}_0^r$ is a left ideal of $\mathcal{A}$.

(ii) For all $x, a \in \mathcal{A}$ and $k \in K_0^l$, we have $(k \cdot a) \cdot x = 0 \cdot x = 0 \Rightarrow k \cdot a \in \mathcal{A}_0^l \Rightarrow \mathcal{A}_0^l$ is a right ideal of $\mathcal{A}$. Similarly, for all $x, a \in \mathcal{A}$ and $k \in K_0^r$, we have $x \cdot (a \cdot k) = x \cdot 0 = 0 \Rightarrow a \cdot k \in \mathcal{A}_0^r \Rightarrow \mathcal{A}_0^r$ is a left ideal of $\mathcal{A}$.

In both K_0^l and K_0^r, $\alpha(k) = 0$ which is in both K_0^l and K_0^r. It follows that they are hom-ideals of $\mathcal{A}$ on the respective side. □

Corollary 2 *$\mathcal{A}_{00}$ is both a subalgebra and two-sided ideal of $\mathcal{A}$. K_{00} is both a hom-subalgebra and two-sided hom-ideal of $\mathcal{A}$.*

Proof $\mathcal{A}_{00} = \mathcal{A}_0^l \cap \mathcal{A}_0^r$. $\mathcal{A}_{00}$ is a two-sided ideal by intersection of ideals. Let $a, a' \in \mathcal{A}_{00}$, then $(a \cdot a') \cdot x = 0 \cdot x = 0$ and $x \cdot (a \cdot a') = x \cdot 0 = 0$ for all $x \in \mathcal{A}$; thus, it is closed under product and a subalgebra of $\mathcal{A}$.

$K_{00} = K_0^l \cap K_0^r$, so K_{00} is a two-sided hom-ideal by intersection of hom-ideals. Let $k, k' \in K_{00}$, then $(k \cdot k') \cdot x = 0 \cdot x = 0$ and $x \cdot (k \cdot k') = x \cdot 0 = 0$ for all $x \in \mathcal{A}$, and $\alpha(k) = 0 \in K_{00}$ for all $k \in K_{00}$; thus, it is a hom-subalgebra of $\mathcal{A}$.

□

We use the ideals appearing upon decomposition of $\mathcal{A}$ and $\ker \alpha$ to study (hom-)simplicity. A hom-algebra $(\mathcal{A}, \alpha)$ is (hom-)simple if it contains no proper (hom-)ideals, left (hom-)simple if it contains no proper left (hom-)ideals, and right (hom-)simple if it contains no proper right (hom-)ideals. Note that simple $\Rightarrow$ hom-simple.

Corollary 3 *Let $(\mathcal{A}, \alpha)$ be left simple, then only one of the following statements holds:*

(i) *There are no global right zero divisors in $\mathcal{A}$. Set-wise, $\mathcal{A} = \{0\} \sqcup \mathcal{A}_{01} \sqcup \mathcal{A}_{11}$.*
(ii) *The product on $\mathcal{A}$ is trivial.*

Let $(\mathcal{A}, \alpha)$ be right simple, then only one of the following statements holds:

(i) *There are no global left zero divisors in $\mathcal{A}$. Set-wise, $\mathcal{A} = \{0\} \sqcup \mathcal{A}_{10} \sqcup \mathcal{A}_{11}$.*
(ii) *The product on $\mathcal{A}$ is trivial.*

Let $(\mathcal{A}, \alpha)$ be simple, then $\mathcal{A}_0^l = \mathcal{A}_0^r$ and only one of these holds:

(i) *There are no global zero divisors in $\mathcal{A}$. Set-wise, $\mathcal{A} = \{0\} \sqcup \mathcal{A}_{11}$.*
(ii) *The product on $\mathcal{A}$ is trivial.*

Proof If $\mathcal{A}$ is left simple, then every left ideal is either $\{0\}$ or $\mathcal{A}$. In the first case, $\mathcal{A}_0^r = \{0\}$ so $\mathcal{A}$ does not contain global right zero divisors. Set-wise, $\mathcal{A}_{00} \sqcup \mathcal{A}_{10} = \{0\} \Rightarrow \mathcal{A} = \{0\} \sqcup \mathcal{A}_{01} \sqcup \mathcal{A}_{11}$. In the second case, $\mathcal{A}_0^r = \mathcal{A} \Rightarrow x \cdot a = 0\ \forall a \in \mathcal{A}_0^r,\ x \in \mathcal{A} \Rightarrow x \cdot a = 0\ \forall x, a \in \mathcal{A}$. If $\mathcal{A}$ is right simple, then every right ideal is either $\{0\}$ or $\mathcal{A}$. In the first case, $\mathcal{A}_0^l = \{0\}$ so $\mathcal{A}$ does not contain global left zero divisors. Set-wise, $\mathcal{A}_{00} \sqcup \mathcal{A}_{01} = \{0\} \Rightarrow \mathcal{A} = \{0\} \sqcup \mathcal{A}_{10} \sqcup \mathcal{A}_{11}$. In the second case, $\mathcal{A}_0^l = \mathcal{A} \Rightarrow a \cdot x = 0\ \forall a \in \mathcal{A}_0^l,\ x \in \mathcal{A} \Rightarrow a \cdot x = 0\ \forall a, x \in \mathcal{A}$. If $\mathcal{A}$ is two-sided simple, then it is both left and right hom-simple. Either $\mathcal{A}_0^l = \{0\} = \mathcal{A}_0^r, \mathcal{A}_0^l = \mathcal{A} = \mathcal{A}_0^r$ or one of each. If one is $\mathcal{A}$ and the other is $\{0\}$, then there are no global zero divisors on one side in $\mathcal{A}$ and the product is identically 0, which is a contradiction.

If $\mathcal{A}_0^l = \{0\} = \mathcal{A}_0^r$, then $\mathcal{A}$ does not contain global zero divisors on either side. Set-wise, $\mathcal{A}_{00} \sqcup \mathcal{A}_{10} = \mathcal{A}_{00} \sqcup \mathcal{A}_{01} = \{0\} \Rightarrow \mathcal{A}_{01} = \mathcal{A}_{10} = \{0\} \Rightarrow \mathcal{A} = \{0\} \sqcup \mathcal{A}_{11}$. If $\mathcal{A}_0^l = \mathcal{A} = \mathcal{A}_0^r$, then $a \cdot x = 0 = x \cdot a \ \forall a, x \in \mathcal{A}$. Conclusion follows. □

In the field of hom-algebras, we study structures with nontrivial product and twisting map. This makes the counter-reciprocal to this result particularly interesting.

Corollary 4 *Any hom-algebra with a left (right) global zero divisor on one side is not right (left) simple. A hom-algebra with a two-sided global zero divisor is not simple.*

Remark 6 Note that this does not prevent such a hom-algebra from being hom-simple. For example, a non-simple hom-algebra where all ideals are not hom is hom-simple. So far our attempts to construct such an algebra have been unfruitful. This aligns with much of the current literature, where plenty of hom-associative algebras are either multiplicative with non-injective α – and thus not simple, hom-unital – in which case all ideals are hom or a Yau twist of an associative algebra, where ideals in the original algebra are related to hom-ideals in the twist. This phenomenon regarding simplicity will be illustrated in Sect. 4.

Up next we apply the same ideas to the decomposition of $\ker \alpha$ and examine hom-simplicity. Item (i) on each statement follows the same reasoning as Corollary 3.

Corollary 5 *Let $(\mathcal{A}, \alpha)$ be left hom-simple, then only one of the following statements holds:*

(i) *There are no global right zero divisors in* $\ker \alpha$. *Set-wise,* $\ker \alpha = \{0\} \sqcup K_{01} \sqcup K_{11}$.
(ii) *α is the zero map.*

Let $(\mathcal{A}, \alpha)$ be right hom-simple, then only one of the following statements holds:

(i) *There are no global left zero divisors in* $\ker \alpha$. *Set-wise,* $\ker \alpha = \{0\} \sqcup K_{10} \sqcup K_{11}$.
(ii) *α is the zero map.*

Let $(\mathcal{A}, \alpha)$ be hom-simple, then $K_0^l = K_0^r$ and only one of these holds:

(i) *There are no global zero divisors in* $\ker \alpha$. *Set-wise,* $\ker \alpha = \{0\} \sqcup K_{11}$.
(ii) *α is the zero map.*

Proof If $\mathcal{A}$ is left simple, then every left ideal is either $\{0\}$ or $\mathcal{A}$. In the first case, $K_0^r = \{0\}$ so $\ker \alpha$ does not contain global right zero divisors. Set-wise, $K_0^r = K_{00} \sqcup K_{10} = \{0\} \Rightarrow \ker \alpha = \{0\} \sqcup K_{01} \sqcup K_{11}$. In the second case, $K_0^r = \mathcal{A} \Rightarrow \ker \alpha = \mathcal{A} \Rightarrow \alpha(a) = 0 \ \forall a \in \mathcal{A}$. If $\mathcal{A}$ is right simple, then every right ideal is either $\{0\}$ or $\mathcal{A}$. In the first case, $K_0^l = \{0\}$ so $\ker \alpha$ does not contain global left zero divisors. Set-wise, $K_0^l = K_{00} \sqcup K_{01} = \{0\} \Rightarrow \ker \alpha = \{0\} \sqcup K_{10} \sqcup K_{11}$. In the second case, $K_0^l = \mathcal{A} \Rightarrow \ker \alpha = \mathcal{A} \Rightarrow \alpha(a) = 0 \ \forall a \in \mathcal{A}$. If $\mathcal{A}$ is two-sided hom-simple, then it is both left and right hom-simple. Either $K_0^l = \{0\} = K_0^r$, $K_0^l = \mathcal{A} = K_0^r$ or one of each. If one is $\mathcal{A}$ and the other is $\{0\}$, then there are no global zero divisors on one side in $\ker\alpha$ and α is identically 0, which is a contradiction. If $K_0^l = \{0\} = K_0^r$, then $\ker\alpha$ does not contain global zero divisors on either side. Set-wise, $K_0^r = K_0^l = \{0\} \Rightarrow K_{00} \sqcup K_{01} \sqcup K_{10} = \{0\} \Rightarrow \ker\alpha = \{0\} \sqcup K_{11}$. If $K_0^l = \mathcal{A} = K_0^r$, then $\ker\alpha = \mathcal{A} \Rightarrow \alpha \equiv 0$. Conclusion follows. □

Hom-associativity can be used to find zero divisors within $\alpha(\mathcal{A})$. We apply the following proposition, adapted from [1].

Proposition 3 ([1], Proposition 6) *Let $(\mathcal{A}, \cdot, \alpha)$ be a hom-associative algebra. Then, for all $r \in \mathcal{A}$*

$$(\ker\alpha \cdot \mathcal{A}) \cdot \alpha(r) = \{0\}, \ \alpha(r) \cdot (\mathcal{A} \cdot \ker\alpha) = \{0\}.$$

Proposition 4 *Let $(\mathcal{A}, \alpha)$ be a hom-associative algebra, α non-injective:*

(i) *If $\ker\alpha \cdot \mathcal{A}$ contains a nonzero element, then $\alpha(\mathcal{A}) \subseteq \mathcal{R}_0(\mathcal{A})$. In particular, $\alpha(\mathcal{A}) \subseteq Ann^r_{\mathcal{A}}(\ker\alpha \cdot \mathcal{A})$.*
(ii) *If $\mathcal{A} \cdot \ker\alpha$ contains a nonzero element, then $\alpha(\mathcal{A}) \subseteq \mathcal{L}_0(\mathcal{A})$. In particular, $\alpha(\mathcal{A}) \subseteq Ann^l_{\mathcal{A}}(\mathcal{A} \cdot \ker\alpha)$.*
(iii) *If both $\ker\alpha \cdot \mathcal{A}$ and $\mathcal{A} \cdot \ker\alpha$ contain a nonzero element, then $\alpha(\mathcal{A}) \subseteq (\mathcal{L}_0(\mathcal{A}) \cap \mathcal{R}_0(\mathcal{A}))$. In particular, $\alpha(\mathcal{A}) \subseteq (Ann^r_{\mathcal{A}}(\ker\alpha \cdot \mathcal{A}) \cap Ann^l_{\mathcal{A}}(\mathcal{A} \cdot \ker\alpha))$.*

Proof If $\ker\alpha \cdot \mathcal{A}$ contains a nonzero element, then by Proposition 3 we have $(\ker\alpha \cdot \mathcal{A}) \cdot \alpha(r) = \{0\}$ for all $r \in \mathcal{A}$. It follows that $\alpha(r) \in \mathcal{R}_0(\mathcal{A})$; thus, $\alpha(\mathcal{A}) \subseteq \mathcal{R}_0(\mathcal{A})$, and in particular $\alpha(\mathcal{A}) \subseteq Ann^r_{\mathcal{A}}(\ker\alpha \cdot \mathcal{A})$.

If $\mathcal{A} \cdot \ker\alpha$ contains a nonzero element, then by Proposition 3 we have $\alpha(r) \cdot (\mathcal{A} \cdot \ker\alpha) = \{0\}$ for all $r \in \mathcal{A}$. It follows that $\alpha(r) \in \mathcal{L}_0(\mathcal{A})$, thus $\alpha(\mathcal{A}) \subseteq \mathcal{L}_0(\mathcal{A})$ and $\alpha(\mathcal{A}) \subseteq Ann^l_{\mathcal{A}}(\mathcal{A} \cdot \ker\alpha)$.

If both $\ker\alpha \cdot \mathcal{A}$ and $\mathcal{A} \cdot \ker\alpha$ contain a nonzero element, then both properties hold and $\alpha(\mathcal{A}) \subseteq (Ann^r_{\mathcal{A}}(\ker\alpha \cdot \mathcal{A}) \cap Ann^l_{\mathcal{A}}(\mathcal{A} \cdot \ker\alpha))$. □

Remark 7 This result implies that the dimension of $\alpha(\mathcal{A})$ is at most that of $Ann^r_{\mathcal{A}}(\ker\alpha \cdot \mathcal{A})$, $Ann^l_{\mathcal{A}}(\mathcal{A} \cdot \ker\alpha)$ or $Ann^r_{\mathcal{A}}(\ker\alpha \cdot \mathcal{A}) \cap Ann^l_{\mathcal{A}}(\mathcal{A} \cdot \ker\alpha)$.

Elements in K_0 are global zero divisors. By definition of K_0^l and K_0^r (see Definition 13), we have $K_0^l \cdot \mathcal{A} = \{0\} = \mathcal{A} \cdot K_0^r$; thus, $\ker\alpha \cdot \mathcal{A} = K_1^l \cdot \mathcal{A}$ and $\mathcal{A} \cdot \ker\alpha = \mathcal{A} \cdot K_1^r$. That is, any element $k \in \ker\alpha$ such that $k \cdot \mathcal{A} \neq \{0\}$ or $\mathcal{A} \cdot k \neq \{0\}$ (if any) must be in K_1 (on the corresponding side). Products $K_1^l \cdot \mathcal{A}$ and $\mathcal{A} \cdot K_1^r$ are nontrivial if and only if K_1^l and K_1^r are non-empty.

Corollary 6 *Let $(\mathcal{A}, \alpha)$ be a hom-associative algebra, α non-injective:*

(i) *If K_1^l is non-empty, then $\alpha(\mathcal{A}) \subseteq \mathcal{R}_0(\mathcal{A})$. In particular, $\alpha(\mathcal{A}) \subseteq Ann^r_{\mathcal{A}}(K_1^l \cdot \mathcal{A})$.*
(ii) *If K_1^r is non-empty, then $\alpha(\mathcal{A}) \subseteq \mathcal{L}_0(\mathcal{A})$. In particular, $\alpha(\mathcal{A}) \subseteq Ann^l_{\mathcal{A}}(\mathcal{A} \cdot K_1^r)$.*
(iii) *If K_1^l and K_1^r are non-empty, then $\alpha(\mathcal{A}) \subseteq (\mathcal{L}_0(\mathcal{A}) \cap \mathcal{R}_0(\mathcal{A}))$. In particular, $\alpha(\mathcal{A}) \subseteq (Ann^r_{\mathcal{A}}(K_1^l \cdot \mathcal{A}) \cap Ann^l_{\mathcal{A}}(\mathcal{A} \cdot K_1^r))$. This property holds, in particular, if K_{11} is non-empty.*

The last property comes from considering that $K_{11} = K_1^l \cap K_1^r$.

In previous sections we have established that image by α of regular elements is in many cases a zero divisor. Here, $\ker \alpha$ having a single element that does not cancel $\mathcal{A}$ on both sides ensures that α maps all $\mathcal{A}$ into zero divisors. The counter-reciprocal to this proposition illustrates the influence of regular elements in the structure of $\ker \alpha$.

Corollary 7 *Let* $(\mathcal{A}, \alpha)$ *be a hom-associative algebra such that* α *is not the zero map. If there is a right-regular (resp. left-regular) element in* $\alpha(\mathcal{A})$*, then* K_1^l *(resp.* K_1^r*) is empty.*

In the following equivalent cases, $\alpha(r) \in \mathcal{R}_0(\mathcal{A})$ *for all* $r \in \mathcal{A}$ *right-regular:*

(i) K_1^l *is non-empty.*
(ii) *At least one between* K_{10} *and* K_{11} *is non-empty.*

In the following equivalent cases, $\alpha(r) \in \mathcal{L}_0(\mathcal{A})$ *for all* $r \in \mathcal{A}$ *left-regular:*

(i) K_1^r *is non-empty.*
(ii) *At least one between* K_{01} *and* K_{11} *is non-empty.*

In both cases, the subset of one-sided regular elements of $\mathcal{A}$ *is not invariant under* α*.*

The presence of regular elements on $\alpha(\mathcal{A})$ forces $\ker \alpha$ to cancel the entire algebra. If α is surjective, then $\alpha(\mathcal{A}) = \mathcal{A}$ and the following follows naturally.

Corollary 8 *Let* $(\mathcal{A}, \alpha)$ *be hom-associative,* α *be surjective. The following statements hold:*

(i) *If* K_1^l *(resp.* K_1^r*) is non-empty, then all elements on* $\mathcal{A}$ *are right (resp. left) zero divisors.*
(ii) *If there is a right-regular (resp. left-regular) element on* $\mathcal{A}$*, then* K_1^l *(resp.* K_1^r*) is empty.*

Another consideration comes with the dimension of $\ker \alpha$: it bounds the dimension of $\mathcal{R}_0(\mathcal{A})$ in the sense that $\alpha(\mathcal{A})$ is a nontrivial subspace which is completely contained in $\mathcal{R}_0(\mathcal{A})$ – that is, we are not going to find a hom-associative domain with non-empty K_1. Indeed, if $\mathcal{A}$ is a domain and K_1^l is non-empty, then $\alpha(\mathcal{A}) \subseteq \mathcal{R}_0(\mathcal{A}) = \{0\}$ and α becomes the zero map.

For nonzero α, $\ker \alpha = K_0^l$. Elements in K_0^l are zero divisors; thus, $K_0^l = \{0\} \Rightarrow \ker \alpha = \{0\}$. The conclusion is known, but still interesting how we arrive to it using the structure of $\ker \alpha$ in relation to zero divisors.

Corollary 9 *The twisting map of every hom-associative domain is injective.*

Remark 8 Domains are considered in the literature as commutative unital rings with the cancellation property. This result only applies to the absence of zero divisors; unitality and commutativity are not required and may invoke additional restrictions.

These considerations over K_1^l and K_1^r can be used to discriminate the presence of regular elements in $\ker\alpha$.

Theorem 5 *If* $(\mathcal{A}, \alpha)$ *is a hom-associative algebra, then:*

(i) *If* $K_1^l = \emptyset$*, then there are no right-regular elements in* $\ker\alpha$.
(ii) *If* $K_1^r = \emptyset$*, then there are no left-regular elements in* $\ker\alpha$.
(iii) *If* $K_1^l = \emptyset = K_1^r$*, then there are no regular elements in* $\ker\alpha$.

Proof Any regular element in $\ker\alpha$ belongs to K_1^l (right), K_1^r (left) or K_{11} (two-sided). □

The following examples illustrate the set-theoretical split of $\ker\alpha$ and Proposition 6. We compute the subspaces K_{ij}, K_1^l, K_1^r and the product spaces $K_1^l \cdot \mathcal{A}$ and $\mathcal{A} \cdot K_1^r$ on the three-dimensional hom-associative algebras which are not of associative type [10], together with the corresponding annihilators and $\alpha(\mathcal{A})$. The notation $\langle v \rangle$ denotes the linear span of v. A detailed description of these algebras can be found in Sect. 4.

4 Hom-Associative Algebras Not of Associative Type in Dimension 3

Multiplicative hom-associative algebras of dimensions 2 and 3 are classified in[10]. We examine the action of $\mathcal{L}_0$ and $\mathcal{R}_0$ on hom-associative algebras of dimension 3 and compute $\ker\alpha$ in terms of subsets K_{ij}. We consider those algebras which are not of associative type: according to [10, Theorem 3.6 and Proposition 3.7], these hom-algebras are A_1^3, A_2^3, A_4^3, A_5^3, A_6^3 and A_{11}^3. For any $i \in \mathbb{Z}_{\geq 0}$, let $p_i \in \mathbb{F}$. Any product relations and images by α of elements not stated explicitly are 0 and α is given by its coordinate matrix in the basis $\{e_1, e_2, e_3\}$. The notation $\langle e_i \rangle$ is used to denote the linear span of element e_i.

In finite dimension, any simple hom-associative algebra is of associative type [10, Theorem 3.5]. It follows that none of these algebras are simple nor hom-simple. In particular, as they are all multiplicative with non-injective α, $\ker\alpha$ is a two-sided hom-ideal [10, Lemma 3.3].

Example 7 (Hom-associative Algebra A_1^3)

Product relations:

$$e_1 \cdot e_1 = e_1$$
$$e_2 \cdot e_2 = e_2 + e_3$$
$$e_2 \cdot e_3 = e_2 + e_3$$
$$e_3 \cdot e_2 = e_2 + e_3$$
$$e_3 \cdot e_3 = e_2 + e_3$$

Twisting map:

$$\alpha = \begin{pmatrix} 1 & 0 & 0 \\ 0 & 0 & 0 \\ 0 & 0 & 0 \end{pmatrix} \text{ (idempotent)}$$

$$R_{e_1} = \begin{pmatrix} 1 & 0 & 0 \\ 0 & 0 & 0 \\ 0 & 0 & 0 \end{pmatrix}, \ker(R_{e_1}) = \langle e_2, e_3 \rangle \qquad L_{e_1} = \begin{pmatrix} 1 & 0 & 0 \\ 0 & 0 & 0 \\ 0 & 0 & 0 \end{pmatrix}, \ker(L_{e_1}) = \langle e_2, e_3 \rangle$$

$$R_{e_2} = \begin{pmatrix} 0 & 0 & 0 \\ 0 & 1 & 1 \\ 0 & 1 & 1 \end{pmatrix}, \ker(R_{e_2}) = \langle e_1, e_3 - e_2 \rangle \quad L_{e_2} = \begin{pmatrix} 0 & 0 & 0 \\ 0 & 1 & 1 \\ 0 & 1 & 1 \end{pmatrix}, \ker(L_{e_2}) = \langle e_1, e_3 - e_2 \rangle$$

$$R_{e_3} = \begin{pmatrix} 0 & 0 & 0 \\ 0 & 1 & 1 \\ 0 & 1 & 1 \end{pmatrix}, \ker(R_{e_3}) = \langle e_1, e_3 - e_2 \rangle \quad L_{e_3} = \begin{pmatrix} 0 & 0 & 0 \\ 0 & 1 & 1 \\ 0 & 1 & 1 \end{pmatrix}, \ker(L_{e_3}) = \langle e_1, e_3 - e_2 \rangle.$$

Global left zero divisors are given by the subspace $Ann^l_{\mathcal{A}}(\mathcal{A}) = \ker(R_{e_1}) \cap \ker(R_{e_2}) \cap \ker(R_{e_3}) = \langle e_3 - e_2 \rangle$. This algebra is commutative, so $Ann^r_{\mathcal{A}}(\mathcal{A}) = Ann^l_{\mathcal{A}}(\mathcal{A}) = \langle e_3 - e_2 \rangle$ as well, all zero divisors are two-sided.

In terms of sets, $\mathcal{A}_{00} = \langle e_3 - e_2 \rangle$ and we have $\mathcal{A} = \mathcal{A}_{00} \sqcup \mathcal{A}_{11}$. By Corollary 4, A^3_1 is not simple. Furthermore, $\mathcal{L}_0(\mathcal{A}) = \mathcal{A}$, $\mathcal{R}_0(\mathcal{A}) = \mathcal{A}$. In addition, $\ker \alpha = \langle e_2, e_3 \rangle$, which is of type $K_{00} \sqcup K_{11}$ with $K_{00} = \langle e_3 - e_2 \rangle$. By Corollary 5, this algebra is not hom-simple.

Example 8 (Hom-associative Algebra A^3_2)

Product relations:	**Twisting map**:
$e_1 \cdot e_1 = p_1 \cdot e_1$ $e_2 \cdot e_2 = p_2 \cdot e_2$ $e_3 \cdot e_3 = p_3 \cdot e_3$	$\alpha = \begin{pmatrix} 1 & 0 & 0 \\ 0 & 1 & 0 \\ 0 & 0 & 0 \end{pmatrix}$ (idempotent)

$$R_{e_1} = \begin{pmatrix} p_1 & 0 & 0 \\ 0 & 0 & 0 \\ 0 & 0 & 0 \end{pmatrix}, \ker(R_{e_1}) = \langle e_2, e_3 \rangle \quad L_{e_1} = \begin{pmatrix} p_1 & 0 & 0 \\ 0 & 0 & 0 \\ 0 & 0 & 0 \end{pmatrix}, \ker(L_{e_1}) = \langle e_2, e_3 \rangle$$

$$R_{e_2} = \begin{pmatrix} 0 & 0 & 0 \\ 0 & p_2 & 0 \\ 0 & 0 & 0 \end{pmatrix}, \ker(R_{e_2}) = \langle e_1, e_3 \rangle \quad L_{e_2} = \begin{pmatrix} 0 & 0 & 0 \\ 0 & p_2 & 0 \\ 0 & 0 & 0 \end{pmatrix}, \ker(L_{e_2}) = \langle e_1, e_3 \rangle$$

$$R_{e_3} = \begin{pmatrix} 0 & 0 & 0 \\ 0 & 0 & 0 \\ 0 & 0 & p_3 \end{pmatrix}, \ker(R_{e_3}) = \langle e_1, e_2 \rangle \quad L_{e_3} = \begin{pmatrix} 0 & 0 & 0 \\ 0 & 0 & 0 \\ 0 & 0 & p_3 \end{pmatrix}, \ker(L_{e_3}) = \langle e_1, e_2 \rangle.$$

Global left zero divisors are given by the subspace $Ann^l_{\mathcal{A}}(\mathcal{A}) = \ker(R_{e_1}) \cap \ker(R_{e_2}) \cap \ker(R_{e_3})$. We show the different possible subspaces:

(i) $Ann^l_{\mathcal{A}}(\mathcal{A}) = \{0\}$ if $p_1, p_2, p_3 \neq 0$,

(ii) $Ann^l_{\mathcal{A}}(\mathcal{A}) = \langle e_i \rangle$ if $p_i = 0$,
(iii) $Ann^l_{\mathcal{A}}(\mathcal{A}) = \langle e_i, e_j \rangle$ if $p_i = p_j = 0$.

This algebra is commutative, and thus $Ann^r_{\mathcal{A}}(\mathcal{A}) = Ann^l_{\mathcal{A}}(\mathcal{A})$, that is, all these global zero divisors are two-sided. Also, $\mathcal{A} = \mathcal{A}_{00} \sqcup \mathcal{A}_{11} = Ann^l_{\mathcal{A}}(\mathcal{A}) \sqcup \mathcal{A}_{11}$ in all cases. According to Corollary 4, if we have at least one $p_i = 0$, then A^3_2 is not simple. If $p_1, p_2, p_3 \neq 0$, it is enough to compute $\ker\alpha = \langle e_3 \rangle$. It is an ideal, since α is multiplicative, and thus A^3_2 is not simple.

In all cases, the sets of zero divisors are given by $\mathcal{L}_0(\mathcal{A}) = \mathcal{A}$, $\mathcal{R}_0(\mathcal{A}) = \mathcal{A}$. There are no regular elements on either side. In addition, $\ker\alpha = \langle e_3 \rangle$, which is of type $\{0\} \sqcup K_{11}$ where $K_{11} = \langle e_3 \rangle \backslash \{0\}$ unless $p_3 = 0$, in which case $\ker\alpha$ is of type K_{00} with $K_{00} = \langle e_3 \rangle$ instead.

Example 9 (Hom-associative Algebra A^3_4.) We assume $p_i \neq p_j, i \neq j$.

Product relations:

$$
\begin{aligned}
e_1 \cdot e_2 &= p_1 \cdot e_1 \\
e_1 \cdot e_3 &= p_2 \cdot e_1 \\
e_2 \cdot e_2 &= p_3 \cdot e_1 \\
e_2 \cdot e_3 &= p_4 \cdot e_1 \\
e_3 \cdot e_1 &= p_5 \cdot e_1 \\
e_3 \cdot e_2 &= p_4 \cdot e_1 \\
e_3 \cdot e_3 &= p_6 \cdot e_1
\end{aligned}
$$

Twisting map:

$$\alpha = \begin{pmatrix} 0 & 1 & 0 \\ 0 & 0 & 0 \\ 0 & 0 & 0 \end{pmatrix} \text{ (2-nilpotent)}$$

$$R_{e_1} = \begin{pmatrix} 0 & 0 & p_5 \\ 0 & 0 & 0 \\ 0 & 0 & 0 \end{pmatrix}, \ \ker(R_{e_1}) = \langle e_1, e_2 \rangle,$$

$$L_{e_1} = \begin{pmatrix} 0 & p_1 & p_2 \\ 0 & 0 & 0 \\ 0 & 0 & 0 \end{pmatrix}, \ \ker(L_{e_1}) = \langle e_1, p_1 \cdot e_3 - p_2 \cdot e_2 \rangle$$

$$R_{e_2} = \begin{pmatrix} p_1 & p_3 & p_4 \\ 0 & 0 & 0 \\ 0 & 0 & 0 \end{pmatrix}, \ \ker(R_{e_2}) = \langle p_1 \cdot e_2 - p_3 \cdot e_1, p_1 \cdot e_3 - p_4 \cdot e_1 \rangle$$

$$L_{e_2} = \begin{pmatrix} 0 & p_3 & p_4 \\ 0 & 0 & 0 \\ 0 & 0 & 0 \end{pmatrix}, \ \ker(L_{e_2}) = \langle e_1, p_3 \cdot e_3 - p_4 \cdot e_2 \rangle$$

$$R_{e_3} = \begin{pmatrix} p_2 & p_4 & p_6 \\ 0 & 0 & 0 \\ 0 & 0 & 0 \end{pmatrix}, \ \ker(R_{e_3}) = \langle p_2 \cdot e_2 - p_4 \cdot e_1, p_2 \cdot e_3 - p_6 \cdot e_1 \rangle$$

$$L_{e_3} = \begin{pmatrix} p_5 & p_4 & p_6 \\ 0 & 0 & 0 \\ 0 & 0 & 0 \end{pmatrix}, \ \ker(L_{e_3}) = \langle p_5 \cdot e_3 - p_6 \cdot e_1, p_4 \cdot e_3 - p_6 \cdot e_2 \rangle.$$

Global left zero divisors are given by the subspace $Ann^l_{\mathcal{A}}(\mathcal{A}) = \ker(R_{e_1}) \cap \ker(R_{e_2}) \cap \ker(R_{e_3})$. This subspace is, if $p_5 \neq 0$:

(i) $Ann^l_{\mathcal{A}}(\mathcal{A}) = \langle e_1 \rangle$ if $p_1 = p_2 = 0$,
(ii) $Ann^l_{\mathcal{A}}(\mathcal{A}) = \langle e_2 \rangle$ if $p_3 = p_4 = 0$,
(iii) $Ann^l_{\mathcal{A}}(\mathcal{A}) = \langle e_2 - e_1 \rangle$ if $p_1 = p_3, \ p_2 = p_4$,
(iv) $Ann^l_{\mathcal{A}}(\mathcal{A}) = \langle \frac{p_3}{p_1} e_2 - e_1 \rangle$ if $p_1 = p_2, \ p_3 = p_4$,
(v) $Ann^l_{\mathcal{A}}(\mathcal{A}) = \{0\}$ otherwise.

Global right zero divisors are given by the subspace $Ann^r_{\mathcal{A}}(\mathcal{A}) = \ker(L_{e_1}) \cap \ker(L_{e_2}) \cap \ker(L_{e_3})$. This subspace is:

(i) $Ann^r_{\mathcal{A}}(\mathcal{A}) = \langle \frac{p_5}{p_6} e_3 - e_1 \rangle$ if $p_2 = p_4 = 0, \ p_6 \neq 0$,
(ii) $Ann^r_{\mathcal{A}}(\mathcal{A}) = \langle e_3 \rangle$ if $p_2 = p_4 = p_6 = 0$,
(iii) $Ann^r_{\mathcal{A}}(\mathcal{A}) = \langle e_1 \rangle$ if $p_1 = p_5 = 0$,
(iv) $Ann^r_{\mathcal{A}}(\mathcal{A}) = \langle e_2 \rangle$ if $p_1 = p_3 = p_4 = 0$,
(v) $Ann^r_{\mathcal{A}}(\mathcal{A}) = \langle e_1, e_2 \rangle$ if $p_1 = p_3 = p_4 = p_5 = 0$,
(vi) $Ann^r_{\mathcal{A}}(\mathcal{A}) = \langle \frac{p_6}{p_4} e_2 - \frac{p_6}{p_5} e_1 \rangle$ if $p_1 = 0, \ p_4, p_5, p_6 \neq 0$,
(vii) $Ann^r_{\mathcal{A}}(\mathcal{A}) = \langle e_2 - e_1 \rangle$ if $p_1 = 0, \ p_4 = p_5 \neq 0, p_6 \neq 0$,
(viii) $Ann^r_{\mathcal{A}}(\mathcal{A}) = \langle e_1, e_3 - e_2 \rangle$ if $p_1 = p_2, \ p_3 = p_4 = p_6 \neq 0, \ p_5 = 0$,
(ix) $Ann^r_{\mathcal{A}}(\mathcal{A}) = \langle e_1 \rangle$ if $p_5 = 0$ unless $p_1 = p_2, p_3 = p_4 = p_6 \neq 0$,

The sets of zero divisors are given by

$$\begin{aligned} \mathcal{L}_0(\mathcal{A}) = \langle e_1, e_2 \rangle \cup \langle p_1 \cdot e_2 - p_3 \cdot e_1, p_1 \cdot e_3 - p_4 \cdot e_1 \rangle \\ \cup \langle p_2 \cdot e_2 - p_4 \cdot e_1, p_2 \cdot e_3 - p_6 \cdot e_1 \rangle \end{aligned}$$

$$\begin{aligned} \mathcal{R}_0(\mathcal{A}) = \langle e_1, p_1 \cdot e_3 - p_2 \cdot e_2 \rangle \cup \langle e_1, p_3 \cdot e_3 - p_4 \cdot e_2 \rangle \\ \cup \langle p_5 \cdot e_3 - p_6 \cdot e_1, p_4 \cdot e_3 - p_6 \cdot e_2 \rangle. \end{aligned}$$

$\mathcal{L}_0(\mathcal{A})$ is a union of three planes in a three-dimensional space. There are, thus, left-regular elements, which are all elements in $\mathcal{A} \backslash \mathcal{L}_0(\mathcal{A})$. The set of these elements is the subset:

$$Reg_l(\mathcal{A}) = \left\{ \begin{pmatrix} a \\ b \\ c \end{pmatrix} \mid \begin{pmatrix} a \\ b \\ c \end{pmatrix} \notin \left\{ \begin{pmatrix} \mu p_4 - \lambda p_3 \\ \lambda p_1 \\ \mu p_1 \end{pmatrix}, \begin{pmatrix} \mu p_6 - \lambda p_4 \\ \lambda p_2 \\ \mu p_2 \end{pmatrix} \right\}, \atop c \neq 0, \lambda, \mu \in \mathbb{F} \right\}$$

where $(a, b, c)^T := a \cdot e_1 + b \cdot e_2 + c \cdot e_3, \ a, b, c \in \mathbb{F}$.

Similarly, all elements in $\mathcal{A} \backslash \mathcal{R}_0(\mathcal{A})$ are right-regular. The set of these elements can be identified with the subset:

$$Reg_r(\mathcal{A}) = \left\{ \begin{pmatrix} a \\ b \\ c \end{pmatrix} \mid \begin{pmatrix} a \\ b \\ c \end{pmatrix} \notin \left\{ \begin{pmatrix} \lambda \\ -\mu p_2 \\ \mu p_1 \end{pmatrix}, \begin{pmatrix} \lambda \\ -\mu p_4 \\ \mu p_3 \end{pmatrix}, \begin{pmatrix} -\lambda p_6 \\ -\mu p_6 \\ \lambda p_5 + \mu p_4 \end{pmatrix} \right\}, \atop \lambda, \mu \in \mathbb{F} \right\}$$

where $(a, b, c)^T := a \cdot e_1 + b \cdot e_2 + c \cdot e_3, \ a, b, c \in \mathbb{F}$.

In addition, $\ker \alpha = \langle e_1, e_3 \rangle$, which is of type $\{0\} \sqcup K_{11}$. This is a two-sided hom-ideal, and thus A_4^3 is neither simple nor hom-simple.

Example 10 (Hom-associative Algebra A_5^3) We assume $p_1, p_2 \neq 0$.

Product relations:

$$e_2 \cdot e_2 = p_1 \cdot e_1$$
$$e_3 \cdot e_3 = p_2 \cdot e_3$$

Twisting map:

$$\alpha = \begin{pmatrix} 1 & 1 & 0 \\ 0 & 1 & 0 \\ 0 & 0 & 0 \end{pmatrix}$$

$$R_{e_1} = \begin{pmatrix} 0 & 0 & 0 \\ 0 & 0 & 0 \\ 0 & 0 & 0 \end{pmatrix}, \ \ker(R_{e_1}) = \mathcal{A} \qquad L_{e_1} = \begin{pmatrix} 0 & 0 & 0 \\ 0 & 0 & 0 \\ 0 & 0 & 0 \end{pmatrix}, \ \ker(L_{e_1}) = \mathcal{A}$$

$$R_{e_2} = \begin{pmatrix} 0 & 0 & 0 \\ 0 & p_1 & 0 \\ 0 & 0 & 0 \end{pmatrix}, \ker(R_{e_2}) = \langle e_1, e_3 \rangle \quad L_{e_2} = \begin{pmatrix} 0 & 0 & 0 \\ 0 & p_1 & 0 \\ 0 & 0 & 0 \end{pmatrix}, \ker(L_{e_2}) = \langle e_1, e_3 \rangle$$

$$R_{e_1} = \begin{pmatrix} 0 & 0 & 0 \\ 0 & 0 & 0 \\ 0 & 0 & p_2 \end{pmatrix}, \ker(R_{e_3}) = \langle e_1, e_2 \rangle \quad L_{e_3} = \begin{pmatrix} 0 & 0 & 0 \\ 0 & 0 & 0 \\ 0 & 0 & p_2 \end{pmatrix}, \ker(L_{e_3}) = \langle e_1, e_2 \rangle.$$

Global left zero divisors are given by the subspace $Ann^l_{\mathcal{A}}(\mathcal{A}) = \ker(R_{e_1}) \cap \ker(R_{e_2}) \cap \ker(R_{e_3})$. This subspace is:

(i) $Ann^l_{\mathcal{A}}(\mathcal{A}) = \langle e_1 \rangle$ if $p_1, p_2 \neq 0$,
(ii) $Ann^l_{\mathcal{A}}(\mathcal{A}) = \langle e_1, e_2 \rangle$ if $p_1 = 0$,
(iii) $Ann^l_{\mathcal{A}}(\mathcal{A}) = \langle e_1, e_3 \rangle$ if $p_2 = 0$.

This algebra is commutative, and thus $Ann^r_{\mathcal{A}}(\mathcal{A}) = Ann^l_{\mathcal{A}}(\mathcal{A})$. In all cases e_1 is a global two-sided zero divisor. According to Corollary 4, A_5^3 is not simple. Set-wise, if $p_1, p_2 \neq 0$, then $\mathcal{A}_{00} = \langle e_1 \rangle$, $\mathcal{A}_{10} = \mathcal{A}_{01} = \emptyset$ and thus $\mathcal{A} = \mathcal{A}_{00} \sqcup \mathcal{A}_{11}$. Sets of zero divisors are, in all cases, $\mathcal{L}_0(\mathcal{A}) = \mathcal{A}$, $\mathcal{R}_0(\mathcal{A}) = \mathcal{A}$. There are no regular elements on either side. In addition, $\ker \alpha = \langle e_3 \rangle$, which is of type $\{0\} \sqcup K_{11}$ and a two-sided hom-ideal. A_5^3 is not hom-simple. If $p_2 = 0$, then $e_3 \in \mathcal{A}_{00}$, and thus $\ker \alpha$ is of type K_{00} instead, with $K_{00} = \langle e_3 \rangle$.

Example 11 (Hom-associative Algebra A_6^3)

Product relations:

$$e_1 \cdot e_2 = e_1$$
$$e_2 \cdot e_2 = e_1$$
$$e_2 \cdot e_3 = e_1$$
$$e_3 \cdot e_2 = e_1$$

Twisting map:

$$\alpha = \begin{pmatrix} 0 & 1 & 0 \\ 0 & 0 & 0 \\ 0 & 0 & 1 \end{pmatrix}$$

$$R_{e_1} = \begin{pmatrix} 0 & 0 & 0 \\ 0 & 0 & 0 \\ 0 & 0 & 0 \end{pmatrix}, \ker(R_{e_1}) = \mathcal{A}, \qquad L_{e_1} = \begin{pmatrix} 0 & 1 & 0 \\ 0 & 0 & 0 \\ 0 & 0 & 0 \end{pmatrix},$$

$$\ker(L_{e_1}) = \langle e_1, e_3 \rangle$$

$$R_{e_3} = \begin{pmatrix} 0 & 1 & 1 \\ 0 & 0 & 0 \\ 0 & 0 & 0 \end{pmatrix}, \ker(R_{e_3}) = \langle e_1, e_3 - e_2 \rangle, \qquad L_{e_3} = \begin{pmatrix} 0 & 1 & 0 \\ 0 & 0 & 0 \\ 0 & 0 & 0 \end{pmatrix},$$

$$\ker(L_{e_3}) = \langle e_1, e_3 \rangle$$

$$R_{e_2} = \begin{pmatrix} 1 & 1 & 1 \\ 0 & 0 & 0 \\ 0 & 0 & 0 \end{pmatrix}, \ker(R_{e_2}) = \langle e_3 - e_1, e_3 - e_2 \rangle, \quad L_{e_2} = \begin{pmatrix} 0 & 1 & 1 \\ 0 & 0 & 0 \\ 0 & 0 & 0 \end{pmatrix},$$

$$\ker(L_{e_2}) = \langle e_1, e_3 - e_2 \rangle.$$

Global left zero divisors are given by the subspace:

$$Ann^l_{\mathcal{A}}(\mathcal{A}) = \ker(R_{e_1}) \cap \ker(R_{e_2}) \cap \ker(R_{e_3}).$$

This subspace is given by $\langle e_3 - e_2 \rangle$. Set-wise, $\mathcal{A}_{01} = \langle e_3 - e_2 \rangle \backslash \{0\}$. Global right zero divisors are given by the subspace:

$$Ann^r_{\mathcal{A}}(\mathcal{A}) = \ker(L_{e_1}) \cap \ker(L_{e_2}) \cap \ker(L_{e_3}).$$

This subspace is given by $\langle e_1 \rangle$. Set-wise, $\mathcal{A}_{10} = \langle e_1 \rangle \backslash \{0\}$. There are no global two-sided zero divisors, and thus $\mathcal{A}_{00} = \{0\}$. We have a decomposition $\mathcal{A} = \{0\} \sqcup \mathcal{A}_{10} \sqcup \mathcal{A}_{01} \sqcup \mathcal{A}_{11}$.

The sets of zero divisors are given by $\mathcal{L}_0(\mathcal{A}) = \mathcal{A}$, $\mathcal{R}_0(\mathcal{A}) = \mathcal{A}$. There are no regular elements on either side. In addition, $\ker\alpha = \langle e_1 \rangle$. In terms of ideals, $\ker\alpha = \langle e_1 \rangle$ is a two-sided hom-ideal, as α is multiplicative, and thus A_6^3 is neither simple nor hom-simple.

Example 12 (Hom-associative Algebra A_{11}^3) We asume $p_i \neq p_j, i \neq j$ and $p_i \neq 0$ for all i.

Product relations: **Twisting map:**

$$e_1 \cdot e_3 = p_1 \cdot e_1 \qquad e_2 \cdot e_3 = p_2 \cdot e_1 \qquad e_3 \cdot e_3 = p_3 \cdot e_1 \qquad \alpha = \begin{pmatrix} 0 & 1 & 0 \\ 0 & 0 & 1 \\ 0 & 0 & 0 \end{pmatrix} \text{ (3-}nilpotent\text{)}$$

$$L_{e_1} = \begin{pmatrix} 0 & 0 & p_1 \\ 0 & 0 & 0 \\ 0 & 0 & 0 \end{pmatrix}, \ \ker(L_{e_1}) = \langle e_1, e_2 \rangle, \quad R_{e_1} = \begin{pmatrix} 0 & 0 & 0 \\ 0 & 0 & 0 \\ 0 & 0 & 0 \end{pmatrix}, \ \ker(R_{e_1}) = \mathcal{A},$$

$$L_{e_2} = \begin{pmatrix} 0 & 0 & p_2 \\ 0 & 0 & 0 \\ 0 & 0 & 0 \end{pmatrix}, \ \ker(L_{e_2}) = \langle e_1, e_2 \rangle, \quad R_{e_2} = \begin{pmatrix} 0 & 0 & 0 \\ 0 & 0 & 0 \\ 0 & 0 & 0 \end{pmatrix}, \ \ker(R_{e_2}) = \mathcal{A},$$

$$L_{e_3} = \begin{pmatrix} 0 & 0 & p_3 \\ 0 & 0 & 0 \\ 0 & 0 & 0 \end{pmatrix}, \ \ker(L_{e_3}) = \langle e_1, e_2 \rangle, \quad R_{e_3} = \begin{pmatrix} p_1 & p_2 & p_3 \\ 0 & 0 & 0 \\ 0 & 0 & 0 \end{pmatrix}, \ \ker(R_{e_3}) =$$

$$= \langle p_1 \cdot e_2 - p_2 \cdot e_1, p_1 \cdot e_3 - p_3 \cdot e_1 \rangle$$

Global left zero divisors are given by the subspace $Ann^l_{\mathcal{A}}(\mathcal{A}) = \ker(R_{e_1}) \cap \ker(R_{e_2}) \cap \ker(R_{e_3}) = \langle p_1 \cdot e_2 - p_2 \cdot e_1, p_1 \cdot e_3 - p_3 \cdot e_1 \rangle = \ker(R_{e_3})$. Global right zero divisors are given by the subspace $Ann^r_{\mathcal{A}}(\mathcal{A}) = \ker(L_{e_1}) \cap \ker(L_{e_2}) \cap \ker(L_{e_3}) = \langle e_1, e_2 \rangle$. Set-wise, $\mathcal{A}_{01} = \ker(R_{e_3}) \backslash \{0\}$ and $\mathcal{A}_{10} = \langle e_1, e_2 \rangle \backslash \{0\}$, unless there are global two-sided zero divisors. These are given by the subspace $\mathcal{A}_{00} = Ann_{\mathcal{A}}(\mathcal{A}) = \langle p_1 \cdot e_2 - p_2 \cdot e_1 \rangle$. Then, we can express $\mathcal{A}_{10} = \langle e_1, e_2 \rangle \backslash \mathcal{A}_{00}$ and $\mathcal{A}_{01} = \langle p_1 \cdot e_2 - p_2 \cdot e_1, p_1 \cdot e_3 - p_3 \cdot e_1 \rangle \backslash \mathcal{A}_{00} = \langle p_1 \cdot e_3 - p_3 \cdot e_1 \rangle \backslash \{0\}$. In this case $\mathcal{A} = \mathcal{A}_{00} \sqcup \mathcal{A}_{01} \sqcup \mathcal{A}_{10} \sqcup \mathcal{A}_{11}$. According to Corollary 4, A_{11}^3 is not simple. Sets of zero divisors are given by $\mathcal{L}_0(\mathcal{A}) = \mathcal{A}$, $\mathcal{R}_0(\mathcal{A}) = \langle e_1, e_2 \rangle$. There are no left-regular elements. The subset of right-regular elements $Reg_r(\mathcal{A})$ can be described as the subset, $Reg_r(\mathcal{A}) = \{(a, b, c)^T \mid c \neq 0\}$, where $(a, b, c)^T := a \cdot e_1 + b \cdot e_2 + c \cdot e_3$, $a, b, c \in \mathbb{F}$. This has a geometrical interpretation: $Reg_r(\mathcal{A})$ can be seen as the remainder of removing the hyperplane $\langle e_1, e_2 \rangle$ from the total space $\mathbb{F}^3$. In addition, $\ker\alpha = \langle e_1 \rangle$, which is of type $\{0\} \sqcup K_{10}$, where $K_{10} = \langle e_1 \rangle \backslash \{0\}$. This is a two-sided hom-ideal and thus A_{11}^3 is neither simple nor hom-simple.

If $p_i = 0$, then $\ker(L_{e_i}) = \mathcal{A}$ instead. If $i = 1, 2$, then e_i is a global two-sided zero divisor, and thus $\langle e_i \rangle \subseteq \mathcal{A}_{00}$.

$$
\begin{aligned}
p_1 = 0 &\Rightarrow \ker(L_{e_1}) = \mathcal{A} \Rightarrow \mathcal{A} = \mathcal{A}_{00} \sqcup \mathcal{A}_{10} \sqcup \mathcal{A}_{11}, \\
&\qquad\qquad \mathcal{A}_{00} = \langle e_1 \rangle, \ \mathcal{A}_{10} = \langle e_2 \rangle \backslash \{0\}, \\
p_2 = 0 &\Rightarrow \ker(L_{e_2}) = \mathcal{A} \Rightarrow \mathcal{A} = \mathcal{A}_{00} \sqcup \mathcal{A}_{10} \sqcup \mathcal{A}_{11}, \\
&\qquad\qquad \mathcal{A}_{00} = \langle e_2 \rangle, \ \mathcal{A}_{10} = \langle e_1 \rangle \backslash \{0\}, \\
p_3 = 0 &\Rightarrow \ker(L_{e_3}) = \mathcal{A} \Rightarrow \mathcal{A} = \{0\} \sqcup \mathcal{A}_{01} \sqcup \mathcal{A}_{10} \sqcup \mathcal{A}_{11}, \\
&\qquad\qquad \mathcal{A}_{01} = \langle e_3 \rangle \backslash \{0\}, \ \mathcal{A}_{10} = (\langle e_1 \rangle \cup \langle e_2 \rangle) \backslash \{0\}.
\end{aligned}
$$

If $p_i = p_j$ for some $i \neq j$, $\ker(R_{e_3})$ is affected:

$$
\begin{aligned}
p_1 = p_2 &\Rightarrow \ker(R_{e_3}) = \langle e_2 - e_1, p_1 \cdot e_3 - p_3 \cdot e_1 \rangle, \\
p_1 = p_3 &\Rightarrow \ker(R_{e_3}) = \langle p_1 \cdot e_2 - p_2 \cdot e_1, e_3 - e_1 \rangle, \\
p_2 = p_3 &\Rightarrow \ker(R_{e_3}) = \langle p_1 \cdot e_2 - p_2 \cdot e_1, p_1 \cdot e_3 - p_2 \cdot e_1 \rangle.
\end{aligned}
$$

This does not affect the structure of $\mathcal{A}$.

A look at these examples reveals that they all have nonzero zero divisors and nontrivial ideals and hom-ideals.

Proposition 5 *Let $(\mathcal{A}, \alpha)$ be a three-dimensional multiplicative hom-associative algebra not of associative type. The following holds:*

1. *$(\mathcal{A}, \alpha)$ has zero divisors.*
2. *$(\mathcal{A}, \alpha)$ is neither simple nor hom-simple.*

A look at the product tables in [10] reveals that hom-associative algebras of associative type have zero products as well. That is, all multiplicative hom-associative algebras of dimension 3 have zero divisors. We can write this in the form of a theorem.

Theorem 6 *There are no multiplicative not-associative hom-associative domains in dimension 3.*

Remark 9 In finite dimension, finding domains can be done in the following direct computational way: since multiplication is bilinear, it is defined by finite number of rules (multiplication table is finite) expressing the products of the basis elements $e_i \cdot e_j$ as finite linear combinations of the same basis elements, we solve the problem of hom-associativity $(e_i \cdot e_j) \cdot \alpha(e_k) - \alpha(e_i) \cdot (e_j \cdot e_k) = 0$ in terms of the matrix corresponding to α. Afterward, introduce the structure constants obtained into the product: if $e_i \cdot e_j = 0$ for some i, j, then the algebra is hom-associative, but not a domain. Otherwise, one considers the problem of zero division between linear combinations of basis elements: $(a_1 e_1 + \cdots + a_n e_n) \cdot (a_1' e_1 + \cdots + a_n' e_n) = 0$. If this system has a nontrivial solution, then the algebra is not a hom-associative domain for the structure constants obtained; otherwise, we obtain a hom-associative domain with product given by the structure constants obtained and twisting map α.

Acknowledgments The authors would like to thank Lars Hellström for his contribution and construction of Example 2 and Per Bäck for some useful comments.

References

1. García Butenegro, G., Kitouni, A., Silvestrov, S.: Divisibility in Hom-algebras, single-element properties in non-associative algebras and twisted derivations. In: Silvestrov, S., Malyarenko, A. (eds) Non-commutative and Non-associative Algebra and Analysis Structures. SPAS 2019. Springer Proceedings in Mathematics and Statistics, vol. 426, pp. 303–337. Springer, Cham (2023)
2. Bäck, P., Richter, J.: Ideals in hom-associative Weyl algebras, arXiv:2403.00104 [math.RA] (2024)
3. Bäck, P., Richter, J., Silvestrov, S.: Hom-associative Ore extensions and weak unitalizations. Int. Electron. J. Algebra **24**, 174–194, (2018)
4. Frégier, Y., Gohr, A.: On unitality conditions for Hom-associative algebras. arXiv:0904.4874[math.RA] (2009)
5. Frégier, Y., Gohr, A., Silvestrov, S.: Unital algebras of hom-associative type and surjective or injective twistings. J. Gen. Lie Theory Appl. **3**(4), 285–295 (2009)
6. Hartwig, J.T., Larsson, D., Silvestrov, S.D.: Deformations of Lie algebras using σ-derivations. J. Algebra, **295**(2), 314-361 (2006) (Preprints in Mathematical Sciences 2003:32, LUTFMA-5036-2003, Centre for Mathematical Sciences, Lund University, 52 pp. (2003))
7. Larsson, D., Silvestrov, S.D.: Quasi-hom-Lie algebras, central extensions and 2-cocycle-like identities. J. Algebra **288**, 321–344 (2005) (Preprints in Mathematical Sciences 2004:3, LUTFMA-5038-2004, Centre for Mathematical Sciences, Lund University, 40 pp, 2004)
8. Larsson, D., Silvestrov, S.D.: Quasi-Lie algebras. In: Fuchs, J., Mickelsson, J., Rozenblioum, G., Stolin, A., Westerberg, A. (eds.) Noncommutative Geometry and Representation Theory in Mathematical Physics, vol. 391, pp. 241–248. American Mathematical Society, Contemporary Mathematics: Providence (2005); (Preprints in Mathematical Sciences 2004:30, LUTFMA-5049-2004, Centre for Mathematical Sciences, Lund University, 11 pp, 2004)
9. Makhlouf, A., Silvestrov, S.D.: Hom-algebra structures. J. Gen. Lie Theory Appl. **2**(2), 51-64 (2008) (Preprints in Mathematical Sciences 2006:10, LUTFMA-5074-2006, Centre for Mathematical Sciences, Lund University (2006))
10. Makhlouf, A., Zahari, A.: Structure and classification of Hom-associative algebras. Acta et Commentationes Universitatis Tartuensis de Mathematica, **24**(1), 79–102, (2020). (arXiv:1906.04969v1)
11. Sigurdsson, G., Silvestrov, S.: Graded quasi-Lie algebras of Witt type. Czechoslovak J. Phys. **56**, 1287–1291 (2006)
12. Yau, D., Hom-algebras and Homology. J. Lie Theory **19**, 409–421 (2009)

Hom-Lie Structure of Generalized $\mathfrak{sl}(2)$-Type

Stephen Mboya, Sergei Silvestrov, Abdennour Kitouni, Elvice Ongong'a, and Jared Ongaro

Abstract This work focuses on the properties and structures of Hom-Lie algebras of generalized $\mathfrak{sl}(2)$-type. We construct classes of linear twisting maps that turn a skew-symmetric algebra of generalized $\mathfrak{sl}(2)$-type into a Hom-Lie algebra, and identify subclasses that result in multiplicative Hom-Lie algebras. We explore ideals, Hom-ideals, subalgebras, and Hom-subalgebras, emphasizing derived series, central descending series, and nilpotence and solvability properties. We also investigate the invariance of these subalgebras under the linear twisting maps and determine whether these subalgebras are weak subalgebras, Hom-subalgebras, weak ideals, or Hom-ideals. In particular, we examine these subalgebras and properties for the subfamilies of non-multiplicative Hom-Lie algebras of generalized $\mathfrak{sl}(2)$-type, highlighting the differences between non-multiplicative and multiplicative cases.

Keywords Hom-Lie algebra · Weak ideal · Hom-ideal · Central descending series · Derived series · Solvable Hom-algebra · Nilpotent Hom-algebras

S. Mboya
Department of Mathematics, University of Nairobi, Nairobi, Kenya

Division of Mathematics and Physics, Mälardalen University, Västerås, Sweden
e-mail: stephen.mboya@mdu.se

S. Silvestrov (✉) · A. Kitouni
Division of Mathematics and Physics, Mälardalen University, Västerås, Sweden
e-mail: sergei.silvestrov@mdu.se

E. Ongong'a
Strathmore Institute of Mathematical Sciences, Strathmore University, Nairobi, Kenya
e-mail: eongonga@strathmore.edu

J. Ongaro
Department of Mathematics, University of Nairobi, Nairobi, Kenya
e-mail: ongaro@uonbi.ac.ke

M. Mitrović, M. N. Hounkonnou (eds.), *Algebra Without Borders*, Mathematics in Mind, https://doi.org/10.1007/978-3-031-86477-3_11

1 Introduction

The subject of Hom-algebra structures has been started by Hartwig, Larsson, and Silvestrov in [4] in 2003. In [4], Hom-Lie algebras and more general quasi-Hom-Lie algebras were introduced while developing a general method for construction of quasi-deformations and discretizations of Lie algebras of vector fields using more general twisted derivations in place of derivatives leading to deformations of Witt- and Virasoro-type algebras based on twisted derivations, motivated by some examples of q-deformed Jacobi identities observed in physics in some q-deformed algebras and their representations by some q-difference operators, discrete modifications of differential calculus, noncommutative differential calculi, and q-deformed versions of homological algebra. The general abstract quasi-Lie algebras and the subclasses of quasi-Hom-Lie algebras and Hom-Lie algebras as well as their general colored (graded) counterparts including color and super quasi-Lie algebras, color and super quasi-Hom-Lie algebras, and color and super Hom-Lie algebras have been introduced by Silvestrov, Larsson, and Sigurdsson in [4, 11, 12, 29] in 2003–2005. Subsequently, various classes of Hom-Lie-admissible algebras have been considered by Makhlouf and Silvestrov in [18] in 2006. In particular, in [18], the Hom-associative algebras have been introduced and shown to be Hom-Lie-admissible, that is, leading to Hom-Lie algebras using commutator map as new product, and in this sense constituting a natural generalization of associative algebras since the associative algebras are Lie-admissible as for any associative algebra the commutator map as new product yields a Lie algebra. In [18], moreover, several other interesting classes of Hom-Lie-admissible algebras generalizing some classes of nonassociative algebras, as well as examples of low-dimensional finite-dimensional Hom-Lie algebras, have been described. Hom-algebra structures are very useful since Hom-algebra structures of a given type include their classical counterparts and open more possibilities for deformations, extensions of cohomological structures, and representations. Since these pioneering works [4, 11–14, 18], Hom-algebra structures have developed into a popular broad area, with an increasing number of publications in various directions (see, e.g., [2, 3, 8, 11, 14, 15, 19–21, 27, 28, 30–33] and references therein).

Hom-Lie algebras are skew-symmetric not necessarily associative algebras together with a linear mapping α satisfying Hom-Jacobi identity, which is a generalization of Jacobi identity. Lie algebras are Hom-Lie algebras, because with $\alpha = Id$ (the identity mapping), the Hom-Jacobi identity becomes the Jacobi identity of Lie algebras. So, if the identity mapping is in the set of all linear mappings α making a given skew-symmetric algebra into a Hom-Lie algebra, then this Hom-algebra is also an ordinary Lie algebra. The investigations on classification of finite-dimensional Hom-Lie algebras have been initiated in [18] considering the systems of polynomial equations involving the structure constants of the bilinear bracket multiplication and the matrix elements of the twisting linear map α. Since then, the classifications of Hom-Lie algebras and other Hom-Lie-type Hom-algebras have been considered in either low dimensions or for some special classes. Multiplicative

3-dimensional Hom-Lie algebras have been classified in [16], more specifically, Hom-Lie algebras which are not Lie algebras. In [17], the special cases of nilpotent and filiform Hom-Lie algebras are studied and classified up to dimension 7. In [26], the algebraic varieties of Hom-Lie algebras over the complex numbers are considered, and it is shown that all 3-dimensional skew-symmetric algebras can be made into Hom-Lie algebras, but this is not true for the 4-dimensional case, and some further properties of the algebraic varieties of Hom-Lie algebras are studied. In [5], the authors consider also the algebraic varieties of Hom-Lie algebras over the complex numbers; they classify Hom-Lie structures with nilpotent twisting maps, as well as address the classification of their orbit closures. In [22], Hom-Lie structures on complex Lie algebras of dimension 4 are investigated. In [6], the classification of 3-dimensional Hom-Lie algebras is considered. The approach here is to first classify skew-symmetric multiplications, while indicating which ones define Lie algebras, and then classify the twisting maps over each skew-symmetric algebra that yield Hom-Lie algebra structure. In [23, 24], the classification of 3-dimensional Hom-Lie algebra is achieved using a different approach. A system of polynomial equations is obtained from the defining identities; the properties of the spaces of all linear mappings that yield a Hom-Lie algebra together with a given skew-symmetric multiplication are studied. A classification of 3-dimensional Hom-Lie algebras for nilpotent twisting maps is achieved by separating non-similar twisting maps, then classifying the multiplications up to isomorphisms of Hom-Lie algebras, that is, linear maps that intertwine both multiplications and twisting maps. The classification of complex multiplicative non-Lie Hom-Lie algebras of dimension 4 associated with a particular linear map has been considered in [1].

This paper aims to explore certain properties and structures of Hom-Lie algebras of generalized $\mathfrak{sl}(2)$-type. We construct classes of linear twisting maps that turn a skew-symmetric algebra of generalized $\mathfrak{sl}(2)$-type into a Hom-Lie algebra and describe the subclasses of such linear maps which yield multiplicative Hom-Lie algebras. We explore some properties of these Hom-algebras related to their ideals, Hom-ideals, subalgebras, and Hom-subalgebras, with emphasis on their derived series, and central descending series, as well as their nilpotence and solvability properties. We investigate the invariance of these subalgebras under the linear twisting maps, and determine whether these subalgebras are weak subalgebras, Hom-subalgebras, weak ideals, or Hom-ideals. In particular, we investigate these subalgebras and properties for the subfamilies of non-multiplicative Hom-Lie algebras of generalized $\mathfrak{sl}(2)$-type indicating the differences between non-multiplicative and multiplicative cases.

The paper is organized as follows. In Sect. 2, we give some general background knowledge necessary for our studies; this includes definitions of Hom-Lie algebras, Hom-subalgebras, and Hom-ideals. In Sect. 3, we construct polynomial equations defining Hom-Jacobi identity in terms of structure constants of the bilinear bracket multiplication, and linear twisting α. Furthermore, we give a system of polynomial equations equivalent to the weak morphism condition for the multiplicative Hom-Lie algebras. In Sect. 4 we confine our studies to 3-dimensional Hom-Lie algebras. Here, we construct all the linear twisting maps so that a bilinear multiplication (bracket) defined by $[e_1, e_2] = \lambda_1 e_2$, $[e_1, e_3] = \lambda_2 e_3$, and $[e_2, e_3] = \lambda_3 e_1$ determines 3-dimensional Hom-Lie algebra. For all different cases of structure constants $\lambda_i, i = 1, 2, 3$, we construct all subfamilies of the linear mappings α for which such skew-symmetric brackets yield a Hom-Lie algebra. Furthermore, we construct subclasses of these twisting maps α such that the resulting Hom-Lie algebras are multiplicative. Section 5 is devoted to the solvability and nilpotency of Hom-Lie algebras. We explore the properties of derived algebras and their invariance under the twisting and summarize the result in Tables 1, 2, and 3 and discussions thereafter.

2 Preliminaries

In this work, we are interested in 3-dimensional algebras with skew-symmetric binary multiplication over a field $\mathbb{K}$ of characteristic zero. We begin by providing some basic definitions and properties of Hom-Lie algebras necessary for our study. All linear spaces considered herein are over $\mathbb{K}$.

Definition 1 (Hom-Lie Algebra, [18]) A Hom-Lie algebra is a triple $(\mathcal{V}, [\cdot,\cdot], \alpha)$ consisting of a linear space $\mathcal{V}$, a bilinear map $[\cdot,\cdot] : \mathcal{V} \times \mathcal{V} \to \mathcal{V}$, and a linear space homomorphism $\alpha : \mathcal{V} \to \mathcal{V}$ satisfying for all $x, y, z \in \mathcal{V}$:

$$[x, y] = -[y, x] \qquad \text{Skew-symmetry identity}$$

$$\sum_{\circlearrowleft(x,y,z)} [\alpha(x), [y, z]] = [\alpha(x), [y, z]] + [\alpha(z), [x, y]] + [\alpha(y), [[z, x]] = 0 \qquad \text{Hom-Jacobi identity}$$

where $\sum_{\circlearrowleft(a,b,c)}$ denotes the summation over cyclic permutation of (a, b, c).

When the twisting α is the identity map, a Hom-Lie algebra becomes a Lie-Algebra.

Definition 2 Let $(\mathcal{V}_1, [\cdot,\cdot], \alpha)$ and $(\mathcal{V}_2, \{\cdot,\cdot\}, \beta)$ be Hom-Lie algebras. A Hom-Lie algebra morphism is a linear mapping $f : (\mathcal{V}_1, [\cdot,\cdot], \alpha) \to (\mathcal{V}_2, \{\cdot,\cdot\}, \beta)$ satisfying:

(i) $f([x, y]) = \{f(x), f(y)\}$ for all $x, y \in \mathcal{V}_1$,
(ii) $f \circ \alpha = \beta \circ f$.

When f satisfies only the first condition (i), we say that f is a weak morphism. A Hom-Lie-algebra $(\mathcal{V}, [\cdot,\cdot], \alpha)$ is said to be multiplicative if α is an algebra morphism, and regular if α is an isomorphism.

We explore some properties of Hom-Lie algebras through their subalgebras and ideals and, therefore, provide the following definitions.

Definition 3 (Hom-Subalgebra) Let $(\mathcal{V}, [\cdot,\cdot], \alpha)$ be a Hom-Lie algebra. A Hom-subalgebra is a subspace $\mathcal{W} \subseteq \mathcal{V}$ that is invariant under the linear α, and closed under bracket multiplication:

$$\alpha(\mathcal{W}) \subseteq \mathcal{W}, \quad \text{(invariance under twisting map)} \tag{2.1}$$

$$[w_1, w_2] \in \mathcal{W} \text{ for all } w_1, w_2 \in \mathcal{W}. \quad \text{(closure under bracket)} \tag{2.2}$$

Definition 4 (Hom-Ideal) Let $(\mathcal{V}, [\cdot,\cdot], \alpha)$ be a Hom-Lie algebra. A Hom-ideal is a subspace $I \subseteq \mathcal{V}$ satisfying

$$\alpha(I) \subseteq I, \quad \text{(invariance under twisting map)} \tag{2.3}$$

$$[\mathcal{V}, x] \subseteq I \text{ for all } x \in I. \quad \text{(weak ideal property)} \tag{2.4}$$

Definition 5 Let $(\mathcal{V}, [\cdot,\cdot], \alpha)$ be a Hom-Lie algebra. A subspace $\mathcal{W} \subseteq \mathcal{V}$ is a weak subalgebra if it satisfies only condition 2.2 in Definition 3, and a weak ideal if it satisfies only condition 2.4 in Definition 4.

Example 13 (Hom-ideal and Weak Ideal)

(1) Let $\mathcal{A} = (\mathcal{V}, [\cdot,\cdot], \alpha)$ and $\mathcal{B} = (\mathcal{W}, \{\cdot,\cdot\}, \alpha)$ be Hom-Lie algebras and $f : \mathcal{A} \to \mathcal{B}$ be Hom-Lie algebra morphism. The subspace $\ker(f)$ is α-invariant, since $f(\alpha(x)) = \alpha(f(x)) = \alpha(0) = 0$ for $x \in \ker(f)$. Furthermore, $\ker(f)$ is a Hom-ideal, since $f([v, x]) = \{f(v), f(x)\} = \{f(v), 0\} = 0$ for $v \in \mathcal{V}$.
(2) The center $Z(\mathcal{V}) = \{z \in \mathcal{V} \mid [\mathcal{V}, z] = 0\}$ of a Hom-Lie algebra $(\mathcal{V}, [\cdot,\cdot], \alpha)$ is a weak ideal, since $[\mathcal{V}, z] \in Z(\mathcal{V})$ for all $z \in Z(\mathcal{V})$.

In the sequel, we express the skew-symmetry condition and the Hom-Jacobi identity in terms of structure constants of bilinear bracket multiplication and linear space endomorphism α.

3 Structure Constants

Given a basis $\{e_1, \ldots, e_n\}$ in an n-dimensional linear space $\mathcal{V}$, the bilinear map $[\cdot,\cdot] : \mathcal{V} \times \mathcal{V}$ and the linear map $\alpha : \mathcal{V} \to \mathcal{V}$ of the Hom-Lie algebra $(\mathcal{V}, [\cdot,\cdot], \alpha)$ can be expressed in terms of the structure constants as

$$\left[e_i, e_j\right] = \sum_{s=1}^{n} C_{i,j}^{s} e_s \text{ and } \alpha(e_i) = \sum_{t=1}^{n} a_{ti} e_t, \text{ for all } C_{i,j}^{s}, a_{ti} \in \mathbb{K},\ 1 \le i, j \le n. \tag{3.1}$$

Substituting (3.1) in the skew-symmetry and Hom-Jacobi identity conditions, we obtain:

(i) **Skew-symmetry**

$$[e_i, e_j] = -[e_j, e_i] \iff \sum_{s=1}^{n} C_{i,j}^{s} e_s = -\sum_{s=1}^{n} C_{j,i}^{s} e_s.$$

(ii) **Hom-Jacobi identity**

$$\begin{aligned}\sum_{\circlearrowleft(i,j,k)} [\alpha(e_i), [e_j, e_k]] &= \sum_{\circlearrowleft(i,j,k)} [\sum_{t=1}^{n} a_{ti} e_t, \sum_{s=1}^{n} C_{j,k}^{s} e_s] \\ &= \sum_{s=1}^{n}\sum_{t=1}^{n}\sum_{\circlearrowleft(i,j,k)} a_{ti} C_{j,k}^{s} [e_t, e_s] = 0 \\ &= \sum_{r=1}^{n}\Big(\sum_{s=1}^{n}\sum_{t=1}^{n}\sum_{\circlearrowleft(i,j,k)} a_{ti} C_{j,k}^{s} C_{t,s}^{r}\Big) e_r = 0. \end{aligned} \tag{3.2}$$

Thus, for Hom-Lie algebras, the Hom-Jacobi identity (3.2) can be written explicitly as a system of polynomial equations:

$$\sum_{s=1}^{n}\sum_{t=1}^{n} \left(a_{ti} C_{j,k}^{s} C_{t,s}^{r} + a_{tk} C_{i,j}^{s} C_{t,s}^{r} + a_{tj} C_{k,i}^{s} C_{t,s}^{r}\right) = 0, \tag{3.3}$$

$$\text{where } C_{i,j}^{k} = -C_{j,i}^{k}, \text{ for all } i, j, k, r = 1, \dots, n.$$

This system of polynomial equations (3.3) can be viewed as linear in the variables $\{a_{ti}\}$ and can be rewritten as

$$\sum_{t=1}^{n} \Big(a_{ti}\Big(\sum_{s=1}^{n} C_{j,k}^{s} C_{t,s}^{r}\Big) + a_{tk}\Big(\sum_{s=1}^{n} C_{i,j}^{s} C_{t,s}^{r}\Big) + a_{tj}\Big(\sum_{s=1}^{n} C_{k,i}^{s} C_{t,s}^{r}\Big)\Big) = 0, \tag{3.4}$$

$$\forall 1 \le i < j < k \le n \text{ and } r = 1, \dots n. \tag{3.5}$$

The Hom-Jacobi identity in (3.4) can be represented in matrix form as $\mathcal{M}A_{\alpha} = 0$, where $\mathcal{M}$ is a $\binom{n}{3} \cdot n \times n^2$ matrix with coefficients associated with the structure constants of the bilinear bracket multiplication $\{(C_{i,j}^{r})_{i<j}\}$, and A_{α} is the column matrix with coefficients associated with the structure constants $\{a_{ti}\}$ of the linear twist α. Therefore, the linear transformation represented by $\mathcal{M}$ is

$$\mathcal{L} : \mathbb{K}^{n^2} \to \mathbb{K}^{\frac{n^2(n-1)(n-2)}{6}}. \tag{3.6}$$

For a given bilinear bracket multiplication, the solution set of (3.4) can be interpreted as the classes of linear map α that turn a skew-symmetric algebra $(\mathcal{V}, [\cdot, \cdot])$ into Hom-Lie algebras $(\mathcal{V}, [\cdot, \cdot], \alpha)$.

By a Hom-Lie structure, we mean the linear subspace of all linear maps α that satisfy the Hom-Jacobi identity for a skew-symmetric algebra, and denote it by

$$\texttt{HomLie}_n = \Big\{\alpha \in \mathrm{End}(\mathcal{V}) \mid \sum_{\circlearrowleft(x,y,z)} [\alpha(x), [y, z]] = 0\Big\}. \tag{3.7}$$

Next, we present the subsets of the linear space of all linear map α in (3.4) such that $(\mathcal{V}, [\cdot, \cdot], \alpha)$ is a multiplicative Hom-Lie algebra. For this construction, the weak morphism condition together with Hom-Jacobi identity hold. The weak morphism condition can be written in terms of structure constants as follows. Let $\{e_1, \dots, e_n\}$ be a basis for $\mathcal{V}$ then,

$$\alpha([e_i, e_j]) - [\alpha(e_i), \alpha(e_j)] = 0 \Leftrightarrow \alpha(\sum_{s=1}^{n} C_{i,j}^s e_s) - [\sum_{t=1}^{n} a_{ti} e_t, \sum_{p=1}^{n} a_{pj} e_p] = 0$$

$$\Leftrightarrow \sum_{s=1}^{n}\sum_{r=1}^{n} a_{rs} C_{i,j}^s e_r - \sum_{t=1}^{n} a_{ti} \sum_{p=1}^{n} a_{pj} \left[e_t, e_p\right] = 0$$

$$\Leftrightarrow \sum_{s=1}^{n}\sum_{r=1}^{n} a_{rs} C_{i,j}^s e_r - \sum_{t=1}^{n} a_{ti} \sum_{p=1}^{n} a_{pj} \sum_{q=1}^{n} C_{t,p}^q e_q = 0, \quad \forall\, 1 \le i < j \le n,$$

$$\Leftrightarrow \sum_{r=1}^{n}\left(\sum_{s=1}^{n} a_{rs} C_{i,j}^s - \sum_{t=1}^{n}\sum_{p=1}^{n} a_{ti} a_{pj} C_{t,p}^r\right) e_r = 0, \quad \forall\, 1 \le i < j \le n,$$

$$\Leftrightarrow \sum_{s=1}^{n} a_{rs} C_{i,j}^s - \sum_{t=1}^{n}\sum_{p=1}^{n} a_{ti} a_{pj} C_{t,p}^r = 0,\ \forall\, 1 \le i < j \le n,\ 1 \le r \le n \tag{3.8}$$

The solution set of (3.3) together with (3.8) gives the subclasses of multiplicative Hom-Lie algebras.

4 3-Dimensional Hom-Lie Algebras

In [18], the authors listed all the classes of bilinear bracket multiplication of the Hom-Lie algebras obtained using a special diagonal linear map α. Here, for a known bilinear bracket multiplication, we construct all classes of Hom-Lie structures, and subclasses of linear space of all linear twisting maps consisting of multiplicative Hom-Lie algebras. In the 3-dimensional case, Hom-Jacobi identity in (3.3) together with skew-symmetry conditions gives, for $r = 1, 2, 3$:

$$\begin{aligned}a_{11}\big(C^3_{2,3}C^r_{1,3}+C^2_{2,3}C^r_{1,2}\big)+a_{12}\big(-C^3_{1,3}C^r_{1,3}-C^2_{1,3}C^r_{1,2}\big)&\\+a_{13}\big(C^3_{1,2}C^r_{1,3}+C^2_{1,2}C^r_{1,2}\big)&\\+a_{21}\big(C^3_{2,3}C^r_{2,3}-C^1_{2,3}C^r_{1,2}\big)+a_{22}\big(C^1_{1,3}C^r_{1,2}-C^3_{1,3}C^r_{2,3}\big)&\\+a_{23}\big(C^3_{1,2}C^r_{2,3}-C^1_{1,2}C^r_{1,2}\big)&\\+a_{31}\big(-C^1_{2,3}C^r_{1,3}-C^2_{2,3}C^r_{2,3}\big)+a_{32}\big(C^2_{1,3}C^r_{2,3}+C^1_{1,3}C^r_{1,3}\big)&\\+a_{33}\big(-C^1_{1,2}C^r_{1,3}-C^2_{1,2}C^r_{2,3}\big)&=0.\end{aligned}\tag{4.1}$$

Thus, substituting $r = 1, 2, 3$, we get:

$$\begin{aligned}a_{11}\big(C^2_{2,3}C^1_{1,2}+C^3_{2,3}C^1_{1,3}\big)+a_{12}\big(-C^3_{1,3}C^1_{1,3}-C^2_{1,3}C^1_{1,2}\big)&\\+a_{13}\big(C^3_{1,2}C^1_{1,3}+C^2_{1,2}C^1_{1,2}\big)+&\\+a_{21}\big(C^3_{2,3}C^1_{2,3}-C^1_{2,3}C^1_{1,2}\big)+a_{22}\big(C^1_{1,3}C^1_{1,2}-C^3_{1,3}C^1_{2,3}\big)&\\+a_{23}\big(C^3_{1,2}C^1_{2,3}-C^1_{1,2}C^1_{1,2}\big)+&\\+a_{31}\big(-C^1_{2,3}C^1_{1,3}-C^2_{2,3}C^1_{2,3}\big)+a_{32}\big(C^2_{1,3}C^1_{2,3}+C^1_{1,3}C^1_{1,3}\big)&\\+a_{33}\big(-C^1_{1,2}C^1_{1,3}-C^2_{1,2}C^1_{2,3}\big)&=0.\end{aligned}\tag{4.2}$$

$$\begin{aligned}a_{11}\big(C^3_{2,3}C^2_{1,3}+C^2_{2,3}C^2_{1,2}\big)+a_{12}\big(-C^3_{1,3}C^2_{1,3}-C^2_{1,3}C^2_{1,2}\big)&\\+a_{13}\big(C^3_{1,2}C^2_{1,3}+C^2_{1,2}C^2_{1,2}\big)+&\\+a_{21}\big(C^3_{2,3}C^2_{2,3}-C^1_{2,3}C^2_{1,2}\big)+a_{22}\big(C^1_{1,3}C^2_{1,2}-C^3_{1,3}C^2_{2,3}\big)&\\+a_{23}\big(C^3_{1,2}C^2_{2,3}-C^1_{1,2}C^2_{1,2}\big)+&\\+a_{31}\big(-C^1_{2,3}C^2_{1,3}-C^2_{2,3}C^2_{2,3}\big)+a_{32}\big(C^2_{1,3}C^2_{2,3}+C^1_{1,3}C^2_{1,3}\big)&\\+a_{33}\big(-C^1_{1,2}C^2_{1,3}-C^2_{1,2}C^2_{2,3}\big)&=0.\end{aligned}\tag{4.3}$$

$$\begin{aligned}a_{11}\big(C^3_{2,3}C^3_{1,3}+C^2_{2,3}C^3_{1,2}\big)+a_{12}\big(-C^3_{1,3}C^3_{1,3}-C^2_{1,3}C^3_{1,2}\big)&\\+a_{13}\big(C^3_{1,2}C^3_{1,3}+C^2_{1,2}C^3_{1,2}\big)+&\\+a_{21}\big(C^3_{2,3}C^3_{2,3}-C^1_{2,3}C^3_{1,2}\big)+a_{22}\big(C^1_{1,3}C^3_{1,2}-C^3_{1,3}C^3_{2,3}\big)&\\+a_{23}\big(C^3_{1,2}C^3_{2,3}-C^1_{1,2}C^3_{1,2}\big)+&\\+a_{31}\big(-C^1_{2,3}C^3_{1,3}-C^2_{2,3}C^3_{2,3}\big)+a_{32}\big(C^2_{1,3}C^3_{2,3}+C^1_{1,3}C^3_{1,3}\big)&\\+a_{33}\big(-C^1_{1,2}C^3_{1,3}-C^2_{1,2}C^3_{2,3}\big)&=0.\end{aligned}\tag{4.4}$$

In this case, the matrix $\mathcal{M}$ represents a linear transformation $\mathcal{L} : \mathbb{K}^9 \to \mathbb{K}^3$. The Hom-Lie algebra is obtained if and only if the matrix of the linear map A_α is in the kernel of $\mathcal{L}$. This follows from the rank-nullity theorem:

$$\dim \mathcal{V} = \dim(\ker\mathcal{L}) + \dim(Im\mathcal{L}) \iff \dim(\ker\mathcal{L}) := 9 - \dim \mathcal{V} \geq 6.$$

Since the $\ker\mathcal{L}$ is nontrivial for any skew-symmetric algebra, there exists a nontrivial element in $\ker\mathcal{L}$. Therefore, we obtain a nontrivial Hom-Lie structure as shown in [25]. Furthermore, in [23], the authors have shown that the space of Hom-Lie structures has dimension, $6 \leq \dim(\texttt{HomLie}_3) \leq 9$.

Theorem 1 ([25]) *Any skew-symmetric* 3*-dimensional algebra is a Hom-Lie algebra.*

Now, we present some results from our computations.

4.1 Hom-Lie Structure

Let $\{e_1, e_2, e_3\}$ be a basis for skew-symmetric algebra $\mathcal{V}$. We construct linear space of all linear maps α such that the bilinear bracket multiplication defined by

$$[e_1, e_2] = \lambda_1 e_2, [e_1, e_3] = \lambda_2 e_3, [e_2, e_3] = \lambda_3 e_1, \quad \forall\, \lambda_1, \lambda_2, \lambda_3 \in \mathbb{K} \tag{4.5}$$

determines 3-dimensional Hom-Lie algebra. The structure constants of the bilinear bracket multiplication in (4.5) is given by

$$C_{i,j}^r = \begin{cases} \lambda_1 & \text{for} \quad i = 1, j = r = 2, \\ \lambda_2 & \text{for} \quad i = 1, j = r = 3, \\ \lambda_3 & \text{for} \quad i = 2, j = 3, r = 1, \\ 0 & \qquad \text{elsewhere.} \end{cases} \tag{4.6}$$

Substituting (4.6) in (4.2), (4.3), and (4.4), we obtain the following system of polynomial equations:

$$\begin{aligned} a_{22}\left(-C_{1,3}^3 C_{2,3}^1\right) + a_{33}\left(-C_{1,2}^2 C_{2,3}^1\right) &= 0 \\ a_{13}\left(C_{1,2}^2 C_{1,2}^2\right) + a_{21}\left(-C_{2,3}^1 C_{1,2}^2\right) &= 0 \\ a_{12}\left(-C_{1,3}^3 C_{1,3}^3\right) + a_{31}\left(-C_{2,3}^1 C_{1,3}^3\right) &= 0 \end{aligned} \iff \begin{cases} -\lambda_2\lambda_3 a_{22} - \lambda_1\lambda_3 a_{33} = 0 \\ \lambda_1^2 a_{13} - \lambda_1\lambda_3 a_{21} = 0 \\ -\lambda_2^2 a_{12} - \lambda_3\lambda_2 a_{31} = 0 \end{cases} \tag{4.7}$$

For each case of $\lambda_{i,\, i=1,2,3}$, we provide classes of Hom-Lie structures, denoted by $[\alpha]$, and comment on the dimensions of each Hom-Lie structure presented.

(a) When $\lambda_1 = 0$, and $\lambda_2, \lambda_3 \neq 0$, then the solution set of (4.7) is given by

$$\{(a_{ij}), 1 \leq i, j \leq 3 \mid a_{22} = 0,\ a_{31} = -\frac{\lambda_2}{\lambda_3} a_{12}\}$$

and the resulting Hom-Lie structure is of dimension 7 given by $[\alpha] = \begin{pmatrix} a_{11} & a_{12} & a_{13} \\ a_{21} & 0 & a_{23} \\ -\frac{\lambda_2}{\lambda_3}a_{12} & a_{32} & a_{33} \end{pmatrix}$.

(b) When $\lambda_2 = 0$ and $\lambda_1, \lambda_3 \neq 0$, the solution set of (4.7) is given by

$$\{(a_{ij}), 1 \leq i, j \leq 3 \mid a_{33} = 0, a_{21} = \frac{\lambda_1}{\lambda_3}a_{13}\}$$

and the resulting Hom-Lie structure is of dimension 7 given by $[\alpha] = \begin{pmatrix} a_{11} & a_{12} & a_{13} \\ \frac{\lambda_1}{\lambda_3}a_{13} & a_{22} & a_{23} \\ a_{31} & a_{32} & 0 \end{pmatrix}$.

(c) When $\lambda_3 = 0$ and $\lambda_1, \lambda_2 \neq 0$, then the solution set of (4.7) is given by

$$\{(a_{ij}), 1 \leq i, j \leq 3 \mid a_{12} = a_{13} = 0\}$$

and the resulting Hom-Lie structure is of dimension 7 given by $[\alpha] = \begin{pmatrix} a_{11} & 0 & 0 \\ a_{21} & a_{22} & a_{23} \\ a_{31} & a_{32} & a_{33}. \end{pmatrix}$

(d) When $\lambda_1, \lambda_2 = 0$ and $\lambda_3 \neq 0$, then every (a_{ij}) is a solution set of (4.7), and the Hom-Lie structure is the whole space $\mathbb{K}^9$ given by the matrix $[\alpha] = \begin{pmatrix} a_{11} & a_{12} & a_{13} \\ a_{21} & a_{22} & a_{23} \\ a_{31} & a_{32} & a_{33} \end{pmatrix}$.

(e) When $\lambda_1, \lambda_3 = 0$ and $\lambda_2 \neq 0$, then the solution set of (4.7) is given by

$$\{(a_{ij}), 1 \leq i, j \leq 3 \mid a_{12} = 0\}$$

and the resulting Hom-Lie structure is of dimension 8 given by the matrix $[\alpha] = \begin{pmatrix} a_{11} & 0 & a_{13} \\ a_{21} & a_{22} & a_{23} \\ a_{31} & a_{32} & a_{33} \end{pmatrix}$.

(f) When $\lambda_2, \lambda_3 = 0$, and $\lambda_1 \neq 0$, then the solution set of (4.7) is given by

$$\{(a_{ij}), 1 \leq i, j \leq 3 \mid a_{13} = 0\}$$

and the resulting Hom-Lie structure is of dimension 8 given by the matrix $[\alpha] = \begin{pmatrix} a_{11} & a_{12} & 0 \\ a_{21} & a_{22} & a_{23} \\ a_{31} & a_{32} & a_{33} \end{pmatrix}$

(g) When $\lambda_1, \lambda_2, \lambda_3 \neq 0$, then the solution set of (4.7) is given by

$$\{(a_{ij}),\ 1 \le i, j \le 3 \mid a_{33} = -\frac{\lambda_2}{\lambda_1} a_{22}, a_{21} = \frac{\lambda_1}{\lambda_3} a_{13}, a_{31} = -\frac{\lambda_2}{\lambda_3} a_{12}\}$$

and the resulting Hom-Lie structure is of dimension 6 given by the matrix $[\alpha] = \begin{pmatrix} a_{11} & a_{12} & a_{13} \\ \frac{\lambda_1}{\lambda_3} a_{13} & a_{22} & a_{23} \\ -\frac{\lambda_2}{\lambda_3} a_{12} & a_{32} & -\frac{\lambda_2}{\lambda_1} a_{22} \end{pmatrix}$.

The following remarks (10) give a trivial relation between Hom-Lie algebras and their corresponding Lie algebras.

Remark 10

(i) The Hom-Lie algebras given by the bracket multiplications in the cases (c), (d), (e), (f), and (g) are moreover Lie algebras since in these cases the identity map is in the corresponding space of Hom-Lie structures.
(ii) The triple $(\mathcal{V}, [\cdot\,, \cdot]\,, \alpha)$ with the linear map α defined by cases (a) and (b) is a Hom-Lie algebra, but the brackets do not define a Lie algebra since in these cases the identity map is not in the corresponding space of Hom-Lie structures.

Example 14 Considering the bracket multiplication defined by

$$[e_1, e_2] = 2e_2, [e_1, e_3] = -2e_3, [e_2, e_3] = e_1$$

and substituting these structure constants in (4.7), we obtain the equation system:

$$a_{22} - a_{33} = 0, 2a_{13} - a_{21} = 0, -2a_{12} + a_{31} = 0, \tag{4.8}$$

with solution $\{(a_{ij}),\ 1 \le i, j \le 3 \mid a_{22} = a_{33},\ a_{21} = 2a_{13},\ a_{31} = 2a_{12}\}$, and thus the Hom-Lie structure linear space $\{[\alpha] = \begin{pmatrix} a_{11} & a_{12} & a_{13} \\ 2a_{13} & a_{22} & a_{23} \\ 2a_{12} & a_{32} & a_{22} \end{pmatrix}\}$ is of dimension 6.

4.2 *Multiplicative Hom-Lie Algebra*

In a 3-dimensional linear space $\mathcal{V}$, the multiplicativity condition of a Hom-Lie algebra, given by the weak morphism condition in (3.8), becomes:

$$\begin{aligned} (a_{11}C_{i,j}^1 + a_{12}C_{i,j}^2 + a_{13}C_{i,j}^3)e_1 + (a_{21}C_{i,j}^1 + a_{22}C_{i,j}^2 + a_{23}C_{i,j}^3)e_2 + \\ (a_{31}C_{i,j}^1 + a_{32}C_{i,j}^2 + a_{33}C_{i,j}^3)e_3 = (a_{3i}a_{2j}C_{3,2}^1 + a_{2i}a_{3j}C_{2,3}^1)e_1 + \\ (a_{2i}a_{1j}C_{2,1}^2 + a_{1i}a_{2j}C_{1,2}^2)e_2 + (a_{3i}a_{1j}C_{3,1}^3 + a_{1i}a_{3j}C_{1,3}^3)e_3 \\ \text{for all } i, j = 1, 2, 3. \end{aligned} \tag{4.9}$$

By the skew-symmetry condition of the bilinear bracket multiplication in (4.5)

$$
\begin{aligned}
&(a_{22}a_{31}C^1_{2,3} - a_{32}a_{21}C^1_{2,3} + a_{12}C^2_{1,2})e_1 + (a_{12}a_{21}C^2_{1,2} - a_{22}a_{11}C^2_{1,2} + a_{22}C^2_{1,2})e_2\\
&\qquad + (a_{12}a_{31}C^3_{1,3} - a_{32}a_{11}C^3_{1,3} + a_{32}C^2_{1,2})e_3 = 0\\
&(a_{23}a_{31}C^1_{2,3} - a_{33}a_{21}C^1_{2,3} + a_{13}C^3_{1,3})e_1 + (a_{13}a_{21}C^2_{1,2} - a_{23}a_{11}C^2_{1,2} + a_{23}C^3_{1,3})e_2\\
&\qquad + (a_{13}a_{31}C^3_{1,3} - a_{33}a_{11}C^3_{1,3} + a_{33}C^3_{1,3})e_3 = 0\\
&(a_{21}a_{32}C^1_{2,3} - a_{31}a_{22}C^1_{2,3} - a_{12}C^2_{1,2})e_1 + (a_{11}a_{22}C^2_{1,2} - a_{21}a_{12}C^2_{1,2} - a_{22}C^2_{1,2})e_2\\
&\qquad + (a_{11}a_{32}C^3_{1,3} - a_{31}a_{12}C^3_{1,3} - a_{32}C^2_{1,2})e_3 = 0\\
&(a_{23}a_{32}C^1_{2,3} - a_{33}a_{22}C^1_{2,3} + a_{11}C^1_{2,3})e_1 + (a_{13}a_{22}C^2_{1,2} - a_{23}a_{12}C^2_{1,2} + a_{21}C^1_{2,3})e_2\\
&\qquad + (a_{13}a_{32}C^3_{1,3} - a_{33}a_{12}C^3_{1,3} + a_{31}C^1_{2,3})e_3 = 0\\
&(a_{21}a_{33}C^1_{2,3} - a_{31}a_{23}C^1_{2,3} - a_{13}C^3_{1,3})e_1 + (a_{11}a_{23}C^2_{1,2} - a_{21}a_{13}C^2_{1,2} - a_{23}C^3_{1,3})e_2\\
&\qquad + (a_{11}a_{33}C^3_{1,3} - a_{31}a_{13}C^3_{1,3} - a_{33}C^3_{1,3})e_3 = 0\\
&(a_{22}a_{33}C^1_{2,3} - a_{32}a_{23}C^1_{2,3} - a_{11}C^1_{2,3})e_1 + (a_{12}a_{23}C^2_{1,2} - a_{22}a_{13}C^2_{1,2} - a_{21}C^1_{2,3})e_2\\
&\qquad + (a_{12}a_{33}C^3_{1,3} - a_{32}a_{13}C^3_{1,3} - a_{31}C^1_{2,3})e_3 = 0.
\end{aligned}
\tag{4.10}
$$

which is equivalent to all coefficients of e_1, e_2, e_3 in linear combinations being zero, that is to a system of polynomial equations. From our computations we give the subclasses of the linear α such that $\mathcal{A} = (\mathcal{V}, [\cdot\,, \cdot]\,, \alpha)$ is a multiplicative Hom-Lie algebra:

1. When $\lambda_1 = 0$, and $\lambda_2, \lambda_3 \neq 0$, then $\mathcal{A}$ has linear twists of the form $[\alpha] = \begin{pmatrix} a_{11} & a_{12} & a_{13} \\ a_{21} & 0 & a_{23} \\ -\frac{\lambda_2}{\lambda_3}a_{12} & a_{32} & a_{33} \end{pmatrix}$ and the corresponding multiplicative $\mathcal{A}$ has a linear twist of the form: $[\alpha] = \begin{pmatrix} 0 & 0 & 0 \\ 0 & 0 & 0 \\ 0 & a_{32} & 0 \end{pmatrix}$.
2. When $\lambda_2 = 0$ and $\lambda_1, \lambda_3 \neq 0$, then $\mathcal{A}$ has the linear twists of the form $[\alpha] = \begin{pmatrix} a_{11} & a_{12} & a_{13} \\ \frac{\lambda_1}{\lambda_3}a_{13} & a_{22} & a_{23} \\ a_{31} & a_{32} & 0 \end{pmatrix}$, and the corresponding multiplicative $\mathcal{A}$ has the linear twists of the form $[\alpha] = \begin{pmatrix} 0 & 0 & 0 \\ 0 & 0 & a_{23} \\ 0 & 0 & 0 \end{pmatrix}$.
3. When $\lambda_3 = 0$ and $\lambda_1, \lambda_2 \neq 0$, then $\mathcal{A}$ has linear twists of the form $[\alpha] = \begin{pmatrix} a_{11} & 0 & 0 \\ a_{21} & a_{22} & a_{23} \\ a_{31} & a_{32} & a_{33} \end{pmatrix}$ and the corresponding multiplicative $\mathcal{A}$ has linear twists:

   (a) If $\lambda_1 \neq \pm\lambda_2$ for all $\lambda_1, \lambda_2 \neq 0$,

$$
\begin{pmatrix} 1 & 0 & 0 \\ a_{21} & a_{22} & 0 \\ a_{31} & 0 & a_{33} \end{pmatrix}, \begin{pmatrix} a_{11} & 0 & 0 \\ a_{21} & 0 & 0 \\ a_{31} & 0 & 0 \end{pmatrix}, \begin{pmatrix} \frac{\lambda_1}{\lambda_2} & 0 & 0 \\ a_{21} & 0 & 0 \\ a_{31} & a_{32} & 0 \end{pmatrix}, \begin{pmatrix} \frac{\lambda_2}{\lambda_1} & 0 & 0 \\ a_{21} & 0 & a_{23} \\ a_{31} & 0 & 0 \end{pmatrix}.
$$

(b) If $\lambda_1 = \lambda_2$ for all $\lambda_1, \lambda_2 \neq 0$,

$$\begin{pmatrix} 1 & 0 & 0 \\ a_{21} & a_{22} & a_{23} \\ a_{31} & a_{32} & a_{33} \end{pmatrix}, \begin{pmatrix} 1 & 0 & 0 \\ a_{21} & a_{22} & 0 \\ a_{31} & a_{32} & a_{33} \end{pmatrix}, \begin{pmatrix} 1 & 0 & 0 \\ a_{21} & a_{22} & a_{23} \\ a_{31} & 0 & a_{33} \end{pmatrix}, \begin{pmatrix} 1 & 0 & 0 \\ a_{21} & 0 & a_{23} \\ a_{31} & a_{32} & 0 \end{pmatrix}.$$

(c) If $\lambda_1 = -\lambda_2$ for all $\lambda_1, \lambda_2 \neq 0$, then $[\alpha] = \begin{pmatrix} -1 & 0 & 0 \\ a_{21} & 0 & a_{23} \\ a_{31} & a_{32} & 0 \end{pmatrix}$.

4. If $\lambda_1, \lambda_2 = 0$ and $\lambda_3 \neq 0$, then $\mathcal{A}$ has a linear map of the form $[\alpha] = \begin{pmatrix} a_{11} & a_{12} & a_{13} \\ a_{21} & a_{22} & a_{23} \\ a_{31} & a_{32} & a_{33} \end{pmatrix}$ and the corresponding multiplicative $\mathcal{A}$ has the linear twisting of the form $[\alpha] = \begin{pmatrix} a_{22}a_{33} - a_{23}a_{32} & a_{12} & a_{13} \\ 0 & a_{22} & a_{23} \\ 0 & a_{32} & a_{33} \end{pmatrix}$.

5. If $\lambda_1, \lambda_3 = 0$ and $\lambda_2 \neq 0$, then $\mathcal{A}$ has the linear twists $[\alpha] = \begin{pmatrix} a_{11} & 0 & a_{13} \\ a_{21} & a_{22} & a_{23} \\ a_{31} & a_{32} & a_{33} \end{pmatrix}$ and the corresponding multiplicative $\mathcal{A}$ has linear twists:

$$\begin{pmatrix} 0 & 0 & 0 \\ a_{21} & a_{22} & 0 \\ a_{31} & a_{32} & 0 \end{pmatrix}, \begin{pmatrix} 1 & 0 & 0 \\ a_{21} & a_{22} & 0 \\ a_{31} & 0 & a_{33} \end{pmatrix}, \begin{pmatrix} a_{11} & 0 & 0 \\ a_{21} & a_{22} & 0 \\ a_{31} & 0 & 0 \end{pmatrix}.$$

6. If $\lambda_2, \lambda_3 = 0$ and $\lambda_1 \neq 0$, then $\mathcal{A}$ has linear twists $\begin{pmatrix} a_{11} & a_{12} & 0 \\ a_{21} & a_{22} & a_{23} \\ a_{31} & a_{32} & a_{33} \end{pmatrix}$ with the following subclasses of twists for the multiplicative $\mathcal{A}$:

$$\begin{pmatrix} 0 & 0 & 0 \\ a_{21} & 0 & a_{23} \\ a_{31} & 0 & a_{33} \end{pmatrix}, \begin{pmatrix} 1 & 0 & 0 \\ a_{21} & a_{22} & 0 \\ a_{31} & 0 & a_{33} \end{pmatrix}, \begin{pmatrix} a_{11} & 0 & 0 \\ a_{21} & 0 & 0 \\ a_{31} & 0 & a_{33} \end{pmatrix}$$

7. If $\lambda_1, \lambda_2, \lambda_3 \neq 0$, then $\mathcal{A}$ has the twists of the form $[\alpha] = \begin{pmatrix} a_{11} & a_{12} & a_{13} \\ \frac{\lambda_1}{\lambda_3}a_{13} & a_{22} & a_{23} \\ -\frac{\lambda_2}{\lambda_3}a_{12} & a_{32} & -\frac{\lambda_2}{\lambda_1}a_{22} \end{pmatrix}$ and the corresponding multiplicative $\mathcal{A}$ has linear twists of the form:

a. When $\lambda_1 \neq \pm\lambda_2$ for all $\lambda_1, \lambda_2, \lambda_3 \neq 0$,

$$[\alpha] = \begin{pmatrix} 1 & 0 & 0 \\ 0 & a_{22} & 0 \\ 0 & 0 & -\frac{\lambda_2}{\lambda_1}a_{22} \end{pmatrix}, \text{ where } a_{22}^2 = -\frac{\lambda_1}{\lambda_2}$$

b. When $\lambda_1 = \lambda_2$, for all $\lambda_1, \lambda_2, \lambda_3 \neq 0$, then the twists are given by $\begin{pmatrix} 1 & 0 & 0 \\ 0 & a_{22} & -\frac{1+a_{22}^2}{a_{32}} \\ 0 & a_{32} & -a_{22} \end{pmatrix}$, for $a_{32} \neq 0$, $\begin{pmatrix} 1 & 0 & 0 \\ 0 & a_{22} & 0 \\ 0 & 0 & -a_{22} \end{pmatrix}$, where $a_{22}^2 = -1$.

c. When $\lambda_1 = -\lambda_2$ for all $\lambda_1, \lambda_2, \lambda_3 \neq 0$, then the twists are given by $\begin{pmatrix} a_{11} & \frac{(-1+a_{11})(1+a_{11})\lambda_3}{2a_{13}\lambda_2} & a_{13} \\ \frac{\lambda_1}{\lambda_3}a_{13} & -\frac{1}{2}(1+a_{11}) & -\frac{a_{13}\lambda_2}{(1+a_{11})\lambda_3} \\ \frac{(1-a_{11})(1+a_{11})}{2a_{13}} & \frac{(-1+a_{11})^2(1+a_{11})\lambda_3}{4a_{13}\lambda_2} & -\frac{1}{2}(1+a_{11}) \end{pmatrix}$, $a_{13} \neq 0, a_{11} \neq -1$,

$\begin{pmatrix} 1 & a_{12} & 0 \\ 0 & -1 & 0 \\ -\frac{\lambda_2}{\lambda_3}a_{12} & -\frac{a_{12}^2\lambda_2}{\lambda_3} & -1 \end{pmatrix}$, $\begin{pmatrix} -1 & 0 & 0 \\ 0 & 0 & \frac{1}{a_{32}} \\ 0 & a_{32} & 0 \end{pmatrix}$ for $a_{32} \neq 0$, $\begin{pmatrix} 1 & 0 & 0 \\ 0 & 1 & 0 \\ 0 & 0 & 1 \end{pmatrix}$

Example 15 (Multiplicative Hom-Lie Algebra of $\mathfrak{sl}(2)$-Type) Consider a Hom-Lie algebra of $\mathfrak{sl}(2)$-type consisting of linear map of the form $[\alpha] = \begin{pmatrix} a_{11} & a_{12} & a_{13} \\ 2a_{13} & a_{22} & a_{23} \\ 2a_{12} & a_{32} & a_{22} \end{pmatrix}$ and the bracket multiplication defined as $[e_1, e_2] = 2e_2$, $[e_1, e_3] = -2e_3$, $[e_2, e_3] = e_1$. Substituting these structure constants in (4.10), we obtain the following system of polynomial equations:

$$\begin{array}{lll} a_{13}a_{32} - a_{22}a_{12} = a_{12} & a_{12}a_{23} - a_{22}a_{13} = a_{13} & a_{22}^2 - a_{32}a_{23} = a_{11} \\ a_{11}a_{22} - 2a_{13}a_{12} = a_{22} & 2a_{13}^2 - a_{11}a_{23} = a_{23} & a_{12}a_{23} - a_{22}a_{13} = a_{13} \\ 2a_{12}^2 - a_{11}a_{32} = a_{32}, & a_{11}a_{22} - 2a_{12}a_{13} = a_{22}, & a_{13}a_{32} - a_{22}a_{12} = a_{12} \end{array} \tag{4.11}$$

The solution set of (4.11) is

$$\left\{ \begin{array}{l|l} & \{a_{12} = \frac{1-a_{11}^2}{4a_{13}}, a_{22} = \frac{-1}{2}(a_{11}+1), a_{23} = \frac{2a_{13}^2}{a_{11}+1}, \\ & \qquad a_{32} = \frac{(a_{11}-1)^2(a_{11}+1)}{8a_{13}^2}\}, \\ a_{ij}, & \{a_{11} = 1, a_{22} = -1, a_{13} = 0,\ a_{32} = a_{12}^2, a_{23} = 0\}, \\ 1 \leq i, j \leq 3 & \{a_{11} = -1, a_{22} = a_{13} = a_{12} = 0, a_{23} = \frac{1}{a_{32}}\}, \\ & \{a_{11} = 1, a_{22} = -1, a_{13} = a_{12} = a_{32} = a_{23} = 0\}, \\ & \{a_{11} = 1, a_{22} = 1, a_{13} = a_{12} = a_{32} = a_{23} = 0\} \end{array} \right\},$$

and therefore, the corresponding multiplicative Hom-Lie algebra of $\mathfrak{sl}(2)$-type has linear twists of the form:

$$\begin{pmatrix} a_{11} & \frac{1-a_{11}^2}{4a_{13}} & a_{13} \\ 2a_{13} & \frac{1}{2}(-a_{11}-1) & \frac{2a_{13}^2}{a_{11}+1} \\ 2(\frac{1-a_{11}^2}{4a_{13}}) & \frac{(a_{11}-1)^2(a_{11}+1)}{8a_{13}^2} & \frac{1}{2}(-a_{11}-1) \end{pmatrix} \text{ for all } a_{13} \neq 0, a_{11} \neq -1$$

$$\begin{pmatrix} 1 & a_{12} & 0 \\ 0 & -1 & 0 \\ 2a_{12} & a_{12}^2 & -1 \end{pmatrix}, \begin{pmatrix} -1 & 0 & 0 \\ 0 & 0 & \frac{1}{a_{32}} \\ 0 & a_{32} & 0 \end{pmatrix} \text{ for all } a_{32} \neq 0\,, \begin{pmatrix} 1 & 0 & 0 \\ 0 & -1 & 0 \\ 0 & 0 & -1 \end{pmatrix}, \begin{pmatrix} 1 & 0 & 0 \\ 0 & 1 & 0 \\ 0 & 0 & 1 \end{pmatrix}$$

The example (15) given above provides classes of regular Hom-Lie algebras of $\mathfrak{sl}(2)$-types.

Remark 11 Multiplicative Hom-Lie algebras arising from simple Lie algebras of $\mathfrak{sl}(2)$-type with nonzero twisting maps are regular Hom-Lie algebras.

5 Solvability and Nilpotency of Hom-Lie Algebras

In this section, we explore some properties of Hom-Lie algebras related to its derived algebras through their ideals and subalgebras. We begin with some definitions and present the notion of sequences of ideals, which generalize central descending series and derived series of ideals.

Definition 6 Let $(\mathcal{V}, [\cdot,\cdot], \alpha)$ be a Hom-Lie algebra, and $I \subseteq \mathcal{V}$ be an ideal of $\mathcal{V}$:

(i) A derived series of an ideal I is defined as

$$D^0(I) = I \text{ and } D^{k+1}(I) = [D^k(I), D^k(I)], \tag{5.1}$$

(ii) A central descending series of I is defined as

$$C^0(I) = I \text{ and } C^{k+1}(I) = \big[C^k(I), I\big]. \tag{5.2}$$

The authors in [10] have proven the following results for cases generalizing to n-Hom-Lie algebras and k-derived series.

Lemma 1 ([10]) *Let $(\mathcal{V}, [\cdot,\cdot], \alpha)$ be a Hom-Lie algebra, and I be an ideal of $\mathcal{V}$ for some $k \in \mathbb{Z}_{\geq 0}$. Then,*

(i) $D^{k+1}(I) \subseteq D^k(I)$,
(ii) $C^{k+1}(I) \subseteq C^k(I)$.

Definition 7 Let $(\mathcal{V}, [\cdot,\cdot], \alpha)$ be a Hom-Lie algebra and $I \subseteq \mathcal{V}$ be a nonzero ideal of $\mathcal{V}$. Then,

(i) I is said to be solvable if there exist some $k \in \mathbb{Z}_{>0}$ such that $D^k(I) = \{0\}$. It is solvable of class k if $D^k(I) = 0$ and $D^{k-1}(I) \neq \{0\}$;
(ii) I is said to be nilpotent if there exist some $k \in \mathbb{Z}_{>0}$ such that $C^k(I) = \{0\}$. It is nilpotent of class k if $C^k(I) = \{0\}$ and $C^{k-1}(I) \neq \{0\}$.

Example 16 Consider a Hom-Lie algebra of $\mathfrak{sl}(2)$-type defined by

$$[e_1, e_2] = 2e_2, [e_1, e_3] = -2e_3, [e_2, e_3] = e_1.$$

Let $I = \text{span}\{e_1, e_2, e_3\}$ be the ideal generated e_1, e_2, e_3. Then,

(i) Central descending series of I is given by $C^k(I) = \text{span}\{e_1, e_2, e_3\},\ k \geq 0$
(ii) Derived series of I is given by $D^k(I) = \text{span}\{e_1, e_2, e_3\},\ k \geq 0$

Therefore, the Hom-Lie algebra of $\mathfrak{sl}(2)$-type is not nilpotent and not solvable.

We present some properties and results about nilpotent and solvable skew-symmetric algebras $\mathcal{A} = \big(\mathcal{V}, [\cdot, \cdot]\big)$, and in particular, we focus on class *two* nilpotent algebras.

Proposition 1 ([7]) *Let $\mathcal{A} = (\mathcal{V}, [\cdot, \cdot])$ be a skew-symmetric algebra. If $\mathcal{A}$ is class 2 nilpotent, then the derived algebra $C^1(\mathcal{A})$ is contained in the center $Z(\mathcal{A})$.*

Lemma 2 ([9]) *Let $\mathcal{A} = (\mathcal{V}, [\cdot, \cdot])$ be a skew-symmetric algebra:*

(i) *If $\mathcal{A}$ is nilpotent, then $Z(\mathcal{A})$ is not trivial.*
(ii) *If* $\dim\ \mathcal{A} = 3$, *then* $\dim\ Z(\mathcal{A}) = 0$ *or* $\dim\ Z(\mathcal{A}) = 1$ *or* $Z(\mathcal{A}) = \mathcal{A}$.

Let $\mathcal{A} = (\mathcal{V}, [\cdot, \cdot])$ be a class 2 nilpotent skew-symmetric algebra, then the decomposition $\mathcal{V} = Z(\mathcal{A}) \oplus \mathcal{W}$ holds for some subspace $\mathcal{W} \subseteq \mathcal{V}$. Suppose that $\dim(\mathcal{A}) = n$ and $\dim Z(\mathcal{A}) = p$ and $\dim C^1(\mathcal{A}) = p_1$. By Proposition 1, the bracket multiplication of $\mathcal{A}$ is determined by

$$\big[e_i, e_j\big] = \sum_{k=1}^{p_1} C_{i,j}^k e_k, \quad p+1 \leq i < j \leq n,\ \big[e_i, e_j\big] = 0,$$
$$1 \leq i \leq p \text{ or } 1 \leq j \leq p$$

where $\{e_1, \ldots, e_{p_1}, \ldots, e_p, \ldots, e_n\} \in \mathcal{A}$ is a basis of $\mathcal{A}$ adapted to the decomposition, and $\{e_1, \ldots, e_p\} \in Z(\mathcal{A})$ is a basis of $Z(\mathcal{A})$ and $\{e_1, \ldots, e_{p_1}\} \in C^1(\mathcal{A})$ is basis of $C^1(\mathcal{A})$. In this sense, every finite-dimensional class 2 nilpotent algebra $\mathcal{A}$ is a trivial extension of a class 2 nilpotent algebra whose center is the derived algebra $C^1(\mathcal{A})$ extended by an abelian ideal, namely, the sub-algebra spanned by $\{e_1, \ldots, e_{p_1}, e_{p+1}, \ldots, e_n\}$ and the ideal spanned by $\{e_{p_1+1}, \ldots, e_p\}$.

Lemma 3 *Let $\mathcal{A} = (\mathcal{V}, [\cdot, \cdot])$ be a skew-symmetric algebra and $\{e_1, \ldots, e_n\}$ be a basis of $\mathcal{V}$:*

(1) *For all $1 \leq k_0, \ldots, k_r \leq n$, $r > 0$, and $1 \leq i \leq r$*

$$\begin{aligned}
&[e_{k_0}, [\dots, [e_{k_i}, [\dots, [e_{k_{r-1}}, e_{k_r}]\dots]]\dots]] \\
&= \sum_{t_0=1}^{n} \Big(\sum_{t_1=1}^{n} \cdots \sum_{t_{r-1}=1}^{n} \Big(C_{k_0,t_1}^{t_0} \cdots C_{k_{r-2},t_{r-1}}^{t_{r-2}} \Big) C_{k_{r-1},k_r}^{t_{r-1}} \Big) e_{t_0} \\
&= \sum_{t_0=1}^{n} \Big(\sum_{t_1=1}^{n} \cdots \sum_{t_{r-1}=1}^{n} \Big(\prod_{j=0}^{r-2} C_{k_j,t_{j+1}}^{t_j} \Big) C_{k_{r-1},k_r}^{t_{r-1}} \Big) e_{t_0},
\end{aligned} \tag{5.3}$$

(2) *For all* $1 \leq k_1, \dots, k_{2^r} \leq n$

$$\begin{aligned}
&[\dots[[e_{k_1}, e_{k_2}], \dots, [e_{k_{2^{r-1}-1}}, e_{k_{2^{r-1}}}]], [[e_{k_{2^{r-1}+1}}, e_{k_{2^{r-1}+2}}], \\
&\qquad\qquad \dots, [e_{k_{2^r-1}}, e_{k_{2^r}}]]\dots] \\
&= \sum_{t_{2^r-2}=1}^{n} \Big(\sum_{t_0=1}^{n} \cdots \sum_{t_{2^r-3}=1}^{n} \Big(\prod_{j=0}^{2^{r-1}-1} C_{k_{2j+1},k_{2j+2}}^{t_j} \Big) \Big(\prod_{p=0}^{2^{r-1}-2} C_{t_{2p},t_{2p+1}}^{t_{2^{r-1}+p}} \Big) \Big) e_{t_{2^r-2}}.
\end{aligned} \tag{5.4}$$

Proof We proceed by induction over $r \in \mathbb{Z}_{>0}$.

(1) For $r = 1$, we have that, $\big[e_{k_0}, e_{k_1}\big] = \sum_{t_0=1}^{n} C_{k_0,k_1}^{t_0} e_{t_0}$, which hold by definition.
Suppose that (5.3) holds for $r = p \geq 2$, that is, for all $1 \leq k_1, \dots, k_p \leq n$,

$$\begin{aligned}
&[e_{k_1}, [\dots, [e_{k_i}, [\dots, [e_{k_p}, e_{k_{p+1}}]\dots]]\dots]] \\
&\qquad = \sum_{t_1=1}^{n} \cdots \sum_{t_p=1}^{n} \Big(\prod_{j=1}^{p-1} C_{k_j,t_{j+1}}^{t_j} \Big) C_{k_p,k_{p+1}}^{t_p} e_{t_1}.
\end{aligned} \tag{5.5}$$

Then, we get:

$$\begin{aligned}
&[e_{k_0}, [e_{k_1}, [\dots, [e_{k_i}, [\dots, [e_{k_p}, e_{k_{p+1}}]\dots]]\dots]]] \\
&= \Big[e_{k_0}, \sum_{t_1=1}^{n} \Big(\sum_{t_2=1}^{n} \cdots \sum_{t_p=1}^{n} \Big(\prod_{j=1}^{p-1} C_{k_j,t_{j+1}}^{t_j} \Big) C_{k_p,k_{p+1}}^{t_p} \Big) e_{t_1} \Big] \\
&= \sum_{t_0=1}^{n} \Big(\sum_{t_1=1}^{n} \cdots \sum_{t_p=1}^{n} \Big(\prod_{j=1}^{p-1} C_{k_j,t_{j+1}}^{t_j} \Big) C_{k_0,t_1}^{t_0} C_{k_p,k_{p+1}}^{t_p} \Big) e_{t_0} \\
&= \sum_{t_0=1}^{n} \Big(\sum_{t_1=1}^{n} \cdots \sum_{t_p=1}^{n} \Big(\prod_{j=0}^{p-1} C_{k_j,t_{j+1}}^{t_j} \Big) C_{k_p,k_{p+1}}^{t_p} \Big) e_{t_0}.
\end{aligned} \tag{5.6}$$

Therefore, by induction (5.3) holds.

(2) For $r = 2$, $[[e_{k_1}, e_{k_2}], [e_{k_3}, e_{k_4}]] = \sum_{t_0=1}^{n} \sum_{t_1=1}^{n} \sum_{t_2=1}^{n} C_{k_1,k_2}^{t_0} C_{k_3,k_4}^{t_1} C_{t_0,t_1}^{t_2} e_{t_2}$ holds.
Suppose that (5.4) holds for some $r \in \mathbb{Z}_{\geq 3}$, that is for all $1 \leq k_1, \dots, k_r \leq 2^r$,

$$[\ldots[[e_{k_1}, e_{k_2}], \ldots, [e_{k_{2^{r-1}-1}}, e_{k_{2^{r-1}}}]], [[e_{k_{2^{r-1}+1}}, e_{k_{2^{r-1}+2}}], \ldots, [e_{k_{2^r-1}}, e_{k_{2^r}}]\ldots] = \sum_{t_{2^r-2}=1}^{n}\Big(\sum_{t_0=1}^{n}\cdots\sum_{t_{2^r-3}=1}^{n}\Big(\prod_{j=0}^{2^{r-1}-1} C^{t_j}_{k_{2j+1},k_{2j+2}}\Big)\Big(\prod_{p=0}^{2^{r-1}-2} C^{t_{2^{r-1}+p}}_{t_{2p},t_{2p+1}}\Big)\Big)e_{t_{2^r-2}},$$

then

$$\big[[[\ldots[[e_{k_1}, e_{k_2}], \ldots, [e_{k_{2^{r-1}-1}}, e_{k_{2^{r-1}}}]\ldots], [\ldots[e_{k_{2^{r-1}+1}}, e_{k_{2^{r-1}+2}}], \ldots, [e_{k_{2^r-1}}, e_{k_{2^r}}]]\ldots]], [[\ldots[[e_{k_{2^r+1}}, e_{k_{2^r+2}}], \ldots, [e_{k_{3(2^r-1)-1}}, e_{k_{3(2^r-1)}}]\ldots], [\ldots[e_{k_{3(2^r-1)+1}}, e_{k_{3(2^r-1)+2}}], \ldots, [e_{k_{2^{r+1}-1}}, e_{k_{2^{r+1}}}]]\ldots]]\big]$$

$$= \Big(\sum_{t_0=1}^{n}\cdots\sum_{t_{2^r-2}=1}^{n}\Big(\prod_{j=0}^{2^{r-1}-1} C^{t_j}_{k_{2j+1},k_{2j+2}}\Big)\Big(\prod_{p=0}^{2^{r-1}-2} C^{t_{2^{r-1}+p}}_{t_{2p},t_{2p+1}}\Big)\Big)$$
$$\Big(\sum_{t_{2^r-1}=1}^{n}\cdots\sum_{t_{2^{r+1}-3}=1}^{n}\Big(\prod_{j=2^{r-1}}^{2^r-1} C^{t_j}_{k_{2j+1},k_{2j+2}}\Big)\Big(\prod_{p=2^{r-1}-1}^{2^r-3} C^{t_{2^{r-1}+p}}_{t_{2p},t_{2p+1}}\Big)\Big)\big[e_{t_{2^r-2}}, e_{t_{2^{r+1}-3}}\big],$$

By the induction hypothesis.

$$= \Big(\sum_{t_0=1}^{n}\cdots\sum_{t_{2^{r-1}-1}=1}^{n}\Big(\prod_{j=0}^{2^{r-1}-1} C^{t_j}_{k_{2j+1},k_{2j+2}}\Big)\Big)$$

$$\Big(\sum_{t_{2^r-1}=1}^{n}\cdots\sum_{t_{2^r-1}=1}^{n}\Big(\prod_{j=2^{r-1}}^{2^r-1} C^{t_j}_{k_{2j+1},k_{2j+2}}\Big)\Big)$$

$$\Big(\sum_{t_{2^r}=1}^{n}\cdots\sum_{t_{2^r+2^{r-2}-1}=1}^{n}\Big(\prod_{p=0}^{2^{r-2}-1} C^{t_{2^r+p}}_{t_{2p},t_{2p+1}}\Big)\Big)$$

$$\Big(\sum_{t_{2^r+2^{r-2}}=1}^{n}\cdots\sum_{t_{2^{r+1}-2^{r-1}-1}=1}^{n}\Big(\prod_{p=2^{r-2}}^{2^{r-1}-1} C^{t_{2^r+p}}_{t_{2p},t_{2p+1}}\Big)\Big)$$

$$\Big(\sum_{t_{2^r+2^{r-1}}=1}^{n}\cdots\sum_{t_{2^r+2^{r-1}+2^{r-3}-1}=1}^{n}\Big(\prod_{p=2^{r-1}}^{2^{r-1}+2^{r-3}-1} C^{t_{2^r+p}}_{t_{2p},t_{2p+1}}\Big)\Big)$$

$$\Big(\sum_{t_{2^r+2^{r-1}+2^{r-3}}=1}^{n}\cdots\sum_{t_{2^{r+1}-2^{r-2}-1}=1}^{n}\Big(\prod_{p=2^{r-1}+2^{r-3}}^{2^{r-1}+2^{r-2}-1} C^{t_{2^r+p}}_{t_{2p},t_{2p+1}}\Big)\Big)$$

$$\Big(\sum_{t_{2^r+2^{r-1}+2^{r-2}}=1}^{n}\cdots\sum_{t_{2^r+2^{r-1}+2^{r-2}+2^{r-4}-1}=1}^{n}\Big(\prod_{p=2^{r-1}+2^{r-2}}^{2^{r-1}+2^{r-2}+2^{r-4}-1}C_{t_{2p},t_{2p+1}}^{t_{2^r+p}}\Big)\Big)$$

$$\Big(\sum_{t_{2^r+2^{r-1}+2^{r-2}+2^{r-4}}=1}^{n}\cdots\sum_{t_{2^{r+1}-2^{r-3}-1}=1}^{n}\Big(\prod_{p=2^{r-1}+2^{r-2}+2^{r-4}}^{2^{r-1}+2^{r-2}+2^{r-3}-1}C_{t_{2p},t_{2p+1}}^{t_{2^r+p}}\Big)\Big)\cdots$$

$$\Big(\sum_{t_{2^r+2^{r-1}+2^{r-2}+\dots+2^2}=1}^{n}\Big)\Big(\sum_{t_{2^r+2^{r-1}+\dots+2^2+1}=1}^{n}\Big)$$

$$[e_{t_{2^r+(2^{r-1}+2^{r-2}+\dots+2^2)}},e_{t_{2^r+(2^{r-1}+2^{r-2}+\dots+2^2+1)}}]$$

$$=\Big(\sum_{t_0=1}^{n}\cdots\sum_{t_{2^r-1}=1}^{n}\Big)\Big(\sum_{t_{2^r}=1}^{n}\cdots\sum_{t_{2^{r+1}-3}=1}^{n}\Big)\Big(\prod_{j=0}^{2^r-1}C_{k_{2j+1},k_{2j+2}}^{t_j}\Big)$$

$$\Big(\prod_{p=0}^{2^{r-1}+2^{r-2}+\dots+2^2+1}C_{t_{2p},t_{2p+1}}^{t_{2^r+p}}\Big)[e_{t_{2^r+(2^{r-1}+2^{r-2}+\dots+2^2)}},e_{t_{2^r+(2^{r-1}+2^{r-2}+\dots+2^2+1)}}]$$

$$=\Big(\sum_{t_0=1}^{n}\cdots\sum_{t_{2^r-1}=1}^{n}\Big)\Big(\sum_{t_{2^r}=1}^{n}\cdots\sum_{t_{2^{r+1}-3}=1}^{n}\Big)\Big(\prod_{j=0}^{2^r-1}C_{k_{2j+1},k_{2j+2}}^{t_j}\Big)\Big(\prod_{p=0}^{2^r-3}C_{t_{2p},t_{2p+1}}^{t_{2^r+p}}\Big)$$

$$[e_{t_{2^{r+1}-4}},e_{t_{2^{r+1}-3}}]$$

$$=\Big(\sum_{t_0=1}^{n}\cdots\sum_{t_{2^{r+1}-3}=1}^{n}\Big)\Big(\prod_{j=0}^{2^r-1}C_{k_{2j+1},k_{2j+2}}^{t_j}\Big)\Big(\prod_{p=0}^{2^r-3}C_{t_{2p},t_{2p+1}}^{t_{2^r+p}}\Big)[e_{t_{2^{r+1}-4}},e_{t_{2^{r+1}-3}}]$$

$$=\Big(\sum_{t_0=1}^{n}\cdots\sum_{t_{2^{r+1}-2}=1}^{n}\Big(\prod_{j=0}^{2^r-1}C_{k_{2j+1},k_{2j+2}}^{t_j}\Big)\Big(\prod_{p=0}^{2^r-3}C_{t_{2p},t_{2p+1}}^{t_{2^r+p}}\Big)\Big)C_{t_{2^{r+1}-4},t_{2^{r+1}-3}}^{t_{2^{r+1}-2}}e_{t_{2^{r+1}-2}}$$

$$=\sum_{t_0=1}^{n}\cdots\sum_{t_{2^{r+1}-2}=1}^{n}\Big(\prod_{j=0}^{2^r-1}C_{k_{2j+1},k_{2j+2}}^{t_j}\Big)\Big(\prod_{p=0}^{2^r-2}C_{t_{2p},t_{2p+1}}^{t_{2^r+p}}\Big)e_{t_{2^{r+1}-2}}$$

$$=\sum_{t_{2^{r+1}-2}=1}^{n}\Big(\sum_{t_0=1}^{n}\cdots\sum_{t_{2^{r+1}-3}=1}^{n}\Big(\prod_{j=0}^{2^r-1}C_{k_{2j+1},k_{2j+2}}^{t_j}\Big)\Big(\prod_{p=0}^{2^r-2}C_{t_{2p},t_{2p+1}}^{t_{2^r+p}}\Big)\Big)e_{t_{2^{r+1}-2}}.$$

Therefore, by induction (5.4) holds.

□

Corollary 1 *If $\mathcal{A}$ is a skew-symmetric algebra, then:*

(1) *$\mathcal{A}$ is nilpotent of class at most r if and only if for all $1 \le k_0, \dots, k_r \le n$, $1 \le t_0 \le n$,*

$$\sum_{t_1=1}^{n} \cdots \sum_{t_{r-1}=1}^{n} \Big(\prod_{j=0}^{r-2} C_{k_j,t_{j+1}}^{t_j}\Big) C_{k_{r-1},k_r}^{t_{r-1}} = 0. \tag{5.7}$$

(2) *$\mathcal{A}$ is solvable of class at most r if and only if for all $1 \le k_1, \dots, k_{2^r} \le n$,*

$$\sum_{t_0=1}^{n} \cdots \sum_{t_{2^r-3}=1}^{n} \Big(\prod_{j=0}^{2^{r-1}-1} C_{k_{2j+1},k_{2j+2}}^{t_j}\Big)\Big(\prod_{p=0}^{2^{r-1}-2} C_{t_{2p},t_{2p+1}}^{t_{2^{r-1}+p}}\Big) = 0. \tag{5.8}$$

Proof

(1) A skew-symmetric algebra $\mathcal{A}$ is nilpotent of class at most r if and only if for all $1 \le k_0, \dots, k_r \le n$, $[e_{k_0}, [\dots, [e_{k_j}, [\dots, [e_{k_{r-1}}, e_{k_r}] \dots]] \dots]] = 0$, which by Lemma 3, (5.3) is equivalent to

$$\sum_{t_1=1}^{n} \cdots \sum_{t_{r-1}=1}^{n} \Big(\prod_{j=0}^{r-2} C_{k_j,t_{j+1}}^{t_j}\Big) C_{k_{r-1},k_r}^{t_{r-1}} = 0,\ 1 \le k_0, \dots, k_r \le n,\ 1 \le t_0 \le n$$

since $\{e_1, \dots, e_n\}$ is a basis of $\mathcal{V}$.

(2) Similarly, a skew-symmetric algebra $\mathcal{A}$ is solvable of class at most r if and only if for all $1 \le k_1, \dots, k_{2^r} \le n$,

$$\Big[\dots[[e_{k_1}, e_{k_2}], \dots, [e_{k_{2^{r-1}-1}}, e_{k_{2^{r-1}}}]],$$
$$[[e_{k_{2^{r-1}+1}}, e_{k_{2^{r-1}+2}}], \dots, [e_{k_{2^r-1}}, e_{k_{2^r}}] \dots\Big] = 0,$$

which by Lemma 3, (5.4) is equivalent to

$$\sum_{t_0=1}^{n} \cdots \sum_{t_{2^r-3}=1}^{n} \Big(\prod_{j=0}^{2^{r-1}-1} C_{k_{2j+1},k_{2j+2}}^{t_j}\Big)\Big(\prod_{p=0}^{2^{r-1}-2} C_{t_{2p},t_{2p+1}}^{t_{2^{r-1}+p}}\Big) = 0$$

since $\{e_1, \dots, e_n\}$ is a basis of $\mathcal{V}$.

□

Remark 12 Let $\mathcal{A}$ be a skew-symmetric algebra:

(i) If $\mathcal{A}$ is nilpotent of class at most r, then the resulting algebraic subvariety in $(C_{i,j}^k)_{i<j}$ variables is determined by n^{r+2} homogeneous polynomial equations of degree r.

(ii) If $\mathcal{A}$ is solvable of class at most r, then the resulting algebraic subvariety in $(C_{i,j}^{k})_{i<j}$ variables is determined by n^{2^r+1} homogeneous polynomial equations of degree $2^r - 1$.

Example 17 Consider a 3-dimensional skew-symmetric algebra $\mathcal{A} = (\mathcal{V}, [\cdot, \cdot])$ and let bracket multiplication be defined by (4.5) then,

(i) $\mathcal{A}$ is class at most 2 nilpotent if and only if $\sum\limits_{t_1=1}^{3} C_{k_0,t_1}^{t_0} C_{k_1,k_2}^{t_1} = 0$ for all $1 \leq k_0, k_1, k_2 \leq 3$, $t_0 = 1, 2, 3$. Specifically, we obtain the following systems of monomial equations:

$$\lambda_1^2=0, -\lambda_1^2=0, \lambda_2^2=0, -\lambda_2^2=0, \lambda_1\lambda_3=0, -\lambda_1\lambda_3=0, \lambda_2\lambda_3 = 0, -\lambda_2\lambda_3 = 0,$$

whose nontrivial solution set is given by $\{\lambda_1 = \lambda_2 = 0\}$.

(ii) $\mathcal{A}$ is class at most 2 solvable if and only if, for all $1 \leq k_1, \ldots, k_4 \leq 3$,

$$\sum_{t_0=1}^{3}\sum_{t_1=1}^{3}\sum_{t_2=1}^{3}\left(C_{k_1,k_2}^{t_0} C_{k_3,k_4}^{t_1} C_{t_0,t_1}^{t_2}\right) = 0.$$

Specifically, we get the following system of monomial equations $\lambda_1\lambda_2\lambda_3 = 0, -\lambda_1\lambda_2\lambda_3 = 0, \lambda_1^2\lambda_3 = 0, -\lambda_1^2\lambda_3 = 0, -\lambda_2^2\lambda_3 = 0, \lambda_2^2\lambda_3 = 0$, and its corresponding nontrivial solution sets are $\{\lambda_1 = \lambda_2 = 0\}, \{\lambda_3 = 0\}, \{\lambda_1 = \lambda_3 = 0\}$ and $\{\lambda_2 = \lambda_3 = 0\}$.

(iii) $\mathcal{A}$ is class at most 3 solvable if and only if

$$\sum_{t_0=1}^{3}\cdots\sum_{t_6=1}^{3}\left(C_{k_1,k_2}^{t_0} C_{k_3,k_4}^{t_1} C_{k_5,k_6}^{t_2} C_{k_7,k_8}^{t_3}\right) C_{t_0,t_1}^{t_4} C_{t_2,t_3}^{t_5} C_{t_4,t_5}^{t_6} = 0$$

for all $1 \leq k_1, \ldots, k_8 \leq 3$. That is, we obtain the following system of monomial equations $\lambda_1^4\lambda_2\lambda_3^2 = 0,\ -\lambda_1^4\lambda_2\lambda_3^2 = 0,\ -\lambda_1\lambda_2^4\lambda_3^2 = 0, \lambda_1\lambda_2^4\lambda_3^2 = 0, -\lambda_1^2\lambda_2^2\lambda_3^2 = 0$, and its corresponding solution sets are $\{\lambda_1 = 0\}$, or $\{\lambda_2 = 0\}$, or $\{\lambda_3 = 0\}$.

Proposition 2 ([9]) *Let $\mathcal{A} = (\mathcal{V}, [\cdot, \cdot], \alpha)$ be a 3-dimensional Hom-Lie algebra, then $\mathcal{A}$ is class 2 nilpotent if and only if* $\dim Z(\mathcal{A}) = 1$ *and* $C^1(\mathcal{A}) = Z(\mathcal{A})$.

Lemma 4 *Let $\mathcal{A} = (\mathcal{V}, [\cdot, \cdot])$ be an n-dimensional skew-symmetric algebra and $\alpha \in End(\mathcal{V})$ be a Hom-Lie structure on $\mathcal{V}$. If $\mathcal{A}$ is class 2 nilpotent, then the linear space of Hom-Lie structures is of dimension n^2.*

Proof We will provide two proofs, one using Hom-Lie algebras axioms and another using structure constants.

Proof 1 (Using axioms of Hom-Lie algebras): Let $\{e_1, \ldots, e_n\}$ be a basis of $\mathcal{V}$ and suppose that $\mathcal{A}$ is class 2 nilpotent. Then, by Lemma 2 and Proposition 2

$$
\begin{aligned}
C^1(\mathcal{V}) \subseteq Z(\mathcal{A}) &\iff [e_j, e_k] \in Z(\mathcal{A}), \forall 1 \le j < k \le n \\
&\iff \big[x, [e_j, e_k]\big] = 0, \forall x \in \mathcal{A}, 1 \le j < k \le n \\
&\Longrightarrow \big[\alpha(e_i), [e_j, e_k]\big] = 0, \ \forall \alpha \in \mathrm{End}(\mathcal{V}), \ 1 \le i < j < k \le n \\
&\Longrightarrow \sum_{\circlearrowleft(e_i, e_j, e_k)} \big[\alpha(e_i), [e_j, e_k]\big] = 0, \\
&\qquad \forall \alpha \in \mathrm{End}(\mathcal{V}), \ 1 \le i < j < k \le n.
\end{aligned}
$$

Proof 2 (Using equations for structure constants): By Lemma 3 we have that, for all $1 \le k_0, k_1, k_2 \le n$

$$
\sum_{t_0=1}^{n} \sum_{t_1=1}^{n} C_{k_0,t_1}^{t_0} C_{k_1,k_2}^{t_1} e_{t_0} = 0 \iff \big[e_{k_0}, [e_{k_1}, e_{k_2}]\big] = 0
$$

$$
\Longrightarrow \big[E_{k_0,i}(e_i), [e_{k_1}, e_{k_2}]\big] = 0, \forall\, 1 \le i \le n
$$

$$
\Longrightarrow \big[\sum_{k_0,i}^{n} a_{k_0,i}\ E_{k_0,i}(e_i), [e_{k_1}, e_{k_2}]\big] = 0, \forall\, 1 \le i \le n, a_{k_o,i} \in \mathbb{K}
$$

$$
\Longrightarrow \sum_{k_0,i}^{n} a_{k_0,i}\big[E_{k_0,i}(e_i), [e_{k_1}, e_{k_2}]\big] = 0, \forall\, 1 \le i \le n, a_{k_o,i} \in \mathbb{K}
$$

$$
\begin{aligned}
\Longrightarrow &\sum_{k_0,i}^{n} a_{k_0,i}\Big[E_{k_0,i}(e_i), [e_{k_1}, e_{k_2}]\big] + \sum_{k_2,k}^{n} a_{k_2,k}\big[E_{k_2,k}(e_k), [e_{k_0}, e_{k_1}]\Big] + \\
&+ \sum_{k_1,j}^{n} a_{k_1,j}\big[E_{k_1,j}(e_j), [e_{k_2}, e_{k_0}]\big] = 0, \forall\, 1 \le i, j, k \le n, \\
&\quad a_{k_o,i}, a_{k_1,j}, a_{k_2,k} \in \mathbb{K}
\end{aligned}
$$

$$
\begin{aligned}
\Longrightarrow &\sum_{k_0,i}^{n} a_{k_0,i}\Big(\sum_{\circlearrowleft(e_i, e_{k_1}, e_{k_2})} \big(\big[\alpha(e_i), [e_{k_1}, e_{k_2}]\big]\big)\Big) = 0, \\
&\qquad \forall \alpha \in \mathrm{End}(\mathcal{V}), 1 \le i \le n, a_{k_o,i} \in \mathbb{K}
\end{aligned}
$$

$$
\Longrightarrow \sum_{\circlearrowleft(e_i, e_{k_1}, e_{k_2})} \big(\big[\alpha(e_i), [e_{k_1}, e_{k_2}]\big]\big) = 0, \forall \alpha \in \mathrm{End}(\mathcal{V}), 1 \le i \le n.
$$

This completes the proof. □

Corollary 2 *Let $\mathcal{A} = (\mathcal{V}, [\cdot, \cdot])$ be 3-dimensional skew-symmetric algebra and $\alpha \in End(\mathcal{V})$ be a Hom-Lie structure on $\mathcal{V}$. If $\mathcal{A}$ is nilpotent, then the linear space of Hom-Lie structures is of dimension* 9.

Proof For 3-dimensional skew-symmetric algebras, nilpotent implies class 2 nilpotent. □

We extend our computations for each case of the structure constants, as illustrated in Example 17, and provide all the derived series and central descending series of this class of algebra $\mathcal{A}$. We classify each subclass considered into nilpotent and solvable classes and draw conclusions as generalized by Lemma 2, Proposition 2, and Corollary 2. Table 1 presents the results of our computation.
The fourth case where the derived algebra is nilpotent illustrates the conditions under which Proposition 2 and Corollary 2 hold. In this case, $C^1(\mathcal{V}) = Z(\mathcal{V}) = \text{span}\{e_1\}$, that is from the commutator relation:

$$[e_1, e_2] = 0, \; [e_1, e_3] = 0, \; [e_2, e_3] = \lambda_3 e_1 \iff e_1 \in Z(\mathcal{V}).$$

Similarly, Proposition 2 holds, and the Hom-Lie structure obtained in case 4 is of dimension 9. From the computation of solvable classes and nilpotent classes, Corollary 1 holds and gives the desired results.

6 Hom-Subalgebras and Hom-Ideals

The following results provide important properties of derived algebras, as shown by the authors in [10]. In particular, we explore conditions on the entries of linear twisting map α of the Hom-Lie algebras, such that for every ideal $I \subseteq \mathcal{V}$, their corresponding derived algebras are Hom-subalgebras and Hom-ideal. Furthermore, we list instances where the corresponding ideals are weak ideals and present cases when such ideals are not Hom-ideals.

Let $\mathcal{W} \subseteq \mathcal{V}$ be a subspace of a $\mathbb{K}$-linear space $\mathcal{V}$, and $\{w_1, \dots w_m\}$ be a basis of $\mathcal{W}$ and $\alpha \in \text{End}(\mathcal{V})$. Then, $\mathcal{W}$ is α-invariant if and only if, for some $\delta_1, \dots, \delta_m \in \mathbb{K}$,

$$\alpha(w_i) = \delta_1 w_1 + \cdots + \delta_m w_m \text{ for all } 1 \le i \le m. \tag{6.1}$$

If $\mathcal{W}$ is the derived algebra of $\mathcal{V}$, then the solution system for (6.1) gives an interesting subclass of linear twisting α with the property that the derived series and central descending series are α-invariant. In the sequel, we explore some properties of such subclasses of linear α and compare them with their corresponding algebra morphisms.

Proposition 3 ([10]) *Let $\mathcal{A} = (\mathcal{V}, [\cdot,\cdot], \alpha)$ be a Hom-Lie algebra, and let $I \subseteq \mathcal{V}$ be a Hom-ideal. Then $D^1(I)$ and $C^1(I)$ are weak ideals of $\mathcal{A}$.*

Proposition 4 ([10]) *Let $\mathcal{A} = (\mathcal{V}, [\cdot,\cdot], \alpha)$ be a Hom-Lie algebra and I be Hom-ideal of $\mathcal{V}$. For all $n \in \mathbb{Z}_{\ge 0}$, $D^n(I)$, and $C^n(I)$ are weak subalgebras of $\mathcal{V}$ and if in addition, $\mathcal{A}$ is multiplicative, then the subspaces $D^n(I)$ and $C^n(I)$ are Hom-subalgebras of $\mathcal{A}$.*

Table 1 Central descending series and Derived series

	Structure Constants	Central descending series of I	Derived Series of I	Class k nilpotent (resp. solvable)
1	$\lambda_1 = 0$, $\lambda_2, \lambda_3 \neq 0$	$C^0(I) = \text{span}\{e_1, e_2, e_3\}$ $C^k(I) = \text{span}\{e_1, e_3\}$ $k \geq 1$	$D^0(I) = \text{span}\{e_1, e_2, e_3\}$ $D^1(I) = \text{span}\{e_1, e_3\}$ $D^2(I) = \text{span}\{e_3\}$ $D^3(I) = 0$	Not nilpotent, Class 3 Solvable
2	$\lambda_2 = 0$, $\lambda_1, \lambda_3 \neq 0$	$C^0(I) = \text{span}\{e_1, e_2, e_3\}$ $C^k(I) = \text{span}\{e_1, e_2\}$ $k \geq 1$	$D^0(I) = \text{span}\{e_1, e_2, e_3\}$ $D^1(I) = \text{span}\{e_1, e_2\}$ $D^2(I) = \text{span}\{e_2\}$ $D^3(I) = 0$	Not nilpotent, Class 3 Solvable
3	$\lambda_3 = 0$, $\lambda_1, \lambda_2 \neq 0$	$C^0(I) = \text{span}\{e_1, e_2, e_3\}$ $C^k(I) = \text{span}\{e_2, e_3\}$ $k \geq 1$	$D^0(I) = \text{span}\{e_1, e_2, e_3\}$ $D^1(I) = \text{span}\{e_2, e_3\}$ $D^2(I) = 0$	Not nilpotent, Class 2 Solvable
4	$\lambda_1, \lambda_2 = 0$, $\lambda_3 \neq 0$	$C^0(I) = \text{span}\{e_1, e_2, e_3\}$ $C^1(I) = \text{span}\{e_1\}$ $C^2(I) = 0$	$D^0(I) = \text{span}\{e_1, e_2, e_3\}$ $D^1(I) = \text{span}\{e_1\}$ $D^2(I) = 0$	Class 2 Nilpotent, Class 2 Solvable
5	$\lambda_1, \lambda_3 = 0$, $\lambda_2 \neq 0$	$C^0(I) = \text{span}\{e_1, e_2, e_3\}$ $C^k(I) = \text{span}\{e_3\}$ $k \geq 1$	$D^0(I) = \text{span}\{e_1, e_2, e_3\}$ $D^1(I) = \text{span}\{e_3\}$ $D^2(I) = 0$	Not nilpotent, Class 2 Solvable
6	$\lambda_2, \lambda_3 = 0$, $\lambda_1 \neq 0$	$C^0(I) = \text{span}\{e_1, e_2, e_3\}$ $C^k(I) = \text{span}\{e_2\}$ $k \geq 1$	$D^0(I) = \text{span}\{e_1, e_2, e_3\}$ $D^1(I) = \text{span}\{e_2\}$ $D^2(I) = 0$	Not nilpotent, Class 2 Solvable
7	λ_1, λ_2, $\lambda_3 \neq 0$	$C^k(I) = \text{span}\{e_1, e_2, e_3\}$ $k \geq 0$	$D^k(I) = \text{span}\{e_1, e_2, e_3\}$ $k \geq 0$	Not nilpotent, Not solvable

Proposition 5 ([10]) *Let $(\mathcal{V}, [\cdot, \cdot], \alpha)$ be a multiplicative Hom-Lie algebra and I be a Hom-ideal of $\mathcal{V}$. If the linear map α is surjective, then for all $n \in \mathbb{Z}_{\geq 0}$, $D^n(I)$ and $C^n(I)$ are Hom-ideals.*

The results in Proposition 5 can be explored by imposing a weaker condition than the algebra morphism condition, and studying classes of such non-multiplicative Hom-Lie algebras such that the derived series and central descending series are α-invariant. We present a detailed conclusion for such classes of Hom-Lie algebras.

6.1 Invariant Subspaces of Non-multiplicative Hom-Lie Algebras

Consider a 3-dimensional $\mathbb{K}$-linear space $\mathcal{V}$ with a basis $\{e_1, e_2, e_3\}$, and the skew-symmetric bracket multiplication defined in (4.5) for all $\lambda_1, \lambda_2, \lambda_3 \in \mathbb{K}$ as

$$[e_1, e_2] = \lambda_1 e_2,$$

$$[e_1, e_3] = \lambda_2 e_3,$$

$$[e_2, e_3] = \lambda_3 e_1.$$

In this section, we explore properties of Hom-Lie algebras related to their ideals and subalgebras. Specifically, our focus is on α-invariant subspaces of Hom-Lie algebras of generalized $\mathfrak{sl}(2)$-type and to study their solvability, nilpotency, and properties related to their derived series and central descending series. Our computations and results are given in Tables 2 and 3 and discussions and examples thereafter.

In the first part, we explore α-invariant subspaces through Hom-subalgebras and Hom-ideals. For each case of Hom-Lie algebra realized by cases on the dimension of $[\mathcal{V}, \mathcal{V}]$, and conditions on $\lambda_{i,\, i=1,2,3}$ realizing it, we analyze the kernels of all linear twisting maps α and establish conditions under which these subspaces, $\ker(\alpha)$, become Hom-ideals. In this context, we observe that these subspaces are inherently α-invariant. Consequently, our focus shifts to determining which among them is closed under the bilinear bracket multiplication of subalgebra condition and Hom-ideal condition (see Definitions 3 and 4), respectively. For every class of Hom-Lie algebra under our consideration, when the $\dim(\ker(\alpha)) = 3$ or $\dim(\ker(\alpha)) = 0$, the subspace $\ker(\alpha)$ becomes a Hom-ideal. These two cases are not interesting; therefore, we exclude them in our discussion and give much more attention to specific cases when $\dim(\ker(\alpha)) = 1$ or 2. For every class of Hom-Lie algebras considered, if $\dim \ker(\alpha) = 1$, then $\ker(\alpha)$ becomes a Hom-subalgebra, since every 1-dimensional subspace is closed under bilinear bracket multiplication conditions defining a Hom-subalgebra. However, these subspaces do not necessarily need to be Hom-ideals.

For each case presented, we discuss conditions on $\lambda_{i,\, i=1,2,3}$ that realize bilinear bracket multiplication of (4.5) and present the corresponding classes of Hom-Lie algebras realized in each case through its linear space of all linear twisting maps α. For every class of algebra considered, we provide the corresponding subsets of the linear space of linear twisting map α with the properties that $\ker(\alpha)$ is a Hom-ideal. We showcase our results in Table 2 consisting of the following parts. In column 2, we list various cases for the dimension of $[\mathcal{V}, \mathcal{V}]$ by stating conditions on λ_i (for $i = 1, 2, 3$) that realize each value. For each case, we also provide the corresponding linear space of all linear maps α (defined by some free parameters a_{ij} and $\lambda_i \in \mathbb{K}$ for $1 \leq i, j \leq 3$), such that $(\mathcal{V}, [\cdot, \cdot], \alpha)$ forms a Hom-Lie algebra. For each of these Hom-Lie structures, we state their dimension, $\dim(\mathtt{HomLie}_3)$. Moreover, for each case and class of Hom-Lie algebras presented in column 2, we provide the corresponding set of morphisms subset of linear twisting maps α that define multiplicative Hom-Lie algebras. In column 3, we present subsets of the space of linear maps α, given in column 2, with the properties that $\ker(\alpha)$ are Hom-subalgebras. Similarly, in column 4, we include all subsets of space of linear twisting maps α for which the $\ker(\alpha)$, given in column 2 are Hom-ideals. For the subclasses presented in column 3 and column 4, it is important to note the cases in which such subspaces, $\ker(\alpha)$, are Hom-ideals. In column 1, we enumerate these cases starting from case 1 to the case 6, and we refer to these classes later in our discussions. For example, in the first row, we consider the case when $\lambda_1 = 0$, and $\lambda_2, \lambda_3 \neq 0$ as indicated in column 2. For this subclass of algebra, the linear space of all linear maps α such that $(\mathcal{V}, [\cdot, \cdot], \alpha)$ is a Hom-Lie algebra is provided in the same column 2; this is followed by the subset(s) of the linear space of linear maps α consisting of algebra morphisms. Here, we list all the subsets of linear maps α such that $(\mathcal{V}, [\cdot, \cdot], \alpha)$ is a multiplicative Hom-Lie algebra. In column 3, we analyze the kernel of this space of linear maps α given in column 2 and list all its subsets with the properties that $\ker(\alpha)$ is a Hom-subalgebra of this class of Hom-Lie algebra. Similarly, in column 4, we give matrices representing the subset of the linear space of linear maps α given in column 2, with the properties that $\ker(\alpha)$ is a Hom-ideal (in this case, we only have one such subset of the linear space of linear maps α). In column 1, we enumerate each subclass of the considered algebra, which will be referred to later in our detailed discussion.

To this end, we conduct an exhaustive comparison of the properties of non-multiplicative Hom-Lie algebras, which consist of α-invariant subspaces of generalized Hom-Lie algebras of $\mathfrak{sl}(2)$-type, with their corresponding multiplicative Hom-Lie algebras.

In case 7 where $\lambda_1, \lambda_2, \lambda_3 \neq 0$, and the bilinear bracket multiplication is defined as in (4.5), the linear space of all linear twisting maps that turns this skew-symmetric algebra into a Hom-Lie algebra is given by $[\alpha] = \begin{pmatrix} a_{11} & a_{12} & a_{13} \\ \frac{\lambda_1}{\lambda_3} a_{13} & a_{22} & a_{23} \\ -\frac{\lambda_2}{\lambda_3} a_{12} & a_{32} & -\frac{\lambda_2}{\lambda_1} a_{22} \end{pmatrix}$.

Table 2 Hom-subalgebras, Hom-ideals, and multiplicative Hom-Lie structures of generalized $\mathfrak{sl}(2)$-type

	Structure Constants, $[\alpha]$ such that $\mathcal{A}$ is a Hom-Lie algebra, and Subclasses of $[\alpha]$ such that $\mathcal{A}$ is multiplicative Hom-Lie algebra	Subclasses of $[\alpha]$ such that $\ker(\alpha)$ are Hom-subalgebras, $\ker(\alpha) \neq \{0\}$, $\ker(\alpha) \neq \mathcal{V}$	Subclasses of $[\alpha]$ such that $\ker(\alpha)$ are Hom-Ideals, and $\ker(\alpha) \neq \{0\}$, $\ker(\alpha) \neq \mathcal{V}$
1	$\lambda_1 = 0$ $\lambda_2, \lambda_3 \neq 0$ $\dim[\mathcal{V}, \mathcal{V}] = 2$, $[\alpha]$ such that $\mathcal{A}$ is a Hom-Lie algebra $\begin{pmatrix} a_{11} & a_{12} & a_{13} \\ a_{21} & 0 & a_{23} \\ -\frac{\lambda_2}{\lambda_3}a_{12} & a_{32} & a_{33} \end{pmatrix}$ $\dim(\mathtt{HomLie}_3) = 7$, Subclasses of $[\alpha]$ such that $\mathcal{A}$ is multiplicative $\begin{pmatrix} 0 & 0 & 0 \\ 0 & 0 & 0 \\ 0 & a_{32} & 0 \end{pmatrix}$ $a_{32} \neq 0$	$\ker(\alpha) = \mathrm{span}\{e_1, e_2\}$ $\begin{pmatrix} 0 & 0 & a_{13} \\ 0 & 0 & a_{23} \\ 0 & 0 & a_{33} \end{pmatrix}$, a_{13}, a_{23}, a_{33} not simultaneously zero $\ker(\alpha) = \mathrm{span}\{e_1, e_3\}$ $\begin{pmatrix} 0 & 0 & 0 \\ 0 & 0 & 0 \\ 0 & a_{32} & 0 \end{pmatrix}$, $a_{32} \neq 0$ $\ker(\alpha) = \mathrm{span}\{e_1\}$ $\begin{pmatrix} 0 & 0 & a_{13} \\ 0 & 0 & a_{23} \\ 0 & a_{32} & a_{33} \end{pmatrix}$, $a_{32} \neq 0$, and a_{13}, a_{23}, a_{33} not simultaneously zero $\ker(\alpha) = \mathrm{span}\{e_2\}$ $\begin{pmatrix} a_{11} & 0 & a_{13} \\ a_{21} & 0 & a_{23} \\ 0 & 0 & a_{33} \end{pmatrix}$ $a_{11} \neq 0, a_{33} \neq 0$, $a_{13}a_{21} - a_{11}a_{23} = 0$ $\begin{pmatrix} 0 & 0 & a_{13} \\ a_{21} & 0 & a_{23} \\ 0 & 0 & a_{33} \end{pmatrix}$ $a_{21} \neq 0$, and a_{13}, a_{23}, a_{33} not simultaneously zero $\ker(\alpha) = \mathrm{span}\{-\frac{a_{12}}{a_{11}}e_1 + e_2\}$ $\begin{pmatrix} a_{11} & a_{12} & a_{13} \\ a_{21} & 0 & a_{23} \\ -\frac{\lambda_2}{\lambda_3}a_{12} & a_{32} & a_{33} \end{pmatrix}$ $a_{11} \neq 0$, $a_{12}a_{21} = 0$, $a_{12}^2\lambda_2 + a_{11}a_{32}\lambda_3 = 0$ $a_{13}a_{21} - a_{11}a_{23} \neq 0$	$\ker(\alpha) = \mathrm{span}\{e_1, e_3\}$ $\begin{pmatrix} 0 & 0 & 0 \\ 0 & 0 & 0 \\ 0 & a_{32} & 0 \end{pmatrix}$ $a_{32} \neq 0$

(continued)

Table 2 (continued)

		$\ker(\alpha) = \mathrm{span}\{\frac{-a_{13}a_{32}\lambda_3+a_{12}a_{33}\lambda_3}{a_{12}^2\lambda_2}e_1 + \frac{-a_{13}}{a_{12}}e_2+e_3\}$ $\begin{pmatrix} 0 & a_{12} & a_{13} \\ 0 & 0 & 0 \\ \frac{-a_{12}\lambda_2}{\lambda_3} & a_{32} & a_{33} \end{pmatrix}$, $a_{12} \neq 0$ $\ker(\alpha) = \mathrm{span}\{\frac{-a_{23}}{a_{21}}e_1 - \frac{a_{21}a_{33}\lambda_3}{a_{21}a_{32}\lambda_3}e_2+e_3\}$ $\begin{pmatrix} 0 & 0 & 0 \\ a_{21} & 0 & a_{23} \\ 0 & a_{32} & a_{33} \end{pmatrix}$, $a_{21} \neq 0, a_{32} \neq 0$ $\ker(\alpha) = \mathrm{span}\{\frac{-a_{23}}{a_{21}}e_1 + \frac{-a_{13}a_{21}+a_{11}a_{23}}{a_{12}a_{21}}e_2+e_3\}$ $\begin{pmatrix} a_{11} & a_{12} & a_{13} \\ a_{21} & 0 & a_{23} \\ \frac{-a_{12}\lambda_2}{\lambda_3} & a_{32} & a_{33} \end{pmatrix}$ $a_{12}, a_{21} \neq 0,$ $a_{12}^2a_{23}\lambda_2 - a_{13}a_{21}a_{32}\lambda_3$ $+a_{11}a_{23}a_{32}\lambda_3$ $+a_{12}a_{21}a_{33}\lambda_3 = 0$ $\ker(\alpha) = \mathrm{span}\{\frac{-a_{13}a_{32}\lambda_3+a_{12}a_{33}\lambda_3}{a_{12}^2\lambda_2+a_{11}a_{32}\lambda_3}e_1 + \frac{-a_{12}a_{13}\lambda_2-a_{11}a_{33}\lambda_3}{a_{12}^2\lambda_2+a_{11}a_{32}\lambda_3}e_2+e_3\}$ $\begin{pmatrix} a_{11} & a_{12} & a_{13} \\ a_{21} & 0 & a_{23} \\ \frac{-a_{12}\lambda_2}{\lambda_3} & a_{32} & a_{33} \end{pmatrix}$ $a_{11} \neq 0,\ a_{12}a_{21} = 0,$ $a_{12}^2\lambda_2 + a_{11}a_{32}\lambda_3 \neq 0,$ $a_{13}a_{21} - a_{11}a_{23} = 0$	
2	$\lambda_2 = 0$ $\lambda_1, \lambda_3 \neq 0$ $\dim[\mathcal{V}, \mathcal{V}] = 2$, $[\alpha]$ such that $\mathcal{A}$ is a Hom-Lie algebra $\begin{pmatrix} a_{11} & a_{12} & a_{13} \\ \frac{\lambda_1}{\lambda_3}a_{13} & a_{22} & a_{23} \\ a_{31} & a_{32} & 0 \end{pmatrix}$ $\dim(\mathtt{HomLie}_3) = 7$,	$\ker(\alpha) = \mathrm{span}\{e_1, e_2\}$ $\begin{pmatrix} 0 & 0 & 0 \\ 0 & 0 & a_{23} \\ 0 & 0 & 0 \end{pmatrix}$, $a_{23} \neq 0$ $\ker(\alpha) = \mathrm{span}\{e_1, e_3\}$ $\begin{pmatrix} 0 & a_{12} & 0 \\ 0 & a_{22} & 0 \\ 0 & a_{32} & 0 \end{pmatrix}$, a_{12}, a_{22}, a_{32} not simultaneously zero $\ker(\alpha) = \mathrm{span}\{e_1\}$ $\begin{pmatrix} 0 & a_{12} & 0 \\ 0 & a_{22} & a_{23} \\ 0 & a_{32} & 0 \end{pmatrix}$,	$\ker(\alpha) = \mathrm{span}\{e_1, e_2\}$ $\begin{pmatrix} 0 & 0 & 0 \\ 0 & 0 & a_{23} \\ 0 & 0 & 0 \end{pmatrix}$, $a_{23} \neq 0$

(continued)

Table 2 (continued)

Subclasses of $[\alpha]$ such that $\mathcal{A}$ is multiplicative	$a_{23} \neq 0$, and a_{12}, a_{22}, a_{32} not simultaneously zero	
$\begin{pmatrix} 0 & 0 & 0 \\ 0 & 0 & a_{23} \\ 0 & 0 & 0 \end{pmatrix}$ $a_{23} \neq 0$.	$\ker(\alpha) = \mathrm{span}\{e_2\}$ $\begin{pmatrix} a_{11} & 0 & a_{13} \\ \frac{\lambda_1}{\lambda_3} a_{13} & 0 & a_{23} \\ a_{31} & 0 & 0 \end{pmatrix}$, $a_{11}, a_{13} \neq 0$, $a_{13}^2 \lambda_1 - a_{11} a_{23} \lambda_3 \neq 0$	
	$\ker(\alpha) = \mathrm{span}\{e_3\}$ $\begin{pmatrix} 0 & a_{12} & 0 \\ 0 & 0 & 0 \\ a_{31} & a_{32} & 0 \end{pmatrix}$, $a_{31} \neq 0$, and a_{12}, a_{32} not simultaneously zero	
	$\ker(\alpha) = \mathrm{span}\{\frac{-a_{32}}{a_{31}} e_1 + e_2\}$ $\begin{pmatrix} 0 & 0 & 0 \\ 0 & 0 & a_{23} \\ a_{31} & a_{32} & 0 \end{pmatrix}$, $a_{31} \neq 0, a_{23} \neq 0$	
	$\ker(\alpha) = \mathrm{span}\{\frac{-a_{12}}{a_{11}} e_1 + e_2\}$ $\begin{pmatrix} a_{11} & a_{12} & a_{13} \\ \frac{a_{13}\lambda_1}{\lambda_3} & a_{22} & a_{23} \\ a_{31} & a_{32} & 0 \end{pmatrix}$ $a_{11}, a_{31}, a_{13} \neq 0$, $a_{12}a_{31} - a_{11}a_{32} = 0$, $a_{13}a_{22} - a_{12}a_{23} = 0$, $a_{12}a_{13}\lambda_1 - a_{11}a_{22}\lambda_3 = 0$, $a_{12}a_{13}a_{32}\lambda_1 - a_{12}a_{22}a_{31}\lambda_3 = 0$, $a_{13}a_{23}a_{32}\lambda_1 - a_{22}a_{23}a_{31}\lambda_3 = 0$, $a_{13}^2 a_{32}\lambda_1 - a_{12}a_{23}a_{31}\lambda_3 = 0$, $a_{13}^2\lambda_1 - a_{11}a_{23}\lambda_3 = 0$, $a_{13}a_{22}a_{32}\lambda_1 - a_{22}^2 a_{31}\lambda_3 = 0$, $a_{13}a_{32}^2\lambda_1 - a_{22}a_{31}a_{32}\lambda_3 = 0$, or $a_{11} \neq 0$, $a_{13}^2\lambda_1 - a_{11}a_{23}\lambda_3 \neq 0$ $a_{12}a_{31} - a_{11}a_{32} = 0$, $a_{12}a_{13}\lambda_1 - a_{11}a_{22}\lambda_3 = 0$, $a_{12}a_{13}a_{32}\lambda_1 - a_{12}a_{22}a_{31}\lambda_3 = 0$	
	$\ker(\alpha) =$ $\mathrm{span}\{\frac{a_{23}a_{32}}{a_{22}a_{31}} e_1 + \frac{-a_{23}}{a_{22}} e_2 + e_3\}$	

(continued)

Table 2 (continued)

	$\begin{pmatrix} 0 & a_{12} & 0 \\ 0 & a_{22} & a_{23} \\ a_{31} & a_{32} & 0 \end{pmatrix}$, $a_{22}, a_{31}, a_{32} \neq 0$, $a_{12}a_{23} = 0$	
	$\ker(\alpha) = \text{span}\{-\frac{a_{22}\lambda_3}{a_{13}\lambda_1} e_1 + e_2\}$ $\begin{pmatrix} 0 & 0 & a_{13} \\ \frac{a_{13}\lambda_1}{\lambda_3} & a_{22} & a_{23} \\ a_{31} & a_{32} & 0 \end{pmatrix}$, $a_{13} \neq 0$, $a_{13}a_{32}\lambda_1 - a_{22}a_{31}\lambda_3 = 0$	
	$\ker(\alpha) =$ $\text{span}\{\frac{a_{13}a_{22}\lambda_3 - a_{12}a_{23}\lambda_3}{a_{12}a_{13}\lambda_1 - a_{11}a_{22}\lambda_3} e_1 +$ $+ \frac{-a_{13}^2\lambda_1 + a_{11}a_{23}\lambda_3}{a_{12}a_{13}\lambda_1 - a_{11}a_{22}\lambda_3} e_2 + e_3\}$ $\begin{pmatrix} a_{11} & a_{12} & a_{13} \\ \frac{a_{13}\lambda_1}{\lambda_3} & a_{22} & a_{23} \\ a_{31} & a_{32} & 0 \end{pmatrix}$, $a_{11} \neq 0$, $a_{12}a_{13}\lambda_1 - a_{11}a_{22}\lambda_3 \neq 0$, $a_{13}^2a_{32}\lambda_1 - a_{13}a_{22}a_{31}\lambda_3$ $+a_{12}a_{23}a_{31}\lambda_3$ $-a_{11}a_{23}a_{32}\lambda_3 = 0$	
	$\ker(\alpha) =$ $\text{span}\{\frac{a_{13}a_{22}\lambda_3 - a_{12}a_{23}\lambda_3}{a_{12}a_{13}\lambda_1} e_1$ $-\frac{a_{13}}{a_{12}} e_2 + e_3\}$ $\begin{pmatrix} 0 & a_{12} & a_{13} \\ \frac{a_{13}\lambda_1}{\lambda_3} & a_{22} & a_{23} \\ a_{31} & a_{32} & 0 \end{pmatrix}$ $a_{13} \neq 0, a_{12} \neq 0$, $a_{13}^2a_{32}\lambda_1 - a_{13}a_{22}a_{31}\lambda_3$ $+a_{12}a_{23}a_{31}\lambda_3 = 0$	
	$\ker(\alpha) =$ $\text{span}\{\frac{a_{13}a_{32}}{a_{12}a_{31} - a_{11}a_{32}} e_1 +$ $-\frac{a_{13}a_{31}}{a_{12}a_{31} - a_{11}a_{32}} e_2 + e_3\}$ $\begin{pmatrix} a_{11} & a_{12} & a_{13} \\ \frac{a_{13}\lambda_1}{\lambda_3} & a_{22} & a_{23} \\ a_{31} & a_{32} & 0 \end{pmatrix}$, $a_{11} \neq 0$, $a_{12}a_{13}\lambda_1 - a_{11}a_{22}\lambda_3 = 0$, $a_{12}a_{31} - a_{11}a_{32} \neq 0$, $a_{13}^2a_{22}\lambda_1 - a_{12}a_{13}a_{23}\lambda_1 = 0$, $a_{13}^2\lambda_1 - a_{11}a_{23}\lambda_3 = 0$	

(continued)

Table 2 (continued)

3	$\lambda_3=0$ $\lambda_1,\lambda_2\neq 0$ $\dim[\mathcal{V},\mathcal{V}]=2$, $[\alpha]$ such that $\mathcal{A}$ is a Hom-Lie algebra $\begin{pmatrix}a_{11}&0&0\\a_{21}&a_{22}&a_{23}\\a_{31}&a_{32}&a_{33}\end{pmatrix}$ $\dim(\mathtt{HomLie}_3)=7$, Subclasses of $[\alpha]$ such that $\mathcal{A}$ is multiplicative $\lambda_1\neq\pm\lambda_2$ $\begin{pmatrix}1&0&0\\a_{21}&a_{22}&0\\a_{31}&0&a_{33}\end{pmatrix}$, $\begin{pmatrix}\frac{\lambda_1}{\lambda_2}&0&0\\a_{21}&0&0\\a_{31}&a_{32}&0\end{pmatrix}$, $\begin{pmatrix}\frac{\lambda_2}{\lambda_1}&0&0\\a_{21}&0&a_{23}\\a_{31}&0&0\end{pmatrix}$, $\begin{pmatrix}a_{11}&0&0\\a_{21}&0&0\\a_{31}&0&0\end{pmatrix}$. $\lambda_1=\lambda_2$ $\begin{pmatrix}1&0&0\\a_{21}&a_{22}&a_{23}\\a_{31}&a_{32}&a_{33}\end{pmatrix}$, $\begin{pmatrix}1&0&0\\a_{21}&a_{22}&0\\a_{31}&a_{32}&a_{33}\end{pmatrix}$, $\begin{pmatrix}1&0&0\\a_{21}&a_{22}&a_{23}\\a_{31}&0&a_{33}\end{pmatrix}$, $\begin{pmatrix}1&0&0\\a_{21}&0&a_{23}\\a_{31}&a_{32}&0\end{pmatrix}$. $\lambda_1=-\lambda_2$ $\begin{pmatrix}-1&0&0\\a_{21}&0&a_{23}\\a_{31}&a_{32}&0\end{pmatrix}$	$\ker(\alpha)=\mathrm{span}\{e_2,\frac{-a_{23}}{a_{21}}e_1+e_3\}$ $\begin{pmatrix}0&0&0\\a_{21}&0&a_{23}\\a_{31}&0&a_{33}\end{pmatrix}$, $a_{21}\neq 0$, $a_{23}a_{31}-a_{21}a_{33}=0$ $\ker(\alpha)=\mathrm{span}\{e_2,\frac{-a_{33}}{a_{31}}e_1+e_3\}$ $\begin{pmatrix}0&0&0\\0&0&0\\a_{31}&0&a_{33}\end{pmatrix}$, $a_{31}\neq 0$ $\ker(\alpha)=\mathrm{span}\{\frac{-a_{22}}{a_{21}}e_1+e_2,e_3\}$ $\begin{pmatrix}0&0&0\\a_{21}&a_{22}&0\\a_{31}&a_{32}&0\end{pmatrix}$, $a_{21}\neq 0$, $a_{22}a_{31}-a_{21}a_{32}=0$ $\ker(\alpha)=\mathrm{span}\{\frac{-a_{32}}{a_{31}}e_1+e_2,e_3\}$ $\begin{pmatrix}0&0&0\\0&0&0\\a_{31}&a_{32}&0\end{pmatrix}$, $a_{31}\neq 0$ $\ker(\alpha)=\mathrm{span}\{e_1,e_2\}$ $\begin{pmatrix}0&0&0\\0&0&a_{23}\\0&0&a_{33}\end{pmatrix}$, $a_{23}\neq 0$ $\ker(\alpha)=\mathrm{span}\{e_1,e_3\}$ $\begin{pmatrix}0&0&0\\0&a_{22}&0\\0&a_{32}&0\end{pmatrix}$, a_{22},a_{32} not simultaneously zero $\ker(\alpha)=\mathrm{span}\{e_2,e_3\}$ $\begin{pmatrix}a_{11}&0&0\\a_{21}&0&0\\a_{31}&0&0\end{pmatrix}$, a_{11},a_{21},a_{31} not simultaneously zero	$\ker(\alpha)=\mathrm{span}\{e_2,e_3\}$ $\begin{pmatrix}a_{11}&0&0\\a_{21}&0&0\\a_{31}&0&0\end{pmatrix}$ a_{11},a_{21},a_{31} not simultaneously zero $\ker(\alpha)=\mathrm{span}\{e_2\}$ $\begin{pmatrix}a_{11}&0&0\\a_{21}&0&a_{23}\\a_{31}&0&a_{33}\end{pmatrix}$ $a_{11}\neq 0$, a_{23},a_{33} not simultaneously zero $\begin{pmatrix}0&0&0\\a_{21}&0&a_{23}\\a_{31}&0&a_{33}\end{pmatrix}$ $a_{21}\neq 0$, $a_{23}a_{31}-a_{21}a_{33}\neq 0$ $\begin{pmatrix}0&0&0\\0&0&a_{23}\\a_{31}&0&a_{33}\end{pmatrix}$ $a_{31}\neq 0, a_{23}\neq 0$ $\ker(\alpha)=\mathrm{span}\{e_3\}$ $\begin{pmatrix}a_{11}&0&0\\a_{21}&a_{22}&0\\a_{31}&a_{32}&0\end{pmatrix}$, $a_{11}\neq 0$, and a_{22},a_{32} not simultaneously zero $\begin{pmatrix}0&0&0\\a_{21}&a_{22}&a_{23}\\a_{31}&a_{32}&a_{33}\end{pmatrix}$ $a_{21}\neq 0$, $a_{22}a_{31}-a_{21}a_{32}\neq 0$, $a_{23}a_{32}-a_{22}a_{33}=0$, $a_{21}a_{33}-a_{23}a_{31}=0$

(continued)

Table 2 (continued)

	$\ker(\alpha) = \text{span}\{e_1\}$ $\begin{pmatrix} 0 & 0 & 0 \\ 0 & a_{22} & a_{23} \\ 0 & a_{32} & a_{33} \end{pmatrix}$ $a_{22} \neq 0,$ $a_{23}a_{32} - a_{22}a_{33} \neq 0$	
	$\begin{pmatrix} 0 & 0 & 0 \\ 0 & 0 & a_{23} \\ 0 & a_{32} & a_{33} \end{pmatrix}$ $a_{32} \neq 0, a_{23} \neq 0$	
	$\ker(\alpha) = \text{span}\{e_2\}$ $\begin{pmatrix} a_{11} & 0 & 0 \\ a_{21} & 0 & a_{23} \\ a_{31} & 0 & a_{33} \end{pmatrix},$ $a_{11} \neq 0,$ a_{23}, a_{33} not simultaneously zero	
	$\ker(\alpha) = \text{span}\{\frac{-a_{23}}{a_{22}} e_2 + e_3\}$ $\begin{pmatrix} a_{11} & 0 & 0 \\ a_{21} & a_{22} & a_{23} \\ a_{31} & a_{32} & a_{33} \end{pmatrix},$ $a_{11}, a_{22} \neq 0,$ $a_{23}a_{32} - a_{22}a_{33} = 0$	
	$\ker(\alpha) = \text{span}\{\frac{-a_{33}}{a_{32}} e_2 + e_3\}$ $\begin{pmatrix} a_{11} & 0 & 0 \\ a_{21} & 0 & 0 \\ a_{31} & a_{32} & a_{33} \end{pmatrix}, a_{11}, a_{32} \neq 0$	
	$\ker(\alpha) = \text{span}\{\frac{-a_{22}}{a_{21}} e_1 + e_2\}$ $\begin{pmatrix} 0 & 0 & 0 \\ a_{21} & a_{22} & a_{23} \\ a_{31} & a_{32} & a_{33} \end{pmatrix},$ $a_{21} \neq 0,$ $a_{22}a_{31} - a_{21}a_{32} = 0,$ $a_{23}a_{31} - a_{21}a_{33} \neq 0$	
	$\ker(\alpha) = \text{span}\{\frac{-a_{32}}{a_{31}} e_1 + e_2\}$ $\begin{pmatrix} 0 & 0 & 0 \\ 0 & 0 & a_{23} \\ a_{31} & a_{32} & a_{33} \end{pmatrix},$ $a_{31} \neq 0, a_{23} \neq 0$	
	$\ker(\alpha) =$ $\text{span}\{\frac{a_{23}a_{32}-a_{22}a_{33}}{a_{22}a_{31}-a_{21}a_{32}} e_1 + \frac{-a_{23}a_{31}++a_{21}a_{33}}{a_{22}a_{31}-a_{21}a_{32}} e_2 + e_3\}$	

(continued)

Table 2 (continued)

		$\begin{pmatrix} 0 & 0 & 0 \\ a_{21} & a_{22} & a_{23} \\ a_{31} & a_{32} & a_{33} \end{pmatrix}$, $a_{21} \neq 0$, $a_{22}a_{31} - a_{21}a_{32} \neq 0$ $\ker(\alpha) =$ $\mathrm{span}\{\frac{a_{23}a_{32}-a_{22}a_{33}}{a_{22}a_{31}} e_1 - \frac{a_{23}}{a_{22}} e_2 + e_3\}$ $\begin{pmatrix} 0 & 0 & 0 \\ 0 & a_{22} & a_{23} \\ a_{31} & a_{32} & a_{33} \end{pmatrix}$, $a_{31}, a_{22} \neq 0$	
4	$\lambda_1, \lambda_2 = 0$ $\lambda_3 \neq 0$ $\dim[\mathcal{V}, \mathcal{V}] = 1$, $[\alpha]$ such that $\mathcal{A}$ is a Hom-Lie algebra $\begin{pmatrix} a_{11} & a_{12} & a_{13} \\ a_{21} & a_{22} & a_{23} \\ a_{31} & a_{32} & a_{33} \end{pmatrix}$ $\dim(\mathtt{HomLie}_3) = 9$, Subclasses of $[\alpha]$ such that $\mathcal{A}$ is multiplicative $\begin{pmatrix} a_{22}a_{33} - a_{23}a_{32} & a_{12} & a_{13} \\ 0 & a_{22} & a_{23} \\ 0 & a_{32} & a_{33} \end{pmatrix}$	$\ker(\alpha) = \mathrm{span}\{e_1, \frac{-a_{13}}{a_{12}} e_2 + e_3\}$ $\begin{pmatrix} 0 & a_{12} & a_{13} \\ 0 & a_{22} & a_{23} \\ 0 & a_{32} & a_{33} \end{pmatrix}$, $a_{12} \neq 0$, $a_{13}a_{22} - a_{12}a_{23} = 0$, $a_{13}a_{23}a_{32} - a_{13}a_{22}a_{33} = 0$, $a_{13}a_{32} - a_{12}a_{33} = 0$ $\ker(\alpha) = \mathrm{span}\{e_1, \frac{-a_{23}}{a_{22}} e_2 + e_3\}$ $\begin{pmatrix} 0 & 0 & 0 \\ 0 & a_{22} & a_{23} \\ 0 & a_{32} & a_{33} \end{pmatrix}$, $a_{22} \neq 0$, $a_{23}a_{32} - a_{22}a_{33} = 0$ $\ker(\alpha) = \mathrm{span}\{e_1, \frac{-a_{33}}{a_{32}} e_2 + e_3\}$ $\begin{pmatrix} 0 & 0 & 0 \\ 0 & 0 & 0 \\ 0 & a_{32} & a_{33} \end{pmatrix}$, $a_{32} \neq 0$ $\ker(\alpha) = \mathrm{span}\{e_1, e_2\}$ $\begin{pmatrix} 0 & 0 & a_{13} \\ 0 & 0 & a_{23} \\ 0 & 0 & a_{33} \end{pmatrix}$, a_{13}, a_{23}, a_{33} not simultaneously zero $\ker(\alpha) = \mathrm{span}\{e_1\}$ $\begin{pmatrix} 0 & a_{12} & a_{13} \\ 0 & a_{22} & a_{23} \\ 0 & a_{32} & a_{33} \end{pmatrix}$, $a_{12} \neq 0$, $a_{13}a_{22} - a_{12}a_{23} \neq 0$, or $a_{12} \neq 0$, $a_{13}a_{22} - a_{12}a_{23} = 0$, $a_{13}a_{32} - a_{12}a_{33} \neq 0$,	$\ker(\alpha) =$ $\mathrm{span}\{e_1, \frac{-a_{13}}{a_{12}} e_2 + e_3\}$ $\begin{pmatrix} 0 & a_{12} & a_{13} \\ 0 & a_{22} & a_{23} \\ 0 & a_{32} & a_{33} \end{pmatrix}$ $a_{12} \neq 0$, $a_{13}a_{22} - a_{12}a_{23} = 0$, $a_{13}a_{23}a_{32} - a_{13}a_{22}a_{33} = 0$, $a_{13}a_{32} - a_{12}a_{33} = 0$ $\ker(\alpha) =$ $\mathrm{span}\{e_1, \frac{-a_{23}}{a_{22}} e_2 + e_3\}$ $\begin{pmatrix} 0 & 0 & 0 \\ 0 & a_{22} & a_{23} \\ 0 & a_{32} & a_{33} \end{pmatrix}$ $a_{22} \neq 0$, $a_{23}a_{32} - a_{22}a_{33} = 0$ $\ker(\alpha) =$ $\mathrm{span}\{e_1, \frac{-a_{33}}{a_{32}} e_2 + e_3\}$ $\begin{pmatrix} 0 & 0 & 0 \\ 0 & 0 & 0 \\ 0 & a_{32} & a_{33} \end{pmatrix}$ $a_{32} \neq 0, a_{13} = 0$ $\ker(\alpha) = \mathrm{span}\{e_1, e_2\}$ $\begin{pmatrix} 0 & 0 & a_{13} \\ 0 & 0 & a_{23} \\ 0 & 0 & a_{33} \end{pmatrix}$ a_{13}, a_{23}, a_{33} not simultaneously zero $\ker(\alpha) = \mathrm{span}\{e_1\}$ $\begin{pmatrix} 0 & a_{12} & a_{13} \\ 0 & a_{22} & a_{23} \\ 0 & a_{32} & a_{33} \end{pmatrix}$,

(continued)

Table 2 (continued)

		$\begin{pmatrix}0&0&a_{13}\\0&a_{22}&a_{23}\\0&a_{32}&a_{33}\end{pmatrix}$, $a_{13}, a_{22} \neq 0$ $\begin{pmatrix}0&0&0\\0&a_{22}&a_{23}\\0&a_{32}&a_{33}\end{pmatrix}$, $a_{22} \neq 0$, $a_{23}a_{32} - a_{22}a_{33} \neq 0$ $\begin{pmatrix}0&0&a_{13}\\0&0&a_{23}\\0&a_{32}&a_{33}\end{pmatrix}$, $a_{32} \neq 0$ and a_{13}, a_{23}, a_{33} not simultaneously zero $\ker(\alpha) = \text{span}\{\frac{-a_{22}}{a_{21}}e_1 + e_2\}$ $\begin{pmatrix}0&0&a_{13}\\a_{21}&a_{22}&a_{23}\\a_{31}&a_{32}&a_{33}\end{pmatrix}$, $a_{22}a_{31} - a_{21}a_{32} = 0$, $a_{13}, a_{21} \neq 0$ $\begin{pmatrix}0&0&0\\a_{21}&a_{22}&a_{23}\\a_{31}&a_{32}&a_{33}\end{pmatrix}$, $a_{21} \neq 0$, $a_{22}a_{31} - a_{21}a_{32} = 0$, $a_{23}a_{31} - a_{21}a_{33} \neq 0$ $\ker(\alpha) = \text{span}\{\frac{-a_{32}}{a_{31}}e_1 + e_2\}$ $\begin{pmatrix}0&0&a_{13}\\0&0&0\\a_{31}&a_{32}&a_{33}\end{pmatrix}$, $a_{13}, a_{31} \neq 0$ $\ker(\alpha) =$ $\text{span}\{\frac{a_{13}a_{32}-a_{12}a_{33}}{a_{12}a_{31}-a_{32}a_{11}}e_1$ $+\frac{-a_{13}a_{31}+a_{33}a_{11}}{a_{12}a_{31}-a_{32}a_{11}}e_2 + e_3\}$ $\begin{pmatrix}a_{11}&a_{12}&a_{13}\\a_{21}&a_{22}&a_{23}\\a_{31}&a_{32}&a_{33}\end{pmatrix}$, $a_{11} \neq 0$, $a_{12}a_{21} - a_{22}a_{11} = 0$, $a_{12}a_{31} - a_{32}a_{11} \neq 0$, $a_{13}a_{21} - a_{23}a_{11} = 0$ $\ker(\alpha) = \text{span}\{\frac{-a_{12}}{a_{11}}e_1 + e_2\}$ $\begin{pmatrix}a_{11}&a_{12}&a_{13}\\a_{21}&a_{22}&a_{23}\\a_{31}&a_{32}&a_{33}\end{pmatrix}$, $a_{11} \neq 0$, $a_{12}a_{21} - a_{22}a_{11} = 0$, $a_{22}a_{31} - a_{21}a_{32} = 0$, $a_{12}a_{31} - a_{32}a_{11} = 0$, $a_{13}a_{21} - a_{23}a_{11} \neq 0$, or	$a_{12} \neq 0$, $a_{13}a_{22} - a_{12}a_{23} \neq 0$ $\begin{pmatrix}0&a_{12}&a_{13}\\0&a_{22}&a_{23}\\0&a_{32}&a_{33}\end{pmatrix}$, $a_{12} \neq 0$, $a_{13}a_{22} - a_{12}a_{23} = 0$, $a_{13}a_{32} - a_{12}a_{33} \neq 0$ $\begin{pmatrix}0&0&a_{13}\\0&a_{22}&a_{23}\\0&a_{32}&a_{33}\end{pmatrix}$, $a_{22} \neq 0, a_{13} \neq 0$ $\begin{pmatrix}0&0&0\\0&a_{22}&a_{23}\\0&a_{32}&a_{33}\end{pmatrix}$, $a_{22} \neq 0$, $a_{23}a_{32} - a_{22}a_{33} \neq 0$ $\begin{pmatrix}0&0&a_{13}\\0&a_{22}&a_{23}\\0&a_{32}&a_{33}\end{pmatrix}$ a_{22}, a_{32}, not simultaneously zero, and a_{13}, a_{23}, a_{33} not simultaneously zero

(continued)

Table 2 (continued)

		$a_{11} \neq 0,$ $a_{12}a_{21} - a_{22}a_{11} = 0,$ $a_{22}a_{31} - a_{21}a_{32} = 0,$ $a_{12}a_{31} - a_{32}a_{11} = 0,$ $a_{13}a_{21} - a_{23}a_{11} = 0,$ $a_{13}a_{31} - a_{33}a_{11} \neq 0$ $\ker(\alpha) =$ $\mathrm{span}\{\frac{a_{23}a_{32}-a_{22}a_{33}}{a_{22}a_{31}} e_1 + \frac{-a_{23}}{a_{22}} e_2 + e_3\}$ $\begin{pmatrix} 0 & a_{12} & a_{13} \\ 0 & a_{22} & a_{23} \\ a_{31} & a_{32} & a_{33} \end{pmatrix}, a_{22}, a_{31} \neq 0,$ $a_{13}a_{22} - a_{12}a_{23} = 0$ $\ker(\alpha) =$ $\mathrm{span}\{\frac{a_{13}a_{32}-a_{12}a_{33}}{a_{12}a_{31}} e_1 + \frac{-a_{13}}{a_{12}} e_2 + e_3\}$ $\begin{pmatrix} 0 & a_{12} & a_{13} \\ 0 & 0 & 0 \\ a_{31} & a_{32} & a_{33} \end{pmatrix}, a_{12}, a_{31} \neq 0$ $\ker(\alpha) =$ $\mathrm{span}\{\frac{a_{13}a_{22}-a_{12}a_{23}}{a_{12}a_{21}-a_{22}a_{11}} e_1 + \frac{-a_{13}a_{21}+a_{23}a_{11}}{a_{12}a_{21}-a_{22}a_{11}} e_2 + e_3\}$ $\begin{pmatrix} a_{11} & a_{12} & a_{13} \\ a_{21} & a_{22} & a_{23} \\ a_{31} & a_{32} & a_{33} \end{pmatrix},$ $a_{11} \neq 0,$ $a_{12}a_{21} - a_{22}a_{11} \neq 0,$ $a_{13}a_{22}a_{31} - a_{12}a_{23}a_{31}$ $-a_{13}a_{21}a_{32} + a_{12}a_{21}a_{33}$ $+a_{23}a_{32}a_{11} - a_{22}a_{33}a_{11} = 0$	
5	$\lambda_1, \lambda_3 = 0$ $\lambda_2 \neq 0$ $\dim[\mathcal{V}, \mathcal{V}] = 1,$ $[\alpha]$ such that $\mathcal{A}$ is a Hom-Lie algebra $\begin{pmatrix} a_{11} & 0 & a_{13} \\ a_{21} & a_{22} & a_{23} \\ a_{31} & a_{32} & a_{33} \end{pmatrix}$ $\dim(\mathtt{HomLie}_3) = 8,$ Subclasses of $[\alpha]$ such that	$\ker(\alpha) = \mathrm{span}\{\frac{-a_{22}}{a_{21}} e_1 + e_2, e_3\}$ $\begin{pmatrix} 0 & 0 & 0 \\ a_{21} & a_{22} & 0 \\ a_{31} & a_{32} & 0 \end{pmatrix},$ $a_{21} \neq 0,$ $a_{22}a_{31} - a_{21}a_{32} = 0$ $\ker(\alpha) = \mathrm{span}\{\frac{-a_{32}}{a_{31}} e_1 + e_2, e_3\}$ $\begin{pmatrix} 0 & 0 & 0 \\ 0 & 0 & 0 \\ a_{31} & a_{32} & 0 \end{pmatrix}, a_{31} \neq 0$ $\ker(\alpha) = \mathrm{span}\{e_2, \frac{-a_{13}}{a_{11}} e_1 + e_3\}$	$\ker(\alpha) =$ $\mathrm{span}\{\frac{-a_{22}}{a_{21}} e_1 + e_2, e_3\}$ $\begin{pmatrix} 0 & 0 & 0 \\ a_{21} & a_{22} & 0 \\ a_{31} & a_{32} & 0 \end{pmatrix}$ $a_{21} \neq 0,$ $a_{22}a_{31} - a_{21}a_{32} = 0$ $\ker(\alpha) =$ $\mathrm{span}\{\frac{-a_{32}}{a_{31}} e_1 + e_2, e_3\}$ $\begin{pmatrix} 0 & 0 & 0 \\ 0 & 0 & 0 \\ a_{31} & a_{32} & 0 \end{pmatrix}$ $a_{31} \neq 0$

(continued)

Table 2 (continued)

	$\mathcal{A}$ is multiplicative $\begin{pmatrix}0&0&0\\a_{21}&a_{22}&0\\a_{31}&a_{32}&0\end{pmatrix}$, $\begin{pmatrix}a_{11}&0&0\\a_{21}&a_{22}&0\\a_{31}&0&0\end{pmatrix}$, $\begin{pmatrix}1&0&0\\a_{21}&a_{22}&0\\a_{31}&0&a_{33}\end{pmatrix}$	$\begin{pmatrix}a_{11}&0&a_{13}\\a_{21}&0&a_{23}\\a_{31}&0&a_{33}\end{pmatrix}$, $a_{11}\neq 0$, $a_{13}a_{21}-a_{11}a_{23}=0$, $a_{13}a_{23}a_{31}-a_{13}a_{21}a_{33}=0$, $a_{13}a_{31}-a_{11}a_{33}=0$ $\ker(\alpha)=\text{span}\{e_2, \frac{-a_{23}}{a_{21}}e_1+e_3\}$ $\begin{pmatrix}0&0&0\\a_{21}&0&a_{23}\\a_{31}&0&a_{33}\end{pmatrix}$, $a_{21}\neq 0$, $a_{23}a_{31}-a_{21}a_{33}=0$ $\ker(\alpha)=\text{span}\{e_2, \frac{-a_{33}}{a_{31}}e_1+e_3\}$ $\begin{pmatrix}0&0&0\\0&0&0\\a_{31}&0&a_{33}\end{pmatrix}$, $a_{31}\neq 0$ $\ker(\alpha)=\text{span}\{e_1, e_2\}$ $\begin{pmatrix}0&0&a_{13}\\0&0&a_{23}\\0&0&a_{33}\end{pmatrix}$, a_{13}, a_{23}, a_{33} are not simultaneously zero $\ker(\alpha)=\text{span}\{e_1, e_3\}$ $\begin{pmatrix}0&0&0\\0&a_{22}&0\\0&a_{32}&0\end{pmatrix}$, a_{22}, a_{32} is not simultaneously zero $\ker(\alpha)=\text{span}\{e_1\}$ $\begin{pmatrix}0&0&a_{13}\\0&a_{22}&a_{23}\\0&a_{32}&a_{33}\end{pmatrix}$, $a_{22}\neq 0, a_{13}\neq 0$ $\begin{pmatrix}0&0&0\\0&a_{22}&a_{23}\\0&a_{32}&a_{33}\end{pmatrix}$, $a_{22}\neq 0$, $a_{23}a_{32}-a_{22}a_{33}\neq 0$ $\begin{pmatrix}0&0&a_{13}\\0&0&a_{23}\\0&a_{32}&a_{33}\end{pmatrix}$, $a_{32}, a_{23}\neq 0$ $\begin{pmatrix}0&0&a_{13}\\0&0&0\\0&a_{32}&a_{33}\end{pmatrix}$ $a_{32}, a_{13}\neq 0$	$\ker(\alpha)=\text{span}\{e_2, e_3\}$ $\begin{pmatrix}a_{11}&0&0\\a_{21}&0&0\\a_{31}&0&0\end{pmatrix}$ $a_{11}\neq 0$ $\ker(\alpha)=\text{span}\{e_1, e_3\}$ $\begin{pmatrix}0&0&0\\0&a_{22}&0\\0&a_{32}&0\end{pmatrix}$ a_{22}, a_{32} is not simultaneously zero $\ker(\alpha)=\text{span}\{e_2\}$ $\begin{pmatrix}a_{11}&0&a_{13}\\a_{21}&0&a_{23}\\a_{31}&0&a_{33}\end{pmatrix}$ $a_{11}\neq 0$, $a_{13}a_{21}-a_{11}a_{23}\neq 0$ $\begin{pmatrix}a_{11}&0&a_{13}\\a_{21}&0&a_{23}\\a_{31}&0&a_{33}\end{pmatrix}$, $a_{11}\neq 0$, $a_{13}a_{21}-a_{11}a_{23}=0$, $a_{13}a_{31}-a_{11}a_{33}\neq 0$ $\begin{pmatrix}0&0&0\\a_{21}&0&a_{23}\\a_{31}&0&a_{33}\end{pmatrix}$ $a_{21}\neq 0$, $a_{23}a_{31}-a_{21}a_{33}\neq 0$ $\begin{pmatrix}0&0&a_{13}\\a_{21}&0&a_{23}\\a_{31}&0&a_{33}\end{pmatrix}$ a_{21}, a_{31} not simultaneously zero, and a_{13}, a_{23}, a_{33} not simultaneously zero $\ker(\alpha)=\text{span}\{e_3\}$ $\begin{pmatrix}0&0&0\\a_{21}&a_{22}&0\\a_{31}&a_{32}&0\end{pmatrix}$ $a_{21}\neq 0$, $a_{22}a_{31}-a_{21}a_{32}\neq 0$

(continued)

Table 2 (continued)

		$\ker(\alpha) = \mathrm{span}\{e_2\}$ $\begin{pmatrix} a_{11} & 0 & a_{13} \\ a_{21} & 0 & a_{23} \\ a_{31} & 0 & a_{33} \end{pmatrix}$, $a_{11} \neq 0$, $a_{13}a_{21} - a_{11}a_{23} \neq 0$ $\begin{pmatrix} a_{11} & 0 & a_{13} \\ a_{21} & 0 & a_{23} \\ a_{31} & 0 & a_{33} \end{pmatrix}$ $a_{11} \neq 0$ $a_{13}a_{21} - a_{11}a_{23} = 0$, $a_{13}a_{31} - a_{11}a_{33} \neq 0$ $\ker(\alpha) = \mathrm{span}\{-\frac{a_{22}}{a_{21}}e_1 + e_2\}$ $\begin{pmatrix} 0 & 0 & a_{13} \\ a_{21} & a_{22} & a_{23} \\ a_{31} & a_{32} & a_{33} \end{pmatrix}$, $a_{21} \neq 0, a_{13} \neq 0$ $a_{22}a_{31} - a_{21}a_{32} = 0$ $\begin{pmatrix} 0 & 0 & 0 \\ a_{21} & a_{22} & a_{23} \\ a_{31} & a_{32} & a_{33} \end{pmatrix}$, $a_{21} \neq 0$, $a_{22}a_{31} - a_{21}a_{32} = 0$, $a_{23}a_{31} - a_{21}a_{33} \neq 0$ $\ker(\alpha) = \mathrm{span}\{-\frac{a_{32}}{a_{31}}e_1 + e_2\}$ $\begin{pmatrix} 0 & 0 & a_{13} \\ 0 & 0 & a_{23} \\ a_{31} & a_{32} & a_{33} \end{pmatrix}$, $a_{31} \neq 0, a_{23} \neq 0$ $\begin{pmatrix} 0 & 0 & a_{13} \\ 0 & 0 & 0 \\ a_{31} & a_{32} & a_{33} \end{pmatrix}$, $a_{31} \neq 0, a_{13} \neq 0$ $\ker(\alpha) = \mathrm{span}\{\frac{-a_{13}}{a_{11}}e_1 + \frac{a_{13}a_{21} - a_{11}a_{23}}{a_{11}a_{22}}e_2 + e_3\}$ $\begin{pmatrix} a_{11} & 0 & a_{13} \\ a_{21} & a_{22} & a_{23} \\ a_{31} & a_{32} & a_{33} \end{pmatrix}$, $a_{11} \neq 0, a_{22} \neq 0$, $a_{13}a_{22}a_{31} - a_{13}a_{21}a_{32}$ $+a_{11}a_{23}a_{32} - a_{11}a_{22}a_{33} = 0$ $\ker(\alpha) = \mathrm{span}\{\frac{-a_{13}}{a_{11}}e_1 + \frac{a_{13}a_{31} - a_{11}a_{33}}{a_{11}a_{32}}e_2 + e_3\}$	$\begin{pmatrix} a_{11} & 0 & 0 \\ a_{21} & a_{22} & 0 \\ a_{31} & a_{32} & 0 \end{pmatrix}$ a_{11}, a_{21}, a_{31} not simultaneously zero, and a_{22}, a_{32}, not simultaneously zero

(continued)

Table 2 (continued)

		$\begin{pmatrix} a_{11} & 0 & a_{13} \\ a_{21} & 0 & a_{23} \\ a_{31} & a_{32} & a_{33} \end{pmatrix}$, $a_{11}, a_{32} \neq 0$, $a_{13}a_{21} - a_{11}a_{23} = 0$ $\ker(\alpha) =$ $\mathrm{span}\{\frac{a_{23}a_{32}-a_{22}a_{33}}{a_{22}a_{31}-a_{21}a_{32}} e_1 + \frac{a_{21}a_{33}-a_{23}a_{31}}{a_{22}a_{31}-a_{21}a_{32}} e_2 + e_3\}$ $\begin{pmatrix} 0 & 0 & a_{13} \\ a_{21} & a_{22} & a_{23} \\ a_{31} & a_{32} & a_{33} \end{pmatrix}$, $a_{21} \neq 0$, $a_{22}a_{31} - a_{21}a_{32} \neq 0$ $\ker(\alpha) =$ $\mathrm{span}\{\frac{a_{23}a_{32}-a_{22}a_{33}}{a_{22}a_{31}} e_1 - \frac{a_{23}}{a_{22}} e_2 + e_3\}$ $\begin{pmatrix} 0 & 0 & 0 \\ 0 & a_{22} & a_{23} \\ a_{31} & a_{32} & a_{33} \end{pmatrix}$ $a_{31} \neq 0, a_{22} \neq 0$	
6	$\lambda_2, \lambda_3 = 0$ $\lambda_1 \neq 0$ $\dim[\mathcal{V}, \mathcal{V}] = 1$, $[\alpha]$ such that $\mathcal{A}$ is a Hom-Lie algebra $\begin{pmatrix} a_{11} & a_{12} & 0 \\ a_{21} & a_{22} & a_{23} \\ a_{31} & a_{32} & a_{33} \end{pmatrix}$ $\dim(\mathtt{HomLie}_3) = 8$, Subclasses of $[\alpha]$ such that $\mathcal{A}$ is multiplicative	$\ker(\alpha) = \mathrm{span}\{\frac{-a_{12}}{a_{11}} e_1 + e_2, e_3\}$ $\begin{pmatrix} a_{11} & a_{12} & 0 \\ a_{21} & a_{22} & 0 \\ a_{31} & a_{32} & 0 \end{pmatrix}$ $a_{11} \neq 0, a_{12}a_{21} - a_{11}a_{22} = 0$, $a_{12}a_{22}a_{31} - a_{12}a_{21}a_{32} = 0$, $a_{12}a_{31} - a_{11}a_{32} = 0$ $\ker(\alpha) = \mathrm{span}\{\frac{-a_{22}}{a_{21}} e_1 + e_2, e_3\}$ $\begin{pmatrix} 0 & 0 & 0 \\ a_{21} & a_{22} & 0 \\ a_{31} & a_{32} & 0 \end{pmatrix}$, $a_{21} \neq 0$, $a_{22}a_{31} - a_{21}a_{32} = 0$ $\ker(\alpha) = \mathrm{span}\{\frac{-a_{32}}{a_{31}} e_1 + e_2, e_3\}$	$\ker(\alpha) = \mathrm{span}\{e_1, e_2\}$ $\begin{pmatrix} 0 & 0 & 0 \\ 0 & 0 & a_{23} \\ 0 & 0 & a_{33} \end{pmatrix}$ a_{23}, a_{33} not simultaneously zero $\ker(\alpha) =$ $\mathrm{span}\{e_2, \frac{-a_{23}}{a_{21}} e_1 + e_3\}$ $\begin{pmatrix} 0 & 0 & 0 \\ a_{21} & 0 & a_{23} \\ a_{31} & 0 & a_{33} \end{pmatrix}$ $a_{21} \neq 0$, $a_{23}a_{31} - a_{21}a_{33} = 0$

(continued)

Table 2 (continued)

	$\begin{pmatrix} 0 & 0 & 0 \\ a_{21} & 0 & a_{23} \\ a_{31} & 0 & a_{33} \end{pmatrix}$ $\begin{pmatrix} 1 & 0 & 0 \\ a_{21} & a_{22} & 0 \\ a_{31} & 0 & a_{33} \end{pmatrix}$ $\begin{pmatrix} a_{11} & 0 & 0 \\ a_{21} & 0 & 0 \\ a_{31} & 0 & a_{33} \end{pmatrix}$	$\begin{pmatrix} 0 & 0 & 0 \\ 0 & 0 & 0 \\ a_{31} & a_{32} & 0 \end{pmatrix}$, $a_{31} \neq 0$ $\ker(\alpha) = \mathrm{span}\{e_2, \frac{-a_{23}}{a_{21}} e_1 + e_3\}$ $\begin{pmatrix} 0 & 0 & 0 \\ a_{21} & 0 & a_{23} \\ a_{31} & 0 & a_{33} \end{pmatrix}$, $a_{21} \neq 0$, $a_{23}a_{31} - a_{21}a_{33} = 0$ $\ker(\alpha) = \mathrm{span}\{e_2, \frac{-a_{33}}{a_{31}} e_1 + e_3\}$ $\begin{pmatrix} 0 & 0 & 0 \\ 0 & 0 & 0 \\ a_{31} & 0 & a_{33} \end{pmatrix}$, $a_{31} \neq 0$ $\ker(\alpha) = \mathrm{span}\{e_1, e_3\}$ $\begin{pmatrix} 0 & a_{12} & 0 \\ 0 & a_{22} & 0 \\ 0 & a_{32} & 0 \end{pmatrix}$, a_{12}, a_{22}, a_{32} not simultaneously zero $\ker(\alpha) = \mathrm{span}\{e_1, e_2\}$ $\begin{pmatrix} 0 & 0 & 0 \\ 0 & 0 & a_{23} \\ 0 & 0 & a_{33} \end{pmatrix}$, a_{23}, a_{33} are not simultaneously zero $\ker(\alpha) = \mathrm{span}\{e_1\}$ $\begin{pmatrix} 0 & a_{12} & 0 \\ 0 & a_{22} & a_{23} \\ 0 & a_{32} & a_{33} \end{pmatrix}$, a_{12}, a_{22}, a_{32} not simultaneously zero, and a_{23}, a_{33} not simultaneously zero $\ker(\alpha) = \mathrm{span}\{\frac{-a_{12}}{a_{11}} e_1 + e_2\}$ $\begin{pmatrix} a_{11} & a_{12} & 0 \\ a_{21} & a_{22} & a_{23} \\ a_{31} & a_{32} & a_{33} \end{pmatrix}$, $a_{11} \neq 0$, $a_{12}a_{21} - a_{11}a_{22} = 0$, $a_{12}a_{22}a_{31} - a_{12}a_{21}a_{32} = 0$, $a_{12}a_{31} - a_{11}a_{32} = 0$, a_{23}, a_{33} not simultaneously zero $\ker(\alpha) = \mathrm{span}\{\frac{-a_{22}}{a_{21}} e_1 + e_2\}$ $\begin{pmatrix} 0 & 0 & 0 \\ a_{21} & a_{22} & a_{23} \\ a_{31} & a_{32} & a_{33} \end{pmatrix}$,	$\ker(\alpha) =$ $\mathrm{span}\{e_2, \frac{-a_{33}}{a_{31}} e_1 + e_3\}$ $\begin{pmatrix} 0 & 0 & 0 \\ 0 & 0 & 0 \\ a_{31} & 0 & a_{33} \end{pmatrix}$ $a_{31} \neq 0$ $\ker(\alpha) = \mathrm{span}\{e_2, e_3\}$ $\begin{pmatrix} a_{11} & 0 & 0 \\ a_{21} & 0 & 0 \\ a_{31} & 0 & 0 \end{pmatrix}$ a_{11}, a_{21}, a_{31} not simultaneously zero $\ker(\alpha) = \mathrm{span}\{e_2\}$ $\begin{pmatrix} 0 & 0 & 0 \\ a_{21} & 0 & a_{23} \\ a_{31} & 0 & a_{33} \end{pmatrix}$ $a_{21} \neq 0$, $a_{23}a_{31} - a_{21}a_{33} \neq 0$ $\begin{pmatrix} a_{11} & 0 & 0 \\ a_{21} & 0 & a_{23} \\ a_{31} & 0 & a_{33} \end{pmatrix}$ a_{11}, a_{21}, a_{31} not simultaneously zero, and a_{23}, a_{33} not simultaneously zero $\ker(\alpha) = \mathrm{span}\{e_3\}$ $\begin{pmatrix} a_{11} & a_{12} & 0 \\ a_{21} & a_{22} & 0 \\ a_{31} & a_{32} & 0 \end{pmatrix}$ $a_{11} \neq 0$, $a_{12}a_{21} - a_{11}a_{22} \neq 0$ $\begin{pmatrix} a_{11} & a_{12} & 0 \\ a_{21} & a_{22} & 0 \\ a_{31} & a_{32} & 0 \end{pmatrix}$ $a_{11} \neq 0$, $a_{12}a_{21} - a_{11}a_{22} = 0$, $a_{12}a_{31} - a_{11}a_{32} \neq 0$, $\begin{pmatrix} 0 & a_{12} & 0 \\ a_{21} & a_{22} & 0 \\ a_{31} & a_{32} & 0 \end{pmatrix}$ $a_{21} \neq 0, a_{12} \neq 0$

(continued)

Table 2 (continued)

		$a_{21} \neq 0$, $a_{22}a_{31} - a_{21}a_{32} = 0$, $a_{23}a_{31} - a_{21}a_{33} \neq 0$ $\ker(\alpha) = \text{span}\{\frac{-a_{32}}{a_{31}} e_1 + e_2\}$ $\begin{pmatrix} 0 & 0 & 0 \\ 0 & 0 & a_{23} \\ a_{31} & a_{32} & a_{33} \end{pmatrix}$, $a_{23} \neq 0, a_{31} \neq 0$ $\ker(\alpha) = \text{span}\{\frac{-a_{33}}{a_{31}} e_1 + e_3\}$ $\begin{pmatrix} 0 & a_{12} & 0 \\ 0 & 0 & 0 \\ a_{31} & a_{32} & a_{33} \end{pmatrix}$, $a_{12} \neq 0, a_{31} \neq 0$ $\ker(\alpha) = \text{span}\{\frac{-a_{23}}{a_{21}} e_1 + e_3\}$ $\begin{pmatrix} 0 & a_{12} & 0 \\ a_{21} & a_{22} & a_{23} \\ a_{31} & a_{32} & a_{33} \end{pmatrix}$, $a_{12} \neq 0, a_{21} \neq 0$, $a_{23}a_{31} - a_{21}a_{33} = 0$ $\ker(\alpha) = \text{span}\{\frac{-a_{12}a_{23}}{a_{12}a_{21}-a_{11}a_{22}} e_1 + \frac{a_{11}a_{23}}{a_{12}a_{21}-a_{11}a_{22}} e_2 + e_3\}$ $\begin{pmatrix} a_{11} & a_{12} & 0 \\ a_{21} & a_{22} & a_{23} \\ a_{31} & a_{32} & a_{33} \end{pmatrix}$, $a_{11} \neq 0$, $a_{12}a_{21} - a_{11}a_{22} \neq 0$, $a_{12}a_{23}a_{31} - a_{11}a_{23}a_{32}$, $-a_{12}a_{21}a_{33} + a_{11}a_{22}a_{33} = 0$ $\ker(\alpha) = \text{span}\{\frac{-a_{12}a_{33}}{a_{12}a_{31}-a_{11}a_{32}} e_1 + \frac{a_{11}a_{33}}{a_{12}a_{31}-a_{11}a_{32}} e_2 + e_3\}$ $\begin{pmatrix} a_{11} & a_{12} & 0 \\ a_{21} & a_{22} & 0 \\ a_{31} & a_{32} & a_{33} \end{pmatrix}$, $a_{11} \neq 0$, $a_{12}a_{21} - a_{11}a_{22} = 0$, $a_{12}a_{31} - a_{11}a_{32} \neq 0$	$\begin{pmatrix} 0 & 0 & 0 \\ a_{21} & a_{22} & 0 \\ a_{31} & a_{32} & 0 \end{pmatrix}$ $a_{21} \neq 0$, $a_{22}a_{31} - a_{21}a_{32} \neq 0$ $\begin{pmatrix} 0 & a_{12} & 0 \\ 0 & a_{22} & 0 \\ a_{31} & a_{32} & 0 \end{pmatrix}$, $a_{31} \neq 0$, a_{12}, a_{22}, a_{32} not simultaneouslyzero

(continued)

Table 2 (continued)

		$\ker(\alpha) = \mathrm{span}\{\frac{a_{23}a_{32}-a_{22}a_{33}}{a_{22}a_{31}-a_{21}a_{32}}e_1 + \frac{-a_{23}a_{31}+a_{21}a_{33}}{a_{22}a_{31}-a_{21}a_{32}}e_2 + e_3\}$ $\begin{pmatrix} 0 & 0 & 0 \\ a_{21} & a_{22} & a_{23} \\ a_{31} & a_{32} & a_{33} \end{pmatrix}, a_{21} \neq 0,$ $a_{22}a_{31} - a_{21}a_{32} \neq 0$ $\ker(\alpha) =$ $\mathrm{span}\{\frac{a_{23}a_{32}-a_{22}a_{33}}{a_{22}a_{31}}e_1 - \frac{a_{23}}{a_{22}}e_2 + e_3\}$ $\begin{pmatrix} 0 & a_{12} & 0 \\ 0 & a_{22} & a_{23} \\ a_{31} & a_{32} & a_{33} \end{pmatrix},$ $a_{22} \neq 0, a_{31} \neq 0,$ $a_{12}a_{23} = 0$	

Now, we present all nontrivial Hom-subalgebras realized as $\ker(\alpha)$ of the Hom-Lie structures above, along with their corresponding subsets of the space of linear maps α that realize them:

(i) $\ker(\alpha) = \mathrm{span}\{e_1, e_2\}$, $\quad [\alpha] = \begin{pmatrix} 0 & 0 & 0 \\ 0 & 0 & a_{23} \\ 0 & 0 & 0 \end{pmatrix}, \ a_{23} \neq 0$

$$[\alpha] = \begin{pmatrix} 0 & 0 & 0 \\ 0 & 0 & 0 \\ 0 & a_{32} & 0 \end{pmatrix}, \ a_{32} \neq 0.$$

(iii) $\ker(\alpha) = \mathrm{span}\{\frac{(a_{13}a_{22}-a_{12}a_{23})\lambda_3}{a_{12}a_{13}\lambda_1 - a_{11}a_{22}\lambda_3}e_1 + \frac{-a_{13}^2\lambda_1 + a_{11}a_{23}\lambda_3}{a_{12}a_{13}\lambda_1 - a_{11}a_{22}\lambda_3}e_2 + e_3\}$,

$$[\alpha] = \begin{pmatrix} a_{11} & a_{12} & a_{13} \\ \frac{\lambda_1}{\lambda_3}a_{13} & a_{22} & a_{23} \\ -\frac{\lambda_2}{\lambda_3}a_{12} & a_{32} & -\frac{\lambda_2}{\lambda_1}a_{22} \end{pmatrix}, \quad \begin{aligned} & a_{11} \neq 0, a_{12}a_{13}\lambda_1 - a_{11}a_{22}\lambda_3 \neq 0, \\ & (a_{13}^2a_{32}\lambda_1 + (2a_{12}a_{13}a_{22} - a_{12}^2a_{23})\lambda_2 \\ & \quad -a_{11}a_{23}a_{32}\lambda_3)\lambda_1 - a_{11}a_{22}^2\lambda_2\lambda_3 = 0, \\ & \text{or} \\ & a_{11} = 0, a_{13} \neq 0, a_{12} \neq 0, \\ & a_{13}^2a_{32}\lambda_1 + (2a_{12}a_{13}a_{22} - a_{12}^2a_{23})\lambda_2 = 0. \end{aligned}$$

(iv) $\ker(\alpha) = \mathrm{span}\{\frac{-a_{13}}{a_{11}}e_1 + e_3\}$,

$$[\alpha] = \begin{pmatrix} a_{11} & a_{12} & a_{13} \\ \frac{\lambda_1}{\lambda_3}a_{13} & a_{22} & a_{23} \\ -\frac{\lambda_2}{\lambda_3}a_{12} & a_{32} & -\frac{\lambda_2}{\lambda_1}a_{22} \end{pmatrix}, \quad \begin{aligned} & a_{11} \neq 0, \\ & a_{12}a_{13}\lambda_1 - a_{11}a_{22}\lambda_3 = 0, \\ & a_{12}^2\lambda_2 + a_{11}a_{32}\lambda_3 \neq 0, \\ & a_{13}(a_{13}a_{22} - a_{12}a_{23}) = 0, \\ & a_{13}^2\lambda_1 - a_{11}a_{23}\lambda_3 = 0 \end{aligned}$$

(v) $\ker(\alpha) = \text{span}\{\frac{-a_{22}\lambda_3}{a_{12}\lambda_1} e_1 + e_3\}, \quad [\alpha] = \begin{pmatrix} 0 & a_{12} & 0 \\ 0 & a_{22} & 0 \\ -\frac{\lambda_2}{\lambda_3} a_{12} & a_{32} & -\frac{\lambda_2}{\lambda_1} a_{22} \end{pmatrix}, \quad a_{12} \neq 0.$

(vi) $\ker(\alpha) = \text{span}\{\frac{-a_{12}}{a_{11}} e_1 + e_2\},$

$$[\alpha] = \begin{pmatrix} a_{11} & a_{12} & a_{13} \\ \frac{\lambda_1}{\lambda_3} a_{13} & a_{22} & a_{23} \\ -\frac{\lambda_2}{\lambda_3} a_{12} & a_{32} & -\frac{\lambda_2}{\lambda_1} a_{22} \end{pmatrix}, \quad \begin{array}{l} a_{11} \neq 0, \\ a_{12}a_{13}a_{32}\lambda_1 + a_{12}^2 a_{22}\lambda_2 = 0, \\ a_{12}a_{13}\lambda_1 - a_{11}a_{22}\lambda_3 = 0, \\ a_{12}^2\lambda_2 + a_{11}a_{32}\lambda_3 = 0, \\ a_{13}^2\lambda_1 - a_{11}a_{23}\lambda_3. \end{array}$$

(vii) $\ker(\alpha) = \text{span}\{\frac{-a_{22}\lambda_3}{a_{13}\lambda_1} e_1 + e_2\}, \quad [\alpha] = \begin{pmatrix} 0 & 0 & a_{13} \\ \frac{\lambda_1}{\lambda_3} a_{13} & a_{22} & a_{23} \\ 0 & 0 & -\frac{\lambda_2}{\lambda_1} a_{22} \end{pmatrix}, \quad a_{13} \neq 0.$

(viii) $\ker(\alpha) = \text{span}\{e_1\}, \quad [\alpha] = \begin{pmatrix} 0 & 0 & 0 \\ 0 & a_{22} & a_{23} \\ 0 & a_{32} & -\frac{\lambda_2}{\lambda_1} a_{22} \end{pmatrix}, \quad \begin{array}{l} a_{22} \neq 0, a_{23}a_{32}\lambda_1 + a_{22}^2\lambda_2 \neq 0, \\ \text{or} \\ a_{22} = 0, a_{32} \neq 0, a_{23} \neq 0. \end{array}$

In case 7, all the subsets of the set of twisting maps α consisting of algebra morphisms are listed in Sect. 4.2. It is noteworthy that, for this specific case, none of the corresponding $\ker(\alpha)$ is a weak ideal, and consequently, none of $\ker(\alpha)$ is Hom-ideal. Therefore, $\ker(\alpha)$ in this case does not provide conclusions about nonsimplicity, or non-Hom-simplicity of the algebra. Moreover, since the derived series and central descending series stabilize at the whole algebra, $\mathcal{V}$, they do not provide any information on nonsimplicity or non-Hom-simplicity of the algebra as well.

Next, we investigate the derived series and central descending series of Hom-Lie algebras of generalized $\mathfrak{sl}(2)$-type and present subsets of space of linear twisting map α, given in column 2 of Table 2, such that the derived series and central descending series are α-invariant subspaces, and investigate whether these derived series and central descending series are weak ideals or not. We refer to column 3 and column 4 of Table 2 to determine if the members of these derived series and central descending series are Hom-ideals. We consider different dimensions of $[\mathcal{V}, \mathcal{V}]$ as presented in Table 2 so that each case presented in Table 3 corresponds to the same cases presented in Table 2. We refer to column 2 of Table 2 for different cases and the corresponding Hom-Lie structures provided. If $\dim([\mathcal{V}, \mathcal{V}]) = 3$ or $\dim([\mathcal{V}, \mathcal{V}]) = 0$, the central descending series and derived series stabilize at $\mathcal{V}$ and $\{0\}$, respectively. Therefore, we exclude these two trivial cases from our discussion and focus on more interesting cases when $\dim([\mathcal{V}, \mathcal{V}]) = 1$ or 2. Our results and computations are presented in Table 3, and discussions thereafter.

In column 2, we present all the derived series and central descending series of the Hom-Lie algebras of generalized $\mathfrak{sl}(2)$-type and state whether the considered algebra in each case is nilpotent (resp. solvable) or not. For each nilpotent (resp. solvable) algebra, we state its nilpotency (resp. solvability) class starting from $k \geq 1$ and give all its nonzero members of central descending series and derived series.

Table 3 Solvability and nilpotency, weak Ideals, and Hom-ideals of Hom-Lie algebras of generalized $\mathfrak{sl}(2)$-type

	Derived Series, Central descending series of $\mathcal{V}$	Subclasses of linear α such that $C^k(\mathcal{V})$, $D^k(\mathcal{V})$ are α- invariant	Hom-Simple or Not-Hom-Simple with respect to $\ker(\alpha)$, $C^k(\mathcal{V})$ and $D^k(\mathcal{V})$
1	Not nilpotent Class 3-Solvable for all $k \geq 1$ $C^k(\mathcal{V}) = \text{span}\{e_1, e_3\} = D^1(\mathcal{V})$ Weak ideal	$a_{21} = a_{23} = 0$ $\begin{pmatrix} a_{11} & a_{12} & a_{13} \\ 0 & 0 & 0 \\ -\frac{\lambda_2}{\lambda_3} a_{12} & a_{32} & a_{33} \end{pmatrix}$	Not Hom-simple
	$D^2(\mathcal{V}) = \text{span}\{e_3\}$ Not weak ideal	$a_{13} = a_{23} = 0$ $\begin{pmatrix} a_{11} & a_{12} & 0 \\ a_{21} & 0 & 0 \\ -\frac{\lambda_2}{\lambda_3} a_{12} & a_{32} & a_{33} \end{pmatrix}$	
2	Not nilpotent Class-3-Solvable For $k \geq 1$ $C^k(\mathcal{V}) = \text{span}\{e_1, e_2\} = D^1(\mathcal{V})$ Weak ideal	$a_{31} = a_{32} = 0$ $\begin{pmatrix} a_{11} & a_{12} & a_{13} \\ \frac{\lambda_1}{\lambda_3} a_{13} & a_{22} & a_{23} \\ 0 & 0 & 0 \end{pmatrix}$	Not Hom-simple
	$D^2(\mathcal{V}) = \text{span}\{e_2\}$ Not weak ideal	$a_{12} = a_{32} = 0$ $\begin{pmatrix} a_{11} & 0 & a_{13} \\ \frac{\lambda_1}{\lambda_3} a_{13} & a_{22} & a_{23} \\ a_{31} & 0 & 0 \end{pmatrix}$	
3	Not nilpotent Class 2-Solvable For $k \geq 1$ $C^k(\mathcal{V}) = \text{span}\{e_2, e_3\} = D^1(\mathcal{V})$ Weak ideal	$\begin{pmatrix} a_{11} & 0 & 0 \\ a_{21} & a_{22} & a_{23} \\ a_{31} & a_{32} & a_{33} \end{pmatrix}$	Not Hom-simple
4	Class-2-nilpotent Class 2-Solvable $C^1(\mathcal{V}) = \text{span}\{e_1\} = D^1(\mathcal{V})$ Weak ideal	$a_{21} = a_{31} = 0$ $\begin{pmatrix} a_{11} & a_{12} & a_{13} \\ 0 & a_{22} & a_{23} \\ 0 & a_{32} & a_{33} \end{pmatrix}$	Not Hom-simple
5	Not nilpotent Class 2-Solvable For $k \geq 1$ $C^k(\mathcal{V}) = \text{span}\{e_3\} = D^1(\mathcal{V})$ Weak ideal	$a_{13} = a_{23} = 0$ $\begin{pmatrix} a_{11} & 0 & 0 \\ a_{21} & a_{22} & 0 \\ a_{31} & a_{32} & a_{33} \end{pmatrix}$	Not Hom-simple
6	Not nilpotent Class 2-Solvable $C^1(\mathcal{V}) = \text{span}\{e_2\} = D^1(\mathcal{V})$ Weak ideal	$a_{12} = a_{32} = 0$ $\begin{pmatrix} a_{11} & 0 & 0 \\ a_{21} & a_{22} & a_{23} \\ a_{31} & 0 & a_{33} \end{pmatrix}$	Not Hom-simple

Furthermore, we classify these nonzero members as either weak ideal or not. In column 3, we give subsets of space of linear twisting map α presented in column 2 of Table 2 with the properties that all derived series and central descending series are α-invariant. Moreover, for each subset of space of linear α presented in this column 3 of Table 3, we state the conditions on the entries of linear α realizing them. Each cell of column 3 is subdivided where necessary, that is, for the cases where $C^k(\mathcal{V})$ is not the same as $D^k(V)$, and in each case, the two subsets of the linear α presented correspond to $C^k(\mathcal{V})$ and $D^2(\mathcal{V})$, respectively. In column 4, it is stated that this class of generalized Hom-Lie algebras of $\mathfrak{sl}(2)$-type is not Hom-simple, as in each case, there exists a Hom-ideal that arises either as the kernel of the linear twist or from the derived series and central descending series.

It is worth comparing the subsets of the space of linear maps, α, presented in column 3 of Table 3 with their corresponding subsets of the space of linear maps α that conform to algebra morphisms, that is, the multiplicative linear maps α presented in column 2 of Table 2. According to Proposition 5, we know that if $(\mathcal{V}, [\cdot,\cdot], \alpha)$ is multiplicative, then, $C^k(\mathcal{V})$ and $D^k(\mathcal{V})$ are α-invariant subspaces. Here, we present some non-multiplicative subsets of the linear space of linear twisting maps, α, with the property that the members of the derived series and the central descending series of an ideal of $\mathcal{V}$ are α-invariant.

Now, we provide some comments on 1-dimensional ideals, central descending series, and derived series of generalized Hom-Lie algebras of $\mathfrak{sl}(2)$-type. In cases 1 and 2, it is noted that there is no 1-dimensional weak ideal. However, in case 3 where $\lambda_3 = 0$, and $\lambda_1, \lambda_2 \neq 0$, we consider a linear subspace $I = \text{span}\{c_1e_1+c_2e_2+c_3e_3\}$, for the weak ideal conditions, our bilinear bracket multiplication on I gives:

$$[e_1, c_1e_1 + c_2e_2 + c_3e_3] = c_2\lambda_1e_2 + c_3\lambda_2e_3,$$

$$[e_2, c_1e_1 + c_2e_2 + c_3e_3] = -c_1\lambda_1e_2, [e_3, c_1e_1 + c_2e_2 + c_3e_3] = -c_1\lambda_2e_3.$$

Then, it follows that $c_2\lambda_1e_2 + c_3\lambda_2e_3 = k_1(c_1e_1 + c_2e_2 + c_3e_3)$ for some $k_1 \in \mathbb{K}$. It is evident that if $k_1 = 0$, then $c_2, c_3 = 0$, and $I = \text{span}\{e_1\}$, which is not a weak ideal for this case. Moreover, if $k_1 \neq 0$, then $c_1 = 0$, and $I = \text{span}\{c_2e_2 + c_3e_3\}$, which is a 1-dimensional weak ideal for this case. Moreover, $-c_1\lambda_1e_2 = k_2(c_1e_1 + c_2e_2 + c_3e_3)$ for all $k_2 \in \mathbb{K}$. We notice that, if $k_2 = 0$, then $c_1 = 0$, which leads to the same case as above, i.e., $I = \text{span}\{c_2e_2 + c_3e_3\}$. For $k_2 \neq 0$, we have $c_1, c_3 = 0$, and $I = \text{span}\{e_2\}$, which is a weak ideal for this case. Furthermore, we have $-c_1\lambda_2e_3 = k_3(c_1e_1+c_2e_2+c_3e_3)$; it is clear that if $k_3 \neq 0$, we obtain another weak ideal $I = \text{span}\{e_3\}$, and if $k_3 = 0$, then $c_1 = 0$ which gives the same case discussed earlier. For this class of Hom-Lie algebras, the weak ideal $I = \text{span}\{c_2e_2 + c_3e_3\}$, where $c_2, c_3 \neq 0$, becomes a Hom-ideal if and only if the linear twisting maps α are of the form $[\alpha] = \begin{pmatrix} a_{11} & 0 & 0 \\ a_{21} & a_{22} & a_{23} \\ a_{31} & a_{32} & a_{22} + \frac{c_3}{c_2}a_{23} - \frac{c_2}{c_3}a_{32} \end{pmatrix}$. Similarly, the other remaining two weak ideals, $I = \mathrm{span}\{e_2\}$ and $I = \mathrm{span}\{e_3\}$, become Hom-ideals if and only if α are of the form $[\alpha] = \begin{pmatrix} a_{11} & 0 & 0 \\ a_{21} & a_{22} & a_{23} \\ a_{31} & 0 & a_{33} \end{pmatrix}$ and $[\alpha] = \begin{pmatrix} a_{11} & 0 & 0 \\ a_{21} & a_{22} & 0 \\ a_{31} & a_{32} & a_{33} \end{pmatrix}$, respectively.

We adapt the same argument for other cases: In case 4 where $\lambda_1, \lambda_2 = 0$, and $\lambda_3 \neq 0$, we obtain $I = \mathrm{span}\{e_1\}$ as the only 1-dimensional weak ideal. Furthermore, this ideal $I = \mathrm{span}\{e_1\}$ becomes a Hom-ideal if the subset of space of the linear twisting map α is of the form given in Table 3, column 3 of case 4. In case 5, there are two 1-dimensional weak ideals: $I = \mathrm{span}\{e_2\}$ and $I = \mathrm{span}\{e_3\}$, each of which becomes a Hom-ideal if the subset of space of the linear twisting map α is of the form $[\alpha] = \begin{pmatrix} a_{11} & 0 & a_{13} \\ a_{21} & a_{22} & a_{23} \\ a_{31} & 0 & a_{33} \end{pmatrix}$ and $[\alpha] = \begin{pmatrix} a_{11} & 0 & 0 \\ a_{21} & a_{22} & 0 \\ a_{31} & a_{32} & a_{33} \end{pmatrix}$, respectively. In case 6 where $\lambda_2, \lambda_3 = 0$ and $\lambda_1 \neq 0$, we obtain two 1-dimensional weak ideals: $I = \mathrm{span}\{e_2\}$, and $I = \mathrm{span}\{e_3\}$. An ideal $I = \mathrm{span}\{e_3\}$ become a Hom-ideal if the subset of linear space of linear twisting maps α is of the form $[\alpha] = \begin{pmatrix} a_{11} & a_{12} & 0 \\ a_{21} & a_{22} & 0 \\ a_{31} & a_{32} & a_{33} \end{pmatrix}$, and $I = \mathrm{span}\{e_2\}$ is a Hom-ideal if the subset of linear twisting maps α is of the form given in Table 3, case 6, column 3.

In cases 1 and 2, the algebra is not nilpotent but is class 3-solvable, as shown in column 2 of Table 3. In case 1, $C^k(\mathcal{V}) = \mathrm{span}\{e_1, e_3\} = D^1(\mathcal{V})$, while $D^2(\mathcal{V}) = \mathrm{span}\{e_3\}$. The corresponding subset of the space of linear α such that $C^k(\mathcal{V}) = \mathrm{span}\{e_1, e_3\} = D^1(\mathcal{V})$ is a Hom-ideal is given by the matrix (containing some free parameters a_{ij}, $1 \leq i, j \leq 3$) in cell one of column 3 in Table 3. For $D^2(\mathcal{V}) = \mathrm{span}\{e_3\}$, the corresponding subsets of the space of linear α such that the subspace $\mathrm{span}\{e_3\}$ is α-invariant is presented by a matrix in cell two of column 3 in Table 3. Furthermore, in column 4, it is stated that the algebra is not Hom-simple. This means that there exists a Hom-ideal, $I = \mathrm{span}\{e_1, e_3\}$ corresponding to subclasses listed in column 4 of Table 2, which makes the algebra in this subcase not Hom-simple. Similarly, case 2 follows the same description, where each subset of the space of linear α given in cell one and cell two corresponds to Hom-Lie structures such that $C^k(\mathcal{V}) = \mathrm{span}\{e_1, e_2\} = D^1(\mathcal{V})$ and $D^2(\mathcal{V}) = \mathrm{span}\{e_2\}$ is Hom-ideal, and α-invariant, respectively. A weak ideal $I = C^k(\mathcal{V}) = \mathrm{span}\{e_1, e_2\} = D^1(\mathcal{V})$ is also a Hom-ideal for a subclass of linear α listed in Table 2, column 4; therefore, the algebra in this subcase is not Hom-simple.

It is shown that, in all cases, $C^1(\mathcal{V})$ and $D^1(\mathcal{V})$ are weak ideals, as stated in Proposition 3. However, in cases 1 and 2, it is shown that $D^2(\mathcal{V}) = \mathrm{span}\{e_3\}$ and $D^2(\mathcal{V}) = \mathrm{span}\{e_2\}$ respectively are not weak ideals. This is an example where the condition and conclusion in Proposition 5 do not hold. It follows from Proposition 5 that in these two cases, there is no surjective linear maps α in the space of linear twisting α consisting of algebra morphisms (see column 2 in Table 2). Moreover, in both cases 1 and 2, the subsets of the linear space of the linear twisting maps α presented in column 3 of Table 3 are neither surjective nor algebra morphisms. Thus, in cases 1 and 2, the conditions of Proposition 4 are not satisfied. This observation indicates the possibility of imposing a weaker condition in Proposition 5 to explore more properties within this class of Hom-Lie algebras. Notably, in both cases 1 and 2, the triple $(\mathcal{V}, [\cdot, \cdot], \alpha)$ defines a Hom-Lie algebra. However, the bilinear bracket multiplication does not define a Lie algebra because, in these cases, the identity map is not within the space of Hom-Lie structures (see column 2 in Table 2). This is equivalent to considering the Jacobi identity in Lie algebras of generalized $\mathfrak{sl}(2)$-type with bracket multiplication defined in (4.5). That is, for some chosen basis $\{e_1, e_2, e_3\}$ for $\mathcal{V}$, we have $[e_1, [e_2, e_3]] + [e_3, [e_1, e_2]] + [e_2, [e_3, e_1]] = 0 \implies \lambda_3(\lambda_1 + \lambda_2)e_1 = 0$. Therefore, (4.5) defines a Lie algebra if and only if $\lambda_3 = 0$ or $\lambda_1 + \lambda_2 = 0$. However, for these cases 1 and 2, none of these conditions on $\lambda_i, i = 1, 2, 3$ hold; thus, we obtain a class of Hom-Lie algebra which does not arise from Lie algebra.

In case 3, we note that this algebra is always not Hom-simple in general, as there exists an ideal $I = \text{span}\{e_2, e_3\}$ that is always α-invariant. In other words, the ideal $\text{span}\{e_2, e_3\} = C^k(\mathcal{V}) = D^1(\mathcal{V})$ for all $k \geq 1$ is a Hom-ideal within this subclass of Hom-Lie algebra.

Acknowledgments Stephen Mboya extends his sincere appreciation to the International Science Program (ISP) at Uppsala University for their financial support in partnership with the Eastern Africa Universities Mathematics Programme (EAUMP), and is grateful to the Division of Mathematics and Physics at the School of Education, Culture, and Communication in Mälardalen University in Västerås, Sweden, for hospitality and for providing an excellent environment for research and research education. In our computations the computer algebra software Mathematica has been helpful.

References

1. Alvarez, M.A., Vera, S.V.: On 4-dimensional Hom-Lie algebras. In: Vlachos, D. (eds.) Mathematical Modeling in Physical Sciences. ICMSQUARE 2023. Springer Proceedings in Mathematics and Statistics, vol 446. Springer, Cham (2024)
2. Ammar, F., Ejbehi, Z., Makhlouf, A.: Cohomology and deformations of Hom-algebras. J. Lie Theory **21**, 813–836 (2011)
3. Benayadi, S., Makhlouf, A.: Hom-Lie algebras with symmetric invariant nondegenerate bilinear forms. J. Geom. Phys. **76**, 38–60 (2014)
4. Hartwig, J.T., Larsson, D., Silvestrov, S.D.: Deformations of Lie algebras using σ-derivations, J. Algebra, **295**(2), 314–361 (2006) (Preprints in Mathematical Sciences 2003:32, LUTFMA-5036-2003, Centre for Mathematical Sciences, Lund University, 52 pp. (2003))
5. Fernández-Culma, E.A., Rojas, N.: On the classification of 3-dimensional complex Hom-Lie algebras. J. Pure Appl. Algebra **227**, 107272 (2023)
6. García-Delgado, R., Salgado, G.: Sánchez-Valenzuela, O.A. On 3-dimensional complex Hom-Lie algebras. J. Algebra **555**, 361–385 (2020)
7. Goze, M., Khakimdjanov, Yu.: Nilpotent and solvable Lie algebras. In: M. Hazewinkel (eds.) Handbook of Algebra, vol. 2, pp. 615–663. Elsevier, Amsterdam (2000)
8. Elchinger, O., Lundengård, K., Makhlouf, A., Silvestrov, S.D.: Brackets with (τ, σ)-derivations and (p, q)-deformations of Witt and Virasoro algebras. Forum Math. **28**(4), 657–673 (2016). (arXiv:1403.6291 [math.QA])

9. Kitouni, A., Silvestrov S.: On properties and classification of a class of 4-dimensional 3-Hom-Lie algebras with a nilpotent twisting map. Axioms **13**(6), 373 (2024). (arXiv:2304.10674 [math.RA])
10. Kitouni, A., Makhlouf, A., Silvestrov, S.: On solvability and nilpotency for n-Hom-Lie algebras and $(n+1)$-Hom-Lie algebras induced by n-Hom-Lie algebras. In: Silvestrov, S., Malyarenko, A., Rancic, M. (eds.) Algebraic Structures and Applications, Springer Proceedings in Mathematics and Statistics, vol. 317, Ch 6, pp. 127–157. Springer, Berlin (2020)
11. Larsson, D., Silvestrov, S.D.: Quasi-hom-Lie algebras, central extensions and 2-cocycle-like identities. J. Algebra **288**, 321–344 (2005) (Preprints in Mathematical Sciences 2004:3, LUTFMA-5038-2004, Centre for Mathematical Sciences, Lund University, 40 pp, 2004)
12. Larsson, D., Silvestrov, S.D.: Quasi-Lie algebras. In: Fuchs, J., Mickelsson, J., Rozenblioum, G., Stolin, A., Westerberg, A. (eds.) Noncommutative Geometry and Representation Theory in Mathematical Physics, vol. 391, pp. 241–248. American Mathematical Society, Contemporary Mathematics, Providence (2005); (Preprints in Mathematical Sciences 2004:30, LUTFMA-5049-2004, Centre for Mathematical Sciences, Lund University, 11 pp, 2004)
13. Larsson, D., Silvestrov, S.D.: Quasi-deformations of $sl_2(\mathbb{F})$ using twisted derivations. Commun. Algebra **35**, 4303–4318 (2007). (Preprint in Mathematical Sciences 2004:26, LUTFMA-5047-2004, Centre for Mathematical Sciences, Lund University (2004). arXiv:math/0506172 [math.RA] (2005))
14. Larsson, D., Sigurdsson, G., Silvestrov, S.D.: Quasi-Lie deformations on the algebra $\mathbb{F}[t]/(t^N)$. J. Gen. Lie Theory Appl. **2**(3), 201–205 (2008)
15. Larsson, D., Silvestrov, S.D.: On generalized N-complexes comming from twisted derivations. In: Silvestrov, S., Paal, E., Abramov, V., Stolin, A. (eds.) Generalized Lie Theory in Mathematics, Physics and Beyond, Ch.7, pp. 81–88. Springer, Berlin (2009)
16. Li, X., Li, Y.: Classification of 3-dimensional multiplicative Hom-Lie algebras. J. Xinyang Norm. Univ. **455**, 427–430 (2012)
17. Makhlouf, A., Mehidi, M.: On classification of filiform Hom-Lie algebras. In: Silvestrov, S., Malyarenko, A., Rančić, M. (eds.) Algebraic Structures and Applications, Springer Proceedings in Mathematics and Statistics, vol. 317, Ch. 8, pp. 189–221. Springer, Cham (2020)
18. Makhlouf, A., Silvestrov, S.D.: Hom-algebra structures. J. Gen. Lie Theory Appl. **2**(2), 51–64 (2008) (Preprints in Mathematical Sciences 2006:10, LUTFMA-5074-2006, Centre for Mathematical Sciences, Lund University (2006). arXiv:math/0609501 [math.RA])
19. Makhlouf, A., Silvestrov, S.: Hom-Lie admissible Hom-coalgebras and Hom-Hopf algebras. In: Silvestrov, S., Paal, E., Abramov, V., Stolin, A. (eds.) Generalized Lie Theory in Mathematics, Physics and Beyond, Ch. 17, pp. 189–206. Springer, Verlin (2009) . (Preprints in Mathematical Sciences 2007:25, LUTFMA-5091-2007, Centre for Mathematical Sciences, Lund University, 13 pp, 2007. arXiv:0709.2413 [math.RA]).
20. Makhlouf, A., Silvestrov, S.D.: Hom-algebras and Hom-coalgebras. J. Algebra Appl. **9**(4), 553–589 (2010) (Preprints in Mathematical Sciences 2008:19, LUTFMA-5103-2008, Centre for Mathematical Sciences, Lund University, 2008. arXiv:0811.0400 [math.RA])
21. Makhlouf, A., Silvestrov, S.: Notes on 1-parameter formal deformations of Hom-associative and Hom-Lie algebras. Forum Math. **22**(4), 715–739 (2010) (Preprints in Mathematical Sciences 2007:31, LUTFMA-5095-2007, Centre for Mathematical Sciences, Lund University, 18 pp, 2007. arXiv:0712.3130v1 [math.RA]).
22. Ongong'a, E., Ongaro, J., Silvestrov, S.: Hom-Lie structures on complex 4-dimensional Lie algebras. In: Dobrev, V. (eds.) Lie Theory and Its Applications in Physics, Springer Proceedings in Mathematics and Statistics, vol. 335, pp. 373–381. Springer, Berlin (2020)
23. Ongong'a, E., Richter, J., Silvestrov, S.D.: Classification of 3-dimensional Hom-Lie algebras. J. Phys. Conf. Ser. **1194**, 012084 (2019)
24. Ongong'a, E., Richter, J., Silvestrov, S.: Classification of low-dimensional Hom-Lie algebras. In: Silvestrov, S., Malyarenko, A., Rančić, M., (eds.) Algebraic Structures and Applications, Springer Proceedings in Mathematics and Statistics, vol. 317, Ch.9, pp. 223–256. Springer, Cham (2020)

25. Remm, E.: 3-Dimensional skew-symmetric algebras and the variety of hom-Lie algebras. Algebra Colloq. **25**(4), 547–566 (2018)
26. Remm, E., Goze, M.: On the algebraic variety of Hom-Lie algebras. arXiv:1706.02484 [math.RA] (2017)
27. Richard, L., Silvestrov, S.D.: Quasi-Lie structure of σ-derivations of $\mathbb{C}[t^{\pm 1}]$, J. Algebra **319**(3), 1285–1304 (2008) (Preprints in mathematical sciences, 2006:12, LUTFMA-5076-2006, Centre for Mathematical Sciences, Lund University (2006). arXiv:math/0608196[math.QA])
28. Richard, L., Silvestrov, S.D.: A note on quasi-Lie and Hom-Lie structures of σ-derivations of $\mathbb{C}[z_1^{\pm 1}, \ldots, z_n^{\pm 1}]$. In: Silvestrov, S., Paal, E., Abramov, V., Stolin, A. (eds.) Generalized Lie Theory in Mathematics, Physics and Beyond, Ch. 22, pp. 257–262. Springer, Berlin (2009)
29. Sigurdsson, G., Silvestrov, S.: Graded quasi-Lie algebras of Witt type. Czechoslovak J. Phys. **56**, 1287–1291 (2006)
30. Sigurdsson, G., Silvestrov, S.: Lie color and hom-Lie algebras of Witt type and their central extensions. In: Silvestrov, S., Paal, E., Abramov, V., Stolin, A. (eds.) Generalized Lie Theory in Mathematics, Physics and Beyond, Ch. 21, pp. 247–255. Springer, Berlin, Heidelberg (2009)
31. Sheng, Y.: Representations of hom-Lie algebras. Algebr. Reprensent. Theory **15**, 1081–1098 (2012)
32. Yau, D.: Enveloping algebras of Hom-Lie algebras. J. Gen. Lie Theory Appl. **2**, 95–108 (2008)
33. Yau, D.: Hom-algebras and homology. J. Lie Theory, **19**, 409–421 (2009)

Index

M. Mitrović, M. N. Hounkonnou (eds.), *Algebra Without Borders*, Mathematics in Mind, https://doi.org/10.1007/978-3-031-86477-3

GPSR Compliance

The European Union's (EU) General Product Safety Regulation (GPSR) is a set of rules that requires consumer products to be safe and our obligations to ensure this.

If you have any concerns about our products, you can contact us on ProductSafety@springernature.com

In case Publisher is established outside the EU, the EU authorized representative is:

Springer Nature Customer Service Center GmbH
Europaplatz 3
69115 Heidelberg, Germany

Batch number: 10755329

Printed by Printforce, the Netherlands